Resilient Cybersecurity

Reconstruct your defense strategy in an evolving cyber world

Mark Dunkerley

Resilient Cybersecurity

Copyright © 2024 Packt Publishing

Senior Publishing Product Manager: Reshma Raman
Acquisition Editor – Peer Reviews: Gaurav Gavas
Project Editor: Meenakshi Vijay
Content Development Editor: Soham Amburle
Copy Editor: Safis Editing
Technical Editor: Kushal Sharma
Proofreader: Safis Editing
Indexer: Pratik Shirodkar
Presentation Designer: Rajesh Shirsath
Developer Relations Marketing Executive: Meghal Patel

First published: September 2024

Production reference: 1170924

Published by Packt Publishing Ltd.
Grosvenor House
11 St Paul's Square
Birmingham
B3 1RB, UK.

ISBN 978-1-83546-251-5

www.packt.com

To my loving family.

— Mark Dunkerley

Contributors

About the author

Mark Dunkerley is a cybersecurity and technology leader with over 20 years of experience working in higher education, healthcare, and Fortune 100 companies. Mark has extensive knowledge in IT architecture and cybersecurity through delivering secure technology solutions and services. He has experience in cloud technologies, vulnerability management, vendor risk management, identity and access management, security operations, security testing, awareness and training, application and data security, incident and response management, regulatory and compliance, and more. Mark holds a master's degree in business administration and has received multiple industry-recognized certifications, has been a keynote speaker, has spoken at multiple events, is a published author, and sits on customer advisory boards.

Thank you to my wife, Robin, and children, Tyne, Isley, and Cambridge, for all your continued support. To my parents for shaping me into the person I am today. To my brother for his ongoing service in the British Army. To anyone I missed, thank you! Without you all, this book would not have been possible.

About the reviewers

Vito Rallo is a cybersecurity expert leading an offensive security research lab with purple team, SecOps, and threat hunting experts. He has worked in cybersecurity and ethical hacking for over 25 years, developing security solutions in the fields of red teaming, OT and IoT, threat-informed and regulated security, incident response, and threat hunting.

As a product security specialist within the IBM X-Force Red global team, Vito delivered pentesting and advanced security services across the EMEA region.

He has also worked in the incident response field as an offensive security and OT expert and director at PwC. He also started Kroll as a managing director.

Vito loves working with the cyber community, and often presents at conferences or roadshows. He is on a mission to innovate the cyber business with integrated solutions across AI, security as code (detection and attack as code), and threat-informed security.

Chintan Gurjar is an cybersecurity expert with over 13 years of experience in the field, specializing in vulnerability management, threat intelligence, penetration testing, and attack surface management. He has a proven track record of working with a diverse range of clients across various industries and countries, showcasing his adaptability and proficiency in managing complex security challenges.

Currently, he serves as the threat and vulnerability manager at M&S in the UK, where he has spearheaded the development and maintenance of comprehensive risk-based threat and vulnerability management strategies. His previous roles include global senior vulnerability management analyst at TikTok and security engineering manager at Tesco, where he designed and implemented robust vulnerability management programs across multiple countries.

His experience includes notable positions such as cybersecurity manager at KPMG in New Zealand, where he planned and delivered comprehensive penetration testing and threat intelligence services, and a security consultant for various SMEs in India. This global exposure has provided Chintan with a unique understanding of international cybersecurity landscapes and practices.

Chintan is a strategic advisor for the CyberPeace Foundation and has authored a course on applied attack surface analysis and reduction for EC-Council. He also holds multiple certifications, including SANS MGT516, OSCP, CEH, CTIA, CCFH, and CCFA.

I've just completed my first stint as a technical reviewer for a book, and I'd like to extend my heartfelt thanks to the author, Mark Dunkerley, and the team at Packt Publishing for this remarkable opportunity. Their support and confidence in my abilities allowed me to lend my expertise to the enriching process of developing this publication.

Additionally, I must express my deep gratitude to my wife, Ankita Kacha, for her immense patience and understanding during the long weekends and evenings I dedicated to this project. Her unwavering support and willingness to listen to endless technical discussions have been invaluable. I truly appreciate her for cheerfully enduring my absorbed state and sharing in my enthusiasm for cybersecurity.

Join our community on Discord!

Read this book alongside other users, Cybersecurity experts, and the author himself.

Ask questions, provide solutions to other readers, chat with the author via Ask Me Anything sessions, and much more. Scan the QR code or visit the link to join the community.

https://packt.link/SecNet

Table of Contents

Chapter 4: Solidifying Your Strategy   101

Chapter 8: Vulnerability Management 299

Chapter 9: User Awareness, Training, and Testing   349

Chapter 12: Operational Technology and the Internet of Things   501

Preface

I'm excited to bring to you *Resilient Cybersecurity: Reconstruct your defense strategy in an evolving cyber world*, which addresses the need for a more robust cybersecurity program for every organization. Every organization should be assessing the current state of their cybersecurity program to ensure that it continues to evolve to meet the needs of today's ongoing cybersecurity threats. We are in a place where organizations still do not have a dedicated cybersecurity program in place. Unfortunately, this is no longer acceptable, and the risk of a major cybersecurity incident or breach increases significantly. Having a mature cybersecurity program in place doesn't guarantee that you will not suffer a major cybersecurity incident or breach, but having a mature program in place will reduce the risk and potential impact of a major cybersecurity incident or breach. More importantly, it will best prepare your organization on how to efficiently respond when a major cybersecurity incident occurs. The reality is, it is only matter of 'when' and not 'if' a major cybersecurity incident or breach occurs.

The idea behind this book is to provide a foundation for your organization's cybersecurity program that is all-inclusive and can serve as a reference for any organization. The hope with this book is that you can take something meaningful away, even if it is just one piece of information that can be applied to support your cybersecurity program whether you are just getting started, or if you already have one in place. The principles in this book may not necessarily be the same as the ones you have in place today, but I am sharing the knowledge I have gained over the years from building a cybersecurity program from the ground up. The end goal is to share as much knowledge as possible with the optimism that we continue to work together and collaborate as one unified front to better protect the confidentiality, integrity, and availability (also known as the CIA triad) of the data and information being stored and accessed within our organizations.

One area we address in more detail is how critical the CISO role has become within the organization, quickly becoming a figure of significance in a very short period of time, and a role that every organization needs to have in place.

The CISO role continues to evolve at a very fast pace from one that traditionally focused more on the technical controls to protect an organization to a much broader risk-based role that needs to interact with every part of the business. With this evolution, we are entering a new generation and era for the CISO with new and expanded responsibilities and expectations, that of the CISO v2.0. As part of this evolving role, the CISO is not only expected to be technical in nature, but more of a business acumen who is integrated into every part of the business and is able to translate technical risk into more quantifiable and business terms for the leadership teams including the **Board of Directors (BoD)**.

It is also important for today's CISO to effectively ensure that accountability for cybersecurity is appropriately distributed across the organization, rather than being solely the responsibility of the CISO. Accountability sits at the top of an organization, more specifically with the executive leadership team and the BoD.

It is important to acknowledge that we have come to a critical point with cybersecurity, and it does not look like it is going to get any easier anytime soon. Threat actors are making substantial profits from cybercrime and businesses have been formed to support these ongoing efforts. With the world we live in becoming more interconnected with the advancement of technology and the internet, preventing these crimes has become extremely complex because of cross-border challenges with differing laws and conflicts. Because of this, we all need to focus on the theme of cybersecurity culture for our users, not just within the organization but for everyone's everyday lives. Cybersecurity should not be an afterthought proceeding forward, but a concept that is engrained in everyone's mindset with everything they do, including their personal lives. With a cybersecurity culture comes a shared responsibility that we all must hold ourselves accountable for. Everyone MUST take responsibility for the protection of the information they are responsible for within an organization in addition to the information they must protect for their personal lives.

Who this book is for

This book focuses on cybersecurity from a program level in which the following roles who build the strategy and execute the program will primarily benefit from reading:

- CISO/CSO
- Other C-Level or executive leaders who overlook cybersecurity
- Directors overlooking cybersecurity
- Program Managers

With this book providing insight into all functions of a cybersecurity program, those who work within cybersecurity and help run the program will benefit from understanding what a comprehensive cybersecurity program involves. This will help provide a better understanding of each of the functions they have to interact with and will enable more productive collaboration across functions:

- Managers overlooking cybersecurity
- Architects involved with cybersecurity
- Engineers involved with cybersecurity
- Administrators involved with cybersecurity
- Analysts involved with cybersecurity
- Project Managers involved with cybersecurity
- Other roles that are part of the cybersecurity program

In addition, those who are new to cybersecurity or still determining what they would like to do within cybersecurity can also benefit from this book. As you look to enter cybersecurity, it is important you understand everything that is involved and what it takes to run a comprehensive program.

In addition, you will learn what is involved with each of the functions to help with any career decisions you are making.

What this book covers

Chapter 1, Current State, begins the book with insights into the current digital world we live in today. It then goes into detail about the current threat landscape, covering different types of attacks, threat actors, and emerging threats. There is also a focus on the use of statistics for your cybersecurity program and the importance of them. Next, we will take a look at some of the skillset challenges we are currently observing within cybersecurity before finishing the chapter with a look into the need to prioritize well-being, a very important topic.

Chapter 2, Setting the Foundations, focuses on the building blocks for your cybersecurity program. As a cybersecurity leader, it will be critical that you understand the business you are working in and are familiar with how to navigate the business. Next, we review finances and where you can expect costs to be incurred within the program. This transitions into the structure overview for the cybersecurity program with an emphasis on the core functions that should be included.

Next, we cover the need to document the cybersecurity organization structure and roles and responsibilities before finishing off the chapter with a review of change management and communication and their importance.

Chapter 3, Building Your Roadmap, provides an in-depth review of the need to build a roadmap for the cybersecurity organization. This includes the need for good program and project management to provide structure around the program. To build efficient roadmaps, you are going to need to better understand the current state of your organization. Once you understand the current state, you can build roadmaps for the immediate short-term (2-4 months), short-term (9-12 months), and long-term (1-3+ years).

Chapter 4, Solidifying Your Strategy, takes us into more details around the importance of a strategy for your cybersecurity program. Within the chapter, a focus on four key strategic areas is covered. The first is around the architecture strategy for your organization, covering multiple different areas, such as modernization, the need to use cloud-based technologies, zero-trust architecture, and more. The next strategy covered the need for a cybersecurity framework and the importance of needing to implement one. We then look at the need to have a strategy around your vendors and product portfolio with an emphasis on reducing this portfolio as much as possible. Finally, we review resource management and the need for a strategy around in-house vs. outsourced resources.

Chapter 5, Cybersecurity Architecture, covers everything architecture for your cybersecurity program. This begins with an overview of the architecture and the importance of embedding the cybersecurity program as part of the broader architecture process. Following this is an in-depth review of the architecture review process and what should be considered within the process from a cybersecurity perspective. Next, we touch upon the foundation of cybersecurity architecture before going into detail on zero-trust architecture, what is involved, and the importance of it. We then finish off the chapter with a detailed review of the technical architecture components, such as network, infrastructure, data, etc.

Chapter 6, Identity and Access Management, focuses on an in-depth review of everything identity and access management. First is an overview of identity and access management with more details about identity, authentication, authorization, and accountability. We then shift our focus to the need to modernize your identity architecture before diving deeper into account and access management, which includes stepping through the identity lifecycle process. We then look at what you need to consider with securing your identities before finishing the chapter with a look into enhanced identity security and protection methods.

Chapter 7, Cybersecurity Operations, takes us through everything involved with cybersecurity operations for your cybersecurity program. To begin the chapter, an overview of cybersecurity operations is provided with the different components involved within this program. Next is a detailed review of the **Security Operations Center (SOC)** with insight into the different operating models. We then go into detail about threat detection and what needs to be considered for this component before reviewing incident management and response, which is not to be overlooked. We then finish off the chapter with a look into the importance of **Business Continuity Planning (BCP)**, **Disaster Recovery Planning (DRP)**, and the **Cybersecurity Incident Response Plan (CIRP)**.

Chapter 8, Vulnerability Management, provides a lot of important information on what needs to be considered as part of your vulnerability program. First, we look at why there is a need for a dedicated vulnerability program and the building blocks required for this program. In the section that follows, there is an emphasis on vulnerability discovery and alerting and what should be considered for this component. Next focuses on the importance of tracking your vulnerabilities and the need to ensure remediation is taking place on time. This leads to update management and email protection considerations as part of your vulnerability management activities. The chapter finishes off with a look into other vulnerability management considerations such as hardware, virtualization, network, and more.

Chapter 9, User Awareness, Training, and Testing, covers everything related to the human element. We begin the chapter with an overview of why this component is so important for the organization. Next, we go into detail on building the foundations for your user awareness, training, and testing program with an emphasis on security culture and maturity. This transitions into user awareness and everything that should be considered with awareness for your users. We then go into detail on what is involved with both user training and testing to ensure a comprehensive approach with your users. We finish the chapter with a look into some other areas that should be considered for your user awareness, training, and testing program, such as gamification, bringing in external speakers, cybersecurity town halls, and more.

Chapter 10, Vendor Risk Management, focuses on everything you need to consider for managing cybersecurity risk with your vendors. We begin with a review of vendor risk management and the different types of risk involved with your vendors, in addition to looking at the current landscape and some statistics. Next, we focus on building your cybersecurity vendor risk management foundation and what should be considered for your program. We then review the need to ensure cybersecurity vendor risk management is integrated across the broader business before covering contract management in more detail, which is an important part of the cybersecurity leadership role.

We finish the chapter with insight into managing your vendors in addition to ongoing and continuous monitoring of your vendors.

Chapter 11, Proactive Services, provides insights into everything you should be considering from a proactive perspective to help reduce risk as much as possible. We begin the chapter with an overview of why we need to implement a proactive services program and the importance of executing these types of services. Next, we take a deeper look into cybersecurity testing and the different types of services that should be considered for your program. This transitions into incident response planning, something that should be in place for every organization. We then move on to reviewing tabletop exercises by providing a detailed overview of what they are and how to execute them. To finish the chapter, we cover other proactive services that can be executed with your proactive services program.

Chapter 12, Operational Technology (OT) and the Internet of Things (IoT), begins with an insight into what exactly OT and IoT are, including what **Industrial Control Systems (ICS)** are and how it fits within OT. We then review why securing both OT and IoT has become so important and the criticality of this technology. We then look at the need for building a dedicated program and what is involved in your OT and IoT programs. Next, we take a deeper look into protecting these environments and what you should consider as part of protecting these environments. We finish off the chapter with a focus on responding to OT and IoT incidents as it will differ from your standard incident response plan. This includes the need to execute tabletop exercises with a theme built around OT and IoT technology.

Chapter 13, Governance Oversight, leads us into the concluding section of the book with an emphasis on **Governance, Risk, and Compliance (GRC)**. In this chapter, we look at the importance of governance for the cybersecurity program. This transitions into the program structure for your GRC program including roles and responsibilities for this program. We then shift our focus over to the need for a GRC application for your organization and what should be included with the GRC application. Next, we go into detail with policies, standards, and processes/procedures for your organization as it relates to cybersecurity. This shifts into ensuring the cybersecurity program is made visible to your leadership team through various communication channels with the need for good and clear reporting. We finish off the chapter with a look into other governance considerations for your governance program.

Chapter 14, Managing Risk, focuses on the importance of risk and everything we need to consider with risk within the cybersecurity program. We begin the chapter with an overview of why risk is so important and how everything we manage within cybersecurity translates back to risk.

This transitions into understanding the different risk types by looking into more detail about how to calculate risk and the different mitigation options for risk. We then transition into a review of risk frameworks and the different frameworks for you to consider for risk management. Next, we look at the importance of tracking risk and the need for a risk register. To finish the chapter, we take a deeper look into the insurance landscape and what is involved with managing cybersecurity insurance.

Chapter 15, *Regulatory and Compliance*, gives us deeper insight into the evolving complex world of regulatory and compliance within cybersecurity. First, we look into the current landscape of regulatory and compliance and how complicated it can be to navigate, especially at a global level. We then cover the importance of building positive relationships with your legal team and the importance of legal expertise within cybersecurity. This transitions into the importance of data protection for your cybersecurity program before going into detail on the need for frameworks and audits for your cybersecurity program. To finish off the chapter, we look into other regulatory and compliance considerations like privacy, data retention, legal hold capabilities, and more.

Chapter 16, *Some Final Thoughts*, brings us to the concluding chapter of the book where we take a closer look at bringing everything together and how the overall program has come together. This transitions into discussing the importance of managing your cybersecurity program as a journey as there will be no destination with this program, it continues to evolve. Next, we look at the top ten considerations you should consider for your cybersecurity program including what I consider the current three top priorities for a cybersecurity program at this time. This takes us into the final section of the chapter where we review observations of what the future may hold with cybersecurity.

To get the most out of this book

Ideally, having knowledge of cybersecurity and its concepts will help as you read the book. This book provides a high-level view of a cybersecurity program, and you will be presented with many different topics within cybersecurity. With this, the focus of each topic will be broad versus a deep technical view. In addition, read this book with an open mind on how a cybersecurity program should look. Every organization is different, and every industry comes with its own set of challenges and uniqueness. This book is meant to serve as a foundation for the core functions that you should be considering for your cybersecurity program based on the current threat landscape. Although this will change over time and as cybersecurity leaders, we need to be dynamic and lead with an open mind.

Download the color images

We also provide a PDF file that has color images of the screenshots/diagrams used in this book. You can download it here: https://packt.link/gbp/9781835462515.

Conventions used

There is one text convention used throughout this book.

Bold: Indicates a new term, an important word, or words that you see on the screen. For instance, words in menus or dialog boxes appear in the text like this. For example: "Select **System info** from the **Administration** panel."

 Warnings or important notes appear like this.

 Tips and tricks appear like this.

Get in touch

Feedback from our readers is always welcome.

General feedback: Email feedback@packtpub.com and mention the book's title in the subject of your message. If you have questions about any aspect of this book, please email us at questions@packtpub.com.

Errata: Although we have taken every care to ensure the accuracy of our content, mistakes do happen. If you have found a mistake in this book, we would be grateful if you reported this to us. Please visit http://www.packtpub.com/submit-errata, click **Submit Errata**, and fill in the form.

Piracy: If you come across any illegal copies of our works in any form on the internet, we would be grateful if you would provide us with the location address or website name. Please contact us at copyright@packtpub.com with a link to the material.

If you are interested in becoming an author: If there is a topic that you have expertise in and you are interested in either writing or contributing to a book, please visit http://authors.packtpub.com.

Share your thoughts

Once you've read *Resilient Cybersecurity, First Edition* we'd love to hear your thoughts! Scan the QR code below to go straight to the Amazon review page for this book and share your feedback.

https://packt.link/r/1835462510

Your review is important to us and the tech community and will help us make sure we're delivering excellent quality content.

Download a free PDF copy of this book

Thanks for purchasing this book!

Do you like to read on the go but are unable to carry your print books everywhere?

Is your eBook purchase not compatible with the device of your choice?

Don't worry, now with every Packt book you get a DRM-free PDF version of that book at no cost.

Read anywhere, any place, on any device. Search, copy, and paste code from your favorite technical books directly into your application.

The perks don't stop there, you can get exclusive access to discounts, newsletters, and great free content in your inbox daily.

Follow these simple steps to get the benefits:

1. Scan the QR code or visit the link below:

https://packt.link/free-ebook/9781835462515

2. Submit your proof of purchase.
3. That's it! We'll send your free PDF and other benefits to your email directly.

1

Current State

There doesn't seem to be a day that goes by that there isn't a new notice of a cybersecurity breach or some form of cybercrime. Cybersecurity incidents have become so prevalent that they are hitting mainstream media on a regular basis because of their impact. Cyber events are not just causing a small inconvenience to organizations, they are causing substantial financial loss (millions), crippling manufacturing operations, damaging reputations, leaking enormous amounts of **Personal Identifiable Information (PII)**, and in some instances, causing organizations to permanently close their doors.

The discussion around cybersecurity and risk has become a critical agenda item on executive leadership teams, board rooms, and within the highest level of governments. And, the unfortunate reality is, it continues to get more challenging as threat actors continue to become more sophisticated. The question from leadership and board members continues to be asked: Are we secure? The simple answer is no. No one is 100% secure in today's digital world and we never will be. There will always be risk. As leaders, it is our responsibility to manage and reduce risk as much as possible. We will never eliminate risk entirely, and it is important that those we work for and report to understand this. As cybersecurity leaders, we must create an environment that balances cybersecurity with business enablement and builds a culture around cybersecurity. This includes the need for full transparency, effective collaboration throughout the organization, and most importantly, trust.

As we hear more news of security breaches like the multiple T-Mobile breaches over the years, the Marriott International breach, the Equifax breach, and the Yahoo breach of 3 billion records, the severity of what we are dealing with is evident and requires our utmost attention. The reality is, we all need to do better.

We not only need to hold ourselves accountable, but also those around us and especially those who are trusted to manage, process, and store our data. This is far from an easy task, especially with the emerging complexity of technologies, an attack surface that continues to widen, and the progression of organized cyber and state-sponsored crime groups with budgets and expertise far exceeding that of most organizations.

Because of this, we need to continue to evolve our cybersecurity programs and strategies to meet the demand of modern-day threats such as AI-driven and supply chain-based attacks. As cybersecurity leaders, this in turn means being innovative, creative, dynamic, and agile. We cannot become complacent with the current state because technology and the world we live in are evolving at a faster pace than we've ever seen. The hope is that this book can help provide the higher-level strategy and insight into a more modern cybersecurity program, whether you already have a program in place or you are looking to build one from the ground up.

As you read through this chapter, you will learn more details about the current state of cybersecurity and the challenges we face as cybersecurity leaders. Specifically, you will learn about the following:

- An evolving digital world
- The current threat landscape
- The importance of statistics
- Skillset challenges
- Prioritizing well-being

An Evolving Digital World

As a cybersecurity leader, it is important to remain current and have a foundational understanding of technology. It is your responsibility to reduce risk within the organization you work for. To do this effectively, you need to understand the technology being used to help make informed decisions on securing that technology. As the digital world continues to evolve at such a fast pace, keeping up with technology isn't the easiest of tasks. But this is a requirement to be more efficient in your role and you must keep yourself up to date. You may be asking yourself how this is possible. This can be accomplished through many channels such as conferences, communities, research and analyst companies like Gartner and Forester, meeting with your strategic suppliers regularly, user groups/forums, business social media channels like LinkedIn, and so on.

Increasing Reliance on Technology

Whether you believe it's fortunate or unfortunate, we are living in a connected world where we have never been so reliant on technology that organizations would not be able to survive without. As our younger generations continue to grow, there is an increased demand for the use of technology in everything that we do in life. It's a scary thought, but look how fast the world has grown within the previous 100 years compared to the overall history of humanity. Technology continues to push the boundaries of innovation, and a significant portion of that change must include the securing of this technology. This has been more evident over the previous 10 years and especially since the world has become a more connected place with the advancement of the internet.

According to the United States Census Bureau, as of the 18th of May, 2024, there is an approximate world population of over 8 billion. Yes, the world population is now over 8 billion:

Figure 1.1: World population clock counter from the United States Census Bureau

Source: https://www.census.gov/popclock/

Let's put things into perspective with the challenges of the increased digital footprint that we face as cybersecurity leaders in today's world. A report published in January 2023, on the *We Are Social Inc.* website, *The Digital 2023*, shows the following statistics:

- There are an estimated 5.44 billion unique mobile phone users and 8.46 billion cellular mobile connections (not including IoT).
- There are an estimated 5.16 billion internet users.
- There are an estimated 4.76 billion active social media users.

Source: https://wearesocial.com/us/blog/2023/01/digital-2023/

To provide a clearer picture of how fast both information technology and cybersecurity have grown within the previous 20 years, the following chart shows the rapid growth of internet usage since the early 1990s. Although the need for technology and cybersecurity professionals was present before this time, you can clearly see from the chart below the impact and demand we are dealing with for increased resources in such a short amount of time.

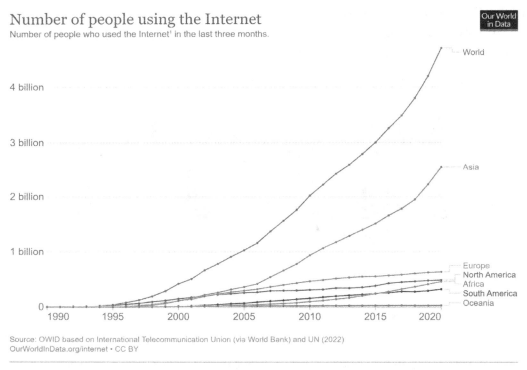

Figure 1.2: Number of people using the internet

Source: The chart above is cited from an article by Hannah Ritchie, Edouard Mathieu, Max Roser, and Esteban Ortiz-Ospina (2023), titled *Internet*. Published online at OurWorldInData.org. Retrieved from https://ourworldindata.org/internet

As we have become more connected throughout the world, traditional borders that separate countries no longer apply when it comes to technology and cybercrime. The laws of one country will not necessarily be applicable when cybercrime occurs from overseas across these different boundaries. This creates a very complex system for holding accountable those who engage in cybercrime and break the law within other countries.

As technology continues to evolve, transform, and innovate at a pace faster than ever before, it has never been more important to ensure that security is considered a core fundamental of this technology. We are not there yet, and we need to continue to push our vendors, technology companies, and ourselves to ensure that a security-first mindset is applied to everything we do with technology moving forward.

Digital Transformation

In recent years, there has been a lot of hype around digital transformation. This has many meanings depending on your organization and the functions within it. In short, at a broader level, digital transformation is the ability to digitally improve your business and/or processes through modern technology with the replacement of legacy systems and antiquated processes.

An example of this includes the shift from a legacy on-premises infrastructure to a modernized cloud-first strategy to support the evolving needs of big data, **Machine Learning (ML)**, **Artificial Intelligence (AI)**, and more. As we take a look back over the previous couple of years, we tend to see hype rise and fall around certain technologies and innovations. For example, in 2021/22, the metaverse was the latest technology everyone was interested in. Shift forward a year and as of this publication, AI is the center of hype, more specifically **generative AI,** also known as GenAI. I'm sure you are all aware of generative AI by now, essentially the next generation of AI that can create new content such as text, images, audio, and video using data it has been provided to learn from. A few examples include ChatGPT, Gemini, and Copilot. *Chapter 7, Cybersecurity Operations*, will cover AI in more detail. We will continue to see these trends and it's important as cybersecurity leaders that we keep close to these trends and continue to educate ourselves as this change continues to occur.

As we continue to digitize and modernize technology, improved security is needed. A simple example of this is the need for some form of endpoint protection tool. Traditionally, this has been an **Anti-Virus (AV)** signature-based tool that is capable of blocking already known threats through known signatures. Unfortunately, in today's world, this type of endpoint protection is no longer appropriate on its own.

Today, the following at minimum needs to be implemented as a replacement for traditional AV signature-based capabilities:

- **Advanced Threat Protection (ATP)** that includes AV and threat protection
- **Endpoint Detection and Response (EDR)**
- Advanced analytics and behavioral monitoring
- Network protection
- Exploit protection

This is just a single example of a specific technology within cybersecurity that has become outdated and it's important you focus on a defense-in-depth strategy using zero-trust principles, which we will cover in more detail in *Chapter 5, Cybersecurity Architecture*. This includes the need to fully understand the role AI is now playing within cybersecurity and the capabilities available. As cybersecurity leaders, we must keep current with the latest cybersecurity technology.

The Evolving Landscape of Cybersecurity

In addition to the ongoing digital transformation activities, there is also an expectation that we can work and access data from anywhere at any time. With the rapid increase of remote work during 2020 and 2021, this model and expectation have been fast-tracked because of COVID. Although many companies are reversing the remote work model and requiring employees to report back to the office, many are resisting and have an expectation of continuing to work remotely, or at least have the flexibility. With this model comes a much larger responsibility from a cybersecurity perspective. As our infrastructure continues to be modernized and shifted to the cloud, so do the cybersecurity requirements. The focus is no longer primarily the network, protecting our data center and devices within a building, but that of the user's identity and, more importantly, the data. Financial gain is the primary motivator for threat actors, and data is the underlying driver for that financial gain. Because of this, it is imperative we provide relevant training and awareness for our users as the technologies evolve and the threat vectors change.

As already stated, attacks are becoming more and more sophisticated every day. There is an ever-growing army of threat actors working around the clock trying to exfiltrate any data they can get their hands on because the cost of private data is very expensive. There has also been a shift in the way bad actors are threatening organizations by looking for weakness in the supply chain and holding companies at ransom. With the advancement of cloud technology, supercomputers, and the reality of quantum computing coming to light, hackers and organized groups now have access to much more powerful systems and are easily able to crack passwords and their hashes much easier, making them obsolete as the only factor of authentication.

No one should be using just passwords anymore; however, the reality is, most still are. The same applies to encryption. The advancement of computers is making algorithms insecure with the ongoing need for stronger encryption. These are just some of the ongoing challenges we are faced with in today's evolving digital world when protecting our assets.

Over the years, cybersecurity has evolved from being a shared role or a role that was non-existent within many companies. Today, well-defined teams and organizational structures exist or are being created to focus solely on cybersecurity. Not only are these teams maturing constantly, but the **Chief Information Security Officer (CISO)** has become a person of significant importance and in some instances may report directly to a **Chief Executive Officer (CEO)** instead of the **Chief Information Officer (CIO)**, the **Chief Technology Officer (CTO)**, or another C-level below the CEO. In addition, we are also observing the CISO being invited to the **Board of Directors (BoD)** quarterly meetings, essentially getting a seat at the table.

Before we move on to the next topic, one additional matter within the digital world that needs mentioning is shadow IT. In short, shadow IT is the setup and use of technology without IT or the security team's approval or knowledge, for example, in a business function like **Human Resources (HR)** or finance. This obviously creates a significant security challenge as technology is being deployed with no standards or best practices in place. This can be a challenge to manage, but it will need to be addressed as part of your role, especially as digital transformation continues to occur across the entire business at a very fast pace.

Now that we have covered the evolving digital world, the next section will take us through the current threat landscape and what to expect in terms of current threats.

The Current Threat Landscape

The threat landscape within the cybersecurity world is extremely diverse and is continually becoming more complex. The task of protecting users, data, and systems is becoming more difficult and requires the progression of even more intelligent tools to keep threat actors out.

Common Cyber Threat Actors

Today, cyber criminals are more sophisticated, and large groups have formed with significant financial backing to support the harmful activities of these groups. The following are common threat actors:

- National governments
- Nation-states

- Terrorists
- **Advanced Persistent Threat (APT)** groups
- Cyber mercenaries
- Cyber arms dealers
- Cyber extortionists
- Spies
- Organized crime groups
- Hacktivists
- Hackers
- Business competitors
- Malicious insiders/internal employees
- Essentially anyone who has some malicious intentions with the use of technology

In addition, with the recent rise of GenAI, ChatGPT has transformed the field of cybersecurity in a very short amount of time. Previously, only highly skilled attackers were able to breach organizations. But with ChatGPT, even less skilled hackers can succeed by using AI in their operations. It's now difficult to judge an attacker's true level of skill during a sophisticated attack.

Types of Cyberattacks

There are many types of cyberattacks in the world today, and this creates a diverse set of challenges for organizations, especially cybersecurity leaders. One of the most common attack methods used today is that of malware. Malware is software or code designed with malicious intent that exploits vulnerabilities found within the system. The following types of threats are considered malware:

- Adware
- Spyware
- Virus (polymorphic, multipartite, macro, or boot sector)
- Worm
- Trojan
- Rootkit
- Bots/botnets
- Ransomware
- Logic bomb

Ransomware in More Detail

With the prevalence of ransomware and the extreme damage it can inflict on an organization, let's review this type of cyberattack in more detail. Ransomware has been around for a long time and the first documented incident occurred in 1989, known as PC Cyborg or the AIDS Trojan. In short, a ransomware attack is where an intruder encrypts data belonging to a user or organization, making it inaccessible. For the user or organization to gain access back to their data, they are held to a ransom in exchange for the decryption keys. The intruders will use many tactics to try and force payment, including threats to leak the data, list the data for sale on the dark web, and erase the backups, to name a few.

As the ransomware business continues to evolve, we are hearing that very mature business models have been put in place to support their efforts to hold organizations to ransom. There is even a ransomware-as-a-service model that allows hackers to subscribe and use the service to commit their own attacks. The latest tactic used by ransomware criminals is double extortion – essentially, exfiltrating the data in addition to encrypting it. This provides additional bargaining power for the threat actors and creates a lot more risk for organizations to handle. Unfortunately, there have been countless ransomware attacks to date that have made the news and they continue to occur often.

A couple of the more notable ransomware attacks include that against *Colonial Pipeline*, one of the largest fuel pipelines in the United States, and *MGM Resorts*, a global entertainment company. Both companies suffered a major impact: Colonial Pipeline was forced to shut down its fuel distribution operations, causing gas shortages for consumers throughout the East Coast of the United States. MGM Resorts encountered major operational challenges for many days and an estimated loss of approximately $100 million.

Other Types of Attacks

In addition to malware, the following table shows other types of attack techniques that can be used to exploit vulnerabilities and that you should be familiar with:

Main Category	Sub-Categories	Description	Examples
Malware	Virus, Worm, Trojan, Ransomware, Adware, Spyware, Bots/Botnets	Malicious software designed to damage, disrupt, or gain unauthorized access to systems.	ILOVEYOU virus, WannaCry ransomware, Mirai botnet

Social Engineering	Phishing, Spear Phishing, Whaling, Vishing, Smishing, BEC, Pretexting, Tailgating, Baiting	Manipulative techniques to trick individuals into divulging confidential information.	CEO fraud, IRS scam calls, lottery scams, tech support scams
Network Attacks	DoS, DDoS, MITM, DNS Tunneling, ARP Spoofing, IP Spoofing, Session Hijacking, Zero-Day Exploits	Disrupting network operations or exploiting network vulnerabilities for malicious purposes.	SYN flood, Wi-Fi evil twin, rogue DHCP server
Web Application Attacks	SQL Injection, XSS, CSRF, RFI, Command Injection, OWASP Top 10	Exploiting web application vulnerabilities to compromise systems or data.	File upload attacks, broken authentication
Exploitation	Zero-Day, Buffer Overflow, Privilege Escalation, RCE	Utilizing software vulnerabilities for unauthorized actions or data breaches.	Heartbleed, Shellshock, Microsoft Exchange Server vulnerabilities
Password Attacks	Brute Force, Dictionary, Credential Stuffing, Rainbow Table, Keylogger, Password Spraying	Techniques aimed at uncovering or bypassing passwords to gain unauthorized access.	John the Ripper, Hydra, Hashcat
Physical Attacks	Tailgating, Shoulder Surfing, Dumpster Diving, Theft, Device Tampering	Direct physical methods to gain unauthorized access or information.	Unauthorized entry, stolen hardware
IoT Attacks	Mirai Botnet, Connected Device Exploits	Targeting IoT devices for unauthorized access or to create botnets.	Unpatched smart home devices, compromised wearable devices

Cryptocurrency-Related	Cryptojacking, Phishing Scams, Exchange Hacks, 51% Attacks	Attacks aimed at cryptocurrencies, including theft, exchange exploitation, and blockchain attacks.	Fake crypto giveaways, compromised exchanges, malware for mining
Other	APT, Insider Threats, Supply Chain Attacks, Mobile Attacks	Diverse attacks including state-sponsored attacks, malicious insiders, and mobile device targeting.	Stuxnet, data theft by employees, SolarWinds attack, SMS-based malware

Supply Chain Challenges

Another attack becoming more common is that against the supply chain, where the threat actors look to compromise a vendor's software or application, which in turn will compromise all its downstream customers. A couple of the more notable include the attack against SolarWinds, a monitoring and performance management tool, and Progress, a company with many solutions including that of MOVEit, a managed file transfer solution. With SolarWinds, threat actors implanted malicious code into their software, which was received by thousands of customers. Once installed, hackers were provided with the ability to infiltrate customer networks. With MOVEit, threat actors took advantage of a zero-day exploit that allowed them to exfiltrate the sensitive data of many companies, the damage of which would continue for many months. In addition to supply chain challenges, there is the need for improved third-party risk management as we need to hold our third parties to a higher level of standard with cybersecurity. Third parties continue to become compromised, potentially putting our data at greater risk and/or impacting the services being provided to us. We will be covering third-party risk in more detail in *Chapter 10, Vendor Risk Management*.

Impact on Organizations

Even more concerning is the case of organizations permanently closing their doors because of a cybersecurity incident. The cybersecurity incident alone may not be the sole reason for the closure of an organization, but it adds an extreme operational and financial burden that an already struggling organization may not be able to recover. Some notable examples recently include that of St. Margaret's Health hospital located in Spring Valley, Illinois.

Although other factors were to blame, a ransomware attack in 2021 that significantly impacted operations was specifically noted. Lincoln College in Illinois is another unfortunate example of the impact of a cyber attack. An institution that was able to survive 157 years finally shut its doors in May 2022. The coronavirus pandemic and a ransomware event were both publicly noted as major events forcing the college to permanently close.

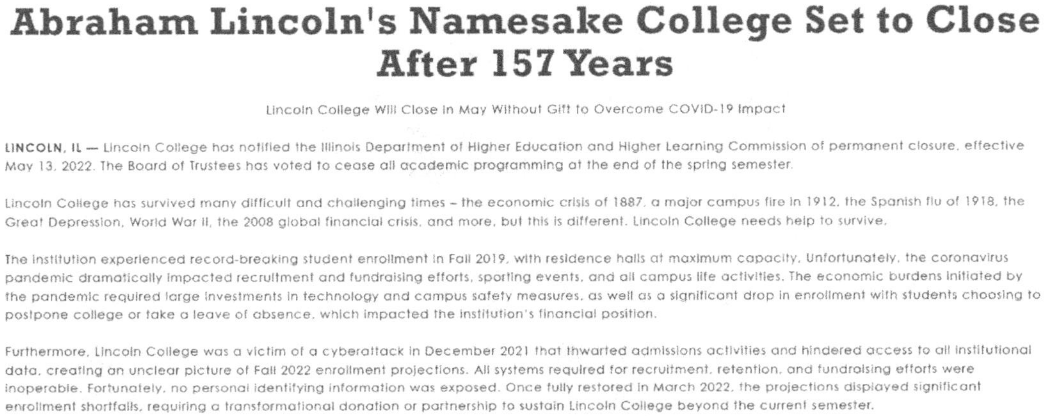

Abraham Lincoln's Namesake College Set to Close After 157 Years

Lincoln College Will Close In May Without Gift to Overcome COVID-19 Impact

LINCOLN, IL — Lincoln College has notified the Illinois Department of Higher Education and Higher Learning Commission of permanent closure, effective May 13, 2022. The Board of Trustees has voted to cease all academic programming at the end of the spring semester.

Lincoln College has survived many difficult and challenging times – the economic crisis of 1887, a major campus fire in 1912, the Spanish flu of 1918, the Great Depression, World War II, the 2008 global financial crisis, and more, but this is different. Lincoln College needs help to survive.

The institution experienced record-breaking student enrollment in Fall 2019, with residence halls at maximum capacity. Unfortunately, the coronavirus pandemic dramatically impacted recruitment and fundraising efforts, sporting events, and all campus life activities. The economic burdens initiated by the pandemic required large investments in technology and campus safety measures, as well as a significant drop in enrollment with students choosing to postpone college or take a leave of absence, which impacted the institution's financial position.

Furthermore, Lincoln College was a victim of a cyberattack in December 2021 that thwarted admissions activities and hindered access to all institutional data, creating an unclear picture of Fall 2022 enrollment projections. All systems required for recruitment, retention, and fundraising efforts were inoperable. Fortunately, no personal identifying information was exposed. Once fully restored in March 2022, the projections displayed significant enrollment shortfalls, requiring a transformational donation or partnership to sustain Lincoln College beyond the current semester.

Figure 1.3: A snippet from Lincon College's home page taken October 2023

Source: `https://lincolncollege.edu`

Another unfortunate example is that of KNP Logistics Group, a UK-based logistics firm that went into administration in September 2023. Along with other challenges mentioned was a ransomware attack that significantly impaired the operations of the firm and the ability to secure the investments needed to continue.

Special Considerations for OT and IoT

Although not applicable to most industries, other challenges that need to be addressed involve continuing to increase the protection of **Operational Technology (OT)** and the **Internet of Things (IoT)**. Managing and securing these technologies efficiently requires a different set of skills. The ability of threat actors to compromise power plants, manufacturing plants, water treatment facilities, internet-connected cars, and more poses a major risk. These types of attacks go beyond the impact of data exfiltration and financial loss; they have the ability to cause significant harm to people. Examples include the ability of a threat actor to take control of systems that could bring down a power plant supplying power to an entire city, take over a power plant and control machinery, or modify the chemicals within a water treatment facility.

These risks cannot be taken lightly, and it is critical that organizations are aware of these risks and ensure cybersecurity is a priority.

Emerging Threats — AI and Beyond

Being a cybersecurity leader requires the ability to be dynamic and up to date as emerging threats continue to evolve at a very fast pace. We need to understand what risk they pose and how to reduce this risk. The most recent emerging threat is that of AI as it becomes more accessible to everyone. Although there are many benefits from using AI, it is already coming with a lot of challenges from a cybersecurity perspective as it is being used to advance cyber threat actors' malicious intents. Unfortunately, AI is already being used to create more advanced attack methods, speed up the ability to create new malware at a rapid pace, impersonate others using deepfake capabilities, and develop and initiate advanced email types of attacks such as sophisticated phishing campaigns with fewer signals (reduced spelling mistakes, more realistic conversation, catered to company culture, etc.). As AI and other technologies continue to evolve, so do our defense mechanisms.

Now that we have covered the current threat landscape, let's move on to the next section, which provides statistics around the reality of what we are dealing with.

The Importance of Statistics

As a leader, I'm big on statistics. I believe it to be one of the more efficient tools within our toolbox to help drive meaningful conversation throughout the leadership team and business in general. Statistics are real facts that show the real picture and allow us to deliver a more realistic story of what we are up against. As you look to justify the need for additional funding for your cybersecurity program, there probably hasn't been an easier time than now with all the statistics and real-life examples of compromises. It is also important to ensure your executive leadership team and board are fully aware of what is happening around us and what impact and implications can occur at any time because of a cybersecurity event. The same applies for user awareness, wherein leveraging statistics and real-life examples provides some very powerful stories that can relate to users to provide better awareness.

Key Reports and Findings

There are countless annual reports being released with great information. I personally reference many of these reports as they provide very useful information on the current state of cybersecurity and the threat landscape.

IBM's Cost of a Data Breach Report

The first is IBM's *Cost of a Data Breach Report*. The 2023 report provided data from 553 organizations affected by data breaches throughout 16 countries and regions within 17 industries. The following provides some data points from the executive summary within the report:

- An all-time high of $4.45 million was reported as the average cost of a data breach, which is a 2.3% increase from $4.35 million in 2022, and a 15.3% increase from $3.86 million in 2020.

- As a result of a breach, 51% of organizations plan to increase investments in security.

- The use of security AI and automation provided an average reduction of a 108-day time frame to identify and contain a breach. A $1.76 million reduction in data breach costs was also reported versus those that didn't use security AI or automation.

- Only 1 in 3 organizations self-identified a breach. Third parties and attackers represent 67% of reported breaches.

- It was noted that organizations experienced an additional cost of $470,000 by not involving law enforcement.

- The healthcare industry continues to report the largest expense from data breaches, a 53.3% increase in breach costs since 2020.

- 82% of breaches included cloud infrastructure.

- Greater levels of incident response planning and testing saved organizations $1.49 million when containing a data breach.

Source: https://www.ibm.com/security/data-breach

Verizon Data Breach Investigation Report (DBIR)

Another great resource is the Verizon DBIR. Like the IBM report, this report is built on a set of real-world data and contains some eye-opening statistics on data breaches. Here are some of the findings from the summary of the 2023 report:

- Within the social engineering category, **Business Email Compromise** (**BEC**) represents more than 50% of incidents.

- 95% of breaches are financially driven.

- The human element is included in 74% of breaches. This is through either human error, privilege misuse, stolen credentials, or social engineering.

- External actor involvement made up 83% of the breaches.

- Stolen credentials, phishing, and exploitation of vulnerabilities are the primary entry points for attackers.
- Ransomware is present in 24% of breaches.

Source: https://www.verizon.com/business/resources/reports/2023/2023-data-breach-investigations-report-dbir.pdf

CISO Perspectives and Challenges

A report providing perspectives from the CISO is Proofpoint's *2023 Voice of the CISO Report*. This report provides insight from 1,600 global CISOs. Some of the highlights provided in this report include:

- A staggering 68% agreed they are at risk of a material breach within 12 months.
- A loss of sensitive information within the previous year was reported by 63%.
- Burnout was reported by 60% within the previous 12 months.
- Personal liability was shared as a concern by 62%.
- 62% responded that cybersecurity expertise at the board level should be a requirement.

Source: https://www.proofpoint.com/us/resources/white-papers/voice-of-the-ciso-report

 Since the topic of personal liability came up in the Proofpoint report, it's important to note an added burden for executives that has recently come to light. There have now been instances of executives being charged with negligence. A few that have made the media include the CSO from Uber, the CEO of Vastaamo, and the CISO of SolarWinds (Case is still pending as of August 2024).

Federal Bureau of Investigation Internet Crime Report

A report I like to reference that includes consumer data statistics is the *Federal Bureau of Investigation Internet Crime Report* released by the **Federal Bureau of Investigation (FBI) Internet Crime Complaint Center (IC3)**. This one is important as it allows us to relate cybersecurity more to our users and their everyday lives in terms of how cybersecurity can impact them and their families personally. This is important and I'll be covering this more in *Chapter 9, Cybersecurity Awareness, Training and Testing*. In 2022, it was stated that more than $10.2 billion in losses were reported from 800,944 complaints.

The following chart taken from the IC3 website represents a very concerning trend and presents a strong message and reminder of what we continue to be challenged with:

Figure 1.4: IC3 complaint statistics over the last 5 years

Source: https://www.ic3.gov/

Additional Resources and Staying Updated

As stated, there are many great reports now being published by various vendors with a lot of great information; there are too many to cover in this book. A few others worth noting are the Microsoft annual *Digital Defense Report* (https://www.microsoft.com/en-us/security/security-insider/microsoft-digital-defense-report-2023), Proofpoint's annual *Board Perspective Report* (https://www.proofpoint.com/us/resources/white-papers/board-perspective-report), and Secureworks, *State of the Threat: A Year in Review Report* (https://www.secureworks.com/resources/rp-state-of-the-threat-2023). These reports are typically annual and a quick search on Google should return the latest report. When using data and statistics as a reference, make sure you confirm there is a good dataset being referenced to provide the output.

Throughout the book, I will continue to reference statistics and data points to help with some of the justification around why certain functions within your program are important.

In addition to the reports being referenced, there are many resources you can utilize to keep up to date with the latest news, which it is vital for leaders to do. There are many sources available to view security news, follow the latest trends, and understand the current best practices. There is no way I could even begin to list all of them, but the following are some general resources to help keep you up to date with the latest news and information:

- **Dark Reading**: `https://www.darkreading.com/`
- **Cyware**: `https://cyware.com/cyber-security-news-articles`
- **SANS Cyber Security Newsletters**: `https://www.sans.org/newsletters/`
- **Cybersecurity Insiders**: `https://www.cybersecurity-insiders.com/`
- **CSO**: `https://www.csoonline.com/`
- **Krebs on Security**: `https://krebsonsecurity.com/`
- **The Hacker News**: `https://thehackernews.com/`
- **Darknet Diaries podcast**: `https://darknetdiaries.com/`
- **Risky Biz podcast**: `https://risky.biz/`

A quick Google search or interaction with ChatGPT will return many additional resources for review. The following is an example of a resource that provides over 50 blogs and websites for reference: `https://heimdalsecurity.com/blog/best-cyber-security-blogs/`.

 We will cover more specific details on threat intelligence and ways to obtain this type of information in *Chapter 7, Cybersecurity Operations*.

Moving on from some general statistics, let's take a closer look at some data around tracking breaches and some places that will provide more awareness on the volume of breaches occurring.

Breaches Continue to Rise

If you follow the news, you are probably aware that there is no shortage of breaches today. They are happening so frequently that it has become a daily occurrence. What is even more concerning is that these are the ones that we hear about; how many do we not hear about?

Identity Theft Resource Center (ITRC)

A great resource to reference is the ITRC. The ITRC is a non-profit in the United States that provides help to victims of identity crime at no cost. In addition to the services provided, it also provides a great source called *notified* to search for any known breaches. To access it, you can visit `https://www.idtheftcenter.org/notified` and search for a breach by time frame, attack vector, and/or company name:

Business Name	Breach Entry Date	
Great Companions (third-party CommerceV3)	10-20-2023	+
Peerstar LLC	10-20-2023	+
Pisenti & Brinker LLP	10-20-2023	+
Revival Animal Health (third-party CommerceV3)	10-20-2023	+
Trust Benefit Technologies, LLC	**10-20-2023**	−

Technology	**Victims Impacted** 2989	**State** California
	Record Type/s Exposed	
Breach Reported 10-19-2023	Sensitive Records Exposed	**Date of Breach** 05-16-2023

Figure 1.5: Sample list of data breaches from notified on ITRC

ITRC Data Breach Reports

In addition to *notified*, ITRC issues quarterly, semi-annual, and annual data breach reports. These reports highlight a lot of eye-opening data relating to breaches that have occurred.

The following diagram shows the *2022 Annual Data Breach Report*, which indicates total compromises, total victims, and the top 10 compromises among other useful data points.

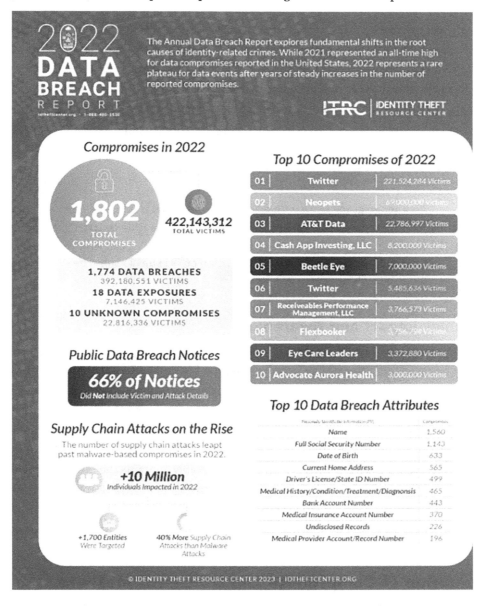

Figure 1.6: The ITRC 2022 Annual Data Breach Report key findings

Source: https://www.idtheftcenter.org/publication/2022-data-breach-report/

Wikipedia's List of Data Breaches

Another good reference for reviewing breaches is Wikipedia's *List of data breaches* page. This is quite a comprehensive list of many of the major breaches referencing back to their sources: https://en.wikipedia.org/wiki/List_of_data_breaches. As you review the breaches on the Wikipedia page and understand how they occurred, you will see a common trend where, for the most part, the breach occurred due to hacking or poor security practices. You might also notice that other common methods of breaches include lost or stolen equipment. These statistics are alarming, and they indicate how critical it is to implement a mature cybersecurity program to reduce risk as much as possible.

 Another great reference with a very powerful visual is provided on Visual Capitalist, which shows the 50 biggest data breaches from 2004 to 2021. The number of records lost from the 50 breaches totals 17.2 billion. You can view the visual here: https://www.visualcapitalist.com/cp/visualizing-the-50-biggest-data-breaches-from-2004-2021/.

Consumer Impact and Awareness

As a consumer, you're probably thinking, "Have I been impacted? And how would I even know if I've been impacted?" Ethically, an organization that has suffered a data breach with your information should inform you. But this is not always the case as many organizations may not be required to notify their customers because of a lack of regulation and/or the data type that has been compromised. I've encountered this firsthand, and I have reached out to organizations stating that I know my information has been compromised to challenge them. If you want to do some research on your own, one resource that probably contains the most comprehensive dataset of compromised information is https://haveibeenpwned.com/. Here you will be able to search the database to see if your email address has been part of a previous breach. You can also sign up for notifications for any breaches using your email address or submit a specific domain to be notified on.

Assumption of Compromise and Defensive Measures

I personally go with the mindset that my data has already been compromised. And there's a high possibility your account information, including passwords, is sitting on the dark web somewhere.

Because of this, we need to be more careful, and look at ways to be better prepared to handle any situation that arises when our personal data is being used for any fraudulent activity.

For example, in the United States, purchasing identity protection as a service to monitor your identity can serve as an insurance policy if you incur any damages. In addition to this, the ability to place your credit reports on hold to prevent bad actors from opening accounts under your name is an example of a defensive approach that you can take to protect your personal identity.

 There are many identity protection plans available today. A couple of notable ones include Norton LifeLock (`https://www.lifelock.com/`) and Aura Identity Guard (`https://www.identityguard.com/`). For those in the United States, you can lock your credit record for free online on each of the credit bureaus' websites: Experian, Equifax, and TransUnion.

As statistics show, we have an extremely challenging road ahead of us as we continue to defend against very mature threat groups throughout the world. And as already stated, even more concerning are organizations that are beginning to close their doors forever because of the added burden of these types of threats.

Skillset Challenges

Let's take a closer look into some of the current skillset challenges we face in the current state as cybersecurity leaders. We will cover multiple different data points along with some of the ways the industry is looking to address the challenges. With these challenges, it is important you are doing everything you can to retain your employees and provide a work environment they want to continue to work in.

Common Cybersecurity Roles

Over the years, many roles that never existed before are appearing within the cybersecurity world, and new skillsets are always needed. The following are some of the more common cybersecurity roles that you can expect to see within a cybersecurity program:

- CISO/CSO
- IT Cybersecurity Manager/Director
- Cybersecurity Program/Project Manager
- Cybersecurity Analyst/Architect/Engineer/Administrator

- Cybersecurity Software/Application Developer/Engineer

- Cryptographer/Cryptologist

- Cybersecurity Consultant/Specialist

- Network Cybersecurity Analyst/Architect/Engineer/Administrator

- Cloud Cybersecurity Analyst/Architect/Engineer/Administrator

- Penetration Tester

- Cybersecurity Auditor

- Governance Manager

Obsolete, Persistent, and Emerging Roles in Cybersecurity

To expand on the roles mentioned above, it is important to understand the evolution of cybersecurity roles to ensure your cybersecurity program remains relevant and up to date. You must continue to assess the current state and ensure your current employees are evolving into newer, more relevant roles. At the same time, when you hire new resources, you need to assess whether they are suitable to support new emerging technologies and threats or not. The following table provides an example of some of the obsolete roles along with those that are currently persistent, with examples of more modern emerging roles that may be needed within your organization to meet today's challenges.

Obsolete Roles in Cybersecurity	Persistent Roles in Cybersecurity	Emerging Roles in Cybersecurity
• Network Administrator • Firewall Administrator • Web Application Security Tester • AV Specialist • IDS/IPS Administrator • Web Server Administrator • Help Desk Technician • Windows Update Administrator	• CISO/CSO • Cybersecurity Analyst/Engineer/Administrator • Incident Responder • Compliance Specialist • Penetration Tester • Cryptographer • SOC Manager • Vulnerability Administrator	• Cloud Cybersecurity Architect • IoT Cybersecurity Specialist • AI/ML Cybersecurity Engineer • Threat Hunter • Deception Technology Specialist • DevSecOps Engineer • Blockchain Cybersecurity Specialist • Cybersecurity Governance Manager

Figure 1.7: Obsolete, persistent, and emerging roles in cybersecurity

High-Level Cybersecurity Organization Structure

As an example, the following shows how the hierarchy in a typical cybersecurity organization may look through an organization chart. Every organization is different, but this will provide you with a basis of what to expect:

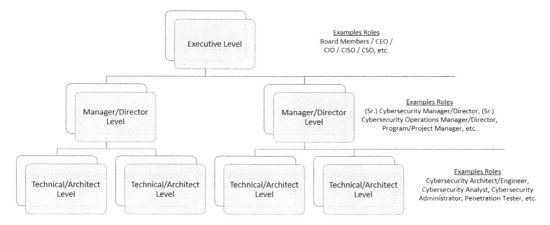

Figure 1.8: Example organization structure

 We will be covering the organization structure in more detail in *Chapter 2, Setting the Foundations.*

Shortage of Cybersecurity Expertise

One major challenge we currently face in the cybersecurity industry is a shortage of the needed expertise within the field. To put things into greater perspective, the IBM *Cost of a Data Breach Report* referenced previously noted that organizations with higher levels of security skills shortages observed an average of $5.36 million in costs from each breach. As you can imagine, only 5 years ago, cybersecurity wasn't necessarily something every organization envisioned as requiring a dedicated team. Fast forward to today, and every organization is frantically looking to build (if they don't already have one) or onboard a **Managed Security Service Provider** (**MSSP**) to meet the demand of the ongoing threats we are continuously dealing with on a day-to-day basis. For some, this is a requirement, while others are reacting from an already experienced cybersecurity incident, and others still are observing the increased risk and growing number of breaches that continue to make headlines.

Regardless of the reason, there is a need to fill millions of open roles in which the majority are skillsets that need some form of expertise to be successful. Acquiring these skillsets doesn't happen overnight, forcing us to re-think the way we hire cybersecurity professionals.

ICS2 2023 Cybersecurity Workforce Study

An *ICS2 2023 Cybersecurity Workforce Study* that surveyed 14,865 global users revealed:

- The global cybersecurity workforce continues to grow, by 8.7% in this year's report.
- The gap in cybersecurity professionals needed also continues to grow with a year-over-year increase of 12.6%.
- It is estimated that the global cybersecurity workforce is approximately 5.4 (4.7 in 2022) million.
- There is a 4 (3.4 in 2022) million worldwide worker gap within cybersecurity.
- Layoffs are not uncommon within cybersecurity with 22% reporting they experienced a layoff.
- The current threat landscape is viewed as the most challenging within the previous 5 years according to 75% of respondents.
- There is a slight drop in job satisfaction among the cybersecurity professionals with a 70% representation (4% drop from last year).
- Cloud computing skills are the most common gap and the hardest to find among qualified employees.
- AI and ML are among the top five skills in demand for the first time.
- It was reported that 39% of respondents know someone who has, or have themselves, been approached by a threat actor for malicious intent.

Source: https://www.isc2.org/research

As you can see from the data provided by **ICS2**, we have a huge challenge ahead of us to fill the current gap within the cybersecurity industry. This gap isn't going to be addressed overnight, and in reality will take years. But the good news is that this has been acknowledged at the highest levels and there are many great initiatives in place to help reduce this gap.

National Cyber Workforce and Education Strategy (NCWES)

The following strategy released by the Biden-Harris Administration known as the NCWES aims to help address the skills gap shortage through partnering with educators, organizations, and government entities: `https://www.whitehouse.gov/briefing-room/statements-releases/2023/07/31/fact-sheet-biden-%E2%81%A0harris-administration-announces-national-cyber-workforce-and-education-strategy-unleashing-americas-cyber-talent/`.

One example of these great initiatives and one listed on the NCWES is from ISC2, who has pledged for 1 million individuals to receive the *Certified in Cybersecurity* certification for free (as of May 2024), which includes both training and the certification exam. This is an incredible initiative and one that I share with others as part of mentoring and those looking to break into the cybersecurity field: `https://www.isc2.org/landing/1mcc`.

In addition to this initiative and the support from many educators, organizations, and government entities is an incredible community of cybersecurity professionals who continue to bring awareness and help educate others to break into the cybersecurity field. There are many forums where collaboration occurs through local and national events, conferences, local chapters, educational institutions, and social platforms like LinkedIn.

Addressing the Talent Gap with Outsourcing

As per the statistics above, there may be several cybersecurity openings within your team. If this is the case, you may want to look at some immediate options to outsource some work until you can appropriately staff your team. As we are all aware, onboarding doesn't happen overnight. With an outsourced approach, you can work to bring in temporary resources as needed to fill the gap as you look for permanent hires. We cover resource management in more detail in *Chapter 4, Solidifying Your Strategy*.

Retaining Top Performers

Another important topic is the ability to hold on to your resources, more specifically your top performers. In recent years we have seen what has been categorized as *The Great Resignation*, where millions of workers have been quitting each month at record numbers in the United States to look for better opportunities, create a better work-life balance, and for other various reasons. Because of this, you need to ensure you take care of your employees, especially your top performers, and create a work environment that supports their needs.

Any good leader knows that losing a good employee can be extremely impactful and the cost of replacing a good worker versus providing additional compensation (as an example) can be substantial. One quote that has always stuck with me is one from Steve Jobs, and I can personally attest to this statement:

> *"A small team of A+ players can run circles around a giant team of B and C players."*

It is important to ensure that your cybersecurity program includes a diverse workforce. The *ICS2 2023 Cybersecurity Workforce Study* referenced above also includes a section on **Diversity, Equity, and Inclusion (DEI)**, which shows diversity within in cybersecurity is moving in the right direction, which is great. Although, more progress is still needed, as women in the under-30 group only represent 26% of the cybersecurity workforce.

Methods of Staying Current

As you look to retain your top performers and provide a work environment where employees want to stay, make sure you are encouraging them to stay up to date, but at the same time provide them a platform and the necessary time to self-educate. For example, the following table provides some methods with which you can allow your employees to update their skills and remain current.

Method	Description	Priority	Why
Cybersecurity Certifications	Study and become certified in industry-recognized certifications. Some of the more common certifications come from ISC2, ISACA, CompTIA, EC-Council, and GIAC.	High	To help with applying for new positions, with promotions, to generally elevate your career, and remain up to date.
Attending Conferences	Participating in cybersecurity conferences to learn about the latest trends, threats, and solutions. Attend industry-specific cybersecurity conferences (e.g., Black Hat, DEF CON, RSA, etc.)	High	Direct exposure to the latest threats, solutions, and networking with experts.

Research and Analyst Companies	Subscribing to research reports and analysis from firms like Gartner, Forrester, or IDC.	High	Market trends, technology evaluations, and strategic recommendations.
Regular Meetings with Suppliers	Engaging with strategic suppliers to understand new technologies and solutions they offer.	High	Insight into product roadmaps, innovative technologies, and potential collaborations.
User Groups/ Forums	Joining cybersecurity user groups and forums to discuss challenges, share insights, and learn.	Medium	To learn from others in the industry, collaborate with like-minded professionals, and grow your personal network.
Business Social Media Channels	Following cybersecurity thought leaders and industry updates on platforms like LinkedIn. Checking dedicated cybersecurity news sites (The Hacker News, Krebs on Security, etc.).	Medium	Breaking news, vulnerability disclosures, and threat analysis. Quick updates, but requires careful filtering of information.
Online Courses and Webinars	Enrolling in cybersecurity courses and attending webinars to acquire new skills and knowledge.	Medium	Validate knowledge, increase credibility, and stay aligned with best practices.
Reading Industry Publications	Keeping up with cybersecurity news, articles, magazines, journals and publications from reputable sources.	Medium	In-depth articles, analysis, and case studies.

It is important we don't overlook the importance of providing time for employees to remain up to date. We often get so busy with projects and operational items that time doesn't allow for these activities. As a leader, you must make time. A very relevant quote I like to reference by Sir Richard Branson is as follows:

> *"Train people well enough so they can leave. Treat them well enough so they don't have to."*

Challenges in the Hiring Process

Switching topics, I continue to observe a lot of feedback and challenges being publicized with the hiring process as it relates to cybersecurity. One area worth mentioning is that of new cybersecurity professionals trying to land their first role encountering unrealistic requirements on many of the entry-level job descriptions, for example, a resume that states entry level but requires 10+ years' experience. You may laugh but they are out there. Another is an issue that not only haunts the cybersecurity industry but is an ongoing issue in general with the overall hiring process being very long, with unrealistic never-ending interviews, and overall being very legacy and frustrating.

Innovative Hiring Practices

As leaders, we have the ability to break down barriers and influence change in this area as we partner closely with our HR leaders. Filling cybersecurity positions doesn't necessarily mean hiring those with current experience. There is an abundance of roles that can be filled by those looking to enter the field who are very smart and hungry to learn. You need to think outside the box to meet your hiring needs, especially by looking within your own organization to those who are familiar with the business, want to learn, and can get the job done. Keep the job descriptions simple and don't list unrealistic requirements. I'm also a believer that you don't need to make a degree a requirement when hiring, as some of the best workers I've had on my teams don't even have a bachelor's degree. For the general hiring process challenges, you may not be able to change the overall application process, but you can change your responsiveness to your applicants, speed up the hiring process, and promote a more agile approach in your hiring to create more efficiency.

Changing the Negative Perception of Cybersecurity

One final item we need to tackle is that of a negative perception by some of the cybersecurity industry. I've heard it firsthand where great talent is considering entering the cybersecurity industry but they decide not to because of the perception that there is a requirement to work non-stop and the stress can be extremely challenging. This is true in some respects, as shown with data in the upcoming section, *Prioritizing Well-Being*. I have also observed this firsthand within the security operations and incident response functions, where there are ongoing fires while dealing with never-ending security incidents. As leaders, we are the only ones who can change this perception to create a more welcoming environment and one that doesn't involve the ongoing demands that are causing long hours and high stress. It is our responsibility to influence change and we need to start now to better protect our team's well-being.

On another note, it is important to make others aware that there are many other functions within cybersecurity that have much less demand, which we will cover throughout the book.

Encouraging Collaboration and Mentorship

As you can see, we have a challenging road ahead of us and one that won't be solved in the short term. I do see a lot of interest from those who haven't worked in cybersecurity and are looking to break into the cybersecurity field. I have mentored and continue to mentor many to help them break into cybersecurity, only to watch them struggle to land a job because of the rigorous requirements and the expectation of hiring only experienced professionals. Remember, we all started somewhere. Let us not see a candidate just based on their experience, but also see how trainable they are, and whether they have a hunger to learn, adapt, and be trained. It is important we create an environment where we allow collaboration, cross-training, and a place where we encourage knowledge sharing and the ability to enhance those around us. Most importantly, building an environment that becomes a place your team wants to work and enjoys working will bring the best out in everyone.

Prioritizing Well-Being

It would be remiss not to touch upon one extremely important topic we continue to hear more about; a topic that we tend to put secondary to everything else in life, and more specifically work. It is that of our own well-being, and the well-being of those who work for us and around us. We have come to live in a world where there's a mindset that has been instilled that we *live to work*, when the reality is we should be *working to live*. As leaders, we need to ensure that we have the health and wellness of those who work for us and those around us at the forefront of our priorities. Cybersecurity can be an extremely demanding field to work in and burnout and mental health are real issues we need to manage head-on as leaders. One topic in particular that needs addressing more is that of mental health, not just in cybersecurity, but in general. There has always been a stigma around mental health, making it difficult for anyone to be open. At the end of the day, the brain is an organ that requires the same care and attention as any other organ within our body. If there is an issue you are struggling with mentally, it should be addressed like any other organ without feeling uncomfortable. Unfortunately, we are hearing more mental health concerns and an increase in mental health issues within the cybersecurity field. Or, we may just be becoming more aware of the situation and others are beginning to speak up. Either way, data is beginning to show that we have an issue that needs addressing.

Data on Well-Being and Burnout

On the flip side, we are beginning to see more data on well-being, mental health, and burnout issues, which allows us to better understand what we are dealing with. As leaders, this allows us to take immediate action to ensure we provide the support and resources needed. To better understand how real the situation is, a simple Google search (or question to ChatGPT) for *mental health articles in cybersecurity* will return countless articles and research to provide a clearer picture of what we are dealing with. I highly encourage all cybersecurity leaders to do this, so you have a better understanding of what challenges we face on this critical topic.

Statistics on Mental Health in Cybersecurity

Let's look at some of the data available.

A study from over 1,000 cybersecurity professionals by Tines in 2022 provided the following:

- 27% say their mental health has declined over the past year.
- Only 54% say their workplace prioritizes mental health.
- 63% say their stress levels have risen over the past year.
- 64% say their work impacts their mental health.
- 51% of respondents have been prescribed medication for their mental health.

Source: `https://www.tines.com/reports/state-of-mental-health-in-cybersecurity`

The 2021 *Global Incident Response Threat Report* from VMware found:

- During the past 12 months, 51 % of respondents experienced extreme stress or burnout.
- Of the 51 %, 65 % said they have considered leaving their job because of it.

Source: `https://blogs.vmware.com/security/2021/08/combating-cybersecurity-burnout-through-self-care-empathy-and-empowerment.html`

Some other article headlines include *Gartner Predicts Nearly Half of Cybersecurity Leaders Will Change Jobs by 2025, 25% of Cybersecurity Leaders Will Pursue Different Roles Entirely Due to Workplace Stress* from a Gartner press release (`https://www.gartner.com/en/newsroom/press-releases/2023-02-22-gartner-predicts-nearly-half-of-cybersecurity-leaders-will-change-jobs-by-2025`), *Concern for cybersecurity workforce mental health is rising* from an article on Healthcare IT News (`https://www.healthcareitnews.com/news/concern-cybersecurity-workforce-mental-health-rising`), and *Ransomware's Relentless Rise Strains Security Teams* from a Mimecast blog.

Strategies for Promoting Well-Being

It is important that we take this data seriously and begin to influence change that provides a positive and healthy work environment for everyone. Of course, there will always be times of stress and the need to work longer hours than normal. But we cannot allow this type of environment to be sustained as it will catch up with us and the burnout will be real. For example, some of the more relevant areas where burnout is more prevalent and that require closer attention include SOCs with a 24x7 operation, vulnerability management with the potential of thousands of vulnerabilities to review each month, and incident response where there is pressure to mitigate and understand if any exfiltration activities have occurred. However, no specific function within cybersecurity is immune to burnout.

Some thoughts to help you prioritize the health and well-being of those on your team and around you include:

- Better understanding the burden of your team.
- Taking responsibility to ensure your employees are well supported.
- Spending extra time checking in on your employees. Maybe schedule weekly touchpoints.
- Ask them how they are doing often and if they need any support.
- Always make yourself available to them and make sure they know about this.
- Respect your employees personal time and be flexible with their schedules by providing compensation time when earned, time away for family events and appointments, allowing remote work options, etc.
- Remind them to take time off and ensure they are not working off-hours.
- Make sure your employees are aware of any HR-related well-being programs or company-sponsored mental health resources and encourage them to take advantage of them.
- Ensure your team is staffed to support each other. Your employees should not always be overworking. If they are, you need to address the resource issue ASAP.
- Provide a safe space for employees to express concerns about workload before burnout occurs.
- Celebrate small wins and show regular appreciation, which will make them feel good and worthy for the company.
- Organize team-building activities to strengthen social connections and enhance team cohesion.
- Lead by example.

Strategies for Individuals and Leaders

Most importantly, you must take care of your own health first. If you are not at your best and taking care of yourself, then you can't be the best for your employees. You need to build good habits for your employees to observe and follow. Some basics of self-care include:

- Getting enough sleep.
- Be active by moving around or getting exercise daily.
- Get into nature.
- Eat a healthy diet.
- Engage in social activities and stay connected.
- Meditation or other mindfulness practices to cultivate mental clarity.
- Learning to say "no" to non-essential requests.
- Activities that bring joy and relaxation, separate from work.
- Manage your stress.
- Have FUN!!!

Make sure your employees are aware that you have their well-being at the front of your mind and share and encourage the items mentioned above. Also, make sure there is a comfortable space for your employees to discuss well-being and encourage the conversations to happen.

A simple phrase to help remind yourself and others:

Step Away > Disconnect > Refresh

Summary

The digital world continues to grow at an incredible pace with new technology and innovation constantly being released, as we observed with the opening section. This transitioned into the current threat landscape and the challenges we face as cybersecurity professionals. In the section that followed, we took a deeper dive into the importance of statistics to ensure that support and buy-in are offered by the business leadership teams. We also reviewed breaches in more detail and saw how they continue to become more prevalent. This shifted into the next section on skillset challenges and the challenges ahead of us with filling the talent gap in cybersecurity. We then finished off the chapter with an often-overlooked priority: the need to prioritize well-being.

In the next chapter, we will be looking into setting the foundations of your cybersecurity program and strategy. Having a solid foundation in place will only set you up for greater success. We will be focusing on the importance of learning about the broader organization and why business relationships are a critical component of being a leader and the overall cybersecurity program. We will briefly touch upon financial management before going into more detail on defining your cybersecurity organization. We will then investigate the building blocks for your cybersecurity program, including a discussion around the importance of risk and how to manage it. We will finish with an overview of change management and the importance of this often-overlooked topic.

Join our community on Discord!

Read this book alongside other users, Cybersecurity experts, and the author himself.

Ask questions, provide solutions to other readers, chat with the author via Ask Me Anything sessions, and much more. Scan the QR code or visit the link to join the community.

`https://packt.link/SecNet`

2

Setting the Foundations

As discussed in the previous chapter, we now understand the current state, which isn't good, and we have a lot of work ahead of us to build or reconstruct our cyber-resilient strategy. Before getting started on this journey, it's important to lay the foundation of what the strategy for your program will look like. To do this, we will cover some very important subjects that will help you better set the foundations to allow greater success for your broader cybersecurity program.

As a cybersecurity leader, it's never been more important to integrate yourself with the broader organization. This means you need to learn the business inside out and build relationships throughout, including the top levels. In your role as a leader, you will need to build positive relationships within every function of your organization including the C-suite. Without these relationships, your success will become an uphill battle. Although you don't need to be a **Chief Financial Officer (CFO)**-level expert, it is critical that you know how to manage finances and speak the same language as the finance team when the need for funds arises, which will be often. You will need to ensure you fully understand the budgetary requirements for the cybersecurity program, especially as the program continues to grow, which will be the case for most at this time.

Whether you are new to the organization or new to your role, you are going to need to review and assess the current cybersecurity posture/state of the organization and determine what changes are needed, if any. Every organization will be different but it's important you understand what is needed to ensure you have the coverage you need to support the broader program. Laying the foundation of your program requires you to clearly define how the program will look and what it entails. This is where you need to define the functions that fall within your program. Finally, some often-overlooked components are change management and communication.

It is important that this is at the forefront of your program for the organization. In this chapter, we will be covering the following topics:

- Learn the business
- What about finances?
- Building blocks for your cybersecurity program
- Defining the cybersecurity organization
- Change management

Learn the Business

The role of a cybersecurity leader within an organization has significantly changed and grown in a short amount of time. The cybersecurity function is no longer a standalone function that has traditionally operated in silo from the rest of the organization. The role of cybersecurity within the organization touches every aspect of your business and is one that requires new evolving qualities. A cybersecurity leader has traditionally been thought of as a technical type of leader, but this role has shifted to become more of a business acumen that can represent risk at the broader organization level. The need to translate and relate technical skills still holds incredible value at this level. Having both these qualities is becoming a requirement for this role. Because of the importance of cybersecurity, we are seeing a significant increase in the need for C-level representation at the executive and boardroom levels, and this role as we know it today is the **Chief Information Security Officer (CISO)**. Many organizations are looking to fill this newly created role.

Understanding the Business Environment

How much you need to learn about the broader business will be determined by whether you are a new leader in the organization or promoted to a leadership role within the organization. In both instances, there will be some level of learning needed. If you already work within the organization, you will have a significant advantage, although you will still have learning ahead of you. As you grow within an organization, you will continue to receive broader responsibilities along with gaining more visibility at a broader level. As this occurs, you gain new peers, and you naturally begin to get more insight into the executive leadership team, which will require learning. Someone new to the organization will have a significant amount of homework to do to get more acquainted with the new organization. You will not only need to learn more about those around you and how they operate, but more importantly how the business and functions operate and all the intricacies that make the organization successful. Unfortunately, this doesn't happen overnight, so patience will be needed as you continue to learn.

Some examples that can help one to learn more about an organization include:

- **Business function deep dive sessions**: Collaborate and connect with leaders from other areas within the organization such as finance, legal, marketing, and sales to understand their mission and vision, challenges and goals, values and culture, decision-making processes, and more.

- **Shadow other business leaders in the organization**: Observe other business leaders to gain insights into their daily operations.

- **Documentation review**: Request and review documents outlining other business function strategies, budgets, **Key Performance Indicators (KPIs)**, and anything else relevant.

- **Business function collaboration**: Engage in cross-functional projects to collaborate with different departments to build meaningful relationships and understand their needs.

- **Stakeholder interviews**: Meet with key stakeholders across different functions to understand their roles, challenges, and any other needs.

- **Business process mapping**: Work with teams to map out critical business processes.

Every organization is different and will hold its own unique set of personalities, especially across different industries. Organizations within the same industries will have similarities but are guaranteed to be different in many ways. The great benefit of being a security practitioner is that no matter which industry you work in, to an extent, cybersecurity is going to be the same across all of them. Unlike most careers, where you are limited by your knowledge within your field and industry, being a cybersecurity professional allows greater opportunities. As you grow into leadership roles, this does change as understanding the business and industry you are working in becomes a critical requirement as a leader. As a cybersecurity leader, you need to continue to learn everything about your organization. This is not a one-time endeavor; this is an ongoing quality that's needed to be successful in leadership. Organizations change often, and you need to evolve with these changes to be efficient.

Embracing Organizational Principles and Culture

As a cybersecurity leader, you must become familiar with and live by some standard principles within a company. Every company will (should) have a mission and vision, which you need to be very familiar with as a leader. In addition, you will need to understand what values the organization follows, the culture it upholds, and the organizational structure it has in place. All these together will be core to your leadership knowledge within any organization. Ensuring your team is fully aware of these principles is also very important as a leader.

You must create an environment of transparency that allows you to build trust not only within your team but the broader organization. This is critical.

In addition to the organization's mission and vision, you will want to create one for the cybersecurity team. This allows the broader organization to better understand what the cybersecurity team's value proposition is within the organization. This is something that can be shared with other function leaders, so they understand what your purpose is within the organization's broader mission and vision. We will provide an example in the *Defining the Cybersecurity Organization* section of this chapter.

As previously mentioned, understanding the organizational structure of the business is critical as you will need to build meaningful relationships throughout the entire business, especially with C-level leaders within the organization. As stated, every organization is different, but to give you an idea of some of the C-levels you can expect to interact with, the following organization chart should help:

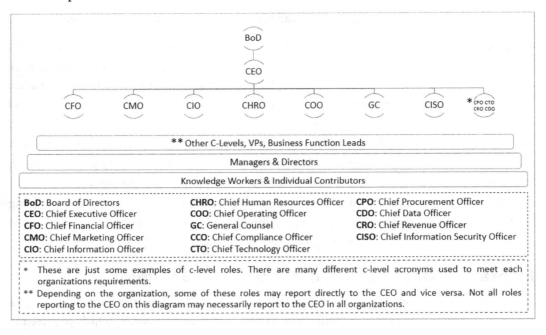

Figure 2.1 An example of C-level roles within an organization

As a reminder, the new version of a CISO in today's modern world needs to be ingrained in every part of the organization to create more value and reduce risk as much as possible. The more you learn about the business and understand the purpose of your organization's existence, the more successful you will be as a leader.

I believe it's time to acknowledge that a new generation of CISOs is upon us, that of the CISO v2.0.

Business Relationships

As you continue to learn about the business and become engrained with the culture of the organization, you will need to put a lot of energy into building relationships throughout the organization. As we've touched upon already, the roles of a cybersecurity leader and CISO continue to evolve and today, the role requires the need to be fully integrated into every part of the organization. This requires the need to meet with and continuously interact with other leaders throughout the organization. You need to consider yourself more of a business partner rather than a security guru.

As a business partner, it is not your place or role to tell the business what they can and cannot do. It is your responsibility to be able to translate the security and technical risk with a given initiative into business risk.

For example, if the marketing team would like to launch a new web application, which, upon reviewing the solution, you identify to have a significant amount of risk because **personally identifiable information (PII)** will be hosted on the application, it is not your place to say no, we can't proceed with this. A good leader and business partner will work with the marketing team to identify the requirements for the proposed solution and guide the team through the correct process to ensure the appropriate architecture and security controls are in place to reduce risk. If there is still a substantial amount of risk, it is your responsibility to present that risk in a way that allows the business and leadership team to make an informed decision on whether to proceed with the proposed web application. What you will find is the business always pushing more on the usability of a solution, which typically means fewer security controls. It will be important to ensure you are able to bring the right balance of usability with the required security controls in place. This is not always an easy conversation, but it is something you will need to work on as a leader as you collaborate with the business. At the end of the day, it is all about being a good partner to build positive relationships and more importantly trust, without compromising the business.

Previously, we discussed the importance of learning the mission and vision of the organization. In addition, you will want to work with the other business functions and/or unit leaders to understand each of their missions and vision. Not every function or unit may have its own mission and vision, but if they do, understanding what each of them stands for within the organization will help you better understand how to support them more efficiently with their needs. Each of these functions and units will most likely have its own values and culture within the organization, which will be important for you to understand.

As you connect with other leaders throughout the organization, make sure you ask them for any documentation that can help you better understand and support their purpose within the organization.

You will also want to better understand the different roles and responsibilities within each of the business functions or units, and how each of them is structured. What layers are involved in each of the structures, for example, executive leadership, senior leadership, middle-level leadership, management/leads, front-line workers, and so on? As you learn more about the organization, you will learn quickly that every function or unit will operate in its unique way. For you to be able to cater to each of them, it will be important for you to have a good understanding of how each function operates to better support their needs. This is a lot of work, which will require time and patience. But the more you build upon your knowledge across the different functions and units, the stronger those business relationships will become.

One of the most important principles I follow, and I would consider this my most important principle, is that of trust. From the day you walk into an organization for the first time, you begin to build trust. Trust is the foundation of being a great leader and it doesn't happen overnight. You cannot buy trust or expect someone to simply start trusting you. Trust is earned over time. Many qualities help build trust within a leader, some of which include:

- Demonstrate competence: Build resiliency, show expertise, and deliver results.
- Communicate effectively: Always be transparent, be an active listener, and deliver meaningful communications.
- Show integrity: Be your authentic self, always be empathetic, take responsibility, and ensure an ethical decision-making process is in place.
- Inclusion and collaboration: Make yourself accessible, ensure inclusive decision making, and celebrate success.
- Adaptability: Always be open to change and ensure continuous learning.

Building trust across the organization will take time, but it will come as you deliver and provide value to the organization. Most importantly, trust can be lost in a heartbeat with one wrongdoing. The reality is we all make mistakes, and the role of a cybersecurity leader comes with a lot of pressure and stress. How we handle these mistakes will determine how that trust is portrayed.

Remember, as a cybersecurity leader or CISO, you are now in a position to support the broader organization and help with its success. Don't create friction across the organization with other functions or units; I see this far too often and it really does make it a challenging and uncomfortable environment to work in.

Be a positive partner and build (and continue to build) positive relationships throughout. This will only provide for more success as a cybersecurity leader and CISO.

Navigating the Business

As a leader, knowing where the resources are within the organization provides a significant advantage as you navigate the complexity of requirements within cybersecurity. To be honest, this applies not just to leaders but to anyone in general. If you have worked in a large enterprise, you will understand that navigating the organization for the correct resources can be a challenge in itself. Approvals, exceptions, funding, requests, and so on, may not fall within one function in your organization. You may also be part of an organization that consists of many smaller organizations, a parent organization, an overseas organization, multiple regional organizations, and so on. This only makes navigating the business even more challenging.

Take, for example, the need to follow up on specific controls on an audit within your organization. Several of these controls may need different parts of the organization to review, provide evidence, and sign off. Another example may be the need to onboard a security tool for an application that is being implemented by a sub-division and the budget to cover this tool will need to come from that sub-division. You will now need to navigate the complexities of that sub-division to discuss and request funding for a new security tool that they may not have considered for their application. Think of a third-party application onboarded by a recently acquired organization that now has a major vulnerability identified. You will possibly need to navigate the complexities of a separate leadership team to work through remediating the vulnerability before any major security incident occurs. Each of these examples demonstrates the need to know how to navigate the business and know where your resources are within the organization. You may not know where to go directly but the more you build positive relationships within the organization and know who and where your resources are, the quicker you will be able to navigate the ongoing challenges that will continue to come your way.

Everything Is About Risk

As the title states, everything within an organization has some form of risk. This applies to the broader organization and not just cybersecurity. There is an amount of risk that needs to be managed.

Managing risk is everyone's responsibility and, for the most part, will not be within our control. Risk needs to be closely monitored in all aspects of the organization. At a high level, some of the more recognized types of risk within an organization include:

Organizational Risk	
	Cybersecurity
	Financial
	Market
	Legal
	Competition
	Social Media
	Operational
	Environmental
	Technology Outages
	Crime
	Supply Chain
	Reputation
	Political
	Compliance
	And More.......

Figure 2.2: Different types of organization risk

There are endless risk factors that an organization must track against and far too many to list here. Unfortunately, cybersecurity risk continues to appear as a top concern for many organizations. At one point in time, this wasn't the case. Today, the risk is real and recognized by most. To provide insight into what organizations are listing as their top business risk, Allianz releases an annual survey of key business risks around the world known as the Allianz Risk Barometer. In 2023, the survey was completed by over 2,700 respondents from 94 different countries and territories. For the second year in a row, cyber incidents were listed as the top risk, the first occurrence of this within the survey. The following image provides the most important business risks identified from the *Alliance Barometer* for 2023.

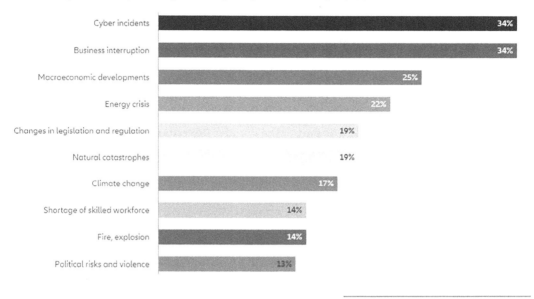

Figure 2.3: Allianz Risk Barometer 2023 (source: `https://commercial.allianz.com/news-and-insights/reports/allianz-risk-barometer.html`*)*

As a cybersecurity leader, everything you manage revolves around risk within cybersecurity. You will, of course, need to be familiar with risk in general and those items listed previously and in the chart provided. With direct oversight of cybersecurity, you will need to learn how to effectively translate cybersecurity risk into business language to help the broader organization make informed decisions. This will be covered in more detail throughout the book and more specifically in *Chapter 14, Managing Risk*. As mentioned previously, it is not our place to tell the business what they should and shouldn't do. Rather, it is our responsibility to provide constructive feedback and provide awareness of the cybersecurity risk with any given solution or idea.

Risk in general should be managed at the broader organizational level. There shouldn't be one individual responsible for accepting risk, especially high or critical risk. Risk needs to be accepted by the business. If you are within a large enterprise, you may have a **Chief Risk Officer (CRO)** with a team and processes in place to manage risk more broadly. Chances are, most don't have this function in place and need to ensure processes, policies, and committees are in place to allow risks to be reviewed and accepted accordingly. This includes any identified high and critical risk going to executive leadership and/or the board for review and approval/denial. Of course, lower-level risks wouldn't make sense going all the way to executive leadership and/or the board, but having processes in place to track, document, and approve the lower-risk items is critical for compliance.

Risk is an integral component of every organization and certainly every cybersecurity program. We have barely touched upon the topic of risk within this chapter, and risk within the cybersecurity program is typically tracked as part of the broader **Governance, Risk, And Compliance (GRC)** function. Because of this, we have a chapter dedicated to risk, *Chapter 14, Managing Risk*.

Now that we have a better idea of what is needed to learn more about the business, we will cover the need to understand the financial requirements for the cybersecurity program.

What about Finances?

Everyone's favorite topic: managing the budget. As you continue to grow your career into management and leadership, you will find the responsibility of financial management continues to become a larger responsibility. The reality is that everything about an organization revolves around the bottom line, otherwise known as the income. If there is no money available, a company simply cannot operate. Part of your role as a cybersecurity leader plays into the larger organization as it relates to finances. You will need to manage a set budget, you will continuously be challenged against that budget, and you will always be asked for ways to cut your budget. It is important that you manage finances efficiently and always be one step ahead when needing to justify the need to retain and potentially increase your budget needs.

Whether your organization is public, nonprofit, or private, you will need to understand financial management at the broader organization level. As a leader and someone who will probably be included in executive and board-level meetings, you will need to understand and speak finance at this level. Be familiar with the organization's financial statements when they are released, learn how to read them, and know what each of them means. Specifically, make sure you understand in detail the income statement, balance sheet, and cash flow statement.

You will also want to build a good working relationship with the **Chief Financial Officer** (**CFO**), as they will be the ones scrutinizing any request for additional funds, so, having them on your good side will only help as you look to gain approval.

 Since this isn't a book on financial management, we won't go into detail on the financial statements and what is included. However, there are many resources available to help you better understand and learn. For example, the United States **Securities and Exchange Commission** (**SEC**) has a *Beginners' Guide to Financial Statement* publication available for review: https://www.sec.gov/reportspubs/ investor-publications/investorpubsbegfinstmtguide.

As a cybersecurity leader running a program, you are probably asking yourself, "How much budget will I need?" For most, there is not an easy answer, as there is not a great deal of historical data to help support how much should be spent on a cybersecurity program. Unfortunately, you won't be able to sit in an executive-level meeting and ask for a budget without providing any justification or baseline data to compare against. This is where you are going to need to do some research and look for some guidance to better assist you when having conversations around budget. Some options include leveraging any current vendors you have on contract that provide consultancy types of services, leveraging advisory types of service such as *Gartner* and *Forrester*, checking with peers from different organizations, checking in with any collaborative and/or advisory groups you are a member of, or even referencing any articles or published reports available on the web.

To get you started, there is a recent report released by *IANS Research* and *Artico Search* named the *2023 Security Budget Benchmark Report*. This report provides data on the cybersecurity budget, is compiled annually, and includes responses from 550 CISOs. Some of the highlights of the report include:

- Although slower than normal, cybersecurity budgets continue to rise.
- 80% indicated budget increases were due to unexpected or non-typical events. An example includes a security incident.
- Of those impacted by a security breach, their budget increased by 18%.
- Staff growth in security was lower than the previous year (31%) but still showed growth with a 16% increase in budget allocation for hiring.
- Staff and compensation continue to be the biggest portion of spending within the cybersecurity budget at 38%.
- The cybersecurity budget consists of 11.6% of the IT budget and continues to rise.

One of the most important statistics to reference is the percentage of cybersecurity spending in comparison to the overall IT budget. All organizations should already have a pretty well-defined IT budget. With something to work with, you can estimate that 11.6% of your IT budget will be needed for cybersecurity. If you have a $20 million IT budget, then you can anticipate your cybersecurity program will need to be around $2.3 million.

In addition to the overall cybersecurity budget, you may be curious about what the breakdown of your spending is within the cybersecurity budget. Although every organization will be different, the same report provides an overview of the breakdown of costs within the cybersecurity program:

- Staff and compensation: 38%
- Off-premises software: 21%
- Outsourcing: 11%
- On-premises software: 9%
- Project: 9%
- Hardware: 6%
- Training and development: 4%
- Discretionary: 3%

 You can access the 2023 Security Budget Benchmark Summary Report here: https://www.iansresearch.com/resources/infosec-content-downloads/ detail/2023-security-budget-benchmark-summary-report.

As part of your cybersecurity program, you're going to need to track your budget and ongoing spending to allow efficient reporting back to the executive leadership team and the board. You will also want to break down that spending into different buckets to more efficiently show where the investments are being made. To track the spending, I suggest two different formats. The first is in the traditional spreadsheet format, such as Microsoft Excel. This allows you to efficiently itemize and track everything with auto calculations as your budget changes; I imagine this is already being done today. The second is in a more visual format that can be more efficiently shared with the executive leadership team and the board. I don't know about you, but I feel spreadsheets are not the most engaging and effective tools for formal presentations. With that, maybe something more simple like the following example could be better for sharing the overall financial spend for the cybersecurity program:

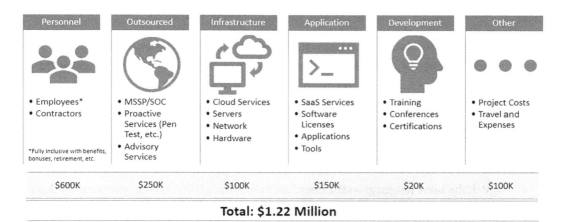

Figure 2.4: Cybersecurity program costs

Like anything within the business world, you are going to be challenged with everything from a financial perspective. Some examples include:

- **Return on investment (ROI)**: You will continuously be challenged on what the ROI is. Like the IT function over the years, you will be scrutinized as a cost center and by your expenses versus generated revenues for the organization.

- **Lack of awareness**: Those outside of cybersecurity will fail to understand the severity of the threat landscape and what is needed to reduce risk as much as possible. This will include your executive leadership team and board of directors. I see this often and it's our responsibility to help close this gap.

- **Competing priorities within the organization**: Other functions with revenue-generating projects may be perceived as a higher priority over cybersecurity projects that don't generate revenue.

As a leader, you will need to ensure you build constructive justification as to why you are requesting additional budget or need to maintain your current budget. Areas that you will need to be proficient in will include tracking financial metrics and KPIs, the ability to conduct cost-benefit analysis and risk assessments, understanding regulatory compliance costs, the consideration and benefits of cybersecurity insurance, the ability to forecast your financial needs, providing long-term financial requirements, and more. A lot of this is going to fall back on the premise of what this book will guide you through.

Some examples of how to navigate through these challenges include:

- **Build your foundation**: Determine the current state of your cybersecurity program using assessments and audits. This will identify gaps to justify the budget for critical and high-risk findings to reduce risk versus the cost of a breach

- **Cost of a major cybersecurity incident**: As I have mentioned, statistics and data are your best friend. Use other organizations that have been impacted by a cybersecurity incident as an example of the cost of a major cybersecurity incident, as discussed in *Chapter 1, Current State.*

- **Speak the same language as the board**: Be able to translate complex cybersecurity threats into business-related terms. As in the previous bullet, highlight the potential financial impact of a major cybersecurity incident.

- **Quantify cybersecurity risk**: Leverage **Cyber Risk Quantification (CRQ)** frameworks to estimate the potential cost of a major cybersecurity incident.

- **Show the ROI**: Show the positives for investing in cybersecurity such as preventing a major cybersecurity incident, maintaining compliance, providing strong security for business continuity, maintaining brand reputation and customer trust, and more.

- **Leverage regulations and compliance**: With the ever-growing regulatory requirements and the need to remain compliant, organizations need to ensure due diligence and that best practices are being put in place to prevent a major cybersecurity incident from occurring.

- **Ensure alignment with business goals**: Demonstrate how a secure environment builds better trust and allows the facilitation of innovation and growth.

- **Benchmarking against others**: Compare both your cybersecurity spending and program maturity with others in the industry to demonstrate the need for additional financial support to remain competitive.

- **Start small if needed**: Propose a phased approach beginning with **Proof Of Concepts (POCs)** to demonstrate the value of solutions before making a large investment.

 CRQ is becoming widely recognized and adopted as a method allowing for more efficient quantification of risk versus the traditional method of qualitative. We are in the early phases, but this is something you need to be reviewing as part of your program. One example of a vendor providing these capabilities is https://www.kovrr.com/.

Hopefully, you can work with your executive leadership team and the board to justify the need for the budgets you are requesting. I've seen it too many times when it has been difficult to receive approval for finances for cybersecurity-related activities until a major event has occurred. Make sure you do everything you can to justify the need for finances without it being a major event. In some sense, you are going to need to become a salesperson, and one with good data to justify the need to reduce risk within your organization. Bringing real data to the table will significantly help as you make your pitch for the needed finances.

A good leader will also identify that they don't necessarily need every tool available on the market, and implementing a new solution may not be worth the cost because of the small amount of risk at hand. This is why it's important that you are able to efficiently assess risk within the organization to make better-informed decisions on your cybersecurity financial needs. One additional thought is taking into consideration the dynamics of cybersecurity. It is not possible to predict everything that may happen throughout the year and there may be a need for an unplanned budget to help with a zero-day finding that needs to be fixed, a cybersecurity incident that requires additional resources to mitigate and restore service, or a new finding on an audit that occurred. Being transparent about these expectations with executive leadership and the finance team is critical. Maybe you need to increase your project funds or ask for additional finances to cover these unknown expenses when they occur, as they will occur.

Building Blocks for Your Cybersecurity Program

To provide additional visibility into how your cybersecurity program looks and what is involved, you will need to provide some form of representation of the program. Within this representation, you will want to highlight the specific functions needed to run your cybersecurity program. This can be done in many ways and may differ per organization and different industries you are in, but having this in place helps provide your leadership teams and the broader organization with better insight into how the cybersecurity program is being executed and operates on a day-to-day basis.

As we work through this book, we will follow the premise of ten core functions that I believe every organization will foundationally need as the building blocks for their cybersecurity program. These core functions are:

- Cybersecurity architecture
- Identity and access management
- Cybersecurity operations
- Cybersecurity awareness, training, and testing

- Vendor risk management
- Vulnerability management
- Proactive services
- Governance oversight
- Managing risk
- Regulatory and compliance

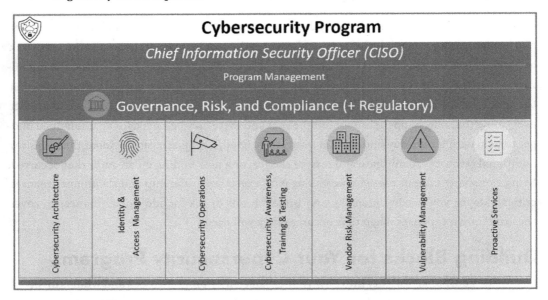

Figure 2.5: Core cybersecurity functions

In addition to the ten referenced functions in the preceding figure, we will cover two additional functions that are not necessarily required by all organizations, **Operational Technology (OT)** and the **Internet of Things (IoT)**. Although we do include OT and IoT as part of the broader cybersecurity architecture function, some specific industries and organizations will need to dedicate personnel and teams to protect this technology.

As we cover each of these functions, you will see that they are not small tasks to manage. Within each function, there are many sub-functions that need committed resources and will take a lot of time to adopt and manage efficiently. You will need to manage your resources appropriately by prioritizing the work ahead. You may be asking how I do this; it all comes back to risk – locate your highest risk within the program and focus on those items first.

Cybersecurity Architecture

The first function we will be covering in more detail is that of cybersecurity architecture. This function I would consider core to your cybersecurity program and one where having highly skilled architects on your team will set the stage for more robust and secure solutions within the organization. In this function, I would probably break it into two distinctive areas: technical and strategic.

From a strategic standpoint, we'll look to focus more on the broader strategy and foundation needed to build the cybersecurity organization. This includes adopting **Zero-trust Architecture (ZTA)**, which allows you to follow a specific architecture and set of principles to help reduce risk as much as possible. Another strategic item is how your broader environment is architected. For example, do you implement a cloud-first, cloud-only architecture model? Do you use one public cloud or many? Do you look to strategize on as few vendors as possible? In addition, you will want to have a solidified **Architecture Review Board (ARB)** process. This allows all solutions to be reviewed and assessed from a risk perspective before being implemented.

When we look at the technical side of the architecture function, this will include anything that needs the cybersecurity team to ensure the services, solutions, and products are secured as much as possible. Security engineering will play a big part in this area by implementing security controls on the devices, within the cloud environments, for email services, and so on.

We will cover cybersecurity architecture in a lot more detail in *Chapter 5, Cybersecurity Architecture*.

Identity and Access Management

The next function is identity and access management, which serves as a critical component of your overall strategy. Identity has taken a front seat as the new perimeter of your environment as we have shifted to a more decentralized, cloud-based model. Not long ago, protecting your data and assets was a much simpler model. Your data center, users, and user devices were all traditionally sitting within your own premises. The protection for this model was a strong network perimeter to protect everything contained inside. Now, with users working from anywhere and the push to public cloud data centers, the identity (username and password) of a user is all a threat actor needs to penetrate your entire environment. Because of this, the protection of your identity needs the utmost attention.

Within this function, there are several principles and technologies that will need to be considered to ensure a robust identity and access management program. This program will also require close collaboration with other departments, such as **Human Resources (HR)**, where the broader provisioning and de-provisioning process will need to be well documented and followed from beginning to end.

Since HR will most likely own the onboarding and offboarding of users, you will need to ensure close coordination to ensure reduced risk and impact to any users.

You will also need defined strategies around your identity and access management with examples such as requiring all applications and **Software-as-a-Service (SaaS)** environments to require **Single Sign-on (SSO)**, otherwise, an exception with approval will be needed. More examples include ensuring no shared user accounts are being used, ensuring there are no privileges on a standard user account, and so on. Most importantly, you will want to ensure all applicable security controls have been enabled and applied to best protect all identities within your environment.

We will cover identity and access management in a lot more detail in *Chapter 6, Identity and Access Management.*

Cybersecurity Operations

This function covers everything operational within the cybersecurity program. Your cybersecurity operations will need to be a 24/7/365 function. If you don't have 24/7/365 operations, you are leaving yourself more vulnerable. The reality is that threat actors are working around the clock, and you can guarantee increased activity during off-hours, weekends, and holidays. If you are in a small organization, running 24/7/365 operations may not seem realistic because of staffing and costs. But there are options available to support this need like outsourcing your **Security Operations Center (SOC)** to a third party, looking at a **Managed Security Services Provider (MSSP)**, or some form of next-generation cloud **Security Information and Event Management (SIEM)** solutions that provides enhanced capabilities.

Cybersecurity operations will be your most active function within the program and the one that will require the most demand to keep running efficiently. The team will be continuously dealing with ongoing alerts and events that will need to be reviewed to determine whether an actual security incident has occurred or not. This function also has the most opportunity for the likes of automation and AI. As thousands to millions of alerts and events continue to flood into your SIEM, you will need to apply effective automation and intelligence to the data to provide more efficiency and the ability to quickly identify high-risk true positive security incidents.

We will cover cybersecurity operations in a lot more detail in *Chapter 7, Cybersecurity Operations.*

Vulnerability Management

In this function, we will dive deeper into vulnerability management within the cybersecurity program. Unfortunately, vulnerability management is a never-ending responsibility. There is simply no light at the end of the tunnel when dealing with vulnerabilities. With technology, there will always be some form of vulnerability that will need to be addressed. As more technology continues to be developed and made available, the more vulnerabilities we will continue to be challenged with. We are continuously challenged with thousands of vulnerabilities monthly, and we need to constantly review and assess whether we are impacted by these vulnerabilities and how we remediate them as quickly as possible.

In the past, vulnerability management has taken more of a passive approach. When updates are released, they are typically pushed into a non-production environment and then finally pushed into the production environment once validation has been completed and sign-off has occurred. This process could take weeks to months depending on the organization. The reason for this has traditionally been to ensure no impact on business applications. Today, we must outweigh the risk of possibly impacting a business application versus the chance of a major security incident occurring because of an unpatched system. Because of this, our strategy must change to get updates pushed out to our devices and applications as quickly as possible. The faster, the better.

In addition, this function will require close collaboration with the broader organization so efficient collaboration is critical. Ensuring you have a well-documented inventory with application owners is a must. As vulnerabilities arise, you are going to need to know who to inform to coordinate mitigating any risk as soon as possible. This will require ongoing teamwork throughout the organization.

We will cover vulnerability management in a lot more detail in *Chapter 8, Vulnerability Management.*

Cybersecurity Awareness, Training, and Testing

Considered one of the most important functions, here, we will be focusing on the human factor within the cybersecurity program. Within this function, everything is about the human element and an area where we must ensure a robust program is in place for your users. It is no longer acceptable to treat this function as a checkbox to ensure an audit or internal requirement has been met. For most, an annual training video and phishing simulation was most likely only executed. Today, this program needs to evolve into much more for our users. Unfortunately, the human element is one of the most vulnerable attack vectors. We can implement all the technical controls in the world, but a simple social engineering attack can bypass them all.

The mindset we need to instill in our users is that cybersecurity is now a shared responsibility throughout the organization. Everyone must be better educated and aware of the threats so they can become their own security experts. Not just for themselves at work, but within their personal lives and for those around them, including their co-workers, family, and friends. To do this, your awareness, training, and testing program is going to need a rehaul if you are treating it as a checkbox to meet some compliance requirements.

We will cover cybersecurity awareness, training, and testing in a lot more detail in *Chapter 9, Cybersecurity Awareness, Training, and Testing.*

Vendor Risk Management

Another function considered one of the most important today is the vendor risk management program. It may be no surprise to you if you follow any cybersecurity news that supply chain types of attacks are happening more frequently. They are an attractive target for threat actors as the compromise of one vendor in the supply chain could impact tens to hundreds to even thousands of downstream customers who may use the impacted vendor. Because of this, the need for a more robust and thorough vendor risk management program is needed by all organizations these days. If you already have a vendor risk management program in place, you may also want to review it for opportunities to modernize to meet today's challenges.

An additional challenge from a cybersecurity perspective is that most of your vendors providing a web or application service will be a SaaS model. This is a big shift from deploying a web or application service within your environment. Your data is now going to be distributed across many different vendor environments, which is going to require a much deeper review of their security controls to ensure they are best protecting your data. The risk of your data getting into the wrong hands significantly increases as we adopt a more SaaS-based architecture.

As you work through your vendor risk management process, you are going to need to collaborate closely with multiple teams to ensure an effective due diligence process has taken place. You'll find yourself needing to work closely with the vendor being onboarded, the business function that is looking to onboard the vendor, the procurement team if one exists, and, most importantly, the legal team. Working through the onboarding process can take time, so, ensuring a well-defined process is in place will be needed to ensure efficiency.

We will cover vendor risk management in a lot more detail in *Chapter 10, Vendor Risk Management.*

Proactive Services

One area that requires its own dedicated function is that of proactive services. These are services that you should be running to identify any current gaps within your program and to help best prepare for when an incident occurs. Proactive services may not be a function that exists within your program today, or it may be part of other functions within your broader cybersecurity program. In today's world, I believe this needs to be a dedicated function as part of the broader program. Unfortunately, as leaders, we must work with the assumption that it is not a matter of "if," but "when" a major cybersecurity incident is going to occur. Data only continues to show this to be true.

The idea behind this program is that you are tracking all the documentation, exercises, and activities that are needed to best evaluate your environment and users. This will include running exercises such as tabletop exercises, threat briefs, and cybersecurity incident response plan reviews and updates. You will also want to ensure the execution of testing is occurring against your environment to identify unknown gaps. These types of activities will include penetration and application type of testing. You will want to bring in an external company to assist with these types of activities even if you have expertise in-house. It's always good to bring an external perspective to the table.

We will cover proactive services in a lot more detail in *Chapter 11, Proactive Services.*

Operations Technology (OT) and the Internet of Things (IoT)

OT and IoT are not included as a core function as part of the cybersecurity program, simply for the reason that these technologies don't impact all organizations. However, these technologies play an integral role within organizations that adopt them. For example, they are part of critical infrastructure that is used to power and operate countries. I would say IoT is probably an exception as there will most likely be some form of footprint in most organizations. But IoT as it relates to OT has become much more prevalent, which is why it is included in this section for a deeper review. These technologies do get mentioned within the cybersecurity architecture sub-functions in the event organizations run a small footprint of any of these.

The reason we need to cover this section separately is that these types of technologies require a completely different set of skill and, in most cases, a separate management plane to operate them. All the other functions fall within the IT scope, which we are very familiar with. As you move into OT and IoT, your traditional IT technologies are not built to manage these technologies. A lot of this technology will be legacy and antiquated, which will require specialized hardening and controls to reduce the risk of compromise.

We will cover OT and IoT in a lot more detail in *Chapter 12, Operational Technology (OT) and the Internet of Things (IoT)*.

Governance, Risk, and Compliance (GRC)

The final functions that we will review will be covered in *Part 3, Bringing It Together*. Here, we will look at how the broader cybersecurity program is governed to ensure everything is being implemented that we commit to, look at how risk is managed for the cybersecurity program, and ensure everything is maintaining compliance.

GRC is a crucial component of the overarching cybersecurity program. This function is sometimes overlooked but is quickly becoming a necessity for all organizations as more audits, regulations, and assessments are bearing down on us. Each of the components of GRC is unique in its own way and because of this distinction, we cover them separately throughout the book. Governance essentially wraps around the entire program to ensure everything is being executed and implemented correctly to reduce risk as much as possible. Here, you will have oversight of the entire program to ensure that the program is well defined and being run as efficiently as possible. This is where we bridge the gap between the cybersecurity program and the broader executive team and board members on how the program is being executed to reduce risk.

Regulatory and compliance is an area that continues to become very challenging to follow and keep up with. Not all organizations necessarily have a regulation requirement, but all organizations do need to follow some form of framework and hold themselves accountable to ensure they comply with that framework. In addition, privacy laws continue to evolve. As more breaches continue to occur, the privacy law space will only become more complex, especially if your organization employs users from multiple states within the United States, and it is even more complex if they reside in different countries. Working with your legal teams or outside council will be required for this function.

As already mentioned, risk management pertains to how we efficiently manage risk for the broader cybersecurity program for the organization. As cybersecurity leaders, we are essentially taking on a broader responsibility of being risk management leaders. How we track and measure risk is not an easy task. We need to continuously review the current threat landscape against our own organization to understand how we prioritize and reduce risk as much as possible.

We will cover GRC in a lot more detail in *Chapter 13, Governance Oversight, Chapter 14, Managing Risk,* and *Chapter 15, Regulatory and Compliance*.

The Cyber Mindmap

By bringing together all the core functions referenced in the preceding sections into one visual, we are able to demonstrate what a comprehensive cybersecurity program entails through a cyber mindmap. Bear in mind this may not be inclusive of everything needed, but it does provide an idea of what we are tasked with as cybersecurity leaders.

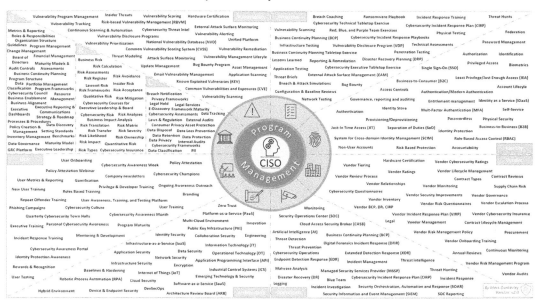

Figure 2.6: The cyber mindmap

The image may be a little small to read all the represented sub-functions, but they will be covered in more detail within each of their respective chapters.

 A great reference of a CISO mind map, which I have personally referenced many times, is that of SANS: `https://pentest.sans.org/security-resources/posters/ciso-mind-map-vulnerability-management-maturity-model/205/download`.

The idea of bringing everything into one image is to demonstrate at a high level what a comprehensive and robust cybersecurity program consists of. It is a visual that can be used as part of your toolbox to share with your leadership team, board members, and other business functions who may not be as well versed with cybersecurity. This image should be used to educate and inform everyone that cybersecurity is not a small task, but rather a massive program that takes dedicated resources and time to implement and manage efficiently.

Remember, each of the sub-functions is not a simple checkbox; they each require a substantial amount of time and attention to implement successfully.

Defining the Cybersecurity Organization

Before going into more detail on the organization's structure, roles, and responsibilities, it is important that you spend some time creating a mission and vision for the cybersecurity organization. As we previously discussed, there will be (or should be) a mission and vision at the organization level and there is a high probability that other functions/units within the organization have their own. Having one for the cybersecurity organization will help the broader organization understand your purpose, values, and goals being delivered for the organization. This is something that can also be shared among other leaders within the organization and something that can be included in any presentations or materials being shared with the broader organization.

A simple cybersecurity mission and vision could look something like this:

Mission

Protect and secure the people, assets, and data at "your company" from the ongoing and evolving threats in today's world.

Vision

A work environment with minimum cybersecurity risk.

 If you want to take this a step further, you could also specify goals that will be used to meet your mission and vision. This will then all tie back to the strategy that will be defined as part of *Chapter 4, Solidifying Your Strategy*.

It is important you clearly define how the cybersecurity function looks within the broader organization. This not only helps your team understand how the program operates, but it also helps provide more visibility to the broader organization and the leadership teams in general. We are also in a place where many companies still don't have a dedicated cybersecurity presence, or the responsibility falls within other teams in the organization, most likely within the IT function. Even if this is the case, you will want to build an organizational structure to identify who is filling each of the cybersecurity roles, even if it is only a partial responsibility of someone within a different function or team. As the need for cybersecurity continues to grow and become a necessity for all organizations, having a documented foundation and an organizational structure will only help as you need to add additional resources and justify the need for additional headcount.

Every organization is different, and there is no one way to build an organizational structure for every cybersecurity program. But it is important you have one in place to allow greater success as you continue to grow and build your team. If you don't have anything defined at this time, a simple way to begin with your cybersecurity organization structure is to follow the outline and ten functions we represent within this book. Following what we covered in the previous section, you can use this as a baseline to build your organizational structure to get started. This will help ensure you fill all the functions with a resource, even if a resource is listed multiple times. The following is an example to get you started:

Figure 2.7: An example cybersecurity organization chart

  Smaller organizations will not be able to staff the represented organization, but it is important that each of the preceding functions are being represented, which will require the need for resources to wear multiple hats, or the need to outsource specific functions.

One additional thought is that of the dynamics within a cybersecurity program. As threats evolve and new technologies are released, so does the need to continue to re-assess the current cybersecurity program and the skillsets needed to efficiently support it.

Roles and Responsibilities

Referencing the organization chart listed previously, you will need to tie back roles and responsibilities for each of these functions. Ensuring your team fully understands what each of them is responsible for within the program will create a much healthier environment to work in. I have seen firsthand what happens when roles and responsibilities have not been well defined, and friction occurs because no one is clear on who owns a specific task. This leads to an uncomfortable atmosphere and, in turn, leads to employees who become less productive and unhappy with their jobs. This is easily solved by documenting and sharing the defined roles with the broader team. It will also help to assign a lead to each of the applications and/or solutions within your portfolio so everyone knows who is ultimately responsible as a point of contact.

The following are some of the more common roles that will cover a variety of the workloads within your cybersecurity program with a brief description of their responsibilities:

- *CISO/CSO*: The CISO or CSO is typically an executive-level role that holds responsibility for the entire cybersecurity program. They define the strategy and provide the pathway to execute the strategy. They are responsible for the building, enablement, and maturing of the cybersecurity program as we cover throughout this book. The CISO may report to the CIO or the CEO (possibly other C-levels) and may begin to take a larger role as part of the **Board of Directors (BoD)**, whether sitting on the board or delivering reports to the board.

- **Director of cybersecurity**: The cybersecurity director of an organization will typically report to the CISO. Their role is to oversee specific functions within the cybersecurity program to ensure they are being executed per the strategy defined. Within a small organization, the director may overlook the entire program. This role will be more of a strategic role providing direction and holding the broader team accountable to ensure projects are being executed and risk is being reduced as much as possible.

- **Manager of cybersecurity**: The cybersecurity manager will typically report to the cybersecurity director or, depending on the size of the organization, the CISO. This role will cover an individual function (or multiple functions) within the cybersecurity program along with the personnel within these teams. The cybersecurity manager will be responsible for ensuring each of the functions they are managing is delivering the strategy defined for each of them. They will tend to have deeper knowledge within the functions they are managing and will typically be more technically inclined.

- **Cybersecurity architect:** The cybersecurity architect role is one of the more seniority technical roles within the program. This role will contain a wealth of knowledge on all the technologies contained within the cybersecurity program. A good architect will also be very familiar with the broader IT technology portfolio as the need to understand how broader integration occurs is critical to provide the best security. This role will typically provide a bigger picture of the intricacies involved with the technologies being used within the organization and how to best protect them. The cybersecurity architect will typically report to a manager or director role.

- **Cybersecurity engineer:** The cybersecurity engineer, on the other hand, will be the role that executes the determined cybersecurity architecture. This role will be the one that deploys, implements, and configures the technologies being used within the cybersecurity program. This is a very senior-level technical role and someone with a very deep technical knowledge. The cybersecurity engineer will typically report to a manager or director role.

- **Cybersecurity administrator:** The cybersecurity administrator will serve as more of an operational type of role with responsibilities of running the day-to-day activities of the cybersecurity technology stack. They will implement changes to the solutions, keep the solutions current with updates, make configuration changes, and so on. As a cybersecurity administrator, you may specialize in specific technologies, or you may have responsibilities over multiple technologies depending on the size of the organization. They typically report to the cybersecurity manager role.

- **Cybersecurity analyst:** I would consider the cybersecurity analyst one of the more entry-level roles within the cybersecurity program. This role will be responsible for a lot of the day-to-day activities that are needed to execute the cybersecurity program. This can include reporting types of activities, data entry types of tasks, console management, and log reviews. Depending on the size of the organization, the cybersecurity analyst may be involved with a single technology or multiple different technologies within the cybersecurity program. They typically report to the cybersecurity manager role.

In short, the management/leadership (although any role can be a leader) positions consist of your manager (entry-level), director (mid-level), and C-level (senior-level) roles. Your technical roles include the architect (designs), engineer (builds), administrator (operates), and analyst (day-to-day activities).

In addition to the generic roles listed previously and depending on the size of your program, you may have more specific roles and responsibilities within your team with more specialized skillsets. Some more specialized roles are software/application security developer/administrator/engineer, network security administrator/engineer, cloud security architect/engineer, penetration tester, security auditor, governance manager, identity and access management administrator, SOC analyst/lead, bug bounty/threat hunter, cryptographer/cryptologist, digital forensics specialist, incident response manager, AI security analyst, and mobile security administrator.

Because of its importance and the need for ongoing education and professional development with how diverse and dynamic the cybersecurity field has become, it's important you are providing a platform for your users to grow. For example, certifications hold a lot of value in the cybersecurity field, so make sure you encourage your employees to pursue these certifications and consider potential employees who have certifications with no degree. Some of the more common certifications come from ISC2, ISACA, CompTIA, EC-Council, and GIAC.

Depending on how large your organization is, many of the specialized skills noted previously may need to be filled within your organization. If you are at a small-/medium-sized organization, you will most likely want to focus on the more generic roles as there will be a need for those on your team to wear multiple hats and oversee many different technologies within the portfolio. Using more generic titles will also help from an administrative perspective; keeping the roles generic will allow for an easier structure to manage and provide growth as you move from an analyst to a senior analyst, to an administrator, to a senior administrator, to an architect, and so forth. In addition, HR tends to favor a simplified model to keep the portfolio of roles simplified and more manageable.

Outsourcing

One additional topic that needs covering is that of in-house resources versus outsourcing or a combination of both. As discussed in the *Skillset Challenges* section of *Chapter 1, Current State*, we are faced with an uphill battle with a significant shortage of cybersecurity professionals. In addition, your organization may be smaller and you can't afford to hire very specialized skillsets or hire enough workers to cover 24/7/365 security operations. In order to support these challenges, there will be a need to look externally to onboard a third-party vendor and/or resources to supplement the specialized expertise as needed and for the ability to staff a 24/7/365 operation.

Although the idea of outsourcing can create fear and provide a negative perception in some instances, you need to look at the idea from a strategic perspective. To do this, you need to assess the model in place, understand the current market along with meeting the organization's goals, including financial obligations. Because of this, and to meet the needs of your cybersecurity organization and requirements, you are almost guaranteed the need to outsource specific functions and skillsets within your program. Unless you have an unlimited budget or you are a Fortune 500 company, you will not be able to staff everything needed to make your program successful. My approach has been to keep the *brains* of your operations in-house and outsource the *moving parts*. In essence, outsource those easier-to-complete repetitive tasks whilst your in-house team manages, designs (architect), builds (engineer), and executes your strategy. As you progress through the book, we will discuss those areas where the need to outsource may make sense for your organization.

 As you do outsource your services, your responsibilities will significantly change to ensure those companies are delivering to your expectations and maintaining compliance. Ensuring clear contracts are in place with well-defined SLAs becomes even more critical. We cover this in more detail in *Chapter 7, Cybersecurity Operations,* and *Chapter 10, Vendor Risk Management.*

As a reminder, every organization is unique and the organization structure, roles, and responsibilities will be different. What I have provided is a very high level that is very basic to get started with. As you re-visit your current organization structure, or if you are just beginning to mature your organization structure, make sure you adapt your cybersecurity program to meet the needs of your organization.

Change Management

One extremely important component that tends to be overlooked frequently is that of change management. Not just within the cybersecurity program, but the broader organization. We tend to find ourselves so busy with the need to move fast that change management is either rushed without thought or bypassed altogether. It is critical that you understand the importance of change management and its place as part of the overall organization and the cybersecurity program. Some benefits that come to mind with good change management include:

- It ensures that testing has been completed, the change is documented, and backout plans are in place.
- It provides an avenue to ensure communication is occurring across the broader organization.

- It allows your support center to be better prepared.

- It reduces risk by following a documented process with the correct controls in place.

- It enforces changes to follow a schedule that everyone agrees to versus implementing a change when you feel like it.

- It allows for scheduling conflicts to be caught ahead of time with other major changes across the organization.

- It provides transparency of what is happening and when to the organization.

- It provides increased collaboration across business functions.

- It builds trust with teams across the broader organization.

- Overall, it allows for a more successful change.

Hopefully, your organization has some form of change control process in place today. If it does, it is important that you collaborate with the team that manages the change management process so you can inject all cybersecurity changes into the broader organization change management process. If there is no change management program in place today, it is highly recommended that one be enabled to provide a more structured and reliable environment.

The following diagram provides an example of a change flow process that you should implement if one doesn't exist within your environment today:

Figure 2.8: Change management flow process

As mentioned, change management is typically part of a larger program, more specifically around service management. One of the more common frameworks to help with change management is the **Information Technology Infrastructure Library (ITIL)**.

 Visit this website to learn more about the ITIL: `https://www.axelos.com/ certifications/itil-service-management/what-is-itil`.

As part of your cybersecurity program, you will have constant change occurring whether you will be deploying new solutions and new capabilities, upgrading current solutions, patching systems, or executing testing. With all these constant changes, your teams must plan them carefully, document them well, and execute on a planned schedule.

Following a process with any and all of these changes needs to be a requirement. If any of these changes go the wrong direction, and they will at times; they can be catastrophic for the organization. Answering questions from your executive leadership team when a change goes wrong is a lot easier when you have followed a process and received the correct approvals from the change management board. Implementing changes without following a process will only create a bad perception of the team, and you even risk impacting any trust gained.

Communications

You will learn quickly that good communication is a must for your cybersecurity program. In general, this applies to any leadership at any level. Your cybersecurity program touches every user in the organization, and the need for effective and efficient communications is not a choice. You need to be reaching out to your users constantly, informing them of updates, new services, and functionality; essentially, any type of change needs to be well communicated. I have seen it many times and even failed myself at times to ensure efficient communications are sent to the users. This is not an easy task and is constant. If you want to build trust with the broader user community, having a good communication program in place is key.

Most organizations will have some form of communications function within their organization. Whether it's a dedicated communications team, a team part of the marketing team, or a team part of the executive leadership team, you will want to build an instant and positive relationship with this team. Collaboration is key with this team. They are the experts when it comes to communications and will (or should) have a well-defined process in place with target audiences already in place. They will also typically have content and graphics experts within the team who can help with providing more professional content as it is communicated. If you don't have a communications team at your disposal, it may be a task you have to manage internally as part of your broader program. Regardless, don't overlook this important step and ensure that the question of "Is communication needed?" comes up with every change. Transparency is a core principle of a good leader, and this can only be done with clear and open communication.

Summary

Ensuring you spend time to implement or review the current foundations in place will only set the program up for greater success. With the dynamics of cybersecurity constantly changing and evolving, you will want to have frequent reviews of the broader program to stay relevant. Being a cybersecurity leader today requires an understanding of the entire organization. You need to learn how to navigate the organization efficiently and build positive relationships throughout.

In addition, translating risk into business terms will only provide greater advantages as you collaborate more with the business. As with any leader, you will need to understand finances and how to manage them. Your program will continue to evolve and the need for funding will be an ongoing requirement as part of your role.

Laying the core building blocks for your cybersecurity program is something that must be completed by all cybersecurity leaders. Having this in place provides greater transparency across the broader organization and allows for easier justification as you provide a vibrant visualization of all the responsibilities needed to run an efficient cybersecurity program. This will also help provide a clear picture of who is responsible for what within the cybersecurity program. As the threat landscape continues to evolve, it is important that we continue to assess and review the building blocks for our cybersecurity programs.

Once you have the program building blocks in place, or once they have been reviewed, you will want to review the current cybersecurity organization structure in place. This will be different for many organizations, especially as cybersecurity can be a shared responsibility within some departments, such as IT. Regardless, you need to define the roles and responsibilities required to operate the program, which will provide clarity to your staff and help alleviate any confusion and friction within the team. Change management and communication are two very important areas that we covered to finish the chapter. These are items that need to be a part of everyone's program and play a critical role in the foundation of your broader program and success within the organization.

In the next chapter, we will be covering road maps and the importance of them as part of your cybersecurity program. Having a road map in place allows better planning and the ability to provide some level of measurement with the program. We will first focus on assessing the current state of the organization, which will differ on whether you are new to the organization or new to the role. We will then cover the importance of a road map and why we should have them in place. Next, we will look at reviewing different road maps that will help as your program evolves. This includes immediate short-term road maps, short-term road maps, and long-term road maps

Join our community on Discord!

Read this book alongside other users, Cybersecurity experts, and the author himself.

Ask questions, provide solutions to other readers, chat with the author via Ask Me Anything sessions, and much more. Scan the QR code or visit the link to join the community.

`https://packt.link/SecNet`

3

Building Your Roadmap

Now that the foundations of your cybersecurity program are understood and in place, the next step is to focus on the program's roadmaps.

If you don't have a well-defined roadmap in place or a roadmap at all, you won't have much structure around your program and a destination for your vision. In short, a roadmap is a plan that guides you on how to get from one place to the next to meet a desired goal. To meet that goal, you will have a pre-defined set of tasks that you will complete to get to your destination. Having a roadmap helps provide a better visual of where your program is today versus where you envision it in the future. It provides clear directions and sets expectations for your executive leadership team and board members. Although your roadmap may have a final destination on paper, the reality is there is no final destination because you will continue to evolve and build new roadmaps as the program grows, matures, and changes with the current threat landscape.

To begin, you must understand the current state of the cybersecurity program you are overlooking. In order to build an efficient roadmap, or roadmaps, you will need to know where to begin. And you can only begin when you know what the current state is with your program; you shouldn't, for example, be purchasing or deploying a new application if you don't know this. This will require you to take a step back and build a roadmap for the program to clearly identify what is needed, allowing you to purchase the new application and ensure you are efficiently staffed to deploy and support it. As we review the importance and benefits of a roadmap, we will take a closer look at the importance of program and project management for your cybersecurity program. Having a well-defined program management foundation in place is a must for greater success.

As we look at different roadmaps for your cybersecurity program, we will focus on three different roadmaps including a current state assessment that will help build upon your foundation, allowing for greater success. There is no right or wrong way to build your roadmap. The example in this chapter is merely a directional guide to get you started with a roadmap that makes sense for your organization. The current state assessment will give you time to assess where the program is at the moment, which will allow the roadmaps that follow to be defined and created more efficiently. The first roadmap will focus on the immediate short-term, which will allow you to identify any quick wins for the cybersecurity program and the organization. It will also be the roadmap that allows you to get better grounded with the current state and the need to tackle any critical or high-risk items if applicable. The second roadmap will be the short-term roadmap, which will include some items that aren't so easy to complete immediately and will need additional time. Here, you will be able to begin maturing the cybersecurity program, but the end goal of this roadmap is to better define the longer-term roadmap to truly build a broader and more mature program. The final roadmap we will cover is the long-term roadmap, which will consist of those items defined from the short-term roadmap that require more time and resources to be efficiently deployed. This roadmap will, essentially, be reviewed and revised every year to meet the current state.

The following will be covered in this chapter:

- The importance of a roadmap
- Assessing the current state
- Immediate short-term impact (2–4 Months)
- Short-Term Impact (5–12 Months)
- Long-Term Impact (1–3+ Years)

The Importance of a Roadmap

Not having a roadmap in place is no different than leaving your home to go on vacation without having any plans. You will be faced with questions such as:

1. Where will you go?
2. How will you get there?
3. Do you have enough money?
4. What will you do when you get there?
5. Will there be availability wherever you go?

A roadmap is essentially a plan for a cybersecurity program that defines how you are going to get to your next destination. As you begin to build your plan, creating roadmaps will provide the guidance, foundation, and structure to efficiently work through what is needed to meet the goals of your plan. In essence, your roadmap will tie back to your mission and vision for the cybersecurity program, which we briefly covered in the *Mission and Vision* section of *Defining the Cybersecurity Organization* in *Chapter 2, Setting the Foundations*. The following table provides a hypothetical overview of similarities between planning a vacation versus building your cybersecurity roadmap. This should help visualize the importance of needing a roadmap and the coordination involved to make it successful, as in everyday life..

Planning a Vacation	Building a Roadmap
Destination Selection	**Mission and Vision**
Where are we going to visit.	Defining the goal of the cybersecurity program by defining what we aim to achieve.
Travel Budget	**Program Budget and Funding**
How much is the trip going to cost for travel, accommodation, activities, etc.?	How much money is needed to fund the cybersecurity program.
Travelling Party	**Resource Management**
How many and who will be travelling.	What resources are needed to build the cybersecurity team.
Itinerary Planning	**Program and Project Management**
Planning a schedule for each day that we are visiting?	Program coordination and oversight of day-to-day project management of all tasks along with scheduling.
Contingency Plans	**Risk Management**
Planning for the unexpected with delays, illness, etc.	Efficiently managing cybersecurity risk for the organization, implementing a cybersecurity policy, etc.
Accommodation and Transport	**Technology and Tools**
How do we get to our destination (flights, bus, train, car, etc.) and where will we stay (hotel, camping, etc.).	Implementing the technology and tools needed to meet your organizational cybersecurity goals.
Documentation	**Policies and Procedures**
Do we need passports, visas, air tickets, hotel reservations, etc.	Creating and maintaining cybersecurity policies, standards, and processes/procedures.
Checklists	**Compliance and Standards**
Building a checklist and to-do list to ensure nothing is missed.	Selecting and implementing a cybersecurity framework and complying with any regulations.
Updates and Modifications	**Continuous Improvement**
Making adjustments based on new information, delays, health issues, etc.	Continuing to review and update the cybersecurity roadmaps based on current state and emerging trends.

Figure 3.1: The similarities between real-life planning and building a cybersecurity roadmap

The following are many of the reasons why you need to spend time focusing on creating a well-defined and robust roadmap for a cybersecurity organization:

- It brings visibility into the cybersecurity organization and provides transparency to the broader organization.
- It's a mechanism that can be used to share and update your executive leadership team along with the board of directors.
- It provides your internal teams with a foundation to follow defined goals in place. It also defines clear expectations for a team.
- It allows you to better define a foundation for the cybersecurity program that will support your annual budget needs.
- It provides you with the ability to efficiently plan projects for the year ahead as part of your program management.
- It clearly defines where the program is today and the journey it will take to meet the goals defined by the program's mission and vision.
- It brings structure to the program.

As we work through this chapter, we will build a roadmap that consists of four separate areas – a current state assessment followed by three roadmaps that will provide you guidance for the next 3 years. The following is a high-level view of the flow:

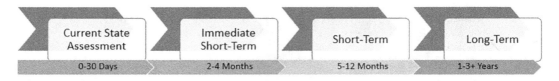

Figure 3.2: High-level workflow of the recommended roadmap journey

One extremely important item that we haven't covered yet is that of program and project management for your cybersecurity program. This is often overlooked but essential to ensure that you efficiently track everything required to run an effective cybersecurity program.

Program and Project Management

One important aspect of your cybersecurity program is that of higher-level program management and its finer details, as it relates to project management. This is far from an easy task, and you'll need to ensure that you align your cybersecurity program and projects with the broader organizational goals, deliverables, and priorities. You'll need to ensure that resources are managed appropriately and not being overworked, that your projects don't conflict with other functions or department projects, that your end users aren't being overwhelmed with too much change at once, and that full transparency is created with the program across the broader organization. This can be a challenging task if there is no centralized program management function for the organization, so you'll need to ensure that you collaborate and communicate effectively with other function leaders throughout the organization with your program priorities. From a hierarchy perspective, your program management is more strategically focused, which will overlook all the projects required to meet the program strategy. At a program level, there will be a significant amount of work that will be underway at the same time, and the need to efficiently track progress and coordinate everything is critical. At a project management level, you will have multiple projects running at the same time, which will require you to resourcefully coordinate your personnel. Scheduling will be critical so that you don't overwork key resources, who will most likely be working on many other projects and tasks already. The following provides a high-level overview of the differences between program and project management.

	Program Management	Project Management
Overview	Manages strategy, aligns with vision, and overlooks multiple projects	Overlooks and manages individual projects
Description	The program management responsibilities focus on oversight of the entire cybersecurity program portfolio	The project management responsibilities focus on the management of many individual projects that make up program management
Scope	Entire program (many projects)	Single project
Length	Long-term	Short-term
Tasks	Strategy management, Stakeholder communication & collaboration, Resource scheduling & conflict management, Project portfolio management, Program status & reporting, Program budget oversight, Risk management for broader program	Manage & oversee individual projects, Resource management with projects, Project timeline management, Budget oversight for projects, Project status & reporting, Risk management within projects
Benefits	Ensures the successful delivery of all objectives for the cybersecurity program	Ensures the successful delivery of individual projects

Figure 3.3: Program management versus project management

As you build your roadmaps, they will fall within the scope of your program management activities for the cybersecurity program. The outcome of your roadmaps will translate into projects that will need to be tracked to completion. As you work from year to year, you will need to plan and build your program for the year ahead. This will require reviewing your roadmap for the specific year and mapping out your projects, meeting each of the goals and deliverables from the defined roadmap for that year. Ideally, having a program or **project manager (PM)** on staff who is familiar with cybersecurity-type projects will be preferred. But this may not be feasible for many. If not, does your organization have a department that can support running your program, such as a **Project Management Office (PMO)**? If so, you will need to work with them to justify the need for resources to support your broader program. If there is no opportunity to bring in a dedicated PM or there is no PMO, you could also consider leveraging a third party, such as a consulting company, who can provide the required expertise. Alternatively, you may need to manage the program yourself, or delegate the responsibility to someone else on the team who can help track the program and all projects within.

Obviously, having someone whose full-time role is a PM will help alleviate a lot of the administration type of work, as they are familiar with and have been trained to manage project management types of tools. If you can bring someone on board, then ensuring they have the relevant background and skill set in project management will be critical. A very common certification for a PM is the **Project Management Professional (PMP)®** from the **Project Management Institute (PMI)**. If you are not familiar with the PMI, you can learn more about them here: https://www.pmi.org/about. If you can't bring dedicated resources in to manage your program, you can always work with your team to see if there is anyone interested in pursuing the PMP certification to gain additional knowledge and add the experience to their résumé.

The program and project management responsibilities shouldn't be overlooked regardless of who runs them.

 Another great learning path (a lot less intense than the PMP) for your current team members who need to manage projects is the *CompTIA Project+*, provided by *CompTIA*. This is a great foundational PM certification that is geared more toward technical professionals. You can learn more here: https://www.comptia.org/certifications/project.

If you are going to need to directly manage both the program and project management yourself, many tools will help easily get you off the ground with managing and tracking the program efficiently. From a program management level, there will be a lot of documentation that needs to be continuously created and updated to represent the broader program. You will find that the likes of Microsoft PowerPoint (bear in mind that PowerPoint comes at a cost) or some other form of presentation application will become one of your most used tools when capturing the broader program and delivering reports to the broader organization, including executive leadership and board members. We will cover reporting in more detail in *Chapter 13, Governance Oversight*. To build your roadmaps, the use of tools like PowerPoint will easily meet your needs. You don't need to be creative but, rather, take advantage of the built-in templates or themes to build your roadmaps in a presentable format. For example, the *Product roadmap infographics poster* shown below is a prebuilt template provided by Microsoft that can be customized to meet your needs. Within PowerPoint, click **File > New**, and then search for your desired template.

Once created, simply add your own information and branding, and you will have a very professional roadmap ready to be shared.

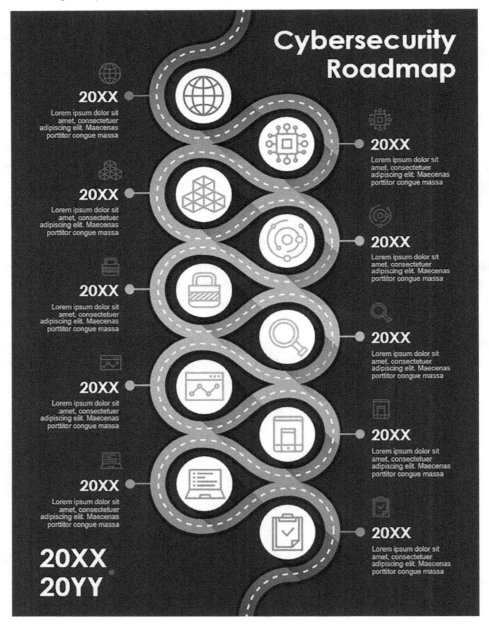

Figure 3.4: Microsoft PowerPoint roadmap template

Direct link to source: `https://create.microsoft.com/en-us/template/product-roadmap-infographics-poster-e91a6422-cb04-4473-9aca-b150e1d949a6`

 If you are not familiar, Microsoft has released AI capabilities within its Office applications with *Microsoft Copilot*. These capabilities do come at an extra cost, and there is a Microsoft PowerPoint-specific Copilot: `https://support.microsoft.com/en-us/copilot-powerpoint`. Simply provide a topic and details to Copilot on what you would like, and it will create a draft presentation.

From a project management perspective, there are many tools and options available to efficiently manage your projects. If your company already strategizes on a specific tool, you'll want to enquire about attaining a license for yourself and the team. Ideally, you will want a SaaS-based tool that will allow you to collaborate on project items more efficiently across the team. If there's nothing available within the organization and you need to procure something, Microsoft has options available to quickly get you up and running if needed. Here are a few of the options available:

1. Microsoft 365 cloud-based project management subscription solutions
2. Microsoft 365 traditional Microsoft Project application-based solutions
3. Microsoft Excel and the built-in templates.
4. Microsoft Planner

The Microsoft Project management solutions (numbers 1 and 2) can be combined. For example, you can purchase the Microsoft Project application and use it to access your cloud-based projects to manage them, instead of using the web **User Interface (UI)**. For number 2, if purchasing a standalone application, you will be limited to only one person having access to the data unless you export and share a static file. You can review this in more detail here: `https://www.microsoft.com/en-us/microsoft-365/planner/microsoft-planner-plans-and-pricing`. If you already use Microsoft Excel or some other similar spreadsheet application, you will be able to capture your projects quickly and easily to get started. Like the above with PowerPoint, you will be able to search for templates that will give you a baseline to get started. Within Excel, click *File > New*, and then search for your desired template. Once created, you can simply customize and add your own information and branding to capture all your annual projects with timelines.

The following is the *Gantt Project Planner* template provided by Microsoft.

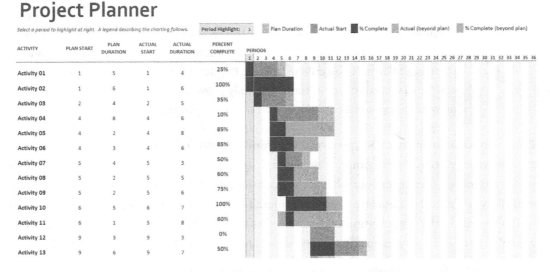

Figure 3.5: Microsoft Excel Project Planner template

Direct link to source: `https://create.microsoft.com/en-us/template/gantt-project-planner-8eab671c-2214-4ce4-b5ee-17b3ad09c5a1`

The tools above are not the only tools available at your disposal; we just share these as an example of how you can quickly get your cybersecurity program up and running. Some other common project management tools include Trello, Asana, and Jira, to name just a few. In addition, there are many common project management techniques you may want to consider with your cybersecurity program, which include:

- Gantt (as shown above)
- Waterfall
- **Work breakdown structure (WBS)**
- Agile methodologies such as Scrum and Kanban

There are many other tools and techniques available to meet your program needs. Make sure you are familiar with what is available within your organization, allowing you to quickly achieve your project management needs.

Assessing the Current State

Assessing the current state will be an ongoing task and is needed every time you look to build your roadmaps. Building your roadmaps is not a one-time activity; they will need to be continuously reviewed to ensure that they meet the needs of an organization to accomplish the final defined goal. There will also be different triggers that initiate the need to review the current state to build your roadmap. The most difficult will be entering an organization as a new leader. Here, you will have a lot to learn, and your focus may be occupied more with learning the business initially. If you are promoted from within the organization to a new leadership role overseeing the cybersecurity program, your focus on the roadmap may be a lot more tactical, as you are both familiar with the technical aspect and the organization as a whole. As a leader already running the cybersecurity program, you may look to build a new roadmap based on an older roadmap coming to an end, or the need to refresh a roadmap because of changing requirements. Whatever the situation, a roadmap will be needed as you continue to evolve and mature the cybersecurity program.

Assessing the current state of both your organization as a whole and the cybersecurity program is needed as you begin to build your roadmaps for the first time. Building a roadmap without knowing the state of the current environment will be very inefficient and potentially provide no value. As mentioned, assessing the current state will differ, depending on whether you are new to the organization or promoted from within. As we journey through the chapter, we will take the approach of building roadmaps from the perspective of someone coming into the leadership role from outside the organization. This will require the most comprehensive approach, providing more details for you to review and ideas to build your own roadmaps. For this phase, we will reference the activities needed as the current state assessment for the roadmaps that will follow. For the current state assessment, you should assign the first month as the time needed to complete the initial assessments of the current state. Assessing the current state can be looked at in three different phases.

Learning the Business

For the first phase, you are going to need to understand your organization and learn the business as quickly as possible. This will help you better understand the culture and provide a better idea of how serious (or not) cybersecurity is within the organization. As you meet with other leaders, you will be able to ask specific questions that will help you better understand how each business function values cybersecurity.

Some questions that can be asked include:

- Are you aware of a formal cybersecurity program within the organization?

- Is cybersecurity awareness part of your program? Is there any cybersecurity training and testing required for your teams? Is it valued?

- Are you aware of any cybersecurity policies or procedures in place?

- Do you see cybersecurity as a priority?

- Is any sensitive data being handled by your team? How do your teams handle sensitive or business data? Do you have an inventory of the data being handled?

- Has your team been involved in any cybersecurity incidents? Are you aware of any cybersecurity incident response procedures?

- Do you own any business applications? Where is the data being hosted? Is there an inventory of applications under your ownership? How do you protect them? Do you run security tests against them?

- Is there any technology and/or infrastructure that your function owns and manages? Is there an inventory of the infrastructure being used?

- Are you required to comply with any audits? Are you impacted by any regulations?

- Have you observed any cybersecurity gaps or risks throughout the organization?

- Is there anything in particular that you expect from the cybersecurity program to support your unit/function?

Understanding whether your organization is publicly traded, regulated, or required to undergo some form of audit will help you determine if the cybersecurity program has maturity in place. In addition, as you learn the organization structure and the different functions and roles, you should be able to easily assess some level of maturity with cybersecurity within the organization. Is there an audit function? Is there a risk management function? Is there a legal team? If these are not in place, there is a high probability you'll have a longer journey maturing your cybersecurity program. For more details, refer back to *Learn the Business* in *Chapter 2, Setting the Foundations*.

Non-Technical Assessment

In the next phase, you should assess the current cybersecurity program from a non-technical perspective, identifying what is currently in place and how the program is run. There are many ways in which this can be completed, including the following:

- Meet with each of your team members to get a better insight into the overall program. Specifically, ask them about any challenges they face and any gaps they have observed.

- Look through current documentation to see how well (or not) the environment is being inventoried. This can include architecture diagrams, build sheets, infrastructure specifications, asset management, user guides, etc.

- Is there a centralized location for user documents, such as how-to guides, awareness material, etc.?

- Is cybersecurity training being executed and are phishing simulations going out to the users? If so, how often?

- Is there a **cybersecurity incident response plan (CIRP)** in place and are tabletop exercises being executed?

- Are cybersecurity policies, processes, and procedures in place for the organization?

- Do contracts include cybersecurity language and is there any formal process around onboarding vendors?

- Is there any intake process or architecture review process for new initiatives, applications, vendors, etc.?

- Does the organization currently hold a cybersecurity insurance policy?

These items shouldn't be difficult to run through to get a quick sense of how the cybersecurity program is being run. This will help you identify any potential low-hanging fruit or "quick wins" as you build your immediate and short-term roadmaps. For example, if tabletop exercises are not being executed, this could be something you execute to get a sense of how well the teams can respond to a cybersecurity incident. If no cybersecurity training is in place, or is only being run annually, you can work toward implementing or enhancing the training schedule for users right away. Just as important, if you identify that there is no cybersecurity insurance policy in place, you can immediately begin the process of onboarding a cybersecurity insurance policy for your organization.

Technical Assessment

In the third phase, you will leverage anything technical in nature that provides any potential data on the current state of the cybersecurity program. Remember, this is not the time to be onboarding vendors or deploying solutions to try and get this data. Here, you will simply look for anything that is already in place to provide a better idea of the state of the cybersecurity program.

Some of the questions to ask can include:

- Are there any vulnerability scans that have been run in the past for review, or is there something already set up that allows you to run a vulnerability scan?

- Are there any audit reports available for review, either within the cybersecurity program or in the broader organization? This could be PCI-DSS, SOC I or II, ISO, etc.

- Have there been any assessments completed that are available for review?

- Have there been any penetration tests against the environment or application-specific testing, with reports available?

- Is there a **Security Operations Center (SOC)** in place? If so, how are incidents being reported, and where are the incident-report documents for review? If not, how are cybersecurity incidents being handled and managed?

- How well is patch management being handled, and what is the time to patch devices when patches are made available?

There's also a chance that you enter an organization where there has been zero cybersecurity presence, and you have been hired to build the program for the first time. In this instance, most of the above likely won't apply, and this is where you would build your roadmap with a greenfield approach. Here, you would begin with the foundations covered in *Chapter 2, Setting the Foundations*, and then prioritize implementing items considered critical and high severity to secure the environment. What these are will differ within each organization, and there is a chance that the IT team may have already deployed some cybersecurity capabilities.

Bringing It Together

Now that we have assessed the current state of the environment, we should have a much better idea of what we need to compile when building roadmaps for a cybersecurity organization. The idea is that the current state assessment is short and should take no more than 30 days to get some immediate data. Having this documented as a preparation phase will quickly allow you to build a more constructive set of roadmaps, which we will cover in the next three sections. These roadmaps are the immediate short-term roadmap, the short-term roadmap, and the long-term roadmap. The following is an example of a template of how you can map out the focus points as you begin your journey to build or reconstruct your defense strategy:

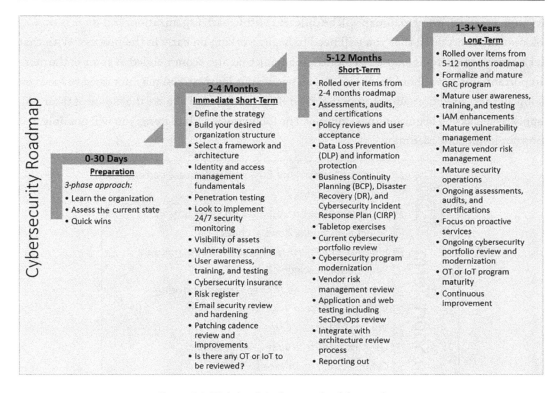

Figure 3.6: High-level draft example of the roadmap

Immediate Short-Term Impact (2–4 Months)

Now that the current state assessment of the program has been completed, you will have a better idea of what is needed for the immediate short-term roadmap. This roadmap will follow your initial review of the environment and consists of a 90-day timeline. The idea here is that you look to make some immediate impact within the program. After your initial review, you may have identified some high-risk items that need to be addressed immediately. There may be some items that are in place but have some gaps that need to be addressed. This immediate roadmap will also give you the time to get better acclimated with the cybersecurity program and to get you moving as quickly as possible. This roadmap will also be a foundation to get some other initiatives in motion as you move into the short-term 5–12 months roadmap. As a new leader, you will quickly observe that accomplishments will not happen overnight, especially within larger organizations with more red tape and processes to navigate. For example, bringing on a new vendor may take months in some companies. Simply getting the process started may be all you can do within this roadmap.

As already stated, every roadmap will be different within each organization, and there is no way of knowing everything that you will need to begin working on early in the process. With that, let's cover several items on the 2–4 months roadmap that are recommended as some of the more important areas you should focus on as you begin your journey. You may not need to focus on some of these or your roadmap may slightly differ, but there is a chance that some of these will apply to you as you begin your new journey. The following is a list of items you will possibly find on your immediate roadmap:

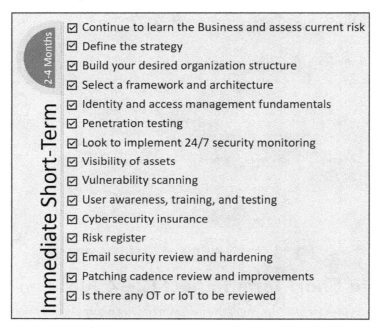

Figure 3.7: Immediate short-term items

Continue To Learn The Business And Assess The Current Risk

This one is important, as we already learned in *Chapter 2, Setting the Foundations*. As you get settled into your new role, you must continue to learn the business and get familiar with all functions within the organization. This includes building meaningful relationships with other leaders throughout the organization. In addition, you need to continuously assess the current cybersecurity risk within the organization, enabling you to better determine the level of overall risk and priorities you need to work toward.

As we journey through the book, we will reference industry-recognized frameworks that will assist you in determining the current risk within the cybersecurity program. For example, cybersecurity frameworks will be covered in *Chapter 4, Solidifying Your Strategy*, and *Chapter 15, Regulatory and Compliance*, and risk frameworks will be covered in more detail in *Chapter 14, Managing Risk*.

Define the Strategy

It is important that you define your strategy early. This includes everything that falls under building or redefining the cybersecurity program, such as laying the building blocks for your program, as covered in *Chapter 2, Setting the Foundations*, understanding the architecture and frameworks available to help create your strategy, how your application portfolio and resource management strategy will look, etc., which will all be covered in *Chapter 4, Solidifying Your Strategy*.

Build your desired Organizational Structure

Here, you will assess the current organizational structure to manage your program and determine if changes are needed. Does the structure need to be modified to meet the defined strategy? Are new resources needed? Do you need to outsource any work? The sooner you get your organization structure in place, the sooner you can begin to focus on everything else that is needed to continue to strengthen the program. We covered this in more detail in *Chapter 2, Setting the Foundations*.

Select a Framework and Architecture

For this task, you will need to determine both a framework to be adopted and a cybersecurity architecture. There are multiple frameworks available, and your organization may already have one implemented as a foundation. If not, you need to determine what makes sense for your organization. The same applies to architecture; you need to select a cybersecurity architecture to support building your program. These will both be covered in *Chapter 4, Solidifying Your Strategy*.

Identity and Access Management Fundamentals

With identity being the new perimeter to your environment, you need to understand how identity is currently set up and being used. For this specific task, you need to review whether there is any immediate risk that needs to be addressed.

For example, how are passwords being managed, what are the requirements, and do they need to be improved? Is **Multi-Factor Authentication (MFA)** being used? If not, a plan will need to be initiated to start enabling it. Is there a separation of accounts for your privileged users? If not, this will also need to be addressed immediately. There is a lot more involved with identity and access management, which will be covered in more detail in *Chapter 6, Identity and Access Management*.

Penetration Testing

If there haven't been any recent penetration tests, you will want to bring in a third party to execute both an internal and an external penetration test immediately. This will help quickly identify any potential critical and high-risk items that can be addressed immediately. If you don't have a vendor on contract to complete this, you will need to work through onboarding a vendor who can complete this for you, which may end up pushing this task into the following roadmap. This will be covered in more detail in *Chapter 11, Proactive Services*.

Look to implement 24/7/365 Security Monitoring

In today's world, you need 24/7/365security monitoring in your organization. If this is not in place today, it may not be something you can immediately implement. What you need to do is assess the current state and determine the path to reach 24/7/365 monitoring. Do you have staff resources in-house or outsource them? This will all be determined by many different factors. We cover this in more detail in *Chapter 7, Cybersecurity Operations*.

Visibility of Assets

Knowing all your assets is critical for your cybersecurity program. As incidents occur, you need to know what has been impacted and how severely. These factors can only be determined when you know what assets are involved and the classification of these assets. This is another task that will not happen very fast if there is nothing in place today. However, you will need to understand what is in place, enabling you to put a plan together to establish a solidified inventory and begin applying classification to everything. We will review this in more detail in *Chapter 13, Governance Oversight*.

Vulnerability Scanning

If there is no vulnerability scanning in place, you will need to quickly look at options to scan your external web-facing environment as quickly as possible. This will allow you to quickly identify any critical or high-related items that will need to be addressed immediately. There are several options available to get this up and running if not currently in place, which we will cover in more detail in *Chapter 8, Vulnerability Management*.

User Awareness, Training, and Testing

One of the most important functions, if not the most important, is awareness, training, and testing for your users. Understanding what is in place, if anything, will help you understand what is needed to ensure that a program gets the attention and resources needed for ongoing user awareness, training, and testing. If nothing is in place, you will want to prioritize onboarding a solution immediately to support this. We will cover this in greater depth in *Chapter 9, Cybersecurity Awareness, Training, and Testing*.

Cybersecurity Insurance

You will want to understand if there is any cybersecurity insurance policy in place for the organization. If not, you will want to make this a high-priority item, seeing what options are available to onboard a policy as soon as possible. As part of assessing your options, you will need to understand the cost of an insurance policy, as well as ensure that your finance team is aligned and a budget can be made available. Onboarding won't happen immediately, but the process will need to begin. Depending on your organization, you may need to own this task, or you may need to partner with someone else in your organization such as the legal team. If cybersecurity insurance is already in place, you will need to better understand the fine details around the policy and how to engage a **Digital Forensic Incident Response (DFIR)** vendor if needed. This will be covered in more detail in *Chapter 14, Managing Risk*.

Risk Register

Since everything we oversee in cybersecurity is about risk, it is very important that we efficiently document identified risks within an environment, and that there is sign-off and acknowledgment when an organization wishes to proceed with risk items identified as high and critical severity. To do this, you need a risk register to formally track an identified risk. If there is no tool in place to track this, you can simply build a risk register using a spreadsheet or search for a template to get you started. This will also be covered in more detail in *Chapter 14, Managing Risk*.

Email Security Review and Hardening

Email is one of the most used tools in many organizations and one that continues to be the most challenging to secure from threat actors. We must do everything we can to harden our email systems as much as possible. Understanding what security mechanisms are set up within your email environment must be well understood, along with knowing what advanced capabilities are available (or not) and are being used. There may be some quick wins with email security to further secure an environment.

If not, you will need to ensure that a plan is in place to review and confirm that everything is set up and configured correctly, and if there is a need for any additional capabilities or tools. We will review email security within *Chapter 5, Cybersecurity Architecture.*

Patching Cadence Review and Improvements

Another common threat vector is via systems and applications that have not been updated with the latest patches. Threat actors continue to exploit systems that have known vulnerabilities because they fail to get updated promptly. Patch management should be a well-defined process, where updates are pushed within a short time of being released. More specifically, you should prioritize **known exploited vulnerabilities (KEV)**. To help with this review, the **Cybersecurity and Infrastructure Security Agency (CISA)** has published a known exploited vulnerabilities catalog, which you should refer to. This will be covered in more detail in *Chapter 8, Vulnerability Management.*

Is there any Operational Technology (OT) or Internet of Things (IoT) to be reviewed?

If there is any OT and/or IoT in your environment, you will need to spend some quality time reviewing and understanding exactly what falls within each of these. If you do have any of these, you'll need to ensure you have the right resources on staff to help efficiently manage these types of technologies. You will first need to build out a comprehensive inventory if one is not already available. From there, you will need to determine how mature the current program is to understand the next steps needed to enhance and mature this program. *Chapter 12, Operational Technology (OT) & the Internet of Things (IoT)*, will cover this in more detail.

Now that we've reviewed the items that you need to focus on within the first 120 days, we will move onto the short-term roadmap, which will take us through to 12 months. Some of the items reviewed in the 2–4 months roadmap will roll over into this roadmap as you continue to evolve the program.

Short-Term Impact (5–12 Months)

Now that you have created the immediate short-term impact roadmap, you can start compiling the roadmap that will take you through to the end of the first year. The first year will be critical for the program as you get your feet off the ground with the cybersecurity program. As mentioned within the *Immediate Short-Term Impact (2–4 Months)* section, there will most likely be several items that will roll over into the short-term roadmap.

You will have identified items that are needed but will take a lot longer to implement because of contractual work, the need for **Request for Proposals (RFPs)**, potential budgetary constraints, etc. Essentially, you'll find a lot of red tape, especially in larger organizations. Because of this, you will need to ensure that you capture your first item on this roadmap as items rolled over from the 2–4 months roadmap. Something else to consider when building this roadmap is that it will not be static. You will need to be dynamic, as changes will occur as you work through your journey with the immediate short-term roadmap. New requirements or priorities will surface that weren't originally captured and need to be added to the roadmap as you progress.

Within the short-term roadmap, you will begin working on some of those items that require additional time and resources to get movement. These items may not necessarily be as high a priority as the immediate roadmap items, but they are nevertheless critical for the overall program. As mentioned, some items will be carried over and ongoing from the previous roadmap, and other items will start to become strategic as you look to improve on capabilities and modernize the overall program. As you become settled with your new role, more familiar with the organization as a whole, and more familiar with the current cybersecurity program, you will be able to start making some impactful changes to the overall cybersecurity program and set the precedence for how the cybersecurity program will be delivered and managed. The following is a list of items you may find on your short-term roadmap:

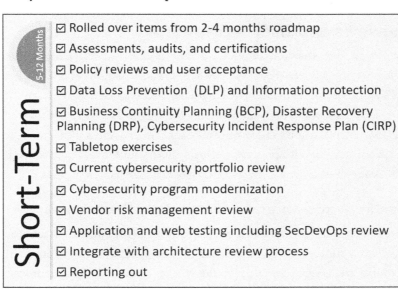

Figure 3.8: Short-term items

Rolled-over Items from the 2–4 Months Roadmap

This will cover any of the items that will roll over from the immediate short-term roadmap. As mentioned, certain tasks will end up requiring a lot more time and resources as you dig deeper into each area. Some items that I'd imagine may roll over into the short-term roadmap include penetration testing, 24/7 monitoring with a SOC, visibility of assets, vulnerability scanning, user awareness, training and testing, cybersecurity insurance, email security, patching, and OT or IoT if applicable.

Assessments, Audits and Certifications

At this point, you should be familiar with any audits or regulatory requirements within the organization. You will know if audits or certifications occur in other parts of the organization and what assessments, audits, or certifications directly impact the cybersecurity program. Based on the immediate short-term roadmap, you should have your cybersecurity framework defined, which will then allow you to begin running an assessment and/or audit using the selected framework. This will be the foundation for your current state and provide the needed baseline to continue maturing your cybersecurity program. Some other audits and certifications you may find within your organization include SOC1 and 2, PCI-DSS, and ISO. These will be covered in more detail in *Chapter 15, Regulatory and Compliance.*

Policy Reviews and User Acceptance

Here, you will need to review and understand what policies are currently in place. Is there a standard cybersecurity policy in place? Is there a retention schedule defined? Is there a data classification policy? Is there any cybersecurity language in the code of business conduct? If there are no documented policies, you will need to begin working on these documents straight away. If there are current policies, you'll need to fully review them and, most likely, make modifications to meet the current state. You'll have to work closely with your HR and legal teams as it relates to policies within your organization. *Chapter 13, Governance Oversight*, will cover this in more detail.

Data Loss Prevention and Information Protection

Data Loss Prevention (DLP) and **Information Protection (IP)** are critical components to help protect your organization's information and data. The premise behind your role is to protect the organization's assets from getting into the wrong hands and being compromised – most importantly, the data you store, manage, and process. You will need to first assess if any DLP or IP technologies are currently in place to determine the next steps you need to take. If nothing is in place, you'll need to quickly create a project to address these needs.

We'll cover this in more detail in *Chapter 5, Cybersecurity Architecture*, and *Chapter 13, Governance Oversight*.

Business Continuity Planning, Disaster Recovery Panning and the Cybersecurity Incident Response Plan

BCP, DRP, and the CIRP are all very important documents that form part of your response activities. As an organization, these plans must be well documented, reviewed at least annually, and circulated with those who will be involved with any of these plans. You may have some of these currently in place, but you'll need to review each of them if they do exist, ensure that you are familiar with them, and make sure that the cybersecurity team is part of each of the plans. For the CIRP, you will directly own this plan and you will need to ensure one is in place if not already. This will be covered in more detail in *Chapter 11, Proactive Services*.

Tabletop Exercises

If your organization hasn't executed any tabletop exercises, you'll need to coordinate and schedule events for both your technical and executive teams. Unfortunately, it's only a matter of 'when' and not 'if' an incident will occur. Being best prepared will only prove to be advantageous, as the need to bring the company back online as quickly as possible will be required. Tabletop exercises help you prepare for these situations and also identify any gaps that will need to be addressed in preparation for an actual incident. *Chapter 11, Proactive Services*, will cover this in more detail.

Current Cybersecurity Portfolio Review

Now that you are a little more settled and beyond the immediate short-term roadmap, you will want to closely assess your current cybersecurity portfolio and all the vendors used to support it. Are there any gaps? Are there any overlaps? Are you using more vendors than you need? Do you need to onboard new vendors to improve any of the cybersecurity functions? These are all questions that you will need to review in more detail as you determine the future portfolio for your cybersecurity program. This will be covered in *Chapter 4, Solidifying Your Strategy*.

Cybersecurity Program and Modernization

As a follow-up to your portfolio review, you will need to assess how mature your cybersecurity program is as it relates to modernization. Is the program heavier on the legacy server type of deployment architecture versus looking at more modernized cloud-based services? Is the program leveraging some of the next-generation technologies and capabilities available to protect your environment? What about AI and automation opportunities?

This will all need to be understood and reviewed further as you review your broader portfolio to determine the future state of your cybersecurity program. We will spend more time on this in *Chapter 4, Solidifying Your Strategy*.

Vendor Risk Management Review

Another hot item within cybersecurity currently is vendor risk management. We are observing a continuous event of vendors becoming compromised with our data. Threat actors target vendors knowing that our information is stored within their environment. They also exploit opportunities within the supply chain of our vendors, which provides them a door into our organizations from vulnerabilities within the vendors' software and applications. Because of this, a lot more due diligence is needed as vendors are onboarded and contracts are renewed, and you need to review current vendors to ensure that they take cybersecurity seriously. We will cover this in a lot more detail in *Chapter 10, Vendor Risk Management*.

Application and Web Testing, including a SecDevOps Review

As you move into the short-term roadmap, you should have full visibility of all your assets from the immediate short-term roadmap activities. With the asset inventory, you will have visibility of your website and web-facing applications. You'll want to understand if any application testing has been completed and if there is any security integrated with these assets. If not, you will want to review testing opportunities, like application penetration testing. You will also want to understand the DevSecOps process if one is in place. This will also require understanding where all the developers are within an organization to provide security best practices to ensure increased protection of all of your in-house applications. We will review this in both *Chapter 5, Cybersecurity Architecture*, and *Chapter 8, Vulnerability Management*.

Integrate with the Architecture Review Process

A critical component of the cybersecurity program is that of architecture – not just cybersecurity architecture but architecture as a whole across an organization. In order to ensure a more secure environment, it is important that cybersecurity resources are fully embedded into the overall architecture review process and that they participate in all architecture reviews.

If cybersecurity is not part of this process, you will need to partner and collaborate with the architecture team to integrate cybersecurity into the process. If there is no architecture review process in place, you will want to work with the correct resources within the organization to begin the process of building one. *Chapter 5, Cybersecurity Architecture*, will cover this in more detail.

Reporting Out

As a cybersecurity leader, you'll find that you will be required to provide more and more reporting on the overall program, general metrics, incidents, audits, assessments, and so on. You will want to begin formalizing templates that you will use to continuously update, as you need to share the overall progress of the cybersecurity program with the broader organization, executive leadership, and the board. The more efficient you are at creating reports and compiling presentations that relate to your intended audience, the more successful you will be at gaining the buy-in needed to keep the program growing and maturing in the right direction. This will be covered more in *Chapter 13, Governance Oversight*.

Now that we've reviewed the items that needed focus on within the first 12 months, we will move on to the long-term roadmap, which will take us through 1–3+ years. Once again, there will be items that roll into this roadmap as you continue to evolve and mature the program.

Long-Term Impact (1—3+ Years)

Like the previous roadmap, you are going to transition into the 1–3+ years roadmap with items that roll over from the short-term roadmap, as they will take longer than hoped and more time will be needed to complete them. Also, a lot will change during the first year, so there will be additional items that will need to be addressed that may not have originally been captured. As stated several time already, this is where you will need to be dynamic as a leader and adapt to the changing environment and requirements around you. In cybersecurity, this happens daily.

You will observe that the 1–3 year long-term roadmap starts naturally evolving into a stabilization phase with reviews and continuous improvement activities. Here, the program will be in a much more mature place, and you should feel more acclimated to what the program has become. Although you are tracking three years into the future with this roadmap, the reality is that no one truly knows what they will be dealing with in the coming years.

Technology changes so fast, and threat actors continue to evolve at such great speed, that our roadmaps will need to be continuously reviewed and updated as we proceed.

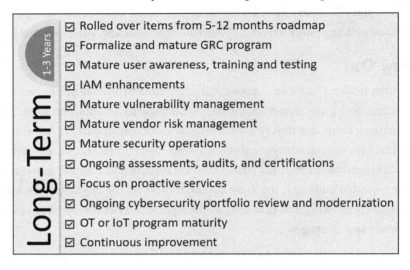

Figure 3.9: Long-term items

Rolled-over Items from the 5–12 Months Roadmap

This will cover any of the items that roll over from the short-term roadmap. With all the competing work and different priorities, certain tasks will end up requiring more time and resources to complete. Some items that I'd imagine may roll over into the long-term roadmap include assessments and/or audits, policy creation, DLP and IP, modernization of the technology stack, vendor risk management, application and web-testing activities (depending on volume), and the possibility of OT and IoT if applicable.

Formalize and Mature the GRC Program

As you continue to gain more visibility into the cybersecurity program, more audits continue to come your way, and regulations continue to evolve, there will be a need to efficiently track everything within the cybersecurity program. This includes an application inventory, the vendors being used, a risk register, audits, policies, security testing, and much more. We need to ensure everything is documented, with full visibility and transparency provided. If there is a major cybersecurity incident, this information becomes critical and should easily be accessible within your **Governance, Risk, and Compliance (GRC)** program as part of any investigation. This will be covered in more detail within *Chapter 13, Governance Oversight, Chapter 14, Managing Risk,* and *Chapter 15, Regulatory and Compliance.*

Mature User Awareness, Training, and Testing

As you head into the second year of your program, it will be time to start maturing each of the functions within your cybersecurity program. At this point, you will have a much clearer picture of where your cybersecurity user awareness, training, and testing program is. Most likely, you will also have some findings or recommendations from any assessments or audits that you have executed so far that will provide specifics on anything that needs improvement. *Chapter 9, Cybersecurity Awareness, Training, and Testing,* will cover this more.

Identity and Access Management (IAM) Enhancement

Another function that is going to need continuous review for enhancement is identity and access management. As you continue to mature and strengthen identity and access management for your users, there will be a need to keep up with the latest technologies and capabilities. This includes the need for identity protection-specific technology, looking to move away from some of the legacy **MFA** technologies to more modern methods, leveraging biometrics, behavioral-based and conditional access capabilities, the move away from passwords to passwordless, using modernized **Business-to-Business (B2B)** capabilities, etc. We will cover this in more detail in *Chapter 6, Identity and Access Management.*

Mature Vulnerability Management

Vulnerability management is another function that will need to be continuously reviewed and enhanced. More and more vulnerabilities will come our way, and we need to continue to improve on how quickly we can identify vulnerabilities within our environment and how quickly we can get them updated once updates become available. Looking beyond the traditional methods within vulnerability management will be critical. Can we automate? Do we need to begin to think about instance patching versus waiting? How do we automatically identify when a **Known Exploited Vulnerability** (**KEV**) impacts one of our services so that it can be remediated immediately? This will be covered in more depth in *Chapter 8, Vulnerability Management.*

Mature Vendor Risk Management

Based on the short-term review of your vendor risk management program, you are going to need to continue to evolve and mature the program. You'll find that an organization as a whole will use a lot of vendors, and you'll need to ensure each of them receives the correct level of review, depending on the risk they pose within your organization. In addition, you'll need to ensure contracts are updated with the relevant cybersecurity language to better protect the organization.

These efforts take a lot of time to work through, especially once you realize how many vendors are in your portfolio. *Chapter 10, Vendor Risk Management*, will cover this in additional detail.

Mature Security Operations

Again, this function will require continuous evaluation and the need to enhance and mature. The security operations team will be the busiest function within your program, and you need to look for ways to relieve the constant pressure and stress as it relates to 24/7/365 operations. Are there more modern approaches that can be evaluated? How can AI and automation continue to help? We cover this in more detail in *Chapter 7, Cybersecurity Operations*.

Ongoing Assessments, Audits, and Certificates

The need for assessments, audits, and certifications will be an ongoing task. You will need to continuously review your current state and ensure that you take the program in the right direction. As the cybersecurity landscape evolves, additional controls will be needed, and you'll need to assess each of these new controls as they are added. This will essentially be an annual task, which will be one of your best measurements to show progress that you can report to the broader organization, executive leadership, and the board. We will cover this in detail within *Chapter 15, Regulatory and Compliance*.

Focus on Proactive Services

Your proactive services function is necessary nowadays. It is important that you are doing everything you can to prevent or reduce the risk of a major incident occurring. The reality is that a security incident is going to happen, so the better prepared you are and the more familiar you are with working through a cybersecurity incident, the quicker you can restore a business and get it back up and running. Tasks within this function include your CIRP, tabletop exercises, penetration testing, threat-hunting exercises, and more. *Chapter 11, Proactive Services*, will cover this in more detail.

Ongoing Cybersecurity Portfolio Review and Modernization

As we covered in the short-term roadmap, the need for a portfolio review of all your current applications and vendors will be an ongoing task as you constantly assess the current threat landscape and market. The same applies to the modernization of your program where applicable. This will be an ongoing task that ensures you keep up to date with the latest threats and allow innovation to occur within the program. This will be covered in more depth in *Chapter 4, Solidifying Your Strategy*.

OT and IoT Program Maturity

If you do have any of these technologies in place, you will need to continue to assess, monitor, and review your program. As previously stated, this program will require dedicated skill sets to manage and protect efficiently. These technologies will consist of a lot of legacy systems that will need additional protection outside of the standard IT security portfolio. It is important they are closely monitored and reviewed for any known vulnerabilities and always kept up to date. *Chapter 12, Operational Technology (OT) & the Internet of Things (IoT)*, will cover this in more detail.

Continuous Improvement

This one goes without saying – the need to continuously review and look for opportunities to enhance and improve the overall program is a must. This needs to be a broader effort with your team to allow input and feedback from a different perspective. What your program looks like this year may need to change next year to meet new challenges. Make sure you have an open forum to allow a conversation around improvement opportunities, as well as the ability to deliver innovation for the cybersecurity program.

This will be the last roadmap as part of the initial development of your roadmaps. This roadmap is a lot more generic than the previous ones because it covers a much broader timeframe, and the effort will be focused more on maturity and enhancements. As you progress year to year, you will need to ensure that a 12-month roadmap is built each year, based on the current state. As you develop your new 12-month roadmap each year, you will want to review and update your long-term 3-year roadmap to reflect the changes made to each of the 12-month roadmaps. This will help keep the long-term roadmap relevant as you continue to report the program to the broader cybersecurity team, executive leadership, and the board.

Summary

As you have seen throughout this chapter, the use for a roadmap will only provide greater success for a broader cybersecurity program. If you are just beginning your role as a new leader, you need to quickly learn the current landscape and what needs to be prioritized. To do this, you will need to build yourself a plan that will support you as you continue to learn the current state and what activities you need to execute next. A roadmap is one of the most efficient ways to accomplish this. Your roadmap will also help as you work through your project items every year. This is where the overall program and project management activities fall within the broader cybersecurity program. Having efficient resources to support the overall program and project-related tasks is needed for a more successful program.

You are going to need to assess the current state of your program and the broader organization to provide more useful and relevant roadmaps. You will have a lot of homework to do here, especially if you are new to an organization. This body of work can also be considered a preparation phase that allows you to collect all the data and information needed to build your roadmaps. Here, you will need to collaborate outside of your own function with business leaders and the organization as a whole to get a better sense of how the program looks from the outside. You will need to work closely with your teams to get a thorough understanding of the current state of the cybersecurity organization to best plan the next steps.

Once the current state has been assessed, you can begin to build out each of your initial roadmaps as you look to build the foundation of your cybersecurity program. The first will be the immediate short-term impact (2–4 months) roadmap where you look to implement some quick wins to begin to strengthen the program. You will also begin to work through any critical or high-priority items that were identified in the current state review. Then, you will build the short-term impact (5–12 months) roadmap, which will focus on the priority items needed to build your program within the first year. You will also continue to work on any items that rolled over from the immediate short-term roadmap. The final roadmap will be the long-term impact (1–3+ years) roadmap, which will focus more on the operational activities by continuing to mature each of the functions. You will also look at continuous improvement opportunities in this roadmap. As you work through each year ahead, you will continue to build out an annual roadmap that captures the items you will work on for that year. This will be an annual exercise.

In the next chapter, we will cover solidifying your strategy. We will investigate the importance of following a strategy and why one is needed. In addition to your broader strategy, we will discuss the need for an architecture strategy for both an organization as a whole and the cybersecurity organization. With this strategy, we will review the need to modernize and what modernization looks like. The need for a framework has been touched upon several times previously, and in the next chapter, we will go into a lot more detail as to why a framework is needed and the options you should look at. The product and vendor portfolio strategy will be reviewed in more detail and some things you should be thinking about before finishing off the chapter, with a more in-depth discussion about resource management and a review of the in-house versus outsourcing options.

Join our community on Discord!

Read this book alongside other users, Cybersecurity experts, and the author himself. Ask questions, provide solutions to other readers, chat with the author via Ask Me Anything sessions, and much more.

Scan the QR code or visit the link to join the community.

`https://packt.link/SecNet`

4

Solidifying Your Strategy

Does your organization have a cybersecurity strategy in place today? Your strategy is an integral part of the foundation of your cybersecurity program. Having a well-defined strategy and ensuring everyone is aware of its importance will only lead to greater success. Your organization will have (or should have) a defined strategy (are you aware of your organization's strategy?) at the broad organization level that supports the reason for the organization's existence. The same applies to all other functions within the organization: there should be a strategy developed to support each of their missions, visions, and goals. In addition to your cybersecurity program strategy, you will also need more specific strategies within your program to help support each of the functions that supports the broader cybersecurity program. You will find that a successful program will have multiple strategies in place. For the cybersecurity program, the core strategies we will focus on throughout this chapter include your architecture strategy, cybersecurity framework strategy, your product and vendor strategy, and your resource management strategy. These strategies are an integral part of the success of the broader cybersecurity program.

First, we will review the importance of having a strategy in place for your program. It is important that you fully understand what a strategy is, why you need to have one, and how it will benefit both the broader organization and the cybersecurity organization. Defining a strategy also allows you to provide more constructive reporting to the executive leadership team and the board members, which will provide confidence that the program has some formality and structure around it. We will then shift over to your strategy around architecture and the cybersecurity program architecture strategy. Without a well-defined cybersecurity program architecture strategy, you will struggle to build an effective program to help reduce overall risk. Another strategy that needs to be implemented is that of a framework for your cybersecurity program.

There are many frameworks available, and you'll need to select one that makes the most sense for your organization. We will review what frameworks are available and provide an overview of a couple of the more prominent cybersecurity frameworks.

One important strategy you'll need to spend some time on is that of your product portfolio and all the applications and services currently used to support your program and protect the environment. Historically, we have onboarded vendor after vendor to meet the needs of specific requirements within the cybersecurity program, but managing all these vendors has become unsustainable and extremely complex. The strategy for this needs to be reviewed and better defined to ensure greater success. Another area of the strategy that will need some focus is your resources. As you build and mature your cybersecurity program, you will find that a broad range of expertise is going to be required. Being a small to medium-sized organization, it will be difficult to hire personnel to support all the requirements within your cybersecurity program. Because of this, you are going to need to strategize around the types of resources you hire in-house to oversee and manage the core functions of the program compared to outsourcing the repetitive day-to-day responsibilities for specific functions, and the need to bring in a third-party vendor or contractor/s to work on more specialized tasks. The following topics will be covered in this chapter:

- The importance of a strategy
- What is your architecture strategy?
- Why a cybersecurity framework?
- Managing your product and vendor portfolio
- Resource management (In-house versus Outsourcing)

The Importance of a Strategy

In simple terms, a strategy is a plan that is used to achieve or meet an overall goal that has been defined. When you define a goal, you need a plan to accomplish it. For example, you set the goal that you want to become certified in cybersecurity. Unfortunately, this just doesn't happen. Ask yourself what is needed to become certified. This is where your strategy comes into play. Your strategy will be the steps needed and the plan you put in place to become certified in cybersecurity. You will want to first set a date you would like to be certified by. Once that date is set, you are going to need to study, which means you will need to purchase study materials. You may need to attend a class, attend study groups, complete practice exams, etc. You will also need to plan when you will study and what material you will use, and you will need to ensure you have covered all material before sitting the exam.

With all this, you are going to need to create a schedule, or roadmap to ensure you accomplish each of your tasks to attain the final goal of becoming cybersecurity certified. The hope is that your strategy will provide you with the foundation to pass your exam and become certified.

From an organizational perspective, the strategy serves the same purpose. To build a successful organization, you will need to build a plan that allows you to meet the mission, vision, and goals defined for the organization. The strategy is the plan created and executed to support the initial creation and the ongoing purpose of your organization. Without a strategy, there will be no set direction for your organization to follow and you will essentially be managing chaos. If there is no plan in place to meet the defined goals, the working environment will become very unsettling, and it will only be a matter of time before the organization no longer exists.

It is important to note that there will be many strategies within an organization. At the organizational level, you will have a broader defined strategy that serves the purpose for the organization's existence. To implement the strategy, supporting goals will need to be defined. You may also find additional organization-level strategies in some instances. As you go down the organization structure, you will find that each function will also (hopefully) have its own strategy, with supporting strategies and goals, and finally, the teams within each function will have their own strategies, with supporting strategies and goals. For example, at a high level, you can expect something like this:

Figure 4.1: Organization, function, and team strategy template

As we shift our focus to a cybersecurity strategy, let's look at a recent example of a strategy released by the US Government in March 2023: the National Cybersecurity Strategy.

The strategy defined is:

> *"to secure the full benefits of a safe and secure digital ecosystem for all Americans."*

To accomplish this strategy, the approach defined is centered around five pillars or supporting goals/strategies:

- Defend critical infrastructure
- Disrupt and dismantle threat actors
- Shape market forces to drive security and resilience
- Invest in a resilient future
- Forge international partnerships to pursue shared goals

MARCH 02, 2023

FACT SHEET: Biden-Harris Administration Announces National Cybersecurity Strategy

Read the full strategy here ↗

Today, the Biden-Harris Administration released the National Cybersecurity Strategy to secure the full benefits of a safe and secure digital ecosystem for all Americans. In this decisive decade, the United States will reimagine cyberspace as a tool to achieve our goals in a way that reflects our values: economic security and prosperity; respect for human rights and fundamental freedoms; trust in our democracy and democratic institutions; and an equitable and diverse society. To realize this vision, we must make fundamental shifts in how the United States allocates roles, responsibilities, and resources in cyberspace.

Figure 4.2: The US Government cybersecurity strategy

There's obviously a lot more detail involved in this strategy, but it should give you a good idea from a cybersecurity perspective of what a broad strategy looks like and the plan to accomplish it.

The fact sheet can be accessed at the following link and the full strategy can be accessed from within the fact sheet: `https://www.whitehouse.gov/briefing-room/statements-releases/2023/03/02/fact-sheet-biden-harris-administration-announces-national-cybersecurity-strategy/`.

As you look to build the strategy for the cybersecurity organization within your organization, you will need to link it back to the mission and vision specified for the program. For example, as we covered in *Chapter 2, Setting the Foundations*, your mission may be to *"Protect and secure the people, assets, and data at <your company> from the ongoing and evolving threats in today's world"* and your vision may be *"A work environment with minimum cybersecurity risk."* Essentially, your strategy is to create a secure work environment for your users by protecting them from the cybersecurity threats faced in today's world.

We are now going to come full circle back to *Chapter 3, Building Your Roadmap*, as one of the best examples of how to achieve your cybersecurity strategy. A roadmap will serve as one of the most efficient tools to build your strategy and meet the objectives of your program. As reviewed in *Chapter 3, Building Your Roadmap*, you should understand the importance of spending the time to build a well-defined roadmap that will deliver the goals for your overall cybersecurity program to meet your strategy. And to deliver the strategy, good program and project management is critical. As touched upon earlier in the chapter, there may be additional strategies to the broader strategy that are needed to efficiently support the broader cybersecurity program.

Now that we've covered the importance of the strategy and how to define the cybersecurity strategy for your organization, we need to execute and implement the strategy and supporting strategies for the cybersecurity organization. This includes your architecture strategy, framework strategy, product vendor strategy, and resource management strategy.

The following provides a visual of the strategies we will focus on:

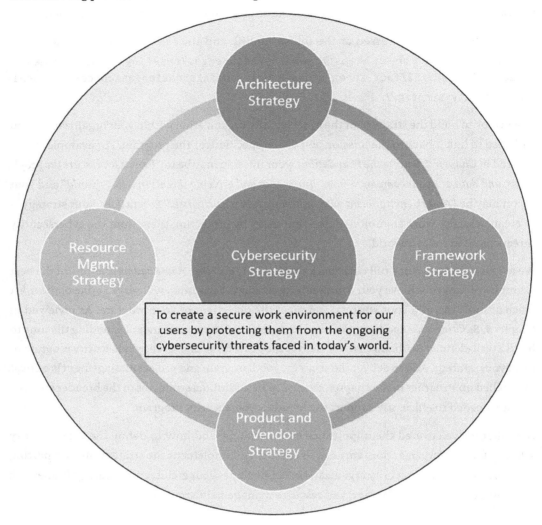

Figure 4.3: Cybersecurity strategy with supporting strategies

As mentioned, we'll be focusing now on the core strategies referenced above as they will be the essential building blocks for minimizing risk and allowing greater success for your cybersecurity program.

What is your Architecture Strategy?

Architecture is a very broad area that goes beyond cybersecurity. It is not a new concept within the organization and, for many, there should be some form of enterprise architecture team and/or strategy in place. Within this section, we will touch upon the enterprise architecture components and the strategy around your cybersecurity architecture.

Architecture Roles

As it relates to enterprise architecture, you will find many different types of architecture roles that will cover many different areas of the technology portfolio. The following table includes many of the architect roles you will come across, and potentially find yourself collaborating with inside of your organization. However, if you work within a small organization, these roles may not exist, or they may be the responsibility of someone who wears multiple hats.

Category	Sub-category	Primary Role	Key Responsibilities	Additional Info
Enterprise Architecture	Enterprise Architect	Oversee the architecture strategy and ensure alignment with the mission, vision, and defined goals.	Develop architecture strategies, policies, and roadmaps. Oversee the architecture framework and ensure compliance.	Familiar with all other architecture areas and provides a leadership role to other architect functions.
Cybersecurity Architecture	Cybersecurity Architect	Design, implement, and oversee the security architecture of the organization and oversee the integration of cybersecurity solutions across the enterprise.	Develop security strategies, policies, and frameworks. Oversee the implementation of security solutions and conduct security assessments.	Works across various domains; involved in risk assessments and incident response planning. Must remain up to date with the threat landscape.

	Compliance Architect	Ensure that security architectures comply with laws, regulations, and standards.	Map regulations to security controls, lead compliance audits, and gap analysis.	Works closely with legal teams; familiar with regulations like GDPR, HIPAA, PCI-DSS.
	Technical Security Architect	Implement and maintain technical security solutions.	Deploy security tools and technologies. Configure and maintain security systems.	Hands-on role; often has a background in systems administration or network engineering.
	Information Security Architect	Protect information assets by establishing security controls and protocols.	Design and implement information-centric security measures, including data classification and encryption.	Knowledgeable about data lifecycle management and information protection best practices.
Network Architecture	Network Architect	Plan and design secure network infrastructure.	Develop secure networking solutions, oversee network segmentation and access controls.	Expertise in networking protocols and network security technologies.
	Network Security Architect	Specialize in securing network communications and data transfer.	Implement firewalls, intrusion detection/ prevention systems, and secure network protocols.	In-depth knowledge of network threats and mitigation techniques.

Software/ Application Architecture	Application Security Architect	Ensure that applications are designed and implemented securely.	Involved in the SDLC to integrate security practices. Perform code reviews and application testing.	Requires programming skills and understanding of secure coding practices.
	Software Architect	Define the structure and behavior of software systems with security in mind.	Architect secure software systems. Align software architecture with business needs and security requirements.	Typically has a strong software development background.
Solutions Architecture	Solutions Architect	Create comprehensive security solutions that meet business requirements.	Translate business requirements into technical solutions. Integrate various security products and services.	Balances technical know-how with business insight.
Cloud Architecture	Cloud Security Architect	Design and secure cloud environments.	Implement security in IaaS, PaaS, and SaaS models. Design cloud network security architectures.	Familiar with CSPs' native security tools and third-party solutions.
	Cloud Architect	Oversee the deployment of applications and infrastructure in the cloud.	Ensure secure migration to the cloud. Optimize cloud services for security and efficiency.	Requires understanding of cloud services and orchestration tools.

Data Architecture	Data Architect	Ensure the security and compliance of data storage and processing.	Design data models. Implement secure database technologies. Oversee data encryption and masking.	Deep understanding of database management systems and big data platforms.
	Data Security Architect	Specialize in protecting data at rest, in use, and in transit.	Develop data-centric security strategies. Secure data environments and workflows.	Expert in data protection regulations and encryption technologies.
Infrastructure Architecture	Infrastructure Architect	Design and secure physical and virtual infrastructure.	Oversee secure deployment of servers, storage, and networking hardware.	Must balance hardware knowledge with virtualization and cloud technologies.
	Infrastructure Security Architect	Focus on securing the foundational IT systems and services.	Design resilient systems to withstand attacks. Implement secure configuration management and patching.	Often requires a blend of traditional IT and modern cloud security expertise.
Identity and Access Management Architecture	IAM Architect	Design systems to manage identities and access controls.	Develop and implement IAM strategies and solutions. Oversee identity provisioning and access governance.	Should be adept in IAM protocols and products.

	Identity Architect	Specialize in defining how identities are managed across systems.	Create identity lifecycle management processes. Integrate IAM with other security and business processes.	Deep knowledge of authentication and authorization mechanisms is required.

Table 4.1: Architectural roles

Alignment with Broader Architecture Strategies

If you work within a larger organization, there will most likely be a dedicated architecture function or an enterprise architecture team. You will also find strategies may already exist within the organization as it relates to architecture within other functions, or within the broader IT function. If so, you'll want to understand and align with any broader architecture strategy defined at a higher level. For example, if there is an enterprise architecture function or team, there will most likely be defined strategies in place you will need to follow. Early alignment will only help move things faster as you need to gain any approvals or work through any processes in place to proceed with projects and tasks on your roadmap.

A Comprehensive View

To ensure a comprehensive view of architecture as it relates to the cybersecurity program, we will take a holistic approach, so you fully understand the entire enterprise architecture landscape. This means that some of the items covered may already exist within your organization, especially in large enterprises. If an enterprise architecture function doesn't exist within the organization, this chapter will provide the guidance needed to ensure architecture is fully understood, and its relationship to the broader cybersecurity program. In addition, it is highly recommended that the core architecture components are defined if they do not exist today. This may be something you need to take ownership of and manage as it will only help support the broader cybersecurity program and reduce the risk across the organization. As briefly covered in *Chapter 2, Setting the Foundations*, we will look at architecture from two different angles, the first from a strategic perspective and the second from a technical perspective.

From a strategic perspective, the areas of focus will include:

- The broader architecture program oversight, more commonly known as enterprise architecture
- **The Architecture Review Board (ARB)**
- The cybersecurity program architecture strategy, such as implementing a cloud-first approach where applicable
- **Zero Trust Architecture (ZTA)**
- Understanding how your vendors fall within the architecture strategy
- Having in-house or outsourced architects and/or a consultancy

As you look at the technical side of architecture, this is where the work becomes more tactical by working with engineers, which will consist of ensuring the applications, systems, services, etc. are all secured based on recommended best practices. Some of the technical architecture areas of focus will include:

- Identity architecture
- Endpoint architecture
- Application and data architecture
- Infrastructure architecture (cloud and traditional)
- Network architecture
- Collaboration architecture (email, voice, IM, etc.)

 Although cybersecurity architecture is considered its own focus area, it will encompass all technical architecture areas as part of its responsibilities. Depending on the size of your organization, your cybersecurity architects may report to the enterprise architecture team or directly to the cybersecurity team. Regardless of where they report, strong collaboration between the cybersecurity architects and architects of the other areas is needed to ensure greater success.

For this chapter, we are going to focus on the architecture strategy for your cybersecurity program and that of enterprise architecture for the broader organization. Although enterprise architecture will most likely be owned by a different function or team, it is important that the cybersecurity team has direct involvement and provides the needed direction as it relates to cybersecurity.

If there is a different strategy across the broader organization from your desired strategy, you will need to work closely with the enterprise architecture team to define the need for your desired cybersecurity architecture strategy and gain approvals to proceed. We will cover the **Architecture Review Board (ARB)**, **Zero Trust Architecture (ZTA)**, and some of the more relevant architecture technical components in *Chapter 5, Cybersecurity Architecture.*

The Need to Modernize

In today's world, there is a need to modernize your technology portfolio to allow greater success and reduced risk. A term you have most likely heard many times in recent years is digital transformation. This is the ability to modernize your technology portfolio and processes from an outdated legacy perspective to a current and modern architecture. For example, one of the more recognized digital transformations to date, and one that transformed the movie rental industry, was Netflix's shift from mail-order DVDs to a cloud digital streaming service using the latest and most modern technology. This is just one example, but there are many more examples of how organizations are transforming their business through digital modernization.

 A quick Google search will provide many examples of digital transformation success stories. Here are a couple for reference, from IBM (`https://www.ibm.com/blog/digital-transformation-examples/`) and McKinsey Digital (`https://www.mckinsey.com/capabilities/mckinsey-digital/how-we-help-clients/rewired-in-action`).

There are many benefits as to why you are going to want to modernize your environment. Some of them include:

- The ability to use the latest and most current technology
- Enables a more innovative culture
- Increased productivity
- Allows for a much-simplified architecture
- It will help reduce risk with improved security
- Improved user experience
- The ability to become much more agile
- The potential for cost savings
- Increased reliability

One of the more common examples of modernization is the shift from a legacy on-premises infrastructure to a modernized cloud-first strategy to support the evolving need for serverless technologies, big data, machine learning, AI, and more. We will be covering cloud-first strategies in more detail later in this section.

The same concept applies to your cybersecurity architecture strategy. It's recommended to simplify where you can. Having disjoined security tools can make maintenance unsustainable and as a result, make you more vulnerable due to their complexities. Because of this, complete a review of what is currently deployed within your environment and set goals to consolidate your security footprint where applicable, which we will cover later, in the *Managing Your Product and Vendor Portfolio* section. Simplicity is key to a successful program. As part of this strategy, it is recommended to modernize to next-generation security tools. Traditional security tools will no longer suffice in the modern world. Next-generation security tools can be enabled and deployed at scale using cloud technologies with limited or zero physical infrastructure under your ownership and are, typically, always kept up to date by the vendor. These tools and services should support a level of automation, use artificial intelligence, analyze big data, and incorporate behavioral analytics. Without these features, organizations will miss out on valuable security insights that can help prevent attacks as opposed to reacting to a major cybersecurity incident.

Core Components

From a cybersecurity architecture perspective, you need to have visibility into all digital assets within the organization to best protect them. Protecting your assets will be very difficult if they are distributed and disconnected from the core architecture. If other business functions within the organization are onboarding their own technologies and services without following the core architecture or process, better known as shadow IT, the risk level of the organization will increase. For example, what if one function decides to onboard **Google Cloud Platform (GCP), another Amazon Web Services (AWS), and another Microsoft Azure?** Your ability to efficiently protect all these environments will be much more difficult than if you had a strategy and architecture that selects one service for the organization to adopt. Also, the cost to maintain multiple environments will significantly increase, in addition to the ongoing operations and management of multiple environments. Because of this, you are going to need to build your strategy around a core set of technologies that are foundational to the organization. The idea here is to simplify as much as possible. This in turn will allow a much more secure environment with reduced costs for the organization.

A unified strategy on a single cloud platform will need to come from the top down. Executive leadership will need to support and enforce this level of strategy. From a cybersecurity perspective, your role is to recommend and influence the need for a unified cloud platform to reduce risk. As with any defined organizational strategy, if there is no support and enforcement from the executive leadership level, it will not be successful.

From an architectural perspective, the following are considered the core technological components to support the technological foundation of the organization:

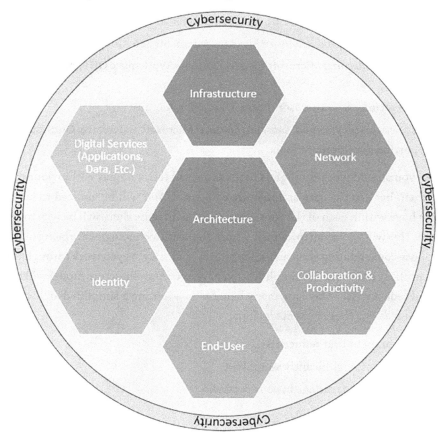

Figure 4.4: Core architecture functions for an organization

You may have noticed the above categories are very similar to that of the zero trust architecture, which will be covered in *Chapter 5, Cybersecurity Architecture*.

Now that you have a better understanding of the core architecture technology stack, you will want to expand on this to capture the core strategic technologies and vendors that can be used to support the needs of the organization. Some examples of the more common vendors you may be familiar with to support each of these architectures are:

- *Infrastructure architecture*: Microsoft Azure, Google Cloud Platform, Amazon Web Services
- *Network architecture*: Cisco, Palo Alto, Juniper, F5, Aruba, Fortinet
- *Collaboration and productivity architecture*: Microsoft 365, Google Workspace
- *End-user architecture*: Microsoft Intune, Omnissa Workspace ONE, Azure Virtual Desktop, Citrix
- *Identity architecture*: Microsoft Entra, Okta, Ping Identity
- *Digital services (application, data) architecture*: Microsoft, Salesforce, OutSystems and other supporting development tools

An exercise you should work through is to inventory all technologies within your organization and map them back to the core architecture categories. You will be amazed to see how many entries you have within each of the core categories. I'd imagine there will be way more than you anticipated. This will allow you to identify where duplicate technologies are being used and allow you to propose consolidating services for long-term efficiency. As you work through this exercise, one of the primary drivers of the simplification of your core architecture model will be increased security and reduced risk. There are many benefits to adopting a simplified architecture model for your organization. These benefits include:

- An environment that reduces risk
- A model that is significantly simplified
- Fewer vendors to manage in your portfolio
- Much simpler integration
- Easier to support and operate
- Cost savings
- Reduced downtime
- Simplified user experience

Although this book is not focused on enterprise architecture and the core technology foundations to support your organization, it is an extremely important topic to understand, and you need to ensure the cybersecurity team is part of the broader enterprise architecture planning and decision-making. Going through this exercise is important as your cybersecurity program will need to be built around the core architecture strategy that is being adopted within your organization. Everything technical has security implications, which is why architecture and cybersecurity must work hand in hand.

Cloud First

As briefly discussed in the The Need to Modernize section, a cloud-first strategy is needed to support the ability to digitally transform and modernize your organization. This requires a shift from what is considered a legacy model, that of a traditional on-premises data center, towards a modern cloud data center. When we reference a cloud data center, this refers to everything from the underlying infrastructure to the ability to use modern applications, the ability to use big data models, AI, machine learning, and so much more. To support these capabilities, the underlying infrastructure is core to the success of modernization. With a traditional on-premises data center, you will not be able to support the requirements for today's business expectations.

Over the years, data centers have changed quite significantly as it relates to the hardware your services run on. In the past, a traditional enterprise data center typically consisted of physical mainframes for storing and accessing information. Data centers during these times were sometimes located on-site or at a separate facility under the management of the organization. As technology has evolved, there has been a shift from the mainframe to server-based data centers. This is where Windows and Linux servers became widely adopted and grew in popularity. Moving beyond standard hardware-based server models is where virtualization technology entered the picture. The ability to run many servers on minimal physical hardware changed the dynamics of the data center significantly. Fast forward to today, and we are in the midst of a major shift to cloud computing. Organizations are slowly moving away from traditional on-premises data centers and moving all their workloads into cloud environments. With cloud data centers, organizations can continue to run traditional servers and services, but the overhead of owning and managing physical infrastructure is greatly reduced or eliminated.

Another major change with a shift to the cloud is the elimination of on-site facility management and physical operations. Building and maintaining a data center is an enormous undertaking that is challenging and comes with substantial cost implications when designing for **highly available (HA)** services and **disaster recovery planning (DRP)** as part of a **business continuity plan (BCP)**. Moving to the cloud changes these dynamics significantly.

Your cost model changes to a subscription-based model with zero upkeep of the physical hardware and facilities. This allows a more robust BCP with the flexibility to spread services across multiple data centers in different locations geographically. This shift also changes the dynamics of physical security for the data center. Traditionally, physical security with access controls, locks, badge readers, and security cameras was all needed. This goes away with the cloud, but how do you ensure the cloud provider is protecting the access and controls? These are all valid concerns and change the way we manage security as opposed to the traditional data center perspective. These concerns will be covered in more detail in *Chapter 10, Vendor Risk Management*. In addition, cloud environments bring a new set of challenges with security in general. How do you ensure your data is safe? How does cloud cybersecurity differ from security in a traditional data center? How does compliance change? Are new skillsets needed? These concerns will be addressed throughout this book.

Let's take a deeper look at the three most common data center models used today, with an emphasis on driving towards a cloud-first model.

On-Premises Data Center

As mentioned previously, the on-premises data center is considered the traditional model. Organizations build out and operate their infrastructure on your business's property or off-site at a separate facility. In this model, you are fully responsible for everything in the physical infrastructure (the building, power, cooling, hardware, security, access, and so on) and everything that runs on the hardware. Let's take a detailed look at the components of an on-premises data center:

- **Initial cost**: Purchasing physical hardware and software licenses, and setting up infrastructure requires an initial investment.

- **Operational cost**: There will be ongoing expenses for maintenance, power, cooling, and upgrades, but you will have a lot of control over optimizing costs.

- **Scalability**: Upgrading or replacing hardware is necessary for scaling, but it can be expensive and time-consuming.

- **Control**: You maintain complete control over the hardware, software, and network configurations, allowing custom solutions.

- **Security**: Full control over security measures requires in-house expertise and resources to manage them effectively.

- **Performance**: Performance is not dependent on external factors, and it can be optimized and tailored to specific needs.

- **Data sovereignty**: Data is stored on–site, giving organizations direct control over the location and compliance of data storage.

- **Maintenance**: Dedicated IT staff are required for maintenance, updates, and trouble-shooting, increasing the workload.

- **Customization**: A high level of customization is possible to meet specific organizational needs.

- **Dependency**: You mainly depend on in-house resources and capabilities.

- **Disaster recovery**: Separate investment in disaster recovery solutions and strategies is required to ensure data integrity.

- **Compliance**: It is easier to ensure compliance with specific industry regulations when all resources are managed internally.

Cloud Data Center

As we look further into the cloud model, it is important to understand public and private cloud offerings. A public cloud is where the services are hosted by the provider and the underlying infrastructure is shared with other organizations. Your environment will be logically separated from other organizations, but the underlying hardware, network, and storage are shared with other subscribers on the same service. A private cloud offering is where the services are hosted in a dedicated environment and only your organization runs on the underlying hardware, network, and storage with physical separation. Determining the appropriate model will most likely be dictated by your organization's industry and compliance requirements. Let's take a detailed look at the components of a cloud data center:

- **Initial cost**: Services offered on a subscription basis have no upfront hardware costs.

- **Operational cost**: The costs for a service may vary according to usage, but the provider will take care of maintenance and upgrades.

- **Scalability**: Technology is highly flexible and resources can be quickly adjusted to meet demand without any physical limitations. This means that it can easily scale up or down as needed to ensure optimal performance and costs.

- **Control**: You have less control over physical infrastructure, depending on the service provider's **Service-Level Agreement (SLA)**.

- **Security**: Security becomes a shared responsibility between the organization and the provider, posing a new set of challenges for the cybersecurity teams.

- **Performance**: This can vary based on internet connectivity and the provider's infrastructure.

- **Data sovereignty**: Data is stored off–site, potentially across multiple locations globally, which can complicate compliance with data protection laws.

- **Maintenance**: The provider handles maintenance, updates, and infrastructure management for the most part, depending on the model you adopt.

- **Customization**: Depending on the model you adopt, there will be limited customization options based on provider offerings.

- **Dependency**: You are dependent on the provider's stability, performance, and security.

- **Disaster recovery**: This often includes more robust and improved disaster recovery solutions due to multiple geographic locations and sites available by the provider.

- **Compliance**: Providers often offer compliance with a wide range of standards, but specific compliance needs must be verified and can be more challenging.

Cloud solutions typically provide three different broad models of primary services available for consumption, as outlined here:

- **Infrastructure as a Service (IaaS)** requires the most involvement from your organization and is operated very similarly to an on-premises virtual environment. The difference is that an organization has no responsibility for physical infrastructure, and the servers, storage, and underlying network fabric are all managed by the hosting provider. You can simply turn on **virtual machines (VMs)** and services as needed.

- **Platform as a Service (PaaS)** provides you with your required platform or service, all bundled together from the cloud provider. Typically, with PaaS, the physical infrastructure, operating system, middleware, and other tools for running services are maintained by the hosting provider. For example, in a traditional IaaS Windows environment for hosting **Internet Information Services (IIS)** or a **Structured Query Language (SQL)** database, you would need to install these components and roles on the operating system. With PaaS, you simply subscribe to the service, and you consume it directly. There is no installation or maintenance of any underlying software to run these apps.

- **Software as a Service** (SaaS) requires the least involvement and essentially provides you with the entire software solution that is ready to be consumed. Although, you will have the ability to configure and customize the service to meet your needs. In addition to what is managed for both the IaaS and PaaS services, the hosting provider also maintains the application itself, including keeping it current and up to date. An example of a SaaS offering is Microsoft Exchange Online, in which your entire Exchange environment is hosted, kept up to date, and managed by Microsoft. You simply consume (and will need to secure) the email services for your organization.

 There are many other "as a service" offerings available on the market. What we have covered here are the three core models that relate to cloud data centers. Some others you may come across are **Desktop as a Service (DaaS), Data-as-a-Service (DaaS), Network as a Service (NaaS), and Database-as-a-Service (DBaaS)**.

Hybrid Data Center

A hybrid model essentially combines the on-premises model with the cloud model, allowing an organization's on-premises deployment to co-exist with cloud services. This model is most likely going to be most suitable for established organizations with existing on-premises data centers simply because they can't easily move to a cloud model overnight or due to compliance requirements (if any exist). What the hybrid model does is provide a pathway from an on-premises data center to the cloud while providing services using both environments. With this model, it is important to provide robust integration tools to allow seamless operation between on-premises and cloud environments. This model also introduces a lot more complexity, so effective management and monitoring will be required. In addition, the cost to run a hybrid data center will increase initially as you migrate your payloads to the cloud. Essentially, the hybrid data center model encompasses both of the components detailed above.

The following image represents each of the models just reviewed.

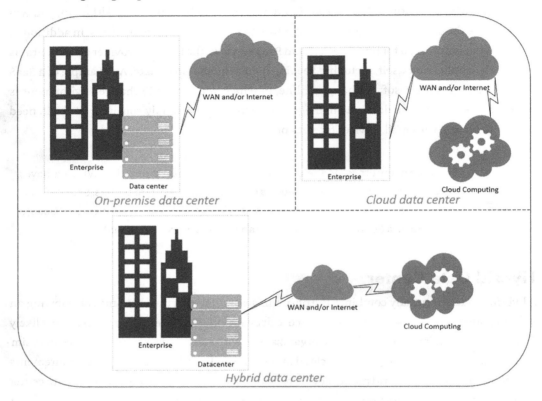

Figure 4.5: Data center models

Depending on your organization, you may have already started your journey to the cloud, you may already be fully in the cloud, or you may not have started yet. If you haven't started your journey, you will need to define a strategy to migrate to the cloud. There are many to choose from and, depending on where you are located in the world, there will be different options. The recommendation is to strategize around one of the major cloud providers for simplicity and reduced risk.

In the past, I've heard the argument in IT that we don't want to be dependent on one vendor or get locked in on price increases. But you need to measure the risk of strategizing on one core cloud provider versus deploying on multiple clouds. Some thoughts that come to mind when strategizing with one cloud provider:

- Security is improved because you only need to secure one environment
- You have a much smaller footprint to manage
- You need less operational support than if you use multiple cloud providers
- You don't need multiple skillsets, unlike with multiple environments
- You can control costs more efficiently

Another example I tend to respond with when I hear the argument of not strategizing on one vendor is "Would you expect your **Human Resources (HR)** team to onboard two different **Human Resource Management Systems (HRMS)** to manage your employees, or would your organization deploy two separate **Enterprise Resource Planning (ERP)** systems. Why should this be any different for your cloud provider?"

 If you already have multiple cloud platforms within your organization, you'll need to work with the leaders throughout the organization, along with executive leadership, to see if there is an opportunity to consolidate and reduce risk within the organization. Consolidation will not occur without support from the very top.

As mentioned previously, the three biggest cloud data centers within the US are AWS, GCP, and Microsoft Azure. From a strategy perspective, you'll want to complete a review, or even a **Request for Proposal (RFP)**, that allows a formal requirements document to be created to determine which of the three providers you should proceed with for your organization. As an example, let's look at the Microsoft cloud platform in more depth as the potential core cloud platform for your organization.

If you select Microsoft, you can build a clear picture of the broader architecture to be implemented based on the three models we have discussed: IaaS, PaaS, and SaaS.

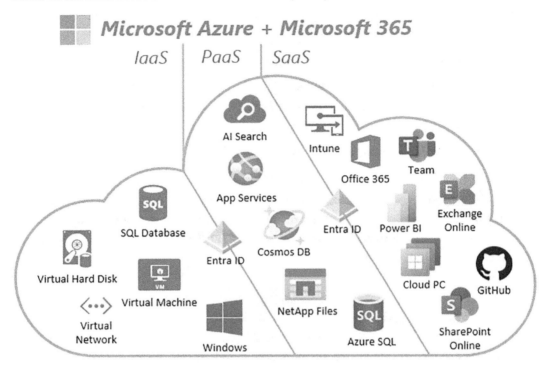

Figure 4.6: High-level Microsoft cloud architecture

 You can use the concept in the image above with any cloud provider to demonstrate the high-level architecture. Each cloud provider may not necessarily provide feature parity, so you'll need to do your due diligence to ensure your requirements are met with the cloud provider you select.

As you start adopting more cloud services, you will want to promote and encourage the use of PaaS and SaaS services as much as possible. The sooner you are able to get away from a legacy server infrastructure that needs never-ending patch management, and all the other risks with physical hardware in an on-premises model, the better. Although PaaS and SaaS still have security challenges and risks, a significant amount of responsibility shifts from you to the service provider. Obviously, ensuring the vendor is doing their due diligence becomes your responsibility.

As we discuss the shift of responsibility (depending on the model you adopt), it is important (as security professionals) that you fully understand your responsibilities and the cloud provider's responsibilities. The model that provides this detail is known as the *shared responsibility model*. This model is similar to a **Responsibility, Accountability, Consult, Inform (RACI)** matrix, just not as detailed. It defines who is responsible for what as it relates to each of the cloud services. The following example is taken from Microsoft and defines who is responsible for what with each of the cloud models. You can use the Microsoft model as a reference for any cloud provider, although there may be some slight differences with other references.

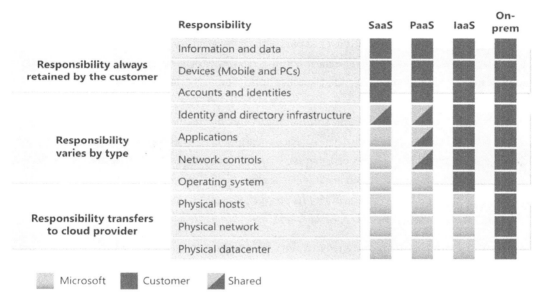

Figure 4.7: The shared responsibility model.
Source: https://learn.microsoft.com/en-us/azure/security/fundamentals/
shared-responsibility

As you transition to PaaS and SaaS services, the need for new and improved security and compliance requirements increases. It is important that you adapt to the changing requirements of these environments, and that you are running ongoing monitoring and assessments to ensure compliance with your standards and any regulations.

As you can observe from the shared responsibility model, as organizations increasingly adopt PaaS and SaaS services, their environments become more dynamic and agile. This will enable your organization to become more innovative in ways that were not possible with traditional IaaS and on-premises deployments. The PaaS and SaaS models will enable the modernization of your business applications, which will provide the benefits mentioned in *The Need to Modernize* section.

Now that we've conducted an in-depth review of your architecture strategy, we will focus on the importance of implementing a cybersecurity framework for your organization.

Why a Cybersecurity Framework?

The next core strategy that we will cover as part of the cybersecurity program is that of frameworks. There's a possibility that your organization may have some form of security framework in place today. If not, it's highly recommended you begin this journey right away to lay the foundation of your security program and strategy. It won't be long before everyone will be required to implement some form of framework at some point. For example, more regulations continue to be enforced, in addition to cybersecurity insurance continually increasing the requirements to retain coverage. Also, if you suffer a major cybersecurity incident and data is exfiltrated, you may find yourself part of a lawsuit in which having a cybersecurity framework will show you were taking due care to reduce risk. If a cybersecurity framework was not being used, you can see where I'm going: the outcome will most likely be very bad.

A cybersecurity framework is designed to build a foundation for your organization's cybersecurity program. As already stated, one of the primary reasons for implementing a cybersecurity framework is to reduce cybersecurity risk as much as possible for your organization. A cybersecurity framework will help cover the basics of everything you need to take into consideration with your cybersecurity program and identify any gaps within cybersecurity for the organization. There are several frameworks available for implementation, and the direction you take will depend on multiple factors, such as your business type, industry requirements, and any required regulations. It is important that you work with your legal, risk management, and privacy teams to understand if there are any specific requirements for your organization. Implementing a cybersecurity framework isn't easy and can in fact be extremely complex, requiring a major investment of time. Implementing a framework won't happen overnight; it will take a lot of planning, many months, and even years to implement correctly. It will be important to think of your framework as a journey that continues to evolve and mature over time. It will also provide transparency and visibility into the cybersecurity program.

A significant benefit of implementing a framework is the ability to provide a well-constructed overview of your cybersecurity program, along with the intended strategy to the executive leadership team and board members. The framework will provide a comprehensive view of what security controls are in place and create a roadmap of improvements that are needed. The framework will also prioritize needs, provide valuable input on missing controls, and provide justification for budget allocation. Ensuring clear transparency and visibility into the cybersecurity program to executive leadership and the board are crucial these days.

To summarize, a framework is essential for the following reasons:

- To lay the foundation of your cybersecurity program
- To identify gaps within your cybersecurity program and identify needed resources
- To define your cybersecurity organization structure
- To reduce the risk for the organization
- To define your cybersecurity strategy for the organization
- To build your short-term and long-term road maps
- To define your budgetary needs for the cybersecurity program
- To provide a visible measurement of your cybersecurity program through metrics and compliance scores
- To show continuous (hopefully) improvement of the cybersecurity program over time
- To measure your success of the cybersecurity program for executive leadership and the board of directors
- To provide transparency to executive leadership, the board of directors, and the broader organization
- To help with the prioritization of work based on identified risks from auditing your implemented framework
- To justify the existence of the cybersecurity function

Let's look at some of the most common and widely adopted cybersecurity frameworks that you may be familiar with:

- **International Organization for Standardization/International Electrotechnical Commission (ISO/IEC)** 27001:2022: https://www.iso.org/standard/27001
- **National Institute of Standards and Technology (NIST)** Cybersecurity Framework (CSF 2.0): https://www.nist.gov/cyberframework
- **Health Information Trust Alliance Common Security Framework (HITRUST CSF)**: https://hitrustalliance.net/product-tool/hitrust-csf

An example of an IT-specific framework is the **Control Objectives for Information and Related Technology (COBIT)**: https://www.isaca.org/resources/cobit

One thing I would like to point out when discussing frameworks is the importance of understanding the difference between a framework, specifically a program framework, and other similar items, such as controls, certifications, regulatory compliance requirements, and acts. I read a lot of posts that share framework recommendations when a lot of them are not actually frameworks. When you think of a program framework, the best way to conceptualize it is to ask, does it cover the entire scope of the cybersecurity program? For example, the NIST Cybersecurity Framework does, but the likes of **System and Organization Controls (SOC)** and **Payment Card Industry Data Security Standard (PCI-DSS)** don't. They typically only cover a specific scope of your environment. Another example is the **Center for Internet Security (CIS)**. You may already be familiar with CIS, and again, you will see CIS listed on a lot of the "most popular framework" lists, although it's not a fully comprehensive framework. Instead, CIS is more of a tactical compilation of controls and guidelines that allow organizations to meet the requirements of a chosen program framework. You can think of CIS as a control framework rather than a program framework.

 Some additional controls to support your selected framework are the NIST SP 800-53 Security and Privacy Controls for Federal Information Systems and Organizations and the ISO/IEC 27002:2022 Information security, cybersecurity, and privacy protection information security controls.

Your industry and location may dictate which framework is to be used, but in general, they can all be used as a foundation in any industry. As an example, healthcare will most likely adopt the HITRUST framework. ISO 27001 will most likely have a global presence over NIST, which was originally created for US government entities.

National Institute of Standards and Technology (NIST)

If you are implementing a framework for the first time, this is where you should start. The NIST Cybersecurity Framework (CSF 2.0) will be a little easier to implement as the first cybersecurity framework for your organization. The NIST Cybersecurity Framework was originally created to improve critical infrastructure cybersecurity within the US.

Although the framework was initially created for critical infrastructure, it has evolved over the years and can be used by any organization of any industry and size. This framework has gained a lot of popularity and is being adopted by many. The NIST Cybersecurity Framework is built around six core functions in the most recent 2.0 publication, which was released in early 2024. Prior to NIST 2.0, there were only five core functions in the 1.1 release. The additional function with NIST 2.0 is Govern.

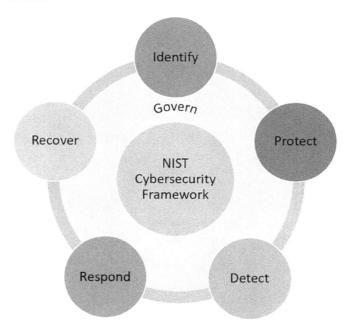

Figure 4.8: NIST 2.0 Cybersecurity Framework

Within these functions are multiple categories that make up the core of the NIST 2.0 Cybersecurity Framework. Then, within each of the categories, there are multiple subcategories that provide additional details on how to meet the requirements for each of the categories. This will allow you to meet the requirements for each of the functions presented above.

The following table presents the functions along with the categories contained within each.

Function	Category
Govern (GV)	Organizational Context Risk Management Strategy Cybersecurity Supply Chain Risk Management Roles, Responsibilities, and Authorities Policies, Processes, and Procedures Oversight
Identify (ID)	Asset Management Risk Assessment Improvement
Protect (PR)	Identity Management, Authentication, and Access Control Awareness and Training Data Security Platform Security Technology Infrastructure Resilience
Detect (DE)	Continuous Monitoring Adverse Event Analysis
Respond (RS)	Incident Management Incident Analysis Incident Response Reporting and Communication Incident Mitigation
Recover (RC)	Incident Recovery Plan Execution Incident Recovery Communication

Figure 4.9: The CSF 2.0 core functions and category names

To take this a step further, let's review a specific category in more detail, along with a single subcategory.

Within the subcategory, you will be provided with a reference that will provide you with a starting point to implement the required control for that subcategory. The following diagram breaks down the Identify function of the NIST 2.0 framework with a selected category, subcategory, implementation examples, and references:

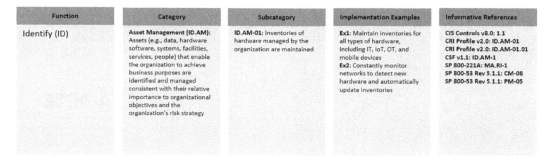

Function	Category	Subcategory	Implementation Examples	Informative References
Identify (ID)	**Asset Management (ID.AM):** Assets (e.g., data, hardware software, systems, facilities, services, people) that enable the organization to achieve business purposes are identified and managed consistent with their relative importance to organizational objectives and the organization's risk strategy	**ID.AM-01:** Inventories of hardware managed by the organization are maintained	**Ex1:** Maintain inventories for all types of hardware, including IT, IoT, OT, and mobile devices **Ex2:** Constantly monitor networks to detect new hardware and automatically update inventories	CIS Controls v8.0: 1.1 CRI Profile v2.0: ID.AM-01 CRI Profile v2.0: ID.AM-01.01 CSF v1.1: ID.AM-1 SP 800-221A: MA.RI-1 SP 800-53 Rev 5.1.1: CM-08 SP 800-53 Rev 5.1.1: PM-05

Figure 4.10: An example NIST 2.0 function with a category, a subcategory, implementation examples, and informative references to meet the subcategory

You can review more details on the informative references by navigating to the link below and clicking **Download (XLSX)** within the **Directly download all the Informative References for CSF 2.0** section: https://www.nist.gov/informative-references.

As you can see from the previous image, the NIST 2.0 Cybersecurity Framework provides guidance and resources that can be used to meet the required controls, to implement a cybersecurity framework efficiently.

International Organization for Standardization (ISO)

We won't go into as much detail as we did for NIST, but a more globally recognized framework is ISO/IEC 27001:2022 for **information security management systems (ISMS)**. More specifically, the ISO/IEC 27001 framework provides the direction needed to establish, implement, maintain, and continuously improve your ISMS.

ISO/IEC 27001 is specifically designed to support risk management, cyber-resilience, and operational excellence.

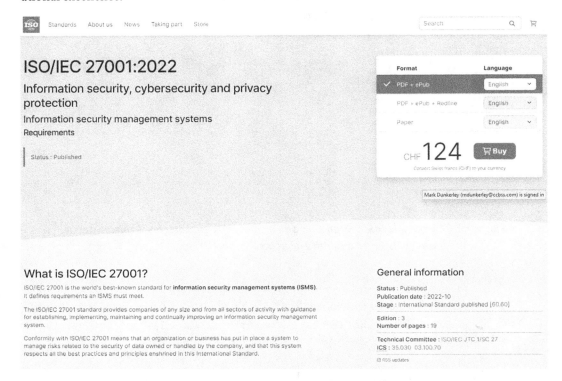

Figure 4.11: The ISO/IEC 27001:2022 webpage.
Link to ISO/IEC 27001:2022: `https://www.iso.org/standard/27001`

You may be wondering which of these frameworks makes the most sense for your organization. The following list provides a high-level guide to the differences:

- ISO/IEC 27001 comes at a cost compared to the NIST Cybersecurity Framework
- ISO/IEC 27001 is more globally recognized, whereas the NIST Cybersecurity Framework originated from the US Government
- You can become formally certified with ISO/IEC 27001, but the NIST Cybersecurity Framework doesn't have a globally recognized certification; however, you can always complete an audit against the NIST to validate the controls
- ISO/IEC 27001 can be considered a little more involved, especially with the certification process, and would better serve more mature organizations than start-ups

With that, if you are at an organization that is still maturing its cybersecurity program or just getting started building its cybersecurity program, the NIST 2.0 Cybersecurity Framework will be the best choice. Once it has been implemented, you will need to engage a third party to audit the controls you are measuring yourself against. We will be covering auditing in more detail in *Chapter 15, Regulatory and Compliance*.

 There are also additional frameworks that focus specifically on risk and privacy. We will review these in more detail in *Chapter 14, Managing Risk*, and *Chapter 15, Regulatory and Compliance*.

If you are unsure of how to get started with a cybersecurity framework, you should consult with a third-party familiar with cybersecurity frameworks or engage a vendor you already have on contract if they provide these types of services. Working with and understanding frameworks can be overwhelming at first so ensure you are reaching out for support to get you started if needed.

To summarize, I can't stress enough the importance of needing a framework for your cybersecurity program nowadays. Implementing a cybersecurity framework will serve as the foundation of your cybersecurity program and will be the core of your overall cybersecurity strategy. I believe this will eventually become a requirement for everyone, so don't delay.

Managing your Product and Vendor Portfolio

The third core strategy that we will be reviewing is your product and vendor cybersecurity portfolio. In the *What is Your Architecture Strategy?* section, we reviewed the broad strategy of consolidating technologies at the organization level. The same applies to your cybersecurity portfolio. We will look at how to best manage your overall cybersecurity portfolio in a world where risk continues to grow. In the past, we needed to onboard one vendor after another to ensure your cybersecurity portfolio had no gaps. This, in some respects, has turned into an environment that has become unmanageable with increased risk within our organizations.

As technology has continued to evolve, you have most likely found the same capabilities being provided by multiple vendors. You need to evaluate your cybersecurity portfolio to optimize it to improve efficiency. Some of the reasons for and benefits of strategizing around your cybersecurity portfolio include:

- Reduced attack surface risk across the organization
- Reduced overall costs (savings on infrastructure, licensing, personnel costs, etc.)
- Less complexity and reduced integration points

- Improved vendor relationships, allowing a more strategic relationship because there will be fewer vendors within the portfolio
- Allows streamlined management, operations, and support
- The ability for your staff to focus on fewer technologies to become more efficient
- Improved security monitoring with a smaller footprint

The good news is that more cybersecurity executives are leaning toward consolidating their security portfolios. A survey of 418 respondents conducted by Gartner in 2022 reported:

- Vendor consolidation is being pursued by 75% of the respondents, which is a significant increase from 29% in 2020
- The same survey found the proportion of organizations working with fewer than 10 security vendors was an impressive 57%

Source: `https://www.gartner.com/en/newsroom/press-releases/2022-09-12-gartner-survey-shows-seventy-five-percent-of-organizations-are-pursuing-security-vendor-consolidation-in-2022`

I'm sure this is not the strategy security vendors like to see, but we don't have much choice because we need to continue to reduce our footprint and reduce risk within our organization as much as possible.

Before you begin to consolidate your portfolio, you are going to need to understand all the vendors, services, applications, etc. in place for the cybersecurity program today. If you haven't already done this, you will need to collect an inventory of all your vendors. Once you have your inventory, map them against each of the 8 (GRC combined as 1) core functions within your portfolio to get a better idea of what all your vendors are being used to support within the program. A high-level example may look like the following, which includes a couple of examples to get you started:

Figure 4.12: Vendor portfolio template

Once you begin to map out your vendors, you will most likely find vendors covering multiple functions within your cybersecurity program, which is good. Your next exercise will be to dive into the details of each of your vendors to understand where there are duplicate technologies and overlaps. This will require a more detailed review of each vendor to understand what you have licensed and what functionalities are being used and are available from each vendor. For example, you may find that you are paying twice for endpoint protection tools, or you may find identity management features are included in one subscription, but you are subscribing to them from another vendor, or there may be functionality with one vendor that can be turned on for a lot less than maintaining another vendor.

This exercise will get complicated, but it will be important as you map out the available functionality from each vendor you have within your portfolio. It will also be important that you are aligning all vendor capabilities with the cybersecurity needs and objectives that we have covered so far. Your cybersecurity framework will help with the needs and objectives of your program as part of your portfolio strategy.

> As you go through this exercise, you may start seeing a trend with one or two more strategic vendors providing functionality for a lot of your portfolio. If this is the case, you will want to focus on these vendors to see how much you can consolidate from other vendors.

An example of how to begin mapping out the functionality of each vendor in more detail can include tracking core needs in a spreadsheet and then building a matrix of whether the functionality is enabled, available, or not available with each vendor. This list will become quite comprehensive.

Functionality	Vendor 1	Vendor 2	Vendor 3
EDR	✓	✕	✕
XDR	✓	✓	✕
SIEM	✕	✕	✓
DLP	—	✓	✓
Firewall	✕	✓	✕
IDM	—	✕	✓
DFIR	✓	—	—
Pen testing	✕	✓	✕
IPS	✓	✓	✕
PAM	✕	—	✓
SIEM	✓	✕	—
Etc.	✕	✓	—

Enabled and being used ✓   Available but not being used —   Not available ✕

Figure 4.13: Mapping services with vendors

> You will find that cybersecurity vendors are always enhancing their portfolios to provide new services in addition to their current services. Make sure you check with your vendors on all the services they have available because they will constantly change. In addition, many larger vendors acquire smaller vendors to continue to grow and increase their portfolios.

As you work through this exercise, you are going to need to understand the licensing model for each vendor you subscribe to, which will quickly get complex. Pricing can also get complex as you'll find that subscription-based models include multiple features, which can make it hard to break down pricing when everything is bundled. As an example, let's take Microsoft's subscription-based model for enterprises. If Microsoft is identified as a core vendor, you will need to understand all the features available within each subscription it provides so you can do an in-depth comparison with other vendors. To get a better understanding of the available Microsoft features, browse to `https://www.microsoft.com/en-us/microsoft-365/enterprise/microsoft365-plans-and-pricing`, scroll down the page until you get to a section that states **Download the full enterprise plans comparison table,** and click on **Get the full comparison table (PDF)** to download and view the details of each of the Microsoft subscriptions. This is a 10-page document (as of August 2024) with a lot of information. It also covers non-security items, so you'll need to work through what is security-related and what is not. The following is an example of one page from the PDF document that shows the complexity of the Microsoft subscription model.

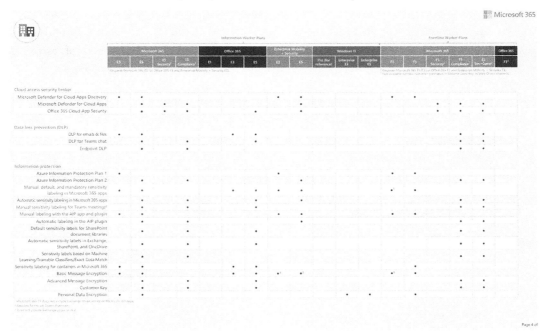

Figure 4.14: Microsoft licensing matrix

Bear in mind that this doesn't include the Azure-specific security services that will also need to be reviewed as part of your matrix. These fall under a different pricing model within Microsoft Azure.

You can access the Azure pricing calculator, which has a section for all security-related products, here: `https://azure.microsoft.com/en-us/pricing/calculator/`. This exercise is going to be needed with all your vendors, and you'll find all of them have some level of complexity with the licensing.

You will want to consolidate your cybersecurity portfolio where applicable, but as demonstrated, this isn't going to be an easy task. This strategy will need to be planned over a long duration, most likely years, as you work through efficiently consolidating vendors into a more streamlined portfolio to reduce risk. Although the strategy will have to be planned over a long duration, it will also have to be examined from time to time based on the changes within the current landscape.

Resource Management (In-House versus Outsourcing)

The last core strategy we will cover is the management of your resources. We briefly discussed skillset challenges in *Chapter 1, Current State*, and touched upon defining the cybersecurity organization in *Chapter 2, Setting the Foundations*, which covered a little on outsourcing and roles and responsibilities. In this section, we will go into more detail about the management of resources and some points to consider as to why you will want to be strategic with your resources.

As stated previously, my preference for resource management has been to keep the *brains* of your operations in-house and to outsource the *moving parts*. In principle, outsource those easy-to-complete repetitive tasks while your in-house team designs (architect), builds (engineer), and executes your strategy. Essentially, you would be looking at a hybrid model for your resources. As much as it would be ideal to bring everything in-house, there will be limitations based on the size of your organization and budget, along with specific tasks that simply can't be completely in-house because of the separation of duties.

With a hybrid model, you will find there are both benefits and challenges with each of these models. Some of the benefits of keeping things in-house include:

- The ability to retain intellectual knowledge of the organization
- Direct oversight of resources
- Increased operational support and turnaround time
- Increased customer satisfaction
- Flexibility without **Statement of Work (SOW)** limitations
- Quicker response times

On the other hand, some of the benefits of outsourcing include:

- The ability to quickly onboard and access resources as needed
- Reduced costs
- The ability to provide 24/7 support
- The ability to scale up and down as economies change
- Access to more specialized resources as needed

Obviously, the downside is the inverse of the benefits provided above for each of these models. For in-house, some of your bigger risk lies with the potential for high turnover and losing any built-up knowledge with your employees. Going through a hiring process can be lengthy, and the time to get new employees in a productive place can take months. You will need to ensure you are doing everything you can to retain employees. Another challenge is that of underperforming employees and not getting the desired results. This will require more effort to manage employees from a leadership perspective, taking you away from more value-added activities. One of the biggest challenges you will face with outsourcing is ensuring your outsourced provider is performing as expected. If not, you will need to lean on the defined SLAs and contracts put in place to hold them accountable. This can become a very demanding task if the outsourced company is under-delivering. Another challenge with outsourcing is the scope of responsibilities you have agreed to. Any work outside of the agreed-upon contract will come at an added cost with the need for an SOW to be created. This can be a very lengthy and challenging process when you need to complete work in a timely manner. As you work with these models over time, you will get a balance between them that will best benefit your organization.

There are many factors that will determine how you address your resource management. A large enterprise with a large budget may hire all resources in-house (other than separation of duty items such as audits, certifications, digital forensics for breaches, etc.) as it can afford to hire specialty skillsets, staff for 24/7 support, etc. On the other hand, a medium-sized organization may have a hybrid model where the core of the team is in-house, and the day-to-day operations to cover 24/7 support and specialty personnel will be outsourced. As a small organization, you may not even be able to afford to hire full-time resources, so you outsource everything to an external vendor to run your entire cybersecurity program. Again, you need to determine what makes the most sense for your organization with resource management.

With that, you are going to want to be strategic about how you staff your internal team or onboard vendors to provide outsourced services. As stated in the previous section, you will want to build a strategy around your cybersecurity product and vendor portfolio, which includes the vendors you chose to outsource any of your program to.

Strategizing on few rather than many vendors to provide resources will provide more efficient operations. If you are familiar with managing vendors in an outsourced model, you will be familiar with the effort that goes into ensuring they are delivering the promised **key performance indicators (KPIs)**. If you are not familiar, just know it can quickly become a full-time responsibility to ensure your outsourced vendors are delivering to expectations. Add multiple vendors and it can quickly become an inefficient model, which will impact your customer satisfaction in the long term.

At a high level, and following the organization structure we've already mentioned, the following organization structure can be used to provide a visual of what roles can be considered in-house and those that can be outsourced:

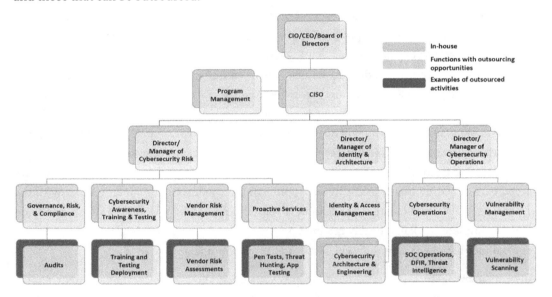

Figure 4.15: Organization chart with in-house and outsourced activities for consideration

Here are some examples and the reasons behind outsourcing specific functions/tasks for your program:

- *Audits* need a separation of duty to ensure the controls you claim are in place are actually in place. You will also need third-party attestation in the event of a cybersecurity incident and any lawsuits that may come your way.

- Vendor Risk Management *assessments* are very repetitive and can be very time-consuming. If all your vendors are well documented, this task can easily be outsourced to vendors.

- *Penetration testing, threat hunting, and application testing activities* can all be completely in-house, but you'll want to bring in outsourced resources to test your environment from a different perspective. It is also good practice to rotate through different vendors if possible.

- Your *SOC operations* will be another area that you may want to look to outsource. Unless you can staff for 24/7/365 operations, outsourcing may be the only option to ensure you have 24/7/365 monitoring.

- *Threat intelligence services* requires a very specialized skillset that you'll most likely need to outsource to get the best value.

- *Digital Forensic Incident Response (DFIR)* is another that you will need to outsource. When an incident occurs, you will need a third party to come in and determine the root cause, even if you are able to determine it in-house. This will also be a requirement for your cybersecurity insurance policy.

- Your *user training and testing activities* can easily be managed and deployed through outsourcing. The creation of the program and content usage will need to be completely in-house, but the day-to-day work can easily be outsourced.

- Ongoing *vulnerability scanning activity* can easily also be outsourced. Once the scope has been defined for where vulnerability scanning needs to occur, this can be managed and executed by a third party on a daily basis.

As mentioned earlier in the chapter, activities that should remain in-house include those areas that require a high-level skillset, require a better understanding of the organization, build positive relationships within the organization, and in some instances could be damaged by being outsourced. For example, anything related to overlooking your identity and management program should be managed in-house. This function will need high-performing resources that are very knowledgeable to ensure the architecture is current and all access management activities follow the process. You'll also want to keep your architecture and engineering activities in-house to ensure the quality of work being delivered can be controlled. Having in-house resources will also allow greater relationship-building throughout the organization. In addition, your management-level and higher positions should always be in-house to ensure that the culture of the organization is understood and that more impactful relationships can be built with the broader organization. This tends to not be possible with outsourcing.

Finally, you will want to map out what is outsourced versus what is being managed in-house, or as a shared responsibility. This will provide a clear picture of the broader cybersecurity organization and of who is responsible for which functionality. You can simply expand upon the exercise completed in the *Managing Your Cybersecurity Portfolio* section with mapping services by adding an in-house/outsourced column, as shown in the following table:

Functionality	In-house/Outsourced	Function	Vendor/Team
EDR	In-house	Architecture	End-User team
XDR	In-house	Architecture	Cloud team
SIEM (SOC)	Outsourced	Cyber Ops	Vendor1
DLP	In-house	Compliance	GRC team
Firewall	In-house	Architecture	Network team
Identity mgmt.	In-house	Identity	IDM team
DFIR	Outsourced	Cyber Ops	Vendor2
Pen testing	Outsourced	Proactive	Vendor3
IPS	In-house	Architecture	Vendor3
PAM	In-house	Identity	IDM team
Threat intel	Outsourced	Cyber Ops	Vendor4
....and so on	Shared	Cyber Ops	Cyber Ops team

Figure 4.16: Mapping in-house and outsourced resources

Now that you have completed the exercise of mapping in-house and outsourced functionalities, you can better evaluate your resource management. This will allow you to determine whether changes need to be made and whether anything can be moved to an outsourced model, or whether it needs to come in-house to improve the overall operations of the organization.

Summary

This chapter has covered everything related to strategy for your cybersecurity program and the importance of having a defined strategy in place. The strategy is what defines the foundation for your cybersecurity program, and without one, you will have no destination to strive for.

As we briefly touched upon, the mechanism that drives and delivers that high-level strategy for your cybersecurity program will be roadmaps, and not just one, but multiple roadmaps for immediate, short-term, and long-term needs. Moving beyond the broad strategy for your cybersecurity organization, we covered in detail four core-level strategies that you need to spend dedicated time solidifying.

The first of the four strategies is your architecture strategy, which we broke down into two distinct areas: technical and strategic. In this chapter, we focused primarily on the architecture strategy for your cybersecurity program and for the broader organization. The additional strategic and technical architecture items will be covered in detail in the next chapter. As part of the architecture for the broader organization, we covered in depth the cloud-first approach, the different types of data centers, and the need to modernize your infrastructure at an organizational level. The next strategy we reviewed was the cybersecurity framework and why you need to strategize around one of them for your organization. We provided some examples of the more common cybersecurity frameworks and went into more detail with NIST and ISO.

The third strategic core item we covered is the strategy around your cybersecurity portfolio and the need to consolidate and simplify to meet the increased risk of today's world. Over the years, we have created a very complex environment with far too many vendors within the portfolio of the cybersecurity program. To reduce risk, you need to revert this complexity and focus on a few key vendors to meet your cybersecurity program needs. The final core strategy is that of resource management and how you best strategize to staff your cybersecurity program, understanding the need to in-house specific functions versus outsourcing and the balance with a hybrid approach for your resource management strategy. Various factors will play into this strategy depending on the size of your organization and how much budget you have available to staff 24/7/365 operations. Other factors include the need for specialized skills that you may not be able to afford on a full-time basis, forcing you to outsource this type of work.

In the next chapter, we will be moving into part two of the book, *The Core*. *The Core* will cover all core functions that make up the cybersecurity program. The first core we will cover is the Cybersecurity Architecture function. We will cover the architecture component of the program in much more detail, including the importance of architecture within the cybersecurity program, what is needed to set the foundation of your architecture program within cybersecurity, what an **Architecture Review Process (ARP)** is, and what is involved in the process. We will also review **Zero Trust Architecture (ZTA)**, which will be covered in detail, and we will finish the chapter by reviewing other architecture items that you need to consider as part of your program, such as email security.

Join our community on Discord!

Read this book alongside other users, Cybersecurity experts, and the author himself. Ask questions, provide solutions to other readers, chat with the author via Ask Me Anything sessions, and much more.

Scan the QR code or visit the link to join the community.

`https://packt.link/SecNet`

5

Cybersecurity Architecture

A critical function that is sometimes overlooked or doesn't always have the involvement of a cybersecurity team across an organization as a whole is that of enterprise architecture. In large organizations, there will typically be an enterprise architecture function or a team of architects that supports the organization as a whole, or you may find architects embedded within different functions of the business. In a smaller organization, there may be no architecture presence, or it falls as a responsibility within the IT function. Regardless of where architecture lives within the organization, there must be a cybersecurity presence for all architecture-related decisions. It is also important that your cybersecurity architecture strategy aligns with your broader architecture strategy and business strategy to securely and efficiently enable your business.

In *Chapter 4, Solidifying Your Strategy*, we discussed the breakdown of architecture into two separate categories, strategic and technical. We focused on the architecture strategy for both the cybersecurity program and broader organization, along with touching upon how the architecture strategy impacts your vendor portfolio and resources. In this chapter, we will cover the remainder of the strategic architecture items along with items that are considered technical.

We will begin the chapter with the importance of architecture and why it is needed within the cybersecurity function, to ensure risk is reduced. A critical component of the architecture program is that of an architecture review process. This is likely a board or committee outside of cybersecurity. If one exists, cybersecurity needs to have a seat at the table to ensure a thorough review of risk has been completed. If one doesn't exist, processes need to be created to ensure a thorough review of all vendors, solutions, and services has been completed.

Then, we will review setting the foundations of your cybersecurity architecture program and review why you need to focus on cybersecurity architecture for your broader program.

This will lead to **Zero -Trust Architecture (ZTA)** and its importance. We will review the ZTA models available, along with building a ZTA roadmap to ensure the adoption of ZTA. We will finish off the chapter with insight into technical architecture components that are required to strengthen and harden your organization. These items include collaboration architecture (email, voice, IM, etc.), application architecture, endpoint architecture, infrastructure architecture, identity architecture, network architecture, and data architecture. More specifically, we will cover the following throughout the chapter:

- Cybersecurity within Architecture
- Architecture Review Process
- Your Cybersecurity Architecture Foundation
- **Zero Trust Architecture (ZTA)**
- Technical Architecture

Cybersecurity within Architecture

First, let's take a high-level look at all the sub-functions that should be addressed as part of cybersecurity architecture. The following image captures much of what the cybersecurity architecture function entails.

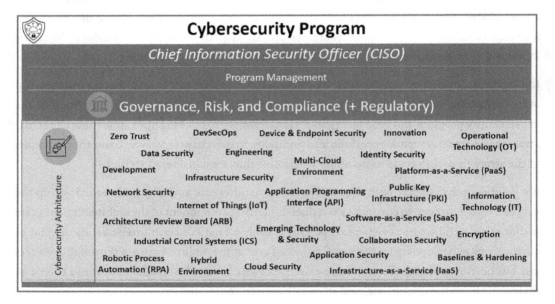

Figure 5.1: Sub-functions of the cybersecurity architecture function

Understanding IT Architecture

Let's take a step back and ensure that we fully understand IT architecture within an organization. IT architecture is defined by *Gartner* as:

> *"A framework and set of guidelines to build new systems. IT architecture is a series of principles, guidelines, or rules used by an enterprise to direct the process of acquiring, building, modifying, and interfacing IT resources throughout the enterprise. These resources can include equipment, software, communications, development methodologies, modeling tools, and organizational structures."*

Source: `https://www.gartner.com/en/information-technology/glossary/architecture`

In short, IT architecture provides standardization to ensure that new vendors who are onboarded, new solutions that are deployed, and the transformation of current services that are modernized all conform to the policies, procedures, and best practices defined by an organization. Although the Gartner definition provides a great overview of IT architecture, I do believe it could be updated to include securing as a process throughout the enterprise. I'm not sure when this definition was created, but it could have been a while back before the need for updated and improved cybersecurity.

Understanding Security Architecture

Now, let's take a look at the definition of security architecture from the **National Institute of Standards and Technology (NIST)** glossary. There are a few definitions from different frameworks, but I tend to lean more toward this definition over the others:

> *An embedded, integral part of the enterprise architecture that describes the structure and behavior for an enterprise's security processes, information security systems, personnel and organizational sub-units, showing their alignment with the enterprise's mission and strategic plans.*

Source: `https://csrc.nist.gov/glossary/term/security_architecture`

Essentially, the definition of architecture from a cybersecurity perspective falls into the broader architecture strategy for an organization, along with aligning with its broader mission and strategy.

Importance of Embedding Cybersecurity within Architecture

Now that we have a better understanding of the broader IT architecture definition along with the cybersecurity definition from NIST, you can see why embedding the role of cybersecurity architecture into a broader architecture program is very important. As new vendors are onboarded, new solutions are built, and current services are modernized, we must ensure that cybersecurity is reviewed, discussed, approved, and embedded early in the process. If cybersecurity is missed as part of the process, solutions will be deployed throughout an organization with significantly increased risk.

Benefits of Embedding Cybersecurity within Architecture

There are many reasons that cybersecurity must be embedded directly into the broader architecture review process from the beginning, including the following:

- Provides a chance to review all the security controls early.
- Ensures that architecture diagrams and documentation are provided with the correct details, specifically relating to cybersecurity.
- Ensures that solutions and vendors are complying with policies and procedures.
- Enforces defined cybersecurity standards.
- Ensures that security testing is conducted and reviewed.
- Allows for a more proactive approach versus reactive.
- Increases collaboration across an organization.
- Provides the ability to collect all relevant information on solutions and vendors for more efficient governance.
- Reduces complexity and overall risk within an organization.
- Ensures that the strategy is being followed.
- Allows for the correct resources to be engaged early.
- Ensures greater success.

Detailed Architecture Review Process

As part of a detailed architecture review process, you are going to need a lot of details in relation to requirements from a cybersecurity perspective. This will include detailed reviews of the solution being proposed and the cybersecurity controls that will be in place. The type of documentation that will need to be reviewed includes technical specifications, architecture documents, build sheets, and any other relevant documentation that can be reviewed through a cybersecurity lens. In addition to the documentation, you will also need the business and/or vendor to complete the detailed requirements of the solution being deployed – for example, what the data type/classification is, whether encryption is used, whether identity can be integrated, what the network controls are, etc. We will cover these in more detail in a later section, *Architecture Review Process*.

Required Architecture Diagrams

On a broader level, you are going to need architecture diagrams that capture the entire environment and some of the more specific technical areas. Having access to these is essential for the cybersecurity team. If they don't exist, they will need to be created as these diagrams are the foundation for understanding the footprint of the environment that needs to be protected. For example, you will want to capture diagrams for the following (if applicable), at a minimum:

- High-level architecture of the environment
- Network architecture
- Application architecture
- Identity architecture
- Device architecture
- Database architecture
- Infrastructure architecture
- Collaboration architecture

The following is an example of the high-level architecture that represents your environment. As you get deeper into each area, your architecture diagrams should contain more detail to help better understand the cybersecurity controls in place. Bear in mind that this high-level architecture will not contain a lot of detail and will need to be customized to your environment.

The idea is that you have something tangible that represents what the basic architecture looks like and can be shared with a broader audience.

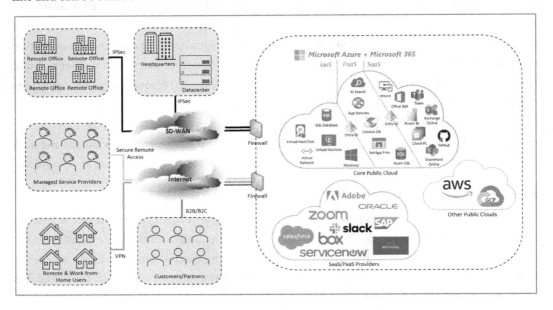

Figure 5.2: Example of a high-level architecture diagram

Cloud Services Architecture

As more continue to adopt cloud services, you will need to be familiar with the architecture for these environments. For example, three of the more common cloud providers each provide an architecture center for reference:

- Microsoft Azure: `https://learn.microsoft.com/en-us/azure/architecture/`
- AWS: `https://aws.amazon.com/architecture/`
- GCP: `https://cloud.google.com/architecture/`

The same should apply to other major cloud providers, including SaaS providers. If a vendor doesn't have architecture reference diagrams available, you'll need to better understand why as this would definitely be a red flag.

If you are running your environment in one cloud provider, becoming familiar with the architecture becomes a much easier task.

Though, the reality is you may have multiple cloud environments and/or a hybrid environment to oversee.

This creates a lot more complexity and challenges with the architecture, as you will need to be familiar with multiple architectures, potentially involving different cloud providers in addition to a legacy, on-premises data center. This is another reason why there is a need to make your strategy as simple as possible.

Cybersecurity Architecture Documentation

In addition to the general architecture, you will want to know if there is any cybersecurity architecture documentation available from a cloud provider, SaaS provider, or vendor that you work with. I'm confident that every organization has some form of cloud provider or SaaS environment within their portfolio these days. With this being the case, it is important that you have access to their cybersecurity architecture diagrams. A great example of this is the Microsoft Cybersecurity Reference Architecture, which you can find here: `https://learn.microsoft.com/en-us/security/adoption/mcra`. This reference architecture provides details on all the cybersecurity technologies and capabilities available with Microsoft. If you use Microsoft products within your portfolio, the following architecture diagram provided in the Microsoft Cybersecurity Reference Architecture material provides all the capabilities that are available from Microsoft to protect your environment. This is a very valuable slide for an organization that strategizes on the Microsoft platform.

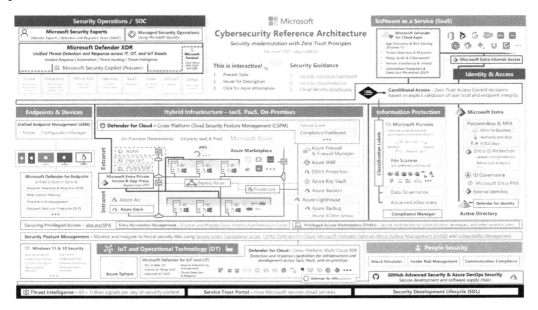

Figure 5.3: Microsoft Cybersecurity Reference Architecture capabilities
Image source: `https://github.com/MicrosoftDocs/security/blob/main/Downloads/mcra-december-2023.pptx?raw=true`

As you can see, there is a lot involved with architecture, and it is important you have a basic understanding of the general architecture requirements along with the cybersecurity architecture needs. Having a good, dedicated cybersecurity architect on your team will make a significant difference within the broader cybersecurity program. Their role serves as a critical one that will be required to partner with the broader architecture function (if one exists) and the business as a whole, as they learn to better understand their needs and ensure cybersecurity is discussed and included at the beginning of any project, not becoming an afterthought. Now that we have a better understanding of the role of cybersecurity in architecture, let's review what an architecture review process entails and what role cybersecurity plays within the broader process.

Architecture Review Process

The overall architecture review process is an important process within any organization in our modern technological world. Every organization should have some form of process in place that allows the correct resources to review all the solutions and vendors deployed within it. If your organization has an architecture review process in place today, make sure you are familiar with it, and ensure that the cybersecurity team has a seat at the table to ensure a thorough cybersecurity review is complete. If there is no process in place currently, or no architecture function within your organization, which may be the case for many smaller to medium-sized organizations, look to build an architecture review process right away. This can even be part of your cybersecurity program, which will allow for greater efficiency and reduced risk.

If there is no architecture review process in place, you will want to begin by forming a committee or assigning someone the responsibility to develop the program needed for the architecture review process. If you already have a cybersecurity architect, they would be the ideal resource to drive the creation of this process. If there is no cybersecurity architect, someone who has more technical acumen would be preferred to get the process up and running. Ideally, you'll want to onboard an architect who specializes in architecture and can manage this process long-term. In addition, you are going to need project management resources – specifically, someone who engages with the business to help with change management and communication to launch the new process.

One very important point is that of buy-in and support from your executive leadership team. If there is no support from the top with this process, you will not be successful with deploying the process throughout the organization as a whole, and you will not receive the participation needed. This process must be adopted by everyone in the organization to be successful, and for that to occur, you will need sponsorship from the very top to ensure enforcement throughout the organization. Before you proceed with building the process, connect with your executive leadership team for approval and support.

As we proceed through this section, we will cover and provide details into what is needed to ensure a successful architecture review process; this includes ensuring the following are in place as part of the broader program:

- The strategy, scope, and goals
- Committees and teams
- Process logistics
- Lifecycle process
- Intake process

The Strategy, Scope, And Goals

As you build your process, the first thing you are going to want to do is create your strategy, scope, and goals/objectives for the architecture review process. We discussed the strategy in detail in *Chapter 4, Solidifying Your Strategy*, so this should simply slide into the strategy defined as part of the process. Next, your goals and objectives should be clearly defined so that everyone is aware of the purpose of the architecture review process and why they need to follow it. Some items that should be listed for the goals and objectives include:

- *Process documentation and governance*
    - Define and document the architecture review process.
    - Provide a governance body for technology deployment and usage within the organization.

- *Compliance and security*

    - Ensure that organizational policies and procedures are followed.

    - Confirm that security controls are in place and effective.

    - Ensure compliance with defined standards, including regulatory and legal requirements.

- *Risk and cost management*

    - Minimize complexity, risk, and costs through proactive risk management strategies.

    - Identify, assess, and prioritize risks regularly.

- *Operational efficiency and performance*

    - Ensure an agile process that provides efficiency for enablement.

    - Set and evaluate performance benchmarks to ensure that the technology meets operational demands.

- *Collaboration and stakeholder engagement*

    - Increase collaboration across the organization.

    - Engage stakeholders in the review process to integrate diverse needs and ensure user-centric outcomes.

- *Sustainability*

    - Incorporate sustainability objectives in technology choices and architectural designs.

    - Design architectures that are modern, scalable, and adaptable to accommodate growth and changes in technology.

Finally, there should be a well-defined scope that provides exactly what initiatives will need to run through the architecture review process. For the most part, this will be all new (including upgraded and modernized) technology and vendors deploying solutions within your environment. You will want to provide specific (and clear) requirements of what must run through the architecture review process so that there is no confusion from the organization as a whole.

Committees and Teams

Then, you will need to ensure your teams and/or committees have been formed to run and manage the architecture review process. Committees and/or teams should include executive sponsorship, architects responsible for each of the core areas, IT representatives from the business function (if they exist), the business owner submitting the request, and project/program managers.

As part of your architecture representation, you will need to ensure you have representation in the following areas:

- The cybersecurity team (including Identity and Access Management) to ensure that all security requirements are reviewed and in place before deployment.

- The IT Support and Operations team to ensure that solutions being deployed have ongoing support, there is a support process in place, the service desk is ready to support the solution, end-user documentation is available, and a **Responsible, Accountable, Consulted, And Informed (RACI)** is developed.

- The collaboration team to review any requirements regarding email, voice, instant messaging, intranets, document libraries, etc.

- The network team to complete a thorough review of all network requirements and ensure anything network-related is set up per best practices and is secure.

- The infrastructure team to support and review any infrastructure requests related to on-premises infrastructure, **Infrastructure as a Service (IaaS)**, **Platform as a Service (PaaS)**, and **Software as a Service (SaaS)** components.

- The end-user team to ensure that any end-user requirements are reviewed, such as device configurations, user application requirements, etc.

- The application and development team to ensure that all application and development work has been completed to the defined standards for the organization and is meeting the defined security standards.

- The OT & IoT team if applicable to review and ensure that all OT & IoT-related work meets the required standards.

A couple of committees to call out are a steering committee that provides the overarching governance and guidance for the overall process. This team will typically include some executive leadership resources, ensuring that the correct support is provided and serving as the liaison with the executive leadership team. The most important committee will be that of the **Architecture Review Board (ARB)**.

This is the team that has been selected to conduct reviews of the items that come through the process. This team has the responsibility to ensure that each item meets the requirements defined for the overall process. There will need to be clear guidelines on the approval process. In essence, if there is an objection from anyone on the ARB, the item shouldn't be approved until everyone is able to gain alignment.

In addition to your committees and teams within the program, you will need to ensure that roles and responsibilities are well defined so that everyone is aware of what is expected of the assigned members.

Process Logistics

As part of the logistics for the architecture review process, you will need to ensure that you implement a foundation to support the process, that a clear meeting schedule is defined, that change management is not overlooked, and integration with the broader program/project management activities for the organization.

Foundation Setup

As you get ready to launch your architecture review process, a foundation will be needed to support the overall process from beginning to end. This includes having somewhere to store all your documents, such as checklists, architecture diagrams, vendor-supplied documentation, etc., having a place to track the items coming through the process with status, and a knowledge base for users to learn about the process, retrieve templates, submit requests, etc. This could be as simple as using your current collaboration tools such as a SharePoint site, leveraging a ticketing system like ServiceNow, and taking advantage of or integrating a project management tool like Microsoft Project, Trello, Asana, or Jira. You can take it a step further and build a custom app that will support the requirements if it makes sense. Most likely, you aren't going to be able to fulfill all your needs with one system, but it is important to ensure that you take advantage of any tools available at your disposal and integrate all components of the process with the foundational setup for more efficient workflows.

Meeting Schedules

There will also be a need to clearly define when the ARB meets to review items that come through the process. Will this be a weekly meeting to cover all items currently active in the queue? Will an ad hoc meeting be set up as requests come through the process? Will you need different meetings to support initial intake versus follow-up and go-live reviews?

As part of these meetings, making sure the correct sponsors and resources are in attendance is very important; this will include your core ARB members, architects responsible for each of the core areas as shown within the committees and teams section, IT representatives from the business function (if they exist), the business owner submitting the request, and any relevant project/program managers. As you need to provide feedback and action items, you will need to ensure that the sponsor and/or organization function owner is in attendance to provide details on the item being reviewed. You'll also want a vendor in the meeting if a third-party solution or application is being reviewed.

Change Management

Another important facet that will need to be managed is communication and any change management activities. Although this should be part of the broader project, they must be discussed, and you need to understand what the impact of any new solution (or upgrade) will have on an organization. Will communication need to be sent to the users? Will documentation need to be created and made available? Will the deployment need to go through a change advisory committee or board? All these questions should be asked as part of the architecture review process.

Project Management

One last item worth noting that will be covered in more detail within the process lifecycle is that of project management. To be effective with the architecture review process throughout an organization, the process should be tied directly into the broader program/project management process. Your architecture review process should essentially become an action from the broader program/project management process that covers the broader organization. That's assuming there is a mature program/project management process in place. The logistics of this integration will require close collaboration with the broader program/project management team to ensure efficiency.

Lifecycle Process

It is important that the architecture review process lifecycle is well documented and understood for those who both manage the process and, more importantly, those who need to run through the process. This lifecycle process provides the steps that a new request needs to follow, from initial intake to final approval/rejection status. To get the final approval, there will be many steps that will need to be followed.

As an example, the following is a high-level process that could be followed:

Figure 5.4: Example of a high-level ARB lifecycle process

When building out your own process, you may want to get more detailed with the decision points and any outputs for each step within the lifecycle. Tools like *Microsoft Visio* enable you to build a much more detailed process lifecycle if needed.

As discussed within the *Process Logistics* section, you will want to ensure the architecture process lifecycle ties directly into the broader program/project management process. If your organization already has a defined program/project management function, there will hopefully already be a process in place to capture all projects that the organization executes on an annual basis. If this is the case, you will be able to partner with the program/project management function to embed the architecture review process directly into the current project lifecycle process. This will allow for a more efficient process and especially help ensure everything gets the opportunity to be reviewed by the architecture team and the correct resources. If there is no formal program/project management function within the organization, you'll want to ensure that basic project management principles are followed, as any new initiative/deployment will essentially need some form of program/project management wrapped around it to allow a more efficient deployment/delivery. The following is an example of how you can tie your architecture review process to the broader program/project management process/lifecycle:

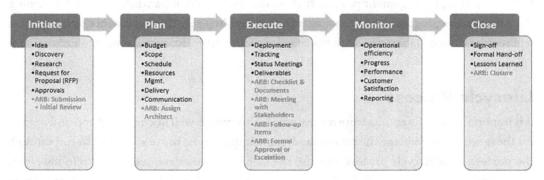

Figure 5.5: Example of a project lifecycle with the ARB lifecycle integrated

Be aware that many organizations may not be sufficiently staffed to have dedicated program/ project management teams within their organization, so it is important that you have resources familiar with program/project management basics. Remember from *Chapter 3, Building Your Roadmap*, that you will need to incorporate this with your broader cybersecurity program to efficiently track everything that is worked on within your program. And trust me – it will not just be a couple of items; you will find that there are dozens of projects/workstreams/tasks all being executed at the same time.

Intake Process

The intake process will be your most important task. Here, you will collect all documentation required to support the review process. This allows you to fully understand the solution being proposed and ensure that all requirements are met from a standardization and cybersecurity perspective. At a minimum, you should collect the following documentation for review:

- A complete checklist
- Architecture documentation
- Vendor-specific documentation

Let's review each of these areas in more detail.

A complete Checklist

Your checklist will need some thought behind it to ensure that it allows you to capture all relevant information needed to complete a thorough review. Your architecture review checklist should contain the following at a minimum.

General Information

In this section, you are going to want to collect information such as the following:

- The name of the solution, vendor, and website to review.
- You will want a description of the solution that is provided and to know if this is a new solution or an upgrade.
- You will want to understand the components (web, app, database, etc.) involved, including the target user base and number of users accessing.

If this is a new vendor, you will also want to confirm that it has been approved from a vendor risk management perspective.

Network Requirements

Here, you are going to want to understand if there are any specific network requirements involved, such as the following:

- Will there be any network access from outside your environment needed, like VPN tunnels, user VPN, remote access, etc.?

- You will want to know how the solution will be accessed (public or internally), whether a public IP is needed, DNS requirements, port configurations, **Quality of Service (QoS)**, etc.

Infrastructure Requirements

For the infrastructure requirements:

- You will want details about the type of infrastructure the solution will be deployed with and whether it is on-prem, cloud (IaaS, PaaS, or SaaS), etc.

- Will a development, staging, or testing environment be needed?

- You will want to confirm where exactly the solution will be hosted and within what regions or countries.

- Understanding what the availability requirements are will be critical, along with **Disaster Recovery Planning (DRP)** requirements with both the **Recovery Time Objectives (RTOs)** and **Recovery Point Objectives (RPOs)**.

Collaboration Requirements

Collaboration requirements will include anything related to email requirements, such as the following:

- Will a mailbox be needed? Will any SMTP or relay setup be needed? Will DKIM, SPF, or DMARC need to be set up and configured?

- You will also want to understand if there will be any collaboration requirements needed with external users through vendors, consultant companies, etc.

- Will there be a need to federate between organizations to allow IM, file-sharing activities, access to intranets, etc.

Application Requirements

For your application requirements, you are going to want to understand if there are any applications to deploy to the users, such as the following:

- Will there be a desktop-type app, mobile app, web app, etc.?

- How can the app be deployed from an enterprise perspective, and what is the update routine of the app?

- You will also want to know about any specific app requirements for deployment, for example, will additional software be needed to support the app, will admin rights be needed to be installed, will any specific customizations of the app be needed, etc.?

Identity and Access Management Requirements

For *Identity and Access Management*, it is very important to ensure that you collect the correct information, such as the following:

- You are going to want to understand how identity will integrate with the solution, preferably via **Single Sign-On (SSO)**.

- Will any user/service accounts be needed?

- What **Multi-Factor Authentication (MFA)** capabilities are available if SSO is not available.

- How will access be managed, and is there **System for Cross-Domain Identity Management (SCIM)** integration?

- Who will manage the identity lifecycle?

- Who will be the privileged users?

- Are there **Role-Based Access Control (RBAC)** capabilities?

- Will there be external users, or B2B or B2C requirements?

Data Requirements

This will consist of getting a better understanding of the data requirements and how data will be handled, such as the following:

- Will any databases or data warehouses be needed?

- How will data be stored, processed, and transmitted?

- What classification will the data be?

- Will data need to be archived?

- Will any retention policies need to be applied?

- Who will be the data owner?

Cybersecurity Requirements

Although cybersecurity considerations will have been captured within other sections, here, you will want to ensure everything else related to cybersecurity is captured, such as the following:

- This will include an understanding of the encryption methods being used.
- Will data be encrypted at rest and in transit?
- Are there any regulatory requirements, such as **Payment Card Industry Data Security Standard (PCI-DSS)**, **Sarbanes-Oxley Act (SOX)**, etc.?
- Are there any privacy or protection laws, such as the **California Consumer Privacy Act (CCPA)** and **General Data Protection Regulation (GDPR)**?
- Will there be a need for privacy agreements?
- Will a **Secure Sockets Layer (SSL)** need to be applied?
- Will data be leaving the environment?
- Will third parties need access?
- What type of reporting and auditing is available?
- Has security testing been completed against the solution?
- Is there security logging or **Security Information And Event Management (SIEM)** integration?
- How long are security logs retained for?

Integration and Automation Requirements

For integration requirements, you are going to want to understand the following:

- Will there be any **Secure File Transfer Protocol (SFTP)** requirements, or will there will be any other mechanism needed for file transfer capabilities?
- Will there be a need to integrate with an API?
- Will there be any data feeds between the environments?

- From an automation perspective, you are going to need to understand what the automation requirements are (if any), how the automations will be set up, and with what accounts and what level of access.

Support and operational requirements

It is important not to forget how the solution will be supported and operationalized once is deployed and ready for production. You don't want to wait until the solution is in production to figure these items out:

- You will need to know who the business owner is.
- What **Service-Level Agreements (SLAs)** are expected?
- Will 24x7x365 support be needed?
- Who will support the solution, and what will the escalation path be?
- Will a war room be needed for initial deployment?
- Are there any licensing considerations?
- Will monitoring be needed?
- Is there a RACI matrix available?

There will be a lot of information that you will need to collect to ensure a complete and thorough review of the solution being deployed. You will need to fully understand the solution being proposed, which means you may need to set aside time with the business function and/or vendor to review the solution in more detail, along with a possible demo.

You will want to ensure that the checklist template is easy to follow and complete. This won't be an easy task, but it is important that all information is provided correctly so that an accurate review can be completed by the architecture team, along with the cybersecurity resources, to ensure minimum risk. For the checklist template, you could look to build an e-form, which will obviously come at more cost to develop and maintain. Alternatively, to get started quickly, a simple Microsoft Word document will give you what you need.

A small example of some of the cybersecurity sections in Microsoft Word format may look like the following:

SECTION 8: CYBERSECURITY REQUIREMENTS

Please complete the following cybersecurity questions to the best of your ability. Please reach out to the assigned architect for any clarification.

What is the data type being stored or transmitted?	○ Public ○ Confidential ○ PII ○ Sensitive PII ○ Internal ○ Restricted ○ Other _____
Regulatory & business compliance	○ SOX ○ PCI ○ PII ○ HIPAA ○ GDPR ○ CCPA ○ Other _____
Are there any California Consumer Privacy Act (CCPA) or General Data Protection Regulation (GDPR) Considerations?	○ Yes ○ No ○ N/A If Yes, has the process to delete requested information been documented? ○ Yes ○ No
Is a privacy agreement required for users?	○ Yes ○ No ○ N/A
Is an SSL be needed?	○ Yes ○ No ○ N/A
Is security event monitoring available?	○ Yes ○ No ○ N/A
How long are security logs retained for?	_____ Months / Years
Does the solution integrate with a SIEM?	○ Yes ○ No ○ N/A Provide additional details: _____
Will data be handles by a third-party?	○ Yes ○ No ○ N/A Provide additional details: _____

Figure 5.6: An example set of questions for an ARB intake form

Architecture documentation

For the architecture documentation, you are going to want a high-level diagram of the solution being proposed, with any data sources, integrations, web interfaces, access controls, etc. This allows a visual review that can help identify where cybersecurity controls have been put in place, as well as identify where there may be gaps in the solution. The following is an example of an architecture diagram that you should request as part of the process:

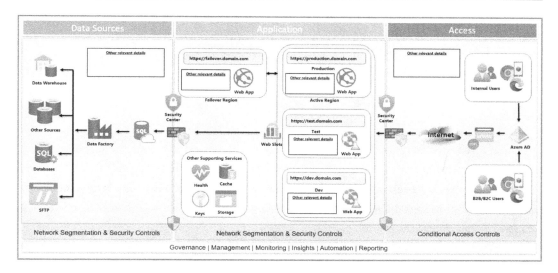

Figure 5.7: Example of a solution architecture diagram

A lot of the time, you will find that you may not get the details you need, or there isn't a clear understanding of what is needed for the architecture team to review. It may be a good idea to have some sample architecture diagrams available for reference; otherwise, your architects may need to spend some time helping to draw out the architecture.

Vendor specific documentation

For any initiative that will require vendor involvement, whether it is a third-party application, standard desktop application, SaaS-type solution, etc., you will need to collect all relevant documentation that relates to the solution. Documents that you will want to collect include any vendor architecture documents, solution overviews, technical specifications, build sheets, requirements, best practices, installation documents, security-related documents, end-user guides, and anything else they have available.

Your architecture review process will need to confirm that any vendors being on-boarded have gone through the correct vendor risk management process. This process will require an in-depth review of the vendor to ensure they have all the correct cybersecurity controls in place. We will cover vendor risk management in more detail in *Chapter 10, Vendor Risk Management*.

One final discussion point before we move into the next section is tracking and keeping a centralized application inventory up to date. As items come through the architecture review process, you will want to ensure you inventory all solutions within the application inventory and keep it up to date with the latest details. We will be covering the application inventory in more detail in *Chapter 13, Governance Oversight*.

Your Cybersecurity Architecture Foundation

Next, we will shift our focus to the architecture strategy that will be used to secure your organization. As you pull the layers back from the broader architecture strategy for the organization, you will find architecture foundations will be in place for your applications, network, infrastructure, data, etc. The same applies to cybersecurity – a well-defined architecture strategy and foundation must be adopted. This ensures that your program follows best practices and defined standards to reduce risk.

Before we discuss the architecture foundation for your cybersecurity program, it is important to touch upon the core principles that form an information security model. In essence, the architecture foundation and strategy that you adopt is what will be used to best protect the core principles. If you are not aware, these principles are known as the **CIA** triad, which stands for **Confidentiality**, **Integrity**, and **Availability**. If you have pursued a security certification, such as the *CISSP* or *Security+* certifications, you will be very familiar with this model. If not, it is recommended that you familiarize yourself with them as a security professional. This book will not go into detail about the CIA triad, but the concepts provided in this book will set a foundation to ensure the confidentiality, integrity, and availability of information within your organization. At a high level, CIA means the following:

- **Confidentiality** involves ensuring that no one other than those who are authorized can access information.

- **Integrity** involves ensuring that the information being protected is original and has not been modified without the correct authorization.

- **Availability** involves ensuring that information is always available when access is needed.

It is also important to note the availability of both enterprise-level architecture frameworks and cybersecurity architecture frameworks you are looking to adopt. Your organization and how large your program is will determine the need to implement a cybersecurity architecture framework (if one is needed). You'll most likely find an enterprise-level architecture framework in place within the architecture function in a large enterprise. Small to mid-size organizations may not have an enterprise architecture framework in place. In this book, we will focus the strategy on a cybersecurity architecture model along with a cybersecurity framework, which will be covered in more detail in *Chapter 15, Regulatory and Compliance*. A few cybersecurity architecture frameworks for your reference include:

- **The Open Group Architecture Framework (TOGAF)**: `https://www.opengroup.org/togaf` (this framework is much broader than cybersecurity and a very common framework amongst enterprise architecture programs)
- **Open Security Architecture (OSA)**: `https://www.opensecurityarchitecture.org/cms/index.php`
- The **Sherwood Applied Business Security Architecture (SABSA)**: `https://sabsa.org/`

If you have researched or looked to see what cybersecurity architecture models are available, you will not find many at your disposal. You will find a lot of concepts and different methodologies that are used to provide a more secure environment. For example, let's take a look at the secure design principles covered in the latest CISSP certification. Within *Domain 3: Security Architecture and Engineering*, the following 11 principles are noted as design principles:

- Threat modeling
- Least privilege
- Defense in depth
- Secure defaults
- Fail securely
- **Separation of Duties (SoD)**
- Keep it simple and small
- Zero trust or trust but verify
- Privacy by design
- Shared responsibility
- Secure access service edge

Source: https://www.isc2.org/certifications/cissp/cissp-certification-exam-outline

There's a high possibility you are familiar with some, if not all, of these principles. Some of these have been around for a while and have been adopted by many cybersecurity programs over the years.

Of all these principles, there is one that provides an architecture model – Zero Trust – and we will base our cybersecurity architecture foundation on the ZTA model within this book. However, many of the principles listed above also fall within ZTA to achieve a comprehensive approach to cybersecurity.

 A more recent cybersecurity architecture model, as defined by Gartner, is the **Cybersecurity Mesh Architecture (CSMA)**: https://www.gartner.com/en/information-technology/glossary/cybersecurity-mesh

In the following section, we will cover the different ZTA models available in more detail, and we will select one to provide more details on how to work toward that ZTA model within your environment. As you build your ZTA roadmap, you will see a transition into the technical architecture component of the program, where you will enable the needed capabilities to meet the requirements of the ZTA model.

Zero-Trust Architecture (ZTA)

In recent years, Zero Trust has gained tremendous momentum, but it has also become a buzzword and marketing term for most security vendors. Taking a step back, the term "zero trust" was used as early as 1994 by Stephen Paul Marsh in his doctoral thesis. The first concepts of zero trust then became realized with the introduction of de-perimeterisation, which was introduced by Jericho Forum (now part of the Open Group). Before this introduction, an organization would protect everything at the perimeter with a much-simplified model. As this model became disrupted, and the need for data to travel outside of this secure perimeter arose, the concept of de-perimeterisation was born, and the need to protect data with additional levels of security controls was needed. The foundation of a Zero Trust model was then founded and popularized by John Kindervag while he was at *Forrester Research Inc.* back in 2009 (https://www.forrester.com/zero-trust/). Since 2009, we have seen the zero-trust concept become more popular over the years as we slowly see more adoption throughout the industry. If you are not clear on what exactly zero trust is, essentially, it is a model where we trust no one until we can validate who they are, who they are meant to be, and whether they are authorized to have access to a system or information. In simple terms, zero trust is well known for the concept of *never trust, always verify*.

If you do some research on zero trust today, you will find the concept being used and referenced by many vendors. You will also see slight differences in the approach of zero trust by different vendors. No matter which vendor you work with, or which approach you take, the core of a zero-trust model will always fall back on the principle of *never trust, always verify*. Effectively implementing a zero trust model requires a multilayered approach with your security strategy, along with the use of the most current and modern technology available. The method of allowing a user to access an environment with only a username and password is long outdated and has proven to be insecure.

Looking at some statistics specifically around zero trust principles, a concerning statistic from the same *IBM Cost of a data breach 2022* report (shared in *Chapter 1, Current State*) stated that only 41% of organizations who participated in the study deployed a ZTA model. Furthermore, those that didn't deploy a zero-trust architecture model incurred a much greater breach cost, with an average addition of $1 million. These concerning numbers clearly state the need for a zero-trust foundation.

The reality is that adoption isn't progressing as fast as it needs to. However, adoption is occurring, according to a report released by CyberRisk Alliance, *The zero-trust dilemma* (https://www.scmagazine.com/whitepaper/the-zero-trust-dilemma), which provides more reality and insight into some of the challenges faced with adopting ZTA. Some highlights from the report include the following:

- Zero trust practices were only implemented by approximately 30% of organizations who participated in the survey.
- Obstacles with implementing ZTA include high costs, complexities with integrating zero trust into existing workflow, and challenges in receiving leadership buy-in.
- Some organizations noted that zero trust is seen as restrictive.
- Larger organizations with mature security teams appear to be making more progress versus smaller organizations with limited resources.

source: https://www.cyberriskalliance.com/press-release/cra-cbir-zero-trust

On a positive note, the report touched upon AI (more specifically, Generative AI) as a driver to efficiently support ZTA by refining ZTA policies, enhancing automation, and real-time insights with privilege access.

With the advancements of Generative AI and its increasing accessibility, it will be critical as a cybersecurity leader to understand how it can benefit a cybersecurity program, especially your ZTA strategy, to support enablement and provide more efficient insight into and protection of your organization's data.

As Generative AI evolves, we will see improved identity protection, network analysis, malicious email detection, and much more. We can expect greater efficiency in automation and enhanced behavioral analytics, which will significantly support the ZTA deployment for an organization. We cover AI in more detail in *Chapter 7, Cybersecurity Operations.*

ZTA Models

As you look to strategize on a zero-trust security model, which will, in turn, become the foundation for your zero trust architecture to best protect your environment, you are going to need to follow some form of deployment and maturity model. A couple of the better maturity models I lean toward for reference are from Microsoft and the **Cybersecurity and Infrastructure Security Agency (CISA)**:

- **Microsoft Zero Trust**: https://learn.microsoft.com/en-us/security/zero-trust/
- **CISA Zero Trust**: https://www.cisa.gov/publication/zero-trust-maturity-model

It is not necessarily a maturity model but the NIST has a zero-trust architecture for reference:

NIST SP 800-207: https://csrc.nist.gov/pubs/sp/800/207/final

Another great resource is from NIST's **National Cybersecurity Center of Excellence (NCCoE)**, *Implementing a Zero Trust Architecture*: https://www.nccoe.nist.gov/projects/implementing-zero-trust-architecture. There are several great resources for reference, including a mapping of the ZTA security requirements with cybersecurity standards and some vendors. All guides are accessible at the NCCoE link provided above.

Both models follow a very similar approach with the pillars and the maturity model used, that can be followed.

CISA Zero Trust Model

The CISA zero trust model is designed to support federal agencies with their ZTA journey, although the model can be applied to any organization. As you can see from the CISA maturity model below, you have five distinct pillars: Identity, Devices, Networks, Applications and Workloads, and Data, with Visibility and Analytics, Automation and Orchestration, and Governance spanning each pillar. For the maturity model, a four-phase approach is used: Traditional, Initial, Advanced, and Optimal.

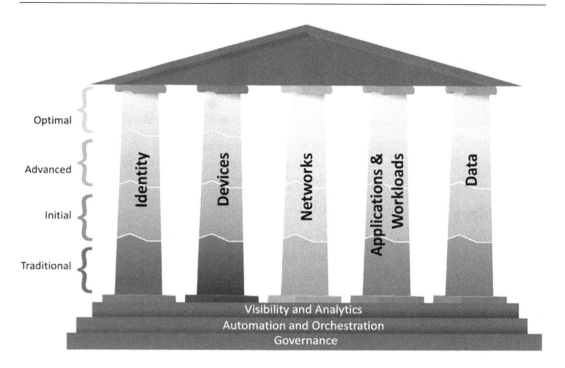

Figure 5.8: The CISA zero trust pilars
Image source: https://www.cisa.gov/sites/default/files/2023-04/CISA_Zero_
Trust_Maturity_Model_Version_2_508c.pdf

Microsoft Zero Trust Model

The Microsoft model is designed for Microsoft-centric customers, but the foundation can be applied to any organization using any technology. Microsoft's model is constantly evolving and being modernized, and some of the references may be from some of their older models, but they still bring value as you build your strategy. As you can see below, the model contains six distinct pillars: Identities, Endpoints, Data, Apps, Infrastructure, and Network, with Visibility, Automation, and Governance spanning each pillar. Microsoft uses the founding principles of verify explicitly, applying least-privilege access and always assuming a breach.

In addition, Microsoft focuses on a three-tier maturity model, using Traditional, Modern, and Advanced as the different measurements for each of the pillars. Obviously, Microsoft ties its model back to its technology stack to meet the requirements of its maturity model, but the model can be followed without needing to use Microsoft technologies. If you are a Microsoft customer, using this model will make more sense than using other models.

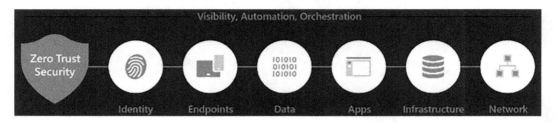

Figure 5.9: The Microsoft zero trust pillars
Image source: `https://learn.microsoft.com/en-us/security/zero-trust/`
`deploy/overview`

ZTA Pillars

For the purpose of this book, we will reference both the CISA and Microsoft ZTA models as we provide an overview of how to proceed with a ZTA model and the strategy that should be followed. From a pillar perspective, I'm going to recommend looking at seven distinct areas. Essentially, we will use the Microsoft pillars and add another pillar, which I feel is very important as part of your ZTA. We will also include Visibility and Analytics, Automation and Orchestration, and Governance to cover the entire architecture model. The following represents the high-level architecture of the recommended zero trust model:

Figure 5.10: The recommended seven pillars for your zero trust architecture

- *Identities* are the new perimeter of zero trust. An identity is something (typically a user) that needs to access an app, data, or some other form of resource. It is critical that identities have multiple layers of protection to prevent unauthorized access.

- *Device and endpoint* protection is an essential component of zero trust. Whether the device is a mobile, laptop, server, IoT, and so on, we need to ensure that recommended baselines are deployed and that devices stay compliant, constantly being scanned for vulnerabilities.

- *Data* is at the core of the zero-trust model. It is ultimately data that intruders look to exfiltrate from your environment. This is the true asset that needs to be protected. Because of this, it is critical that you know where all your data lives, who has access to it, whether it's classified correctly and encrypted, and that you have the correct controls to prevent it from being removed from your environment.

- *Applications and application programming interfaces (APIs)* are gateways to your data. They need to be governed and deployed with best practices to prevent unauthorized access to data, whether intentionally or unintentionally. Ensuring that a business follows enterprise standards is critical to prevent shadow IT.

- *Infrastructure* pertains to everything within your environment that provides the means to store your data and/or run applications such as servers, VMs, appliances, **IaaS**, **PaaS**, and **SaaS**. Preventing unwanted access to your infrastructure is crucial, and you need to ensure that you have best practices and baselines in place, along with effective monitoring.

- A *network* is the medium where your data travels. Once considered the perimeter for defense, this pillar still holds a critical role as part of zero trust to ensure that all data is encrypted during transport, next-generation protection is deployed, micro-segmentation is in place, and ongoing monitoring takes place to detect unauthorized access to your data in transit.

- *Collaboration* technologies include everything that an end user may use to communicate or interact with others within an organization or externally. This includes everything from email, instant messaging, voice, cloud storage, file sharing services, etc. These tools can be easily exploited or used by threat actors to target your users. Continuing to enhance these tools is critical.

The following screenshot shows the seven pillars, with some examples of the technologies and solutions that should be implemented to support your broader zero-trust strategy:

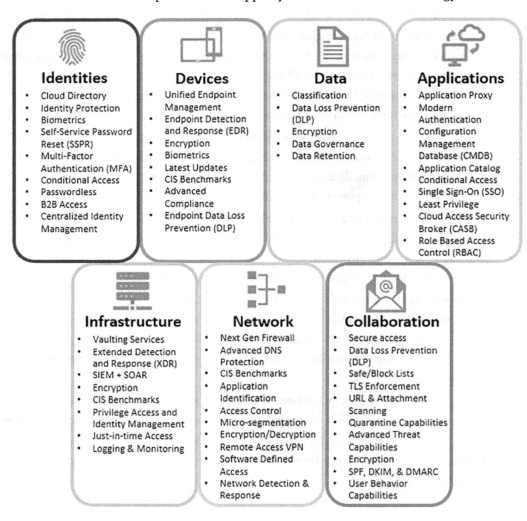

Figure 5.11: The ZTA seven pillars with example technologies

Now that your ZTA model is finalized with the seven pillars to provide for ZTA, next is the transition into the maturity model that will help guide you through your journey in becoming more complaint with zero trust.

ZTA Maturity Model

Now that you have an understanding of what ZTA is and what models are available, including the pillars we will focus our attention on, you will need to have a plan in place to ensure success. Zero trust is not going to occur overnight or within a few weeks. Zero trust is a strategy and a foundation that needs to be planned over months and even years, depending on your organization's maturity. Also, you will not be able to accomplish zero trust with a single vendor. A successful ZTA strategy will require multiple technologies and vendors to be successful. With this, you will need to be able to measure your success by understanding where you are today and what is needed to ensure you advance your maturity. To do this, we will reference both the CISA and Microsoft maturity models. Since I'm a believer in simplicity, we will focus on a three-stage maturity model instead of the four-stage model that CISA provides, although we will reference information from the CISA model to show how your plan and strategy should come together. The following figure provides an example of how to track your ZTA maturity, along with what each of the maturities represents:

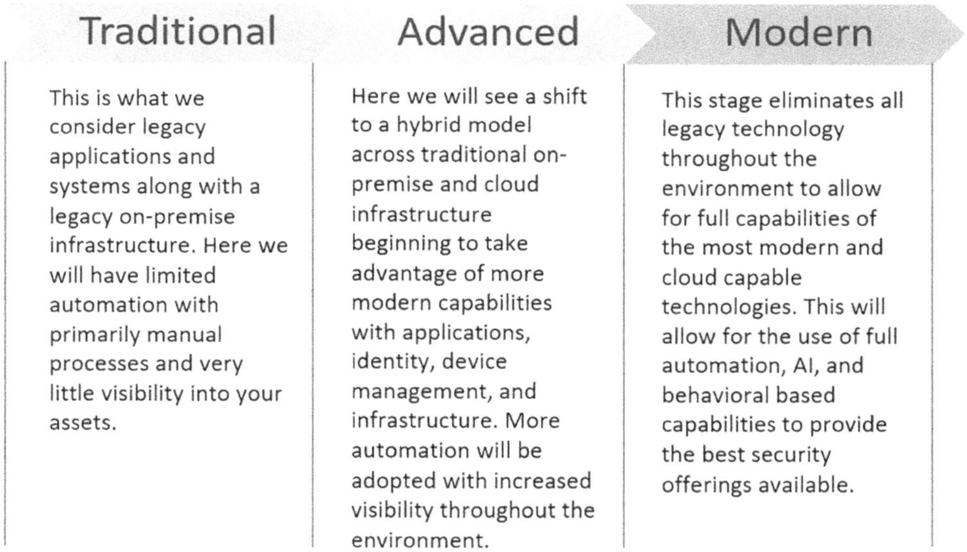

Traditional	Advanced	Modern
This is what we consider legacy applications and systems along with a legacy on-premise infrastructure. Here we will have limited automation with primarily manual processes and very little visibility into your assets.	Here we will see a shift to a hybrid model across traditional on-premise and cloud infrastructure beginning to take advantage of more modern capabilities with applications, identity, device management, and infrastructure. More automation will be adopted with increased visibility throughout the environment.	This stage eliminates all legacy technology throughout the environment to allow for full capabilities of the most modern and cloud capable technologies. This will allow for the use of full automation, AI, and behavioral based capabilities to provide the best security offerings available.

Figure 5.12: The Microsoft ZTA maturity model progression

Expanding on the previous figure, you can create a high-level overview of the maturity of each of the pillars as a reference. Let's take the Devices pillar as an example of how the maturity will work:

Traditional	Advanced	Modern
Traditional on-premise domain joined devices using Group Policy and technologies like Microsoft Endpoint Configuration Manager (formerly SCCM). Legacy security tools such as AV only and no DLP based capabilities.	A hybrid model is in place with cloud-based capabilities being leveraged to mange devices. The ability to shift away from domain joined to cloud joined devices to allow for more efficient policy management with devices.	Fully cloud joined devices leveraging the latest capabilities for device management for both corporate and personal devices. The use of advanced threat detection capabilities with automated information protection and DLP capabilities.

Devices

Figure 5.13: The Microsoft ZTA maturity model applied to Devices

You can now expand this exercise with each of the additional six pillars (Endpoints, Data, Apps, Infrastructure, Network, and Collaboration) to provide a holistic view of your maturity, from traditional to optimal.

> Make sure you review each of the CISA and Microsoft maturity models, as they both provide an overview of what is considered traditional through to the optimal level of maturity.

With the foundation of your maturity model defined, the next step is to build your strategy around accomplishing this transition. Remember, this transformation isn't going to happen overnight, so you'll need to build a roadmap, based on where you are today with the technologies you have in place versus where you would like to be. Accomplishing this journey is not going to be easy, but the better you document your strategy and vision, the easier your journey will become. As an example, here is a high-level visual of how you can build out your roadmap, consisting of the seven pillars previously discussed, with Identities and Devices populated with some examples:

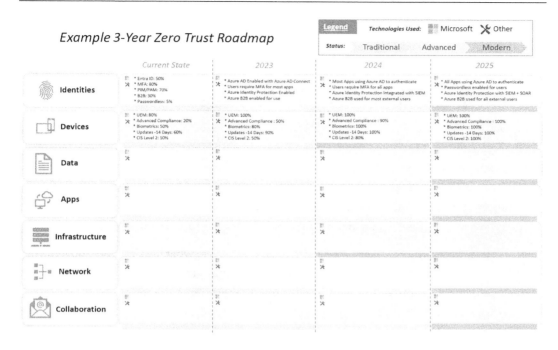

Figure 5.14: ZTA roadmap template

Obviously, you will need additional detailed plans that need to be tracked as you work through your ZTA strategy and roadmap, but this visual should provide a simple way to track your progress, also allowing you to share the progress of the ZTA model with your leadership team.

Microsoft also has a zero-trust maturity assessment that can help provide guidance on how to take the next steps. Once the assessment is complete, you will be provided with recommendations, and you'll have the option to download a playbook to implement zero trust security. There will be references to Microsoft technologies as part of the assessment, but you can easily substitute the recommended Microsoft technologies for any technologies you may already be using: https://www.microsoft.com/en-us/security/business/zero-trust/maturity-model-assessment-tool.

Now that we have the ZTA strategy and roadmap in place, we will dive deeper into the technical side of the architecture and look at some of the technologies and considerations to best protect your environment.

Technical Architecture

As we look into the technical side of the architecture, we will focus on the specifics of protecting the services, solutions, products, applications, data, etc. within an organization. This is where we will implement controls in the environment and where hands-on configurations will occur. Here, we will see a big involvement from the engineering resources, as they will be the ones primarily executing the work needed to ensure the technological controls are implemented, meeting the requirements of the agreed-upon architecture for each of the pillars in the ZTA strategy.

Before we go into detail about each of the seven pillars, let's cover what baselines are and why they are important. Baselines ensure that you implement a minimum recommended set of controls to meet compliance. If you have already implemented some form of framework, you will already be following some set of baselines. If you haven't begun to implement a framework, you will find that baselines will be required as you work through your framework requirements.

Following a Baseline

Security baselining is the practice of implementing a minimum set of standards and configuration within your environment, for example, capturing a minimum configuration for your end-user devices. Building a baseline provides a minimum defined standard that will help ensure a more secure environment as you deploy technology throughout the enterprise. Depending on the size of your organization, baselines could be in the form of checklists or spreadsheets that someone follows to ensure the predefined security controls are in place. A more advanced method includes the use of management tools to layer and enforce baseline configurations. The end-user devices management tools that will help implement those baselines include **Group Policy Objects (GPOs)** and **Mobile Device Management (MDM)**.

As you begin to define and deploy your baselines, you will find that one size does not fit all. You are going to need to document and build them for different use cases. These are some examples of where unique baselines may need to be defined:

- Network devices (switches, routers, firewalls, and so on)
- Windows systems: servers and clients
- Linux/Unix systems
- Storage/file servers

- Database servers

- Web servers

- Application servers

- **Operational Technology (OT)**

- **Internet of Things (IoT)**

Building Baseline Controls

One of the more adopted control services is the **Center for Internet Security (CIS)**. You may already be familiar with CIS, and you will see CIS listed on a lot of the popular framework lists, although it's not a fully comprehensive framework. Instead, CIS is more of a tactical compilation of controls and guidelines that allow organizations to meet the requirements of a chosen framework. You can think of CIS as a control framework rather than a fully comprehensive program framework. You can access CIS here: `https://www.cisecurity.org/`.

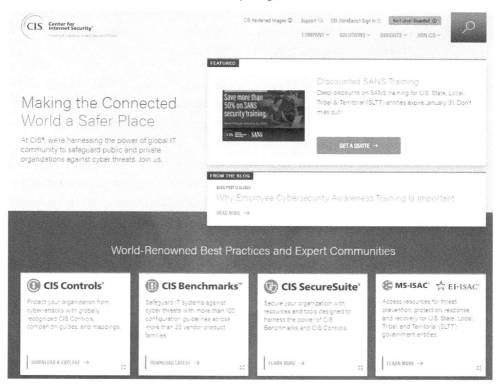

Figure 5.15: The CIS home page

CIS is a non-profit organization comprising a global community to provide protection against the ongoing cybersecurity threat landscape. More specifically, the CIS vision is:

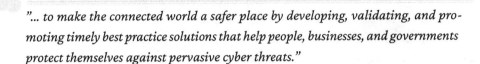

"Leading the global community to secure our ever-changing connected world."

Their mission is:

"... to make the connected world a safer place by developing, validating, and promoting timely best practice solutions that help people, businesses, and governments protect themselves against pervasive cyber threats."

You can learn more about CIS at `https://www.cisecurity.org/about-us/`, and sign up for the MS-ISAC advisories, newsletters, and webinars at `https://learn.cisecurity.org/ms-isac-subscription`.

CIS has an overwhelming number of tools and resources available that can be utilized by any organization of any size. Many of the tools and resources are provided free of charge; you can find more details on all their tools and resources here: `https://www.cisecurity.org/cybersecurity-tools`. More specifically, CIS provides two sets of best practices widely adopted throughout the world: CIS Controls and CIS Benchmarks. As of version 8, the CIS Controls are a broader set of 18 foundational and advanced controls that provide a more comprehensive approach to overall security protection for your organization, whereas the CIS Benchmarks are focused more on the specific hardening of your systems, software, networks, and more.

CIS cybersecurity best practices, which include a link to both the CIS Controls and CIS Benchmarks, can be found at `https://www.cisecurity.org/cybersecurity-best-practices/`.

The list of available benchmarks is extensive and includes categories for the following:

- Cloud Providers
- Desktop Software
- DevSecOps Tools
- Mobile Devices

- Multi-Function Print Devices
- Network Devices
- Operating Systems
- Server Software

Now that we have a better understanding of baseline controls and the importance of how they can support a broader architecture strategy, let's move on to each of the architecture functions, which we will cover from a technical perspective.

Identity Architecture

As previously mentioned, Identity has become the new perimeter of your environment protection. Because it's so important, we have an entire function within the program dedicated to Identity, which will be covered in a lot more detail in *Chapter 6, Identity and Access Management*. Your identity architecture is not as simple as creating a user account and distributing it to the user. This area is very complex, with many moving pieces that need to be taken into account. We won't go into too much detail here, but some of the items that will need to be architected and engineered include some of the foundational identity areas, including identity management provisioning and de-provisioning, access management activities, password management for your users, how authentication will occur, etc. Some of the more advanced technologies you will need to become more familiar with include those that have SSO integration into all your applications (whether internal or SaaS), shifting to full modern authentication away from legacy protocols, focusing on **Role-Based Access Control (RBAC)**, etc. You'll also need to ensure that you have an efficient process for the external users and partners who will need access to your environment. Look at new and modern ways to provide access via **Business-to-Business (B2B)** and **Business-to-Consumer (B2C)** types of technologies.

Most important is securing these identities by leveraging the likes of Privileged Management technologies, the concept of least privilege, **Just-in-Time (JIT)** access, **Just Enough Access (JEA)**, advanced identity protection capabilities, striving for passwordless capabilities, etc. We will cover these in more detail in the next chapter.

Endpoint architecture

Your endpoints include your desktop PCs, laptops, virtual machines, mobile devices, Audio-Visual equipment, printers, IoT, and endpoints within OT. With each of these endpoints comes some form of OS that will need to be secured and hardened.

There is a lot of complexity involved in efficiently managing and securing your endpoints within an organization, and for the most part, this will require dedicated skillsets to fully understand how to architect, engineer, secure, and manage them. Referring back to the *Building Baseline Controls* section, specifically CIS – you will want to ensure that you reference the benchmarks available for your endpoints. There's a high possibility that a benchmark will be available for the endpoints within your environment, and these baselines are highly regarded.

As you look to protect your endpoints efficiently, there are multiple layers that need to be considered. This includes protecting the hardware of the endpoints and/or any virtualization technology needed; you will need to ensure the network components of your endpoints are secured correctly, along with the necessary identity components, to ensure that access to the endpoints is locked down and secured. In addition to these layers, you need to monitor your endpoints to ensure that there are no issues, they meet compliance, and that there are no security issues. Also, you need to provide efficient reporting to ensure that information is made available if any action needs to be taken.

Endpoint Management and Advanced Protection

Another important factor is the management of the endpoints. How do you manage and enforce a policy for your endpoints? You must be able to deploy consistent baselines to your endpoints as per your company policy. This includes all your endpoints, which will require different platforms to be managed. Mostly, you'll be able to use a unified management platform for a lot of your core endpoints like Windows, Apple, Android, Linux, etc. A good example is Microsoft Intune. For some other technologies, like IoT, OT, printers, AV equipment, etc., you will most likely manage on a different platform. We will cover the management of OT & IoT devices in more detail in *Chapter 12, Operational Technology (OT) & the Internet of Things (IoT)*.

From a security standpoint, ensuring robust protection of your endpoints is critical. Some of the basic requirements to secure your endpoint devices include:

- The need for a baseline hardened image or OS.
- Making sure your endpoints are updated as security updates become available.
- Ensuring that encryption is enforced on all endpoints with the efficient management of encryption keys.

- Providing modern login capabilities with passwordless and biometrics, with technologies like Windows Hello for Business.

- The need for compliance enforcement to ensure that defined baselines are met.

- Ensuring that AntiVirus (AV) is installed.

- Leveraging the firewall and network protection features within an OS.

- Taking advantage of account protection features.

- Removing local administrator rights for users.

- Ensuring any other device-specific security is utilized for the best protection.

Looking at some more advanced requirements for your endpoints, you will want to ensure that **Advanced Threat Protection (ATP)** capabilities are used with features such as AV and threat protection, **Endpoint Detection and Response (EDR)**, advanced analytics, and behavioral monitoring. You will want to ensure that any enterprise web browsers such as Microsoft Edge and Google Chrome are secured on a device using a recommended baseline, using technologies to secure your applications with application control capabilities and ensuring that untrusted applications are not being installed, enabling tamper protection technologies to ensure that security features remain enabled, enforcing removable storage policies to prevent the use of USB drives as an example, along with leveraging **Data Loss Prevention (DLP)** and information protection capabilities.

With your endpoints, you also want to ensure that you mitigate against some of the more common attack vectors. This includes protecting against **Adversary-in-the-Middle (AiTM)** or **Man-in-the-Middle (MiTM)** attacks, which take advantage of specific network protocols, including **Link-Local Multicast Name Resolution (LLMNR)**, **NetBIOS Name Service (NBT-NS)**, **Multicast DNS (mDNS)**, **Web Proxy Auto-Discovery Protocols (WPAD)**, the **Server Message Block (SMB)** relay, IPv6 **Domain Name Server (DNS)** spoofing, and the **Address Resolution Protocol (ARP)** cache. Protecting against lateral movement and privilege escalation is critical to prevent resources from being enumerated, protect Kerberos tickets, prevent OS credential dumping, and prevent user access to the registry. One other consideration is reviewing and configuring any privacy settings on the endpoint and applicable applications – for example, ensuring the protection of a camera and mic from unauthorized access, as well as ensuring device location capabilities are not used inappropriately.

Endpoint security is no simple task. To give you an idea of the complexity and what is involved, I co-authored a book with Matt Tumbarello to secure Windows devices called *Mastering Windows Security and Hardening, Second Edition*. This book contains over 800 pages of everything you need to consider when securing a Windows device.

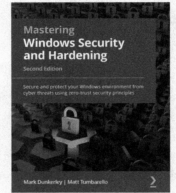

Figure 5.16: Mastering Windows Security and Hardening Second Edition

Here is a link to the book: https://www.amazon.com/Mastering-Windows-Security-Hardening-environment/dp/180323654X/.

Data architecture

The data in your organization is essentially the crown jewels you are protecting. There is nothing more important than ensuring that your data has the correct controls in place and is effectively protected. A major role of a cybersecurity team is to ensure that the appropriate architecture is put in place, along with the correct controls being implemented and enforced to protect the data. You are going to find that there are many locations in which your data is located, including databases, data lakes, data warehouses, cloud storage, file servers, SFTP, etc. You will quickly find that managing your data is not a simple task.

Data Inventory and Classification

As you look to efficiently manage and protect the data within your organization, you are going to need to ensure that you have good inventory management of where everything is, as well as who the data owner is. In the event of a cybersecurity incident, this is going to be critical. As part of inventorying all of your data, you will need to ensure that it's also classified correctly.

You will need to understand what the data type is – **Personal Identifiable Information** (**PII**) versus business-critical versus sensitive versus non-critical, etc. The auditing of all data is also critical, as this will be needed when a cybersecurity incident occurs. When information gets compromised, you need to be able to quickly identify what type it is, where it is located, when it was compromised, and by whom.

Data Protection

Some of the more technical components of data protection are going to include selecting how data will be stored, whether in a database, data warehouse, or somewhere else. Once stored, ensure that each of the data locations has the correct security controls in place, such as locking down access appropriately and making sure the correct access controls are in place. You will need to ensure that encryption is applied to both the data at rest and in motion. Other considerations will include when data needs to be moved, what protocols/technology is used to move the data, and whether it is secure – for example, using data feeds, integration into other systems, API, SFTP, and so on. You will also need to ensure that data availability has been considered if there is a cybersecurity incident. This includes ensuring backups that are resistant to the likes of a ransomware attack. You will also want to ensure that DLP capabilities are applied as an additional layer of protection. This ties back to classification, which will allow you to provide insight into data and apply appropriate controls to it, depending on the classification. Another area to understand is the automation and orchestration of your data. With any automated processes, you will need to fully understand how the data is accessed and moved; typically, a service account is used for these activities, which will need to be updated to more secure methods.

Data Governance and Monitoring

An often overlooked component is the governance of your data. In addition to the auditing of your data, is there good reporting in place to allow easy insight into all your data? Also, are there monitoring capabilities of your data so that you can be informed of any anomalies, and are the alerts integrated with your SIEM and **Security Operations Center** (**SOC**)? You will also need to know who has access to the data and whether there is any external, third-party access to your data. Does this access have the correct documented approvals in place, and how long is that access needed for? Knowing if your data is leaving your organization and where it will be stored is also critical. If a vendor who stores your data is compromised, you will need to ensure that the correct processes are in place for notification. You will also want to ensure that your data follows your **Records Information Management** (**RIM**) schedule.

Holding data longer than needed could create more liability, so you'll need to ensure that data is properly disposed of when no longer needed.

A lot of what is covered within data architecture will be reviewed in more detail within the Governance, Risk, and Regulatory and Compliance chapters in the final section of the book, *Bringing It Together*.

Regulatory and Legal Considerations

One final and very critical component of data management is knowing if there are any regulations and/or privacy-related laws or requirements that apply to your data. This can get very complex, and collaboration with your legal team will be needed. If PII data is part of a compromise, you will have certain obligations relating to the likes of **GDPR and CCPA**. If healthcare information is impacted, there are requirements from the **Health Insurance Portability and Accountability Act (HIPAA)**. A publicly traded company in the US that has been compromised will have reporting requirements under the **Securities and Exchange Commission (SEC)**. There are also requirements from GDPR and CCPA to allow a consumer to request the removal of their information from a company that stores and processes any data of theirs. All these examples enforce the need to know where your data is, what type of data you store and process, ensure that it is correctly audited, and ensure that the correct controls and protections are in place.

Application Architecture

With data being the crown jewels of your organization, think of the application as the conduit to those crown jewels. The application is what allows your users, customers, partners, etc. to access data to accomplish a task. Protecting and securing these applications has continued to become more and more challenging as technology evolves and threat actors continue to look for ways to compromise them. The majority of requests that will go through the architecture review process will most likely be some form of application. Because of this, it will be critical that a thorough review is completed to ensure that an application is secure before it goes into production. Ensuring the right architecture for your applications is very important and will require close collaboration with the Development teams, as they will be the ones building them. You will need to ensure that your developers instill a mindset of security first as they develop any apps within an environment.

Authentication and Access Control

Some of the basic things that will need to be considered from a technical perspective include the authentication method for your applications, whether modern authentication is supported, and the ability to integrate with your **Identity Provider (IdP)** to leverage SSO capabilities.

This will allow you to enforce your enterprise-level identity and access capabilities. If SSO is not supported, how are the identities of the application being managed, what are the password requirements, will MFA be supported, etc.? You will want to move away immediately from any type of legacy authentication that doesn't support MFA.

Update Management

One area that will need ongoing effort is that of update management for your applications. This will be an ongoing task and one that you need to ensure has no user impact or performance degradation when updates are applied. Having a well-defined process in place will be critical, as you will find the need to update your applications will be constant, and it is critical you do so with every new update. This will be covered in more detail in *Chapter 8, Vulnerability Management*.

Infrastructure and Business Continuity

As applications are designed, built, deployed, and enabled, you will need to have a thorough understanding of the underlying infrastructure to support them. Will the applications be deployed using legacy methods with a traditional server model, whether on-premises or in the cloud using IaaS? Or will the app take a more modernized approach, using PaaS and SaaS cloud-based technologies like containerization and an application-type service? Whatever method is selected, you will need to ensure that the correct controls are put in place and that security has been reviewed and approved. In addition to this, ensuring that requirements from the business are met, as this relates to availability and **Disaster Recovery Planning (DRP)** for the application. You will also need to ensure that **the RTO** and **RPO** are understood so that the application continues to meet any **SLAs** that are in place. As part of the application architecture, you will need to understand if a testing, development, or staging environment is also needed, as they will need the same level of security controls as production to ensure minimum exposure.

Advanced Components and Integration

As you look at some of the more advanced components of your applications, you are going to need to ensure there are well-defined policies and processes around access management within the applications. What is the user audience of the application? Are they internal users, external guests, vendors, customers, etc.? Are there RBAC capabilities within the applications, is the concept of least privilege being enforced, how are the privileged users being handled, and how are all these roles being assigned and managed? This will dictate how to correctly architect an application to ensure that more access than is necessary is not provided. Understanding what integrations into the applications is important, as they will all need to be secured.

Is there a need for any APIs or other data connectors in the application? Is there a need to set up any type of automation that will need some form of a service account, or can it leverage more modern capabilities within cloud services, like key vaulting with secrets, etc.? Will there be any database, data lake, or data warehouse integration needed, and how will that integration occur? You will also need to know if there will be any type of web application or interface, and if access is needed over the public web or just internally in the organization. If the application is web-facing, meaning it is accessible over the public internet, the security review will need to be more thorough.

Application Security and Protection

Some of the more specific security items with your applications are going to include the need for application threat protection capabilities. As you deploy more SaaS throughout an organization, you will want to ensure the likes of a **Cloud Access Security Broker (CASB)** are in place. A very important part of the application development and deployment process is that of thorough testing with your applications before they go live. Ensure that the likes of application penetration testing occur, and reference resources like the **Open Web Application Security Project (OWASP)** to ensure that the most common security risks are mitigated before an application goes live. This will be covered in more detail in *Chapter 8, Vulnerability Management*. Another security component that shouldn't be overlooked is that of DevSecOps as part of your application development lifecycle. DevSecOps represents development, security, and operations, and the idea is that security is embedded in every part of the development process for your applications. This is critical in today's world of application development.

Governance of Applications

The final area we will cover is the governance of your applications. Like the governance of your data, there will be a need to understand if there are any regulations and/or privacy-related laws or requirements that apply to your applications. For example, there may be a need to require consumers to accept terms on a website or provide them the ability to opt out of using cookies on their devices. Another important component of governance is having a good inventory of all your applications that tracks all relevant information, such as the application owner, contact information of the application, classification, how often access reviews are needed, whether there are any audit requirements, the accessibility of the application, the audience of the application, and so on. This inventory should be complete, as it will serve as your place of reference when audits arise and any potential cybersecurity incidents are investigated. This will be covered in more detail in *Chapter 13, Governance Oversight*.

The following initiative from CISA is intended to provide a secure-first mindset for technology providers and software developers: https://www.cisa.gov/ securebydesign.

Infrastructure architecture

The infrastructure is essentially everything used to host and allow services to be delivered. This includes everything from a traditional on-premises deployment that includes your physical data center along with all hardware such as servers used to host your applications, data, etc. With a more modern approach, workloads are shifting to cloud architecture with PaaS and SaaS capabilities. We covered the data center models in more detail in *Chapter 4, Solidifying Your Strategy*. Securing your infrastructure is no easy feat, especially since we have shifted from on-premises to cloud infrastructure. Over the years, skillsets have matured to become efficient at protecting the on-premises infrastructure. As we have shifted to the cloud, these skillsets have needed to change and evolve to support cloud environments. The reality is that architecting and engineering cloud environments have become a lot more complex because of the constant addition of new features and ongoing innovation. To make it even more challenging, there are different cloud environments, which means you need more specialized skillsets for each of these cloud environments if you adopt more than one.

Infrastructure Protection

There will be a lot of similarities to the securing of your servers within infrastructure, as noted in the *Endpoint architecture* section. You'll want to ensure you refer to the CIS Benchmarks – more specifically, the OS, Server Software, and Cloud Provider baselines. As you look to protect your infrastructure efficiently, there are multiple layers that need to be considered. This includes protecting the hardware of the infrastructure and/or any virtualization technology needed, and you will need to ensure that the network components of your infrastructure are secured correctly, along with the identity components needed to ensure that access to the infrastructure is locked down and secured. As you move more into a cloud environment, your scope of knowledge will increase, as you will need to become more familiar with and protect containers, microservices, AI, machine learning, IoT, and more. In addition to these layers, you need to monitor your infrastructure to ensure that there are no issues, compliance is being met, and that there are no security issues. Also, you need to provide efficient reporting to ensure that information is made available for action to be taken.

Infrastructure Management

Another important factor is the management of the infrastructure. How do you manage and enforce policy on your infrastructure? It is important that you deploy consistent baselines throughout your infrastructure as per your company policy. Mostly, you'll be able to use a unified management platform for a lot of your core infrastructure if you have strategized on one cloud platform. If not, your management may be distributed, and consistency will become more challenging in addition to a more complex security model.

Infrastructure Security

As you look to protect and secure some of the more traditional services within your infrastructure, like your servers, some of the basic requirements to secure them include the need for a baseline hardened image or OS, making sure that your servers are updated as security updates become available, ensuring encryption is enforced with efficient management of encryption keys, enforcing MFA capabilities, the need for compliance enforcement to ensure that defined baselines are met, ensuring that an AV is installed, leveraging the firewall and network protection features within the OS, taking advantage of account protection features, and ensuring that any other device-specific security measure is utilized for best protection. For all other infrastructure services, a similar approach will be needed where applicable to ensure that they are deployed securely.

Advanced Security Requirements

When looking at some more advanced requirements for your infrastructure, you will want to ensure that ATP capabilities are used, with features including AV and threat protection, EDR, advanced analytics, and behavioral monitoring. As you review your infrastructure portfolio, there will be specialized infrastructure that will need to be secured and hardened differently from other components and may need much tighter scrutiny. Some of these services include the likes of your **Domain Controller (DC)** to protect **Active Directory (AD), Domain Name System (DNS), Dynamic Host Configuration Protocol (DHCP), Public Key Infrastructure (PKI), and Windows Server Update Services (WSUS).** Is there any virtual machine infrastructure, backup infrastructure, monitoring infrastructure, SIEM, **Remote Desktop Protocol (RDP)** services, and so on that need to be hardened and protected? Finally, access management to all these services needs to be closely managed and monitored. Here, you will need to deploy the likes of **Privileged Access Management (PAM)** and **Privileged Identity Management (PIM)** to provide privileged access only when it is needed. This will be covered in more detail in *Chapter 6, Identity and Access Management*.

Business Continuity Planning (BCP) and Disaster Recovery Planning (DRP)

A very important part of your infrastructure architecture is ensuring that a robust disaster recovery plan is in place as part of your broader business continuity planning. As touched upon in the *Application architecture* section, it is important that a business is able to continue if there are any major issues within the infrastructure or a data center. Ensuring a well-defined RTO and RPO are understood and documented for the business is critical. We will cover both BCP and DRP in more detail in *Chapter 7, Cybersecurity Operations*.

Network architecture

Network security has traditionally been at the core of cybersecurity programs, but as previously mentioned, this has shifted. Network security hasn't become any less important, but with the adoption of cloud computing and the rapid acceleration of remote work, security strategies need to shift from a strong focus on the network perimeter to the application, device, and identity areas. A benefit within the networking realm is that network professionals have traditionally been looked at as the cybersecurity experts within an organization. Cybersecurity has naturally been embedded within networking, and there are a lot of cybersecurity awareness and skills readily available within network teams.

Challenges with Network Management

Unfortunately, networking can be a very challenging task for technology and cybersecurity teams. Networks can be very sensitive and commonly take the blame for most outages, without people even knowing the true root cause of an issue. This is simply because most of our data traverses over a network, so it's critical that it performs optimally. If it doesn't, it can bring a business to its knees because of how dependent we have become on the network. In addition to the already challenging task of network operations, we are also concerned with network security. Ensuring that the data we transmit is secure, ensuring that no perpetrators access our network who shouldn't be, preventing traffic that isn't welcome, and ensuring confidential data is isolated are some of the challenges we will encounter.

Network Vulnerabilities

With the network being the conduit through which all your data traverses, it can easily be compromised in many ways. Networks are vulnerable to many attacks, including:

- **Man-in-the Middle (MitM)**
- Traffic/packet sniffing

- Insider threats
- Spoofing
- **Denial-of-Service (DoS)**
- **Distributed Denial-of-Service (DDoS)**
- Port scanning
- DNS compromise

Network Baselines

Referencing back to the *Building baseline controls* section, specifically CIS, you will want to ensure that you reference the benchmarks available for network devices to help protect against these types of attacks. The CIS benchmarks include baselines for Cisco, Palo Alto Networks, the Check Point Firewall, Juniper, Fortinet, F5, Sophos, and the pfSense Firewall. As previously mentioned, these benchmarks are highly regarded and referenced in many frameworks as part of the recommended controls.

Key Technologies for Network Security

Because there are many components involved within a network architecture and the topology can be extremely complex, it is important you have the right personnel and skillsets available on the team to architect and engineer. The following technologies are considered more critical for your enterprise deployment, as they relate to your network security and should be implemented to reduce risk:

- Routers and switches using **Virtual Local Area Networks (VLANs)**
- Next-generation-type firewalls
- A **Virtual Private Network (VPN)** to encrypt connections
- **Intrusion Detection Systems (IDSs)/Intrusion Prevention Systems (IPSs)** to proactively detect and prevent threats
- Wi-Fi with a minimum of **Wi-Fi Protected Access 2 (WPA2)** enterprise security
- **Network Access Control (NAC)** to better manage endpoint access to your network
- Proxy/web content filters to prevent malicious websites
- **Domain Name System Security Extensions (DNSSEC)** to protect your DNS services
- **Public Key Infrastructure (PKI)** to provide digital certificates for encryption
- **Secure Access Service Edge (SASE)**

Also, don't forget to lock down and closely monitor any remote access connectivity for administration, specifically RDP connectivity types.

Network Management

When managing networking equipment, the following should be considered to ensure a more robust and secure environment:

- Always keep the software of your network devices current.
- Enable auditing on devices.
- Integrate authentication using LDAP.
- Use a PAM solution.
- Disable or prevent local account access and change default usernames and passwords.
- Ensure that the management of devices is encrypted (SSH).
- Isolate the management network.
- Don't allow the management of network devices from the internet.
- Ensure that logs are sent to a SIEM to help you detect abnormal activity.

As reviewed within the other areas, governance over your network architecture is critical as you closely manage your access controls, ensuring that you have well-documented policies and procedures, that your network is resilient, and that it can support any defined SLAs in place.

Collaboration Architecture

The final area we will cover is collaboration architecture. I'm not sure why, but collaboration is not included in the more common ZTA models that have been referenced. I believe this is a big gap, which is why I want to ensure that it is covered. Collaboration architecture includes email services, intranet sites, traditional file-sharing technologies using file servers, modern file-sharing capabilities using the cloud, voice technologies, instant messaging, meeting platforms, business social media types of platforms, other communication applications on mobile devices, and so on. These types of technologies are extremely vulnerable and require dedicated expertise to manage and protect. One of the most challenging is that of email security. Phishing continues to be one of the most used attack vectors, and with the advancement of AI technology, it isn't slowing down.

Within the collaboration space, there'll be a need to reduce your attack surface, like any other area, by ensuring you are running the latest version of the tools available, ensuring that you implement security updates as they become available, and looking to eliminate any legacy-type collaboration services.

For example, Microsoft Exchange Server continues to be challenged with ongoing vulnerabilities. Some of the more specific attacks across all the collaboration technologies include malicious attachments, malicious **Uniform Resource Locators (URLs)**, and social engineering techniques. As far as email goes, **Business Email Compromise (BEC)** and phishing, such as spear phishing, whale phishing, vishing, smishing, quishing, and pharming, continue to be extremely challenging to prevent. Looking at the CIS controls, we can see that there are benchmarks available for the two most popular and adopted collaboration platforms, Microsoft 365 and Google Workspace. Within the Cloud Providers section, you will find benchmarks for each of these platforms, which provide a very mature foundation to start from to harden your collaboration services against the aforementioned vulnerabilities and attack vectors.

In order to efficiently secure, deploy, and operate your collaboration technologies, you are going to need someone familiar with and who has experience in managing these types of technologies. You can't set up these technologies and expect them to be secure and run themselves. Threat actors continue to evolve, and so does the need to review and improve the protections against these technologies. As you look to implement your controls to best protect your organization, there are many capabilities that can be enabled and configured, including preventing the sharing capabilities of files with anonymous access and any external users, ensuring that DLP capabilities have been enabled with your intranets, file servers, and cloud sharing platforms, ensuring that there are tightened access controls and policies for all collaboration services, locking down external access to only allow those organizations you want to access your environment for collaboration, ensuring that modern authentication has been enabled across all collaboration services, enabling file and URL scanning capabilities across all technologies, and ensuring that MFA is enforced for your users.

As we focus more on the email-specific services that need protecting, you will find that your resources spend most of their time protecting this environment over other collaboration technologies. Items to consider in securing email services include managing attachment types, ensuring attachment and URL scanning is enabled, ensuring that spam and anti-phishing policies are leveraged, enabling quarantine feature capabilities, taking advantage of any spoofed domain capabilities, ensuring the efficient management of transport rules, enforcing DLP capabilities and ensuring that encryption is enabled and enforced as needed, ensuring that TLS capabilities are enforced between trusted partners, ensuring that **Sender Policy Framework (SPF)**, **DomainKeys Identified Mail (DKIM)**, and **Domain-based Message Authentication, Reporting, and Conformance (DMARC)** are all implemented and used per the best practices, and ensuring any advanced threat protection capabilities are enabled for your email services.

Summary

We have covered a lot in this chapter, and as you can see, architecture plays a very important role as part of a broader cybersecurity program. It is important that you have a foundational understanding of architecture and that you have the right skillsets on your team to represent architecture within the cybersecurity function. Ensuring a simplified, well-designed, architected foundation will significantly reduce risk within your organization. We covered in detail the need for a well-defined architecture review process to ensure that technology isn't deployed without prior approval. As part of this process, you will need to ensure that the program has a defined strategy, scope, and goals, along with a committee to look over the program and approval process. As part of the process, you will want to ensure that the intake process allows for a thorough review, ensuring that there are no gaps in any solutions that are deployed. Simultaneously, it is critical that the process is agile to prevent a business from bypassing any processes in place.

We then reviewed the foundation of your cybersecurity architecture and briefly covered the CIA triad: Confidentiality, Integrity, and Availability. This was followed by a review of some cybersecurity architecture frameworks before we reviewed some of the architecture and engineering principles provided by the CISSP. From these principles, we then focused our attention on the principle of zero trust and shifted our focus to the ZTA model. Here, we looked at the history of zero trust and how the models we know today were founded. There are a couple of more prevalent ZTA models as we reviewed the specific pillars that should be part of your ZTA foundation: Identities, Endpoints, Applications, Data, Infrastructure, Network, and Collaboration. We then reviewed the importance of a maturity model, with an example of how to implement one for your organization.

We then finished off the chapter with a deeper dive into the technical architecture of your cybersecurity program. As you focus on the technical aspects, engineering becomes critical, with the need for skilled personnel who understand the technology and are capable of deploying your solutions with a security-first mindset. We covered the need to deploy a technology baseline within your organization – more specifically, by referencing the well-adopted CIS benchmarks and controls. We then finished off the chapter with a more detailed review of each of the seven different areas that map back to the ZTA pillars reviewed previously.

In the next chapter, we will review *Identity and Access Management* in more detail. In this chapter, we will provide an overview of what is involved in Identity and access management. We will go into detail about the need to modernize your identity strategy and move away from legacy infrastructure and architecture with your identity and access management.

Then, we will cover the basic and advanced account and access management principles needed to efficiently run your identity and access management program. In the final sections of the chapter, we will look at what is needed to secure your identities, before finishing off the chapter by reviewing some of the enhanced identity security capabilities you should consider.

Join our community on Discord!

Read this book alongside other users, Cybersecurity experts, and the author himself. Ask questions, provide solutions to other readers, chat with the author via Ask Me Anything sessions, and much more.

Scan the QR code or visit the link to join the community.

https://packt.link/SecNet

6

Identity and Access Management

You have already heard me say a couple of times that identity is now considered the new perimeter of your organization. Because of this, the need for a mature identity and access management program is a necessity. Identity and access management needs a dedicated function within the cybersecurity program because of how important this area has become. Through the ongoing shift to work from anywhere at any time and with data centers continuing to shift to the cloud, users can now access their corporate information from anywhere over the internet. A simple breach of identity will allow an intruder to log in and access information for that compromised user, then laterally move through the network to create a lot more damage. Because of this, we need to revisit the traditional authentication methods and add enhanced protection to our identity and access model.

We will begin the chapter with an overview of identity and access management and many of the components required to build a robust program. Here, we will review **Identification, Authentication, Authorization, and Accountability (IAAA)**. Next, we will review the need to modernize your identity program with the need to shift away from legacy and outdated methods that you may be using today. We will follow with a detailed review of account and access management within the organization, beginning with some basic concepts and then progressing to more advanced concepts.

One of the more important topics in this chapter will be around securing your identities to ensure you minimize the risk of them being compromised.

This will consist of reviewing some of the current technologies to best protect your identities and prevent threat actors from compromising them. We will then move into the last section of the chapter, which will consist of enhanced identity security and some of the more current protection methods that should be used to best protect your identity today. We will also look into the protection of your privileged identities and some best practices to protect them from getting into the wrong hands.

This chapter will include the following topics:

- Identity and access management overview
- Modernizing your identity architecture
- Account and access management
- Securing your identities
- Enhanced identity security

Identity and Access Management Overview

First, let's take a high-level look at all the sub-functions that should be addressed as part of identity and access management. The following image captures much of what the identity and access management function entails.

Figure 6.1: Sub-functions of the identity and access management function

Today, identity can be somewhat considered as the foundation for cybersecurity within your organization. Although there are other methods of compromising data, simply gaining access to a user or administrative account can be destructive. If an intruder compromises an account, they now acquire the same account level of access across all systems and data. All this can take place without anyone being alerted. It is very important that you are rigid with your identity and access policies. The role of least privilege is a must!

This is a concept where you only receive the required access to data based on your job function. For example, when a user account is set up for someone in HR managing the payroll, they will only receive the access needed to complete the HR payroll responsibilities. They will not need access to other HR roles, such as recruiting or benefits. More importantly, they will not need any access to data in other areas of the business, such as finance. We will cover this in more detail in the *Authorization* section below and *Securing Your Identities* later in this chapter.

Essentially, if you don't need access, you don't get it. In addition, ensuring that you use separate user accounts from administrative, or privileged accounts, or using capabilities like privileged access and identity management capabilities. There should never be permanent elevated or privileged permissions added to your regular user accounts where general day-to-day tasks are performed. In today's world, passwords have become obsolete and do not provide the level of security needed anymore. You need to start implementing multi-factor and passwordless capabilities as soon as possible, along with more modern technologies to protect your users and your organization's data.

Let's look at the foundation or framework of what comprises a solid identity and access management program. If you have studied for your **Certified Information Systems Security Professional (CISSP)** certification or some other security-related exam, you will be familiar with **IAAA** or just **Authentication, Authorization, and Accountability (AAA)**.

Here, we will use IAAA to emphasize identity as the first component of the process because, without an identity, the preceding components cannot exist.

Figure 6.2: Sub-functions of the identity and access management function

Next, let's review each of the components in more detail to better understand the purpose of each. We will begin with the first required component, which is identification.

Identification

The Identification portion of your access relates to something that identifies who you are. Simply put, it is a unique identifier that you enter to let a system know that it is you who is trying to access it. A simple example could be a username, an email address, or an employee ID number. Traditionally, within a Windows environment, you may be familiar with sAMAccountName, which was essentially a username that only worked within your corporate network. Using this outside your corporate network was not possible, as there was no unique identifying factor that accompanied the username on places such as the internet. Today, you will need to adopt the **User Principal Name (UPN)** method for your identity to best support internet and cloud-based technologies. The UPN appends a domain that you own to the end of your username to provide a unique or immutable identity no matter where you are and how you access it. An example of a UPN identity could be *markdunkerley@myorganization.com*.

 Identities can come in various forms, such as a standard user account, a contractor account, an external account, a bot account, or a service account.

Next, let's look at authentication to confirm you are the correct person looking to gain access to something.

Authentication

After you have entered your identity, you will typically be presented with some form of authentication to validate that you are the person who should be gaining access to the system. The most common form of this is entering a password. As part of authentication, there are four different methods to authenticate your identity that are worth mentioning:

- **Type 1** is something you know. This is currently the most common and widely adopted method of authentication. Some examples include a password, PIN, or passphrase.

 A passphrase is recommended over a password today as it can be created much longer while being a lot easier to remember than a password.

- **Type 2** is something you have. This authentication method consists of something you have with you to confirm it is you. Some examples include a hard token, a soft token on your phone, a certificate on your device, or a smart card. The Microsoft and Google Authenticator apps are a good example of this.

- **Type 3** is something you are. This authentication method is commonly known as biometrics, or something physical that is used to authenticate you. Some examples of this include your fingerprints, and facial or iris scans. Examples include Windows Hello and Apple's Face ID and Touch ID.

- **Type 4** is location-based. This authentication method can be extremely powerful with the advancement of cloud computing and AI technologies. Authenticating based on location allows you to authenticate users based on a defined geolocation. Think of a company that only conducts business within the United States (**US**). Why would they expect anyone to access their servers from outside the country? Anyone accessing them from outside the US is expected to be an intruder and should be blocked. In Azure, Conditional Access can provide this level of authentication. This is a very simplified example and threat actors can easily bypass this control by setting up infrastructure in the US. But this same principle applies to only allowing access if you are in an office or on a VPN for added protection.

Ensure you aren't using legacy authentication such as basic authentication. Make sure you are using modern authentication protocols such as **Security Assertion Markup Language (SAML)**.

In addition to the different methods of authentication, you will want to ensure that risk-based authentication is implemented within your organization. This in some way could be considered Type 5 authentication, although it is being used more as a defensive approach to prevent authorization with any detected anomalies. We will cover this more in the *Enhanced Identity Security* section further in the chapter. Next, let's look at authorizing your identity against the system to gain access to data.

Authorization

Once you have been authenticated into your environment or system, you will need to access data or systems that you have been authorized to access. Authorization is a method in which permissions have been added to your identity to allow access to some information or system. Just because you have been authenticated doesn't mean that you are authorized to access data or a system. This is where the **Principle of Least Privilege (PoLP)** comes into play, and as a best practice, there should be no authorization added to a new identity by default. Authorization isn't an easy task, especially as an organization grows. Ensuring that someone only has access to what they are authorized to do can be a full-time role, especially ensuring that authorization is updated and removed as roles change. To better help with authorization and to define your users' job functions, a well-defined access control model should be in place to implement **Role-Based Access Control (RBAC)**, using identity management tools and centralized management of users and groups that use a cloud identity store, such as *Microsoft Entra ID* or a traditional identity store, like on-premises **Active Directory (AD)**.

Now that the identity has been authenticated and authorized to your system, let's look at what accountability is in terms of actions and data accessed inside an environment.

Accountability

The last function is known as accountability, also referred to as accounting. This is the function that allows us to ensure that an identity or user is not abusing their rights or the authorization provided. Here, we need to ensure that we track all the activities of our users and ensure auditing is in place to hold anyone accountable for any misuse of their identity and access.

This is also a critical process as part of any investigative and forensic work that is carried out for cybersecurity incidents or compromised accounts. Knowing exactly where an account has been authenticated and what data it has accessed must be captured and audited for historical purposes.

As you can see, the full scope of an identity and access management program takes great effort and time to implement correctly. With the identity of your users being at the forefront and serving as the key to your data and systems, organizations must implement a well-rounded and robust identity and access management program. As we discussed earlier in this book, ensure your policies and standards have leadership support to ensure your program is effective.

Next, we will review account and access management and what is involved in the identity management lifecycle.

Modernizing Your Identity Architecture

To take advantage of the latest capabilities and security features for identity management, you are going to need to modernize your identity architecture. This won't be a simple task but it will be important that you understand the current state of your current identity architecture so you can build a plan to ensure you are allowing for the modernization of your identity architecture. This will allow you to support more modern capabilities being made available by **Software-as-a-Service (SaaS)** providers and modern applications. In turn, this will allow for the best protection against the continued advancement of threat actors.

Identity and Access Management Statistics

To better understand the current state of identity and access management, let's first take a look at some statistics that support the need to ensure you are modernizing your identity architecture to best protect your users. The *Identity Defined Security Alliance* along with *Dimensional Research* provided the following data from their *2023 Trends in Securing Digital Identities* whitepaper of over 500 participants working in organizations with more than 1,000 employees:

- Managing and securing digital identities is the number 1 priority for 17% of businesses and a top 3 priority for 44%.
- In the previous year, identity-related incidents occurred from (all that applied):
    - Phishing (62%)
    - Brute force attacks (31%)
    - Socially engineered passwords (30%)

- Compromised privileged identity (28%)

- Stolen credentials (28%)

- Third-party attacks (27%)

- Insider attacks (22%)

- Man-in-the-middle attacks (18%)

- No incident to report (10%)

- The following direct impact on business results was reported because of an identity-related incident in the previous year (all that applied):

  - Costs to recover from breach (39%)

  - Significant distraction form core business (33%)

  - Negative impact on reputation (25%)

  - Loss of revenue (21%)

  - Customer attrition (20%)

  - Lawsuit or other legal action (17%)

  - No impact (33%)

Source: `https://www.idsalliance.org/white-paper/2023-trends-in-securing-digital-identities/`

In addition, the 2023 *Verizon DBIR* report (`https://www.verizon.com/business/resources/reports/dbir/`) referenced in *Chapter 1, Current State*, provided some insightful data related to identities:

- Credentials were involved in 49% of breaches.

- The human element is included in 74% of breaches. This is through either human error, privilege misuse, stolen credentials, or social engineering.

- Stolen credentials, phishing, and exploitation of vulnerabilities are the primary entry points for attackers.

As you can see from the statistics referenced above, identity and access management plays a pivotal role in protecting organizations from compromise. More concerning is that the *2023 Trends in Securing Digital Identities* whitepaper referenced above found that 90% of its participants suffered an identity-related cybersecurity incident.

Although modernizing your identity architecture is not going to eliminate a cybersecurity event from occurring with your identities, it will allow you to significantly decrease risk and improve your resiliency, as escalation of privileges and abuse of authorization will drastically reduce by taking advantage of more modern and improved capabilities.

Legacy Identity and Access Management

When we think of legacy in terms of identity and access management, this comprises older on-premises directory stores, older protocols, manual processes, and a lack of support for modern capabilities, such as **Single Sign-On (SSO)**, to name a few. Running a legacy identity architecture can become very complex with the need for many different solutions, multiple integrations, and significant infrastructure requirements, such as the need for many on-premises servers, and so on.

To consider a few examples, the following is a list of commonalities when looking at legacy identity and access management architecture:

- Traditional on-premises identity stores such as AD
- On-premises server-based identity management tools
- Legacy protocols such as basic authentication, such as NTLM, LDAP, CHAP, EAP, RADIUS, Kerberos, and so on
- Allowing username and password-only access
- Limited SSO capabilities
- Leveraging old password types of requirements (simple and only eight-character length)
- Complex federation
- The need for dedicated user accounts for external entities
- Limited security features such as lack of user risk or behavioral capabilities
- Antiquated HR systems
- Manually provisioning/deprovisioning to **Identity Management (IdM)** systems
- The use of service and shared accounts

With a legacy identity and access management architecture in place, you are going to be lacking the capabilities that a modern identity and access management architecture provides to more efficiently protect your identities.

Based on the previous list that touches upon some of those legacy capabilities, the following list provides some of the challenges and lack of capabilities you can expect with a legacy identity and access management architecture:

- Limited integration with your external partners.
- Lack of next-generation protection capabilities such as passwordless.
- Limited identity protection capabilities.
- Decentralized access controls across the enterprise.
- Very complex environment to manage, maintain, operate, and keep current.
- Skill sets will be harder to hire the longer you keep legacy infrastructure.
- Vendors will slowly stop supporting older technologies (end of life).
- Increased costs to maintain legacy infrastructure.
- The need to build custom code to meet business requirements.
- Limited or complex integration with cloud environments.
- Complex business continuity and highly available models.
- Modern apps and SaaS environments may not support legacy identity models.
- Inefficient account life cycle management.

With that, let's shift our focus over to looking at the need to modernize your identity and access management architecture and what is required to modernize your identities.

The Need to Modernize

To allow your business to maintain a competitive advantage and remain current, the need for a modern identity architecture is a necessity. By shifting to a modernized identity and access management model, your organization's capabilities and user experience will increase significantly. The following capabilities and technologies will allow your organization to transform to the needed identity and access management architecture to meet your business needs and support the ongoing need for your digital transformation journey:

- Supports zero-trust concepts and architecture.
- Modernized cloud-based HR systems.
- Cloud-based identity stores such as Microsoft Entra ID.
- Full integration and automation between HR systems and identity stores.
- Leverages modern authentication to support the latest authentication capabilities, such as the following:

- **Multi-Factor Authentication (MFA)** including phish-resistant methods
- Passwordless
- Biometrics
- Software (authenticator app) and hardware-based tokens
- Certificates
- FIDO2

- Advanced identity protection capabilities that leverage behavioral capabilities, ML, and AI.
- Modern cloud-based capabilities that support SSO, **System for Cross-Domain Identity Management (SCIM)**, SAML, OpenID Connect, **Web Services Federation (WS-Fed)**, and OAuth.
- Well-defined RBAC is in place and being enforced.
- Advanced federation capabilities with modern B2B and B2C capabilities
- Privilege access and least privilege capabilities using Privilege access and least privilege capabilities using PRivileged Identity Management (PIM), Privileged Access Management (PAM), Just-In-Time(JIT) Access, and Just Enough Access (JEA) are being enforced. are being enforced.
- End-user self-service capabilities.
- Supports conditional access policy capabilities.

As you plan your journey to modernized architecture, you need to remember that this will not happen overnight. It will not even happen within days or weeks. This type of transformation is going to take months to years to transition through. Your legacy identity infrastructures include years and years of compounded systems and capabilities that will be intertwined with endless on-premises infrastructure and applications. Another dependency with moving to a modern identity model is that of legacy applications. There will be old applications that are being used by the business that have been custom-developed to use your legacy infrastructure. Modernizing or re-wiring these applications will not be easy and every one of them will be unique.

Identity and Access Management Modernization Strategy

To modernize your identity and access management program, you are going to need a strategy behind your modernization along with a roadmap of how you plan to accomplish this journey. Based on what we covered in previous chapters, the **Zero Trust Architecture (ZTA)** strategy is a perfect place to begin when building your journey to a modern identity architecture.

You don't want to be onboarding multiple different vendors that will create more complexity. First, you'll need to build the path needed to get from where you are today to that final modern architecture. To do this, the following is an example that references both the Microsoft and CISA ZTA maturity models that we covered in *Chapter 5, Cybersecurity Architecture*.

Figure 6.3: Identity and access management ZTA maturity strategy

> You will also want to consider your strategy as it relates to your product portfolio for your identity and access management needs. You'll want to simplify and keep the portfolio to a minimum while focusing on one primary vendor for your core identity needs.

Once you define your strategy, you are going to need to execute against the strategy to ensure you achieve your end goal, which is where your roadmap will come into play.

Identity and Access Management Modernization Roadmap

Next, you'll want to build a more tactical roadmap, as demonstrated in *Chapter 5, Cybersecurity Architecture*. This roadmap will capture where you are today with the current identity and access management infrastructure and capabilities and where you'd like to be in the future with transformation to a modern architecture. The following is a more detailed example of how your roadmap may look if you use Microsoft Entra ID:

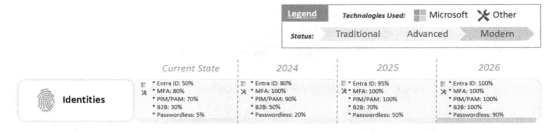

Figure 6.4: Identity and access management ZTA tactical maturity roadmap

Obviously, these are higher-level tasks that will each become individual projects to carry out the transformation required to get to the end state. As already stated, these activities will take time and will not come easily. You will need to ensure you have the right resources to support the transition, plan carefully, and execute with precision. Identity-related services can be very impactful on a user's day-to-day activities so it will be important that the user experience is fully understood, and thorough change management and communication activities are also occurring. The good news is once you make it through the transition, the user experience will significantly transform if executed correctly. Next, let's take a closer look at account and access management activities.

Account and Access Management

To ensure success with your account and access management program, it must start with well-defined policies that enforce the correct standards and procedures throughout the life cycle process. Any mismanagement of this process can result in significant damage to your organization due to the critical nature of identity in cybersecurity. Some examples of this include not applying the **PoLP** to a user account, untimely terminations, or not managing and auditing your privileged accounts. One method to help with any mismanagement or manual errors is implementing automation wherever possible. The more you can automate, the fewer errors will occur. There will always be a need for manual processes, whether it be the initial input of user data, physical validation that the information is correct, or applying an end date to a user, but the more you can reduce these processes, the less error prone they will be.

Every organization is structured differently, but for the most part, you will likely have accounts categorized as **Full-Time Employees (FTEs)** or associates, contractors, vendors, guests, or service accounts and service principals. For each of these categories, you will need separate policies and procedures for their management. Your FTE accounts may be managed differently from your contractor accounts, as well as your vendor and guest accounts. Ensuring all these accounts are managed correctly will require close collaboration between HR, the cybersecurity team (including the identity management team), and the hiring or reporting managers of these accounts.

Never use shared accounts within your environment. All users who access your environment should have an account assigned that identifies them. Without an assigned identity, individual accountability is difficult. If you do have shared accounts, work to remove them immediately.

Identity Life Cycle Process

One of the most important tasks with your identity and access management program is the management and life cycle of accounts and the auditing of the access they have. The whole life cycle process in account and access management may involve multiple teams to make the process efficient and successful. Likely, there's a chance that multiple systems and tools are involved in life cycle management, including manual human processes that increase vulnerability due to poor housekeeping and being error prone. Account and access management is a complex process and only becomes more challenging through the ongoing expansion of the application portfolio, as well as the shift to the cloud. A typical account and access management program may involve resources from HR, the cybersecurity team (including the identity and access management team), technical operations teams, hiring managers, and potentially others. You must familiarize yourself with your identity life cycle process so you can identify any potential gaps to minimize risk and challenge the need for a modernized identity life cycle process if old legacy methods are in place.

Let's look at a typical account and access life cycle scenario that starts with HR as the source of authority for identity. Since integration between HR systems and a cloud-based **Identity Provider (IdP)** is relatively new, there's a high possibility you are still operating a legacy account life cycle process, which will look somewhat like the following:

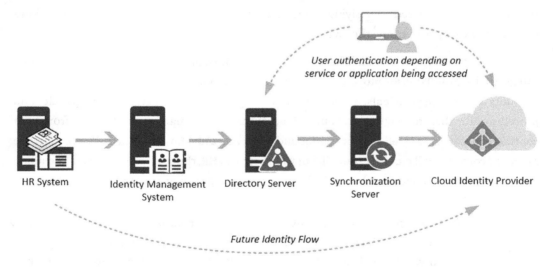

Figure 6.5: Identity account life cycle

As noted in the diagram above, the ideal future architecture will be to provision directly into the cloud IdP from the HR system, thus eliminating a lot of complexity.

As HR system integration with the cloud becomes more accessible and adopted, this architecture will significantly simplify. In the meantime, let's review each of the phases in the identity life cycle process.

HR and Identity Management

Your identity management life cycle needs to start somewhere, and that is most likely with your HR department. This is where your employees start their digital journey within your organization. Once an employee has accepted a position, their digital profile will be created. HR software is very specialized and is typically independent of the core IT identity services. Within an HR system, some of the items you can expect to manage are your personal information, time off, payroll, performance, the employees you manage, and employee training. One of the primary challenges with your HR system is efficiently integrating the application into your core identity services.

 There are many HR platforms available on the market today. Some examples include SAP SuccessFactors, Workday Human Resource Management, Oracle PeopleSoft, and Microsoft Dynamics 365 Human Resources.

Today, when you look at your core identity service, it's highly probable that it will consist of an on-premises identity store, such as **AD**, that's likely synchronized to a cloud identity store, like Microsoft Entra ID, for example. This is known as a hybrid identity model. One of the challenges with the traditional deployment using a legacy on-premises identity store is the limitations on delivering a robust and modernized identity service beyond basic account management. Having the ability for employees to request contractor accounts, manage their own directory groups, update profile information, use a self-service portal, manage the account life cycle process, and build automation are not native features of legacy directory services. This is where an identity management solution will typically sit between your HR source and on-premises identity store to supply the features needed to provide the additional capabilities.

One solution that has been around for a while is **Microsoft Identity Manager (MIM)**, which is the latest edition of **Forefront Identity Manager (FIM)**. MIM is an extremely powerful management tool and serves as a critical component of the overall identity life cycle.

MIM deployments can be highly complex, depending on the level of customization needed. Some other alternatives to MIM for traditional identity management with an on-premises identity store include solutions from *IBM, SailPoint, One Identity, and OKTA*.

 The HR system will contain both **Personally Identifiable Information (PII)** and sensitive PII, so you must work closely with the HR team to ensure that data is properly secured and encrypted both in transit and at rest.

As cloud adoption increases, an ideal scenario is to directly integrate your HR system with a cloud IdP, such as Microsoft Entra ID. For example, Workday and SAP SuccessFactors currently support directly provisioning into Microsoft Entra ID. However, you will need to check the prerequisites from your HR system, and you'll need to be using their latest SaaS offerings to support integration. Unfortunately, this model is still in its infancy and is going to require a major effort to make this transition. There will be many complexities with on-premises dependencies that will include legacy on-premises applications, traditional file shares, legacy authentication, and so on. Although migration is possible and should be part of your current roadmap, the shift isn't going to happen overnight and will need a lot of preparation with the right resources to support it.

Next, we will review the core identity store used to support your identity life cycle.

Directory Services

In the previous section, we reviewed the beginning of the identity life cycle with the HR system as the truth source, which then feeds its data into, or is integrated with, some form of identity management system. This brings us to the next step in the life cycle, which is that of your identity store. Your identity store, or directory service, can be thought of as Microsoft AD. AD is an on-premises hierarchical directory that stores objects, such as user accounts, passwords, user information, computer objects, and security groups. This is where the user objects will be active and enabled for accessing the IT systems within your environment.

Referring back to the overall identity life cycle, once a record has been established in your identity management solution (MIM as an example) based on the HR feed, it will then provision an account in your directory service, or AD, as already referenced for the user being onboarded. The object that has been created in the identity management solution isn't the account that the user will use but is typically set up as the authoritative source of the AD object. What this means is that if the object within the identity management solution is set to be termed because the HR system sent a term instruction, the AD object will be disabled.

Depending on your configuration, re-enabling the account directly within AD will eventually revert it to a disabled state, since the identity management solution is authoritative. This is extremely powerful and exactly how the process should work by providing HR with the authority to have accounts created for new employees and have them termed as users leave. It also helps to prevent anyone with access to AD from creating and enabling accounts that they shouldn't. All requests should filter through the identity management solution, for better control and accountability, as it will serve as the centralized place for all identity and access requests.

> Protecting your active identities is critical, and your identity store needs to be at the core of this protection. If you use Microsoft AD, the following are best practices for securing AD: `https://learn.microsoft.com/en-us/windows-server/identity/ad-ds/plan/security-best-practices/best-practices-for-securing-active-directory`. We go into more detail on protecting your identities in the *Securing Your Identities* section later in the chapter.

For on-premises identities to work in a cloud directory, a copy will need to be synchronized to support the identity and authentication requirements of the modern world. To support this, a synchronization tool will be needed between your on-premises directory and the cloud provider. This will enable what's known as the hybrid identity for your organization.

Hybrid Identities

How you implement your hybrid identity strategy will depend on your cloud provider. Regardless of which cloud provider you use, you will want to ensure there is an authoritative source that allows the synchronization of user details, including passwords, so management isn't required with two separate environments to update account properties. As you work in a hybrid environment, you'll need to consider that users will need to continue to authenticate against the traditional directory services for applications that are still housed within a traditional on-premises environment and applications that have been modernized and running in the cloud will authenticate using the cloud IdP. The end goal will be to fully move all applications to a cloud-based model where your authentication will all occur from the cloud IdP. This, unfortunately, will not happen overnight and will potentially take many years for organizations to transition.

> Depending on your cloud provider, there may be a way to federate authentication back to legacy on-premises applications. This may get complex depending on how old the applications are and which authentication protocols are used.

Referencing back to the Microsoft architecture, this can be accomplished using Microsoft's Entra Connect, which provides synchronization for all or any selected objects, **Organizational Units (OUs)**, and user attributes from your AD into Azure AD. With Entra Connect, you can provide a single identity for your user to access both on-premises resources as well as cloud-based resources. Entra Connect supports the synchronization of users, user attributes, security groups, and other on-premises objects. Other cloud providers like AWS and GCP provide similar capabilities to connect to each of their respective clouds:

- *AWS Directory Service*: https://docs.aws.amazon.com/directoryservice/latest/admin-guide/what_is.html
- *Google Cloud Directory Sync*: https://support.google.com/a/answer/106368?hl=en

It will be critical that security best practices are applied to the hybrid model approach no matter which cloud environment you use hybrid identities with. If you work with multiple cloud providers, your scope for securing your identity will increase significantly along with the complexity of your identity environment.

As briefly mentioned previously, one important decision that will need to be made is around your password management between your on-premises directory services and cloud IdP. For the most part, you can expect two supported models for your hybrid identities: cloud authentication and federated authentication. While there are many considerations when choosing the authentication model, in cloud authentication, the cloud provider will handle the sign-in, whereas, with the federated model, the sign-in is passed back to the on-premises directory services or some form of federation services intermediary. Your choice of which you proceed with will depend on your environment, the cloud provider you are using, and what support capabilities are available. My recommendation would be to leverage the cloud authentication model where you will be able to begin taking advantage of more advanced cloud identity protection capabilities.

With Entra Connect, you are presented with a couple of options when using cloud authentication as the primary method for authentication. The same principles should apply to other cloud providers in which you will need to confirm the options available. The options available with Entra Connect are to use pass-through authentication to on-premises so that users can use the same password both in the cloud and on-premises environment or to synchronization the password hash (known as password hash synchronization) to the cloud IdP. Let's review the difference between pass-through authentication and password hash synchronization.

With Microsoft's version of password hash sync, the actual passwords of your user accounts are not synchronized. For optimal security of your user passwords, a hash of the password with a per-user salt is synchronized.

- **Pass-through authentication** validates passwords between on-premises and the cloud. If you wish to enforce on-premises security and password policies and do not want passwords to be stored within the cloud IdP authentication service, choose pass-through authentication.

- **Password hash synchronization** synchronizes the stored on-premises password hash to the cloud IdP authentication service. Authentication takes place against the cloud IdP, not against the on-premises directory services. Whenever a password is changed on-premises, the password is synced (in a timely manner) to the cloud IdP. If password writeback is configured, then a user can change the password from the cloud IdP and it syncs back to the on-premises directory service. Some on-premises password policies may still apply to the cloud identity if configured for password hash sync depending on which cloud IdP you integrate with. For example, with Entra Connect, an on-premises password complexity policy will override any policies configured in Microsoft Entra ID. Additionally, any on-premises expiration policy will not affect the cloud identity, as the default configuration is set to never expire. If a user's password expires on-premises, cloud services will not be interrupted. It will be important you know how the password policies apply within your cloud environments.

Again, if multiple cloud environments are being used within your organization, the hybrid identity model will quickly become complex and challenging to manage efficiently. You will need to ensure you have the right resources in place to architect and engineer the most efficient and secure hybrid model for your users.

Cloud Identities

Your cloud IdP is essentially the next generation of identity management and security for your users; you can think of it as a modern directory service. The following are the identity management services available from each of the three larger cloud providers:

- *Microsoft Entra ID*: `https://www.microsoft.com/en-us/security/business/identity-access/microsoft-entra-id`

- *AWS Identity Services*: `https://aws.amazon.com/identity/`

- *Google Cloud Identity*: `https://cloud.google.com/identity?hl=en`

For example, if you adopt Microsoft Entra ID, the cloud resources that a user will log into using their cloud identity include Exchange Online, OneDrive for Business, SharePoint Online, Windows (if Entra ID Joined), and so on. The user will not know the difference between the directory services account and the cloud account, which is optimal for the user experience.

Once the identity is synchronized to the cloud, this completes the life cycle of a user's identity from the system of record. While there may be variances in the products between vendors, the life cycle framework and policy provided can be applied using any identity service or vendor. To recap, the following workflow should be referenced to process your identities:

HR > identity management system > traditional directory service > identity synchronization > modern directory services

It's also critical to ensure the off-boarding process works correctly throughout this life cycle. If properly configured, when HR initiates a termination, the account should be disabled throughout all directories. Any failures in this process can allow a user to access resources after they have left the organization, which can cause a serious security risk.

Make sure you are fully aware of synchronization times within each of your identity systems. Terminating a user within HR can take time to synchronize downstream through your systems. For immediate action, you may need to manually intervene to ensure identities are correctly disabled in a timely fashion.

The account life cycle can be very complex, and mismanagement of this process can easily create vulnerabilities. Some best practices with this architecture include working closely with your HR team, ensuring all traffic, feeds, and integrations are encrypted, minimizing the number of privileged identities, and enforcing MFA on all accounts. Be sure to enable auditing to hold accountability and add safeguards to ensure the identity management portal is secure and not accessible over the internet if possible.

Group Management and RBAC

In addition to your identity life cycle is your group management strategy. Within a traditional on-premises directory, you will have most likely managed access to the environment and applications using security groups. Although this significantly simplified the access model by eliminating the need to add individual users to every location or application in which access is needed, the concept of group sprawl has occurred over the years where you are most likely dealing with hundreds, if not thousands, of groups within your environment.

Where this becomes challenging is with needed governance, more specifically around auditing and ensuring everyone has access to the correct groups and users don't have access to locations or applications they shouldn't. Groups aren't going to go away any time soon so it's critical that a well-defined governance model is in place with a policy and process that ensures groups are being managed, reviewed, and audited often for compliance.

 Make sure users are removed from groups as they change roles so their accounts aren't over-provisioned with more permissions than they need to do their jobs. Ideally, having this automated as part of RBAC will significantly help. The reality is you are going to need good processes in place to ensure a thorough audit is complete regularly with user permissions.

A more efficient approach that should be considered if not already being used is that of **RBAC**. Essentially, RBAC provides the user access based on their role within the organization. Roles are built and provided the access and permissions needed for a user to complete their job responsibilities. This significantly simplifies the model by only needing to apply a role to a user versus needing to apply to different groups. This also simplifies the process when users move into a different role within the organization. Once the updated role is applied to the new user, the old access and permissions will be removed while receiving the newly applied access and permissions. However, you may need to manage cross-function roles as users transition into new roles or are needed to support other areas of the business. Essentially, a user may have multiple roles assigned in certain circumstances. Implementing an RBAC model will not be easy; you will need to work with your HR teams to align roles throughout the organization with the permissions they need to accomplish their job. In a large organization, this will require time and planning. Once in place, your access model will be in a much better place.

RBAC doesn't just apply to your users based on their role. RBAC is also applied to your privileged users and those who administer your environment. The cloud environment you operate out of will most likely have RBAC roles for all administrative activity. For example, Microsoft has a magnitude of roles across all its cloud services. RBAC roles are available with Exchange Online, Teams, SharePoint, Entra ID, Security and Compliance, and more. It will be critical that you leverage RBAC as part of your cloud management model to simplify the governance of access within the cloud. In addition, these roles should feed into your privileged management model that allows users to request access to these roles as needed, and not permanently. If you aren't using RBAC for your privileged users, prioritize this today to simplify your privileged access model.

Another access control model is that of **Attribute-Based Access Control** (ABAC). This model is like RBAC but instead of applying roles, attributes are used to apply the access and permissions needed.

Next, let's take a look at service accounts to better understand what they are used for and the risk that comes with them.

Service Accounts

Over the years, there has been a need for service accounts to be used to provide access to applications and databases, for automation to be enabled, scripts to be configured, and more. In some situations, there is a possibility that users leveraged their own accounts for such setups. Although there haven't been many options available for most configurations other than using a service account, especially with traditional on-premises directory services deployment, there is a need to modernize and move away from password-based service accounts. Even more so, there shouldn't be any user accounts set up as a service account. Another major concern with service accounts over the years is the over-provisioning of access. Providing a service account with more access than needed can be catastrophic when a breach occurs, especially if these accounts have been provided admin-level access. You must ensure service accounts are only provided with the access they need.

Looking to modernize these types of accounts is important as the use of username and password-based accounts is simply not secure anymore. They can easily become compromised with a threat actor being able to access all data that a service account has access to. In many instances, service accounts will have access to a lot more information than needed. Moving beyond password-based service accounts, you need to look at capabilities that allow the use of certificate-based authentication. Another preferred method over a traditional service account would be that of secret keys that are stored centrally and managed in some form of a vault where the secret can be stored and rotated securely. In some of the cloud providers, these may be better known as service principal types of accounts.

 One major concern with a password or secret key type of account is the use of it with bad development practices when the password or secret key is viewable in plaintext within the production code. The same applies to **Application Programming Interfaces (APIs)**. This ties back into the DevSecOps process and ensures developers have a security-first mindset.

Another type of account that you may be familiar with is that of a bot account.

Although you may be familiar with the term from social media and fake accounts trying to manipulate others, bot accounts are also used in the enterprise essentially for the same purpose but for legitimate purposes. Think of **Robotic Process Automation (RPA)** where bot accounts are set up to manipulate and automate user activities. In essence, these are, in some aspect, a service account that has been automated to run a repetitive process. Like service accounts, these accounts will have been provisioned access to data to complete their actions, so a lot of risk comes with managing and protecting these accounts.

External Access

Most, if not all, reading this will have external users accessing your environment. Whether it be a customer, vendor, partner, or any other external entity, there will need to be a model and process in place to provide the needed access to your environment. When looking at a traditional model with an on-premises directory service accessing traditional on-premises applications, providing access to external partners was far from an easy task. For most organizations, setting up an identity within their local directory service was probably standard practice. This meant the need for local accounts to be set up, managed, and governed within your environment, the same as with your employee and contractor accounts. This is far from an ideal model with a lot of risk involved. If your organization was better staffed with the right resources, you may have set up some form of federated application or service to support the ability to allow an external party to use their own identity from the connecting organization. The issue with this is the involvement, complexity, and effort to set up the federation between both parties. The more external vendors you work with, the more integration you will need to set up and systems to manage.

As we continue to shift to modernized cloud models with more adoption of SaaS-based services, new and more efficient ways are needed to provide access to our services along with the governance of these identities. The most current and most efficient capabilities available are **Business-to-Business (B2B)** identity capabilities. These capabilities are a game-changer within the enterprise identity space. In short, the B2B model is more focused on a vendor or partner type of relationship that may require access to your internal resources or the need to interact and collaborate more efficiently.

 There is also the concept of **Business-to-Consumer (B2C)** identity capabilities, which is intended for consumer type of users for commerce types of interaction that are separate from your organization's IdP. Typically, this will allow users to leverage their personal email addresses or social media type of accounts as an option to authenticate.

The B2B model greatly simplifies this process by allowing you to invite a guest user into your environment and provide them with access to your applications and services. Once invited into your organization, the guest user then authenticates using their current organization's work identity to access approved applications hosted in your environment. There is no need to provision new user accounts or identities within your environment with this model. The best part of this process is that the user's identity is managed and maintained by their hosting organization, which is a significant improvement from an account management perspective. Although housekeeping will be required to clean up stale accounts, once the external user is disabled within their hosting environment, they can no longer access your environment using that identity, which is a major advantage. B2B models will typically support collaboration with organizations through external identity partners, such as Facebook, Google, Microsoft accounts, and other SAML/WS-Fed IdPs. Another benefit of this model is the ability to apply the same access policies as you do internally with the likes of MFA, control policies, life cycle capabilities, and so on.

Privileged Access

Privileged access is any elevated access provided to a standard user account. This can be viewed in two separate risk buckets. The first is that of your technical staff and developers who have elevated access to manage and operate the environment. For example, a domain admin or a SQL admin. The second bucket can be thought of as your users' elevated access needed to accomplish their responsibilities. An example could be someone in the HR department who has access to all users' sensitive PII, or a user in the financial department who has access to all account and routing information for the organization. Either way, both pose a significant risk to the organization.

Unfortunately, privilege access continues to be abused in both instances whether a threat actor gains access to a privileged account on the technical team or insider threats where an employee takes advantage of the elevated access they have been provided to complete their responsibilities. There is a lot involved in managing privileged access and gone are the days of over-provisioning and providing more access than needed to users. The same applies to domain admin-level or cloud admin-level access; the accounts with the highest level of privilege need to be managed very carefully with minimum users having access to this privilege.

As part of privileged management within your environment, the following will need to be considered:

- Local administration access with capabilities such as **Local Admin Password Solution (LAPS)**

- **Privileged Access Management (PAM)**, **Privileged Identity Management (PIM)**, and **Just-in-Time (JIT) capabilities**
- **PoLP** and **Just-Enough Access (JEA)**
- **Separation of Duties (SoD)**
- Insider threats

We will review each of these in more detail within the *Securing Your Identity* section.

Governance, Reporting, and Auditing

A very important function of your account and access management administration is that of governance, reporting, and auditing. As much as it would be great to have everything automated, the reality is that the human element needs to be involved to oversee the overall process and execute additional due diligence to reduce risk as much as possible. In addition, there will be a need to build, review, publish, and continually update policies and procedures for account and access management. There will also be many different teams involved in the identity life cycle that will require close collaboration to ensure efficiency. You can expect close collaboration with the HR and IT teams at a minimum.

As you can see, there is a lot of complexity involved with account and access management, which only makes operating the identity life cycle more challenging. Ensuring your users' accounts are working as intended and they have the correct permissions to execute their daily functions requires ongoing support. As users are onboarded and offboarded, roles change, and new applications are onboarded, your identity operations will need to be involved to ensure the identities are being provisioned and de-provisioned correctly, that access is being applied correctly with new roles, and new access is made available as new applications are onboarded. All this operational work requires a team that fully understands the identity requirements and, at the same time, has a security-first mindset with the role they fulfill.

From a governance and risk perspective, you will need to pay close attention to the provisioning process. I recommend you set up some time to review the offboarding process in more detail. This process can quickly become very complex depending on the different scenarios when users are offboarded. The one that comes with a higher risk is that of users who are being terminated on the spot. You will need to ensure you are embedded in the process for these types of terminations so you can ensure that the termination process for these users is efficient. A typical timeline for an account to terminate could be hours or longer in some instances, depending on the timing of systems working through their workflows.

With a termination on the spot, access needs to be terminated immediately to prevent any initial reaction of malicious intent from a disgruntled employee. This process will need special attention to ensure all access is terminated immediately, including the use of their badge to get back into the building.

 It is also important to make sure that employees are aware that they should not be allowing anyone into the building who doesn't have an active badge, otherwise known as tailgating. This will prevent a potential disgruntled employee from gaining access to the building and causing harm. Tailgating should be included in your access management or cybersecurity policy and users should be trained on this.

A more modern approach to governing your identities is that of identity entitlement management. Entitlement management allows for a much more efficient management of your identity life cycle with automation capabilities. As your identity and access management footprint continues to grow, entitlement management provides capabilities that allow you to streamline the management of access to groups and applications for both internal and external users. Automated access can be provided based on specific properties a user may have on their account, and the same applies as those properties change. For example, a user who is part of a specific business unit with a tagged attribute on their account may receive access to specific groups. If the user moves to a different business unit, their account will be updated with a new tagged attribute on their account, which will, in turn, remove access to the old groups and apply any new groups needed. Also, look for capabilities that allow the expiration of access in situations with external users.

Finally, let's not forget auditing and reporting. You will need to conduct frequent audits within your organization of account and access management. The process of provisioning and de-provisioning users begins with the human element, and there will be mistakes because of this. More specifically, the offboarding process could be missed in certain situations. For example, a user gives their manager notice and departs the organization, but the manager fails to process the termination within the system although HR has already been informed causing the account to remain active once the user has departed the organization (ideally, the account would be terminated as part of HR process but, for this example, let's assume the process is different and the account deactivation was missed). Ensuring all accounts have been terminated that should have been will require ongoing auditing to reconcile HR data with the identity system data. The same will apply to access to applications and data throughout the organization; there will be a need to execute ongoing audits to confirm access isn't applied where it shouldn't be.

There will be many different areas that will need to be reviewed and audited frequently. In addition, if you have any audit requirements, such as SOC1 Type 2, for example, you will need to provide evidence of access management audit and reviews. There will also be an ongoing need for reporting to be provided throughout the organization as communications are needed, user access reviews are needed, reconciliation of user accounts is needed, and so on. For any reporting coming out of your identity systems, make sure they are encrypted and accessed securely only by those who need them.

Securing Your Identities

In this section, we will shift our focus toward best practices and recommendations for securing the identities within your environment. As already mentioned, your identities are the new perimeter of your network and the keys to all your data. As leaders, you need to commit the time and resources needed to ensure the identities within your environment are protected as best as possible.

If you do a quick Google search on "password statistics," you will be provided with an overwhelming amount of data on the challenges we face with protecting our users and their behaviors. Traditional authentication methods, specifically password-only authentication, are irrelevant these days. We have an enormous task ahead of us to continue pushing for the modernization of our authentication and to move away from the dependency on passwords. This effort also includes the awareness needed for our users to prevent them from continuing to use bad habits with their accounts and password usage. Once again, referencing back to the 2023 Verizon DBIR report, some statistics relevant to identity management include:

- 49% of all breaches reported involved the use of credentials.
- The primary method used as an entry point into organizations was through stolen credentials.
- 86% of breaches against basic web applications involved the use of stolen credentials.

As you look to protect your infrastructure from malicious intent, you will want to ensure you are implementing recommended baselines to protect your environment. Your internal infrastructure will be most vulnerable if a threat actor can get access. Once a threat actor gains access to your internal network, the damages can be catastrophic as exploitation of other users can quickly allow for lateral movement throughout the environment. Because of this, ensure you are referencing back to the CIS Controls that will provide the guidance needed to harden your server infrastructure, as well as directory services, for example, any **Domain Controllers** (**DCs**) you have in place. As a reminder, you can access the latest CIS benchmarks here: `https://www.cisecurity.org/cis-benchmarks/`.

The CIS benchmarks will apply to a lot of your identity and access management services so make sure your teams are familiar with the available benchmarks and taking advantage of them. You should at least be striving for CIS Level 1 with anything applicable within your environment.

Let's move into ways we can best protect the identities within our environment by first looking into directory services protection and best practices.

Directory Services Protection

For those of you who still have a traditional directory service infrastructure in an on-premises environment, which I imagine will be most, must have the right protection in place to reduce risk when a threat actor infiltrates your environment. Once a threat actor gains access to your directory services, they essentially have the keys to the kingdom for your traditional on-premises infrastructure, which will also provide a door into your cloud environment as you are most likely synching your directory services to your cloud IdP. Make sure you are following best practices with the directory services solution you have in place, as covered earlier in the *Directory Services* section.

In addition, CIS has AD-specific controls that can be applied to ensure your directory service is hardened per best practices. We have already covered CIS a couple of times. In addition to CIS, Microsoft also has its own Windows security baselines from the Microsoft **Security Compliance Toolkit (SCT)**, which provides recommended configurations to harden your Windows systems, including a DC. If you aren't using Microsoft, ensure you follow up with the vendor you are using to review any best practices they have available.

For access controls within a traditional directory services identity store, such as AD, Microsoft has traditionally recommended using a tiered model for privileged access. This model essentially separates your roles into tiers preventing lower-level accounts from being able to access higher-level accounts in the event of a compromise. However, this model has been replaced with the Microsoft enterprise access model, which incorporates the features from the tiered model. The enterprise model essentially covers the entire enterprise going beyond traditional infrastructure into the cloud. You can review more details on the model here: `https://learn.microsoft.com/en-us/security/privileged-access-workstations/overview`.

If you are an organization that does not use Microsoft technologies, the Microsoft enterprise access model can be applied generically to ensure you are following best practices to protect your identity infrastructure.

This model also emphasizes a focus on the privileged access strategy model, which we will briefly cover next.

Privileged Access Strategy

The privileged access strategy follows zero-trust principles to provide the best protection for your privileged accounts across all your identity sources throughout your environment. There are four initiatives to follow with the implementation of this strategy, as outlined here:

- End-to-end session security.
- Protect and monitor identity systems.
- Mitigate lateral traversal.
- Rapid threat response.

 More information on the privileged access strategy can be found in this article: https://learn.microsoft.com/en-us/security/privileged-access -workstations/privileged-access-strategy.

To help accomplish the privileged access strategy within your environment, Microsoft advises using the **Rapid Modernization Plan** (**RAMP**). RAMP is built on a roadmap that provides the steps needed to implement the recommended controls. The first section of the roadmap focuses on separate and managed privileged accounts, and is outlined here:

- Ensure you have emergency access accounts in the event of an emergency.
- Enable and implement privileged management capabilities.
- Identity and audit all privileged accounts and limit who needs privileged accounts.
- Ensure separate accounts for on-premises and cloud; ensure no mailbox on administrator accounts.
- Enable and configure identity protection capabilities.

The second section focuses on how to improve the credential management experience and is outlined here:

- Enable **Self-Service Password Reset** (**SSPR**) and enable the combined security information registration experience.
- Require MFA or allow passwordless on all privileged user accounts.
- Block legacy authentication protocols.

- Disable the ability for users to consent to applications.

- Enable identity protection capabilities and ensure notifications are set up to review and clean up alerts.

The last section covers administrator workstations' deployment and is outlined here:

- Set up dedicated workstations for privileged users to log in and use with their privileged accounts.

You can learn more about RAMP here: `https://learn.microsoft.com/en-us/security/privileged-access-workstations/security-rapid-modernization-plan`.

Setting up the preceding recommendations requires time and investment to implement correctly.

It is easy to fall back and apply elevated access to standard users and over-permission accounts to meet deadlines and so on. Don't fall into this trap! Ensure you spend time implementing these recommendations correctly from the ground up and clearly define the account provisioning process.

One consideration is ensuring a clean backup of your identity infrastructure in the event of ransomware. If threat actors compromise your environment and your backups because they aren't secure, you will have a catastrophe on your hands. Backups will be covered in more detail in *Chapter 7, Cybersecurity Operations*.

Next, let's look into password management for your environment.

Password Management

In today's world, the use of passwords is becoming obsolete. Due to major breaches and the advancement of technology, it's not unrealistic to assume passwords associated with our accounts (including old used passwords that can become seeds for different variations) have been leaked and are sitting on the dark web for sale. In addition, bad password management leads to the reuse of passwords on other accounts or with slight variations that are easy to crack. What does this mean? Essentially, someone with your password will be able to easily access all your accounts, especially if no additional security protections are in place. It's commonplace to see attackers use the same username and passwords within services that haven't been breached to try and gain access.

This type of attack is known as credential stuffing, and unfortunately, it is working!

The reality is passwords are going to be around for a while. Not ideal, but as security professionals, it is a risk we are going to need to continue to manage. Most importantly, it is a risk we need to continue to share and educate our users on as password hygiene for many needs improving. One of the challenges we face is that the original password recommendations we have all followed for many years are now obsolete, making our passwords more vulnerable because of these practices. Traditionally, password policy recommendations included changing them every 90 days, using 8 characters as a minimum, and ensuring complexity. With new research coming to light, so have new recommendations for password management. This is due to the advancement of cracking tools, password leaks on the dark web, predictable passwords due to frequent changes, and users writing them down due to complexity.

One reference I always like to share is the work by *Hive Systems* on brute-forcing passwords. The analysis is primarily based on MD5 hashing for their outputs with the hope they start seeing more use of stronger hashing with **PBKDF2** and **bcrypt,** in which their research will shift toward providing outputs on these hashes, which will change the outcome.

The following chart is one of their latest using the same hardware, **Graphics Processing Units (GPUs),** that is used for ChatGPT to brute-force passwords with MD5 hashes:

USING CHATGPT HARDWARE TO BRUTE FORCE YOUR PASSWORD IN 2023

Number of Characters	Numbers Only	Lowercase Letters	Upper and Lowercase Letters	Numbers, Upper and Lowercase Letters	Numbers, Upper and Lowercase Letters, Symbols
4	Instantly	Instantly	Instantly	Instantly	Instantly
5	Instantly	Instantly	Instantly	Instantly	Instantly
6	Instantly	Instantly	Instantly	Instantly	Instantly
7	Instantly	Instantly	Instantly	Instantly	Instantly
8	Instantly	Instantly	Instantly	Instantly	1 secs
9	Instantly	Instantly	4 secs	21 secs	1 mins
10	Instantly	Instantly	4 mins	22 mins	1 hours
11	Instantly	6 secs	3 hours	22 hours	4 days
12	Instantly	2 mins	7 days	2 months	8 months
13	Instantly	1 hours	12 months	10 years	47 years
14	Instantly	1 days	52 years	608 years	3k years
15	2 secs	4 weeks	2k years	37k years	232k years
16	15 secs	2 years	140k years	2m years	16m years
17	3 mins	56 years	7m years	144m years	1bn years
18	26 mins	1k years	378m years	8bn years	79bn years

HIVE SYSTEMS

› Learn how we made this table at **hivesystems.io/password**

Figure 6.6: Hive Systems Password Brute Force
source: https://www.hivesystems.com/blog/are-your-passwords-in-the-green-2023

Without going into too much detail on the above chart, it should serve as a guiding principle and educational tool as to how passwords should be created and managed.

Another powerful reference I like to provide awareness with is that of *Have I Been Pwned* (https://haveibeenpwned.com/) and the fact that over 14 billion accounts as of September 2024 have been identified as compromised from their discovery and research.

The risk we have ahead of us with password management is real. Based on experience, recommendations from others, and seeing frameworks/controls also changing their recommendations, the following should be implemented and recommended for your users:

- Consider using passphrases over passwords.
- Use a minimum of 15 characters (20 or more for privileged accounts).
- Remove the periodic requirements to change passwords and only change them in the event of account compromise.
- Ban common passwords through a centralized banned password list.
- Use a unique password for each account.
- Use a password manager (more in the next section).
- Go passwordless (more in the *Enhanced Identity Security* section).

An important capability to consider is that of password self-service for your users: the ability to allow your users to reset, change, or update their own passwords. Thus, taking away the capability from the service desk, who are prone to social engineering attacks that have been observed to have been successful with many breaches. If you look to enable these types of capabilities, make sure the correct validations are in place before allowing a password to be reset. Simply asking for some answers to questions is not secure. Additional controls will need to be in place to validate a user before resetting.

If you do require your service desk to reset passwords, make sure you have a well-defined and robust process in place. There must be enhanced validations in place to confirm a user calling in is who they claim to be. At the same time, there should be processes in place when resetting the password on behalf of a user. Spending time on hardening this process will be time well invested.

Password Vaulting

I purposely added password vaulting right after password management. If we are requiring our users to use longer, more complex passwords, they are going to need a tool to support the capability. Writing down your passwords on a notepad or storing them in Word or Excel is not the answer. There is no way any user can use best practices with their password management without the support of a tool. The tool to support this reequipment is that of a password vault, or password manager as you may be more familiar with. Obviously, there is a lot of controversy over specific types of password vaults, which has caused a lot of hesitancy in using them. More specifically, with cloud-based password vaults, and for good reason. A few of the more notable reasons are your passwords being stored in someone else's infrastructure in which you are reliant upon their security being strong. Another is the attention we have seen in the past with cloud-based password vaults becoming compromised, and the third is that they are all a common target for threat actors because of the amount of information that can be gained. However, there are alternatives and more secure methods of password vaulting that need to be considered.

I would categorize password vaulting into three different categories:

- **Cloud-Based (SaaS)**: As already mentioned, these are password vaults that are hosted within a vendor's environment in some form of a SaaS service most likely. This allows easy access from anywhere. Some examples include **LastPass**, **1Password**, and **Keeper**, but there are many more.

- **Self-Hosted**: With this model, you will deploy the vaulting infrastructure within your own environment. This allows you to implement your own security controls and significantly reduce the footprint from a public cloud-based (SaaS) model. Examples of vendors with this type of model include **BitWarden**, **Psosno**, and **Passbolt**.

- **Offline Password Vault**: Here, your password vaults will be fully offline and stored on your local computer or cloud storage such as OneDrive or Box to allow access across multiple devices. Some examples of these vaults include **Enpass**, **KeePass**, and **KeePassXC**.

Of the three options, the offline password vault provides the least risk with a significantly reduced footprint because vaults are distributed to each user and not stored centrally.

MFA

Today, MFA should not be an option but a requirement for every account that a user logs in with. According to the *Microsoft Digital Defense Report 2023* (https://www.microsoft.com/en-us/ security/security-insider/microsoft-digital-defense-report-2023), the risk of compromise can be reduced by 99.2% when using MFA according to a recent study with actual attack data from Microsoft Entra. Although MFA is not 100% resilient and there are ways that it is currently being compromised with SIM swapping type of capabilities, MFA fatigue type of tactics, and social engineering, enabling this capability will significantly reduce the risk of compromise for your users.

Some of the more common methods for MFA include:

Authentication Method	Recommendation
Security questions	Not Recommended
A call to your phone where you will need to press a #	Not Recommended
A text message to your phone with a verification code	Not Recommended
An email to your mailbox	Not Recommended
A verification code provided by a hardware token using Time-Based One-Time Password (TOTP)	Recommended
Security keys with biometric capabilities	Recommended
Mobile Application: TOTP code you enter on your phone	Recommended
Mobile Application: Push notification for you to click approve	Not Recommended
Mobile Application: Number matching by entering a number provided by the app	Recommended

Figure 6.7.png: MFA recommended methods

 Although several are listed as not recommended, if you are not able to implement the recommended options, having something in place is better than having no MFA. Just ensure you have a roadmap to attain the recommended methods.

If you are not requiring MFA for your users today, I would make this one of your top priorities and begin working on it straight away. If it is enabled, review the current requirements, and look for ways to continue to strengthen your MFA requirements as threat actors continue to evolve by compromising MFA. You may have heard the term phishing-resistant MFA. This is a newer concept that aims to prevent the current MFA bypass techniques as referenced above.

By using phishing-resistant MFA, you will be opting for a much more secure method of using MFA. We will cover this in more detail along with passwordless (which should be your ultimate goal) in the *Enhanced Identity Security* section.

Non-User Accounts

As we briefly covered in the *Account and Access Management* section, there is an ongoing need for non-user types of accounts, such as service accounts, service principal, bot accounts, and so on. The governance of these accounts is no easy task as you need to understand what the accounts have access to, the data type they are accessing, who is managing the accounts, password management, and so on. As the accounts are issued to the business, keeping track of them is very difficult. Even more concerning is that these accounts typically have privileged access within certain systems so if they get into the wrong hands, significant damage can be done. Some best practices to consider with service accounts, bot accounts, and/or service principals include:

- If possible, leverage certificate-based authentication. If not possible, use secret key-based service principals over standard username and password-based service accounts.

- Use one service account, bot account, or service principal for each use case; never reuse accounts or principals for multiple applications, databases, and so on. Do not allow the use of these accounts for anything other than the purpose they were set up for.

- Follow the best practice of least privilege, never apply more permissions than needed, and especially don't apply any admin-level permissions. Trust me, the request will come through many times.

- If you are using password-based service accounts, ensure long complex passwords are being used and they are being changed on a schedule. If possible, use vaulting capabilities with automation to manage service accounts.

- If using a service principal with secret keys, leverage a vault to manage and auto-rotate your secrets.

- Ensure no one is using a personal or admin account as a service or bot account.

- You will need to keep an active and up-to-date inventory of all service accounts and service principals, including what access they have and their purpose.

- Execute frequent audits on all service accounts and service principals to ensure they are still needed, or the same level of access is needed.

- Audit and monitor all sign-on and access activity for these accounts.

- Remove any service accounts or service principals that are no longer needed.

- Monitor expirations of passwords, secrets, and certificates to ensure no impact on services.

- Ensure SecDevOps best practices are being used by all developers to prevent plaintext passwords or secrets within any code.

Next, we will review SSO and the benefits of adopting SSO throughout your organization.

SSO

SSO is a technology that has been around for a while and is something you should already be familiar with. SSO provides the ability for your users to only have to use one identity to log into applications that support SSO with your IdP. Most modern IdPs will support current methods, such as SAML 2.0, to authenticate with applications. As more applications are being onboarded, more specifically, SaaS-based applications, you are going to need to integrate SSO with each of these applications and SaaS vendors.

Having SSO in your environment provides significant improvement from both a security and user experience perspective. Without SSO, users would need to manage usernames and passwords for every unique application they access. As we've already discussed, password hygiene is far from optimal. In addition to passwords, there'd be a requirement to have to configure MFA for every application you need users to access. With SSO, users will have only one password and MFA will only need to be configured once with your primary IdP.

From an identity management perspective, the governance around managing tens to hundreds of application user identities would not be realistic and would take an army to efficiently manage and govern. Think of the auditing requirements and the process that would be needed to onboard, offboard, and apply the correct access to all these applications individually within the application. With an SSO solution, those tens to hundreds of applications now become one centralized identity store with consistent configuration and requirements. There will still be a need to potentially manage roles within the application locally, depending on provisioning capabilities and how much can be automated. Every application will be different.

Since we touched upon provisioning, an additional capability that should be considered in addition to your SSO integration is SCIM. This is a standard for provisioning users and groups directly into applications from your IdP, which will allow for centralized management of your user provisioning into the application along with group management. This will also significantly help reduce risk as you can manage both the users and groups centrally more efficiently within your IdP.

SSO also comes with its own risk when using one identity to access many applications. Therefore, it is important to ensure strong access and management policies are in place with your SSO setup and configuration. It will be important that you have firm compliance in place with your SSO configuration, specifically enforcing MFA and even leveraging advanced capabilities that require stronger authentication for certain apps depending on their classification. Make sure you fully understand the capabilities of your IdP when it comes to SSO and the capabilities that can be used to reduce risk as much as possible.

Privileged Accounts

As we already discussed in the *Account and Access Management* section, privileged access is any elevated access needed beyond a basic user account setup and is typically needed to manage the infrastructure or support business applications and services. These accounts are what threat actors look for once they are in your environment. The more access they can gain, the more damage they can do, and the more data they can exfiltrate. For this reason, privileged access must be managed and secured with the strictest controls available. Let's review ways to help better protect these accounts.

Local Administrative Access

Local administrative access can occur in many different places within your environment. The first is that of your end-user devices and users who request local admin access on their local devices. Today, there should be no users with local administrative access on their devices. This creates a significant risk within the environment and there should be processes in place that allow users to complete their responsibilities without the need for local administrative access on their work device. Another is having local administrative access on servers within your environment. Again, this is something that should not be granted to any user and there should be no one using the local admin account to log into any servers.

 Ensure you review built-in administrator and/or guest accounts on user devices and servers. It is recommended these are disabled to prevent compromise of these accounts.

The following are some reasons that will help with justification as to why local administrative access should not be allowed on user devices, or at a minimum, not allowing permanent local administrative access:

- The ability for users to install any application increases the risk of malware being installed. The same applies if a malicious actor gets access to the device.

- Inadvertently (social engineering as an example) providing access to an external threat actor who can quickly use the elevated permission to compromise the environment.

- If a malicious link is clicked on or a malicious attachment is opened, there are no protections to prevent installation.

- Allows a malicious attacker to disable endpoint or other security software on a device.

- Many other tasks could benefit a malicious attacker, including modifying registry keys, managing system services, allowing the use of port scanning tools or other tools to exploit the network, uninstalling patches, and so on.

 Although a different concept with Linux and that of the root user, the same principles apply by ensuring the privileged accounts are not being misused and following best practices to reduce risk.

There are solutions available to assist with users' needs with elevated access on their devices as well as the ability to manage local accounts. For example, Microsoft has a capability known as **LAPS** that allows the management of local administrative accounts on Windows devices, and there are third-party PAM tools, such as Delinea, CyberArk, or BeyondTrust, that provide capabilities to manage local administrative accounts along with user access and permissions more efficiently.

Another important consideration regarding local admin access is its necessity within business applications and any SaaS applications owned by the business. You will find that as the business deploys applications and services, there will be a need for local administrator-level access on these applications or SaaS-provided services. It will be critical that those who manage these environments and have access to these privileged accounts are following best practices and being closely monitored. In addition, frequent auditing must be conducted on these accounts to confirm access is still needed and additional administrator accounts haven't been provisioned without following the correct process with approvals.

PAM, PIM, and JIT

PAM, PIM, and JIT are a must in today's environments. As the need to manage multiple on-premises-based roles, cloud admin-level roles, application admin-level roles continues to increase, we need much more efficient methods to protect these roles versus that of the past where permanent privileged permissions were most likely assigned to user accounts.

Today, there shouldn't be permanent access provided to any elevated roles within your environment. There needs to be an approval workflow process that allows users to request access to perform a specific task as needed. Access should only be applied for the duration needed and auditing of all activity must be captured. Anything less will only increase your risk.

With a PAM solution, think of privileged management primarily for your legacy on-premises infrastructure within a domain environment. Although you can store and manage passwords/secrets within PAM that are used within your cloud, the cloud environment becomes more challenging to manage because of RBAC-based roles versus traditional security groups within an on-premises domain. In essence, a PAM solution is a requirement today to reduce the risk of compromise against your privilege accounts.

Although the terminology for PIM comes primarily from Entra ID, think of PIM as the same as PAM but for your cloud environment. No matter what cloud environment you operate within, you will need some form of privilege management solution to assist with efficiently managing your roles within your cloud environment. As mentioned above with the PAM benefits, the same applies to PIM. A concept used by both PAM and PIM is that of JIT. JIT provides the ability to provide access to an account or assign a role to an identity that allows the needed administrative action to be taken for a specified amount of time. This prevents the need to permanently apply permissions to user accounts or the need to create secondary accounts with elevated permissions permanently assigned.

When looking to implement a privileged management access solution for your environment, the following capabilities should be considered. This isn't an exhaustive list but includes some of the more important requirements:

- The ability to discover accounts for all systems, devices, and applications. Examples of systems and devices include Windows, Linux, and networking equipment.
- Local and service account management.
- Capable of managing credentials including password rotation.
- The solution can provide workflow approval with auditing capabilities.
- **Remote Desktop Protocol (RDP)/Secure Shell (SSH)** capabilities to systems and capable of keeping credentials secure by not allowing them to be read or copied.
- Notifications when new accounts are created outside the standard process.
- The ability to monitor and audit all activity with the ability to record remote sessions.
- **Security Information and Event Management (SIEM)** integration.
- IdP integration to leverage SSO authentication and MFA.

Ensure your PAM solution requires stronger authentication than your standard policy when accessing. If an account gets compromised that has access to the PAM solution, you want to ensure additional verification is taking place before anyone can access it. The recommendation is phish-resistant level authentication, which is covered later in the chapter within the *Enhanced Identity Security* section.

An important consideration is that of emergency or what you may be more familiar with as break-glass administrative accounts. There could be several scenarios in which these accounts may be needed such as being locked out of your PAM solution, enforcing policies in your cloud that accidentally lock specific elevated accounts from gaining access, or a ransomware event that impacts access to your accounts. Whatever the reason, you'll need to have break-glass accounts available for certain components and applications within your environment. Storing these within your PAM will not work, so you'll need a strategy around how you manage, store, and protect these accounts in the event they are ever needed.

Let's take a look at the **PoLP** and **JEA**.

PoLP and JEA

One of the core principles of zero trust is that of least privilege access, or you may have seen it referenced as the PoLP. This is a core principle in that you only receive the access needed on your user account to execute the responsibilities for your assigned job role. There should be no additional access or permission added unless you require it for your role. JEA follows the same principle, by only providing the access needed to accomplish something you have been assigned to do.

Over the years, this most likely hasn't been the case in many organizations. We have all been there; a new project is initiated, and we need to move fast. To get things moving as quickly as possible, infrastructure needs to be deployed, services need to be provisioned, databases need to be set up, and so on. This quickly turns into the need to provide more permissions to users than necessarily needed to prevent delay. Once the project is complete, the clean-up tasks tend to get forgotten, leaving user accounts with permissions that are no longer needed. The same applies within the business as users change roles and move into different departments. The access tends to stay with users as they move around, providing their accounts with way more access to data than needed.

One of the biggest risks is that of your technical support staff; simply providing them the highest level of privileged access to their user account for convenience is extremely risky and should never be done.

It is important that the mindset of least privilege and JEA are being applied at all times. As threat actors continue to compromise organizations, they are looking for those with privileged access. Once they get access to a privileged account, the damage can be destructive if a user account they can compromise has access to the entire environment. You need to ensure the footprint of your user account access is minimized as much as possible. One way to simplify the management of least privilege and JEA is that of RBAC in which users are simply assigned to a role versus applying multiple permissions. This will help keep the access management a lot cleaner. In addition, the importance of governance around your access provisioning is critical. You must track an inventory of all access that has been provided and ensure auditing is being complete. In addition, clean-up should occur frequently if not automated.

SoD

SoD holds a critical role and one that may not have been enforced as much as it should have in the past. Within the financial industry, SoD is widely adopted in the corporate world to help prevent fraud, and more importantly, to ensure compliance with the **Sarbanes-Oxley Act (SOX)**. For example, having SoD within the payroll function and ensuring that there is not a single person responsible for issuing, approving, and distributing checks. There should be someone preparing the checks for distribution and someone else signing and approving them for distribution. In cybersecurity, the same concept should be applied, specifically as privileged types of tasks need to be complete within the environment.

As it relates to cybersecurity, the SoD process should prevent any one user from making changes within the environment or accessing data they don't initially have access to without prior approval. For example, an administrator who needs to make some global changes within the cloud environment shouldn't be able to make any changes without going through an approval process. This approval process should at a minimum require a second person to review and physically approve the request for the administrator to make the changes. Once approved, the correct permissions should be applied, or the user should be able to check out a secondary account that allows the changes to be complete. In addition, all work being completed should be audited to ensure the correct changes are being applied.

Another method of reducing risk, especially if you can't deploy technologies to support the SoD process, is to use separate accounts for normal user activity versus privileged activity. A best practice in general is to set up a separate account with privileged access that is used for administrative purposes only, and with enhanced protection. This prevents a threat actor from gaining more access than needed in the event the administrator's standard user account becomes compromised.

By enforcing SoD within your environment for your technical staff, you will significantly reduce risk by preventing a standard compromised account from gaining access to more data than needed. If the correct processes are in place with the SoD model, a threat actor should be prevented from additional lateral movement within the environment beyond what the compromised user account has access to. To ensure efficiency with the SoD model, technologies such as PAM and PIM will need to be deployed correctly.

Insider Threats

You may be thinking, "Why have insider threats been listed within privileged access?". In reality, any user who has been provided additional access to their standard user account can be considered a privileged user. Take, for example, the HR team; depending on the role, they will have access to your sensitive PII. As part of their role, there is a high possibility that their standard user account has been provisioned the needed access as part of their day-to-day duties and there will most likely be no SoD type of accounts within the business functions. This leaves users vulnerable to threat actors who will have access to all information that the user has access to if compromised. Even more concerning is the misuse of these privileges by the users themselves, better known as an insider threat.

As already referenced in the 2023 Verizon DBIR report, the human element is included in 74% of breaches. This is through either human error, privilege misuse, stolen credentials, or social engineering. When referencing privilege misuse, most incidents reference internal threat actors as the source. The report notes that financial gain is the primary motivation with personal data being the most common target. A little older than the Verizon report is the *2022 Ponemon Institute Cost of Insider Threats: Global Report* (https://www.proofpoint.com/us/resources/threat-reports/cost-of-insider-threats), which identified that 26% (a total of 6,803 incidents were represented) of the incidents reported were caused by malicious insiders and that the average cost of an incident was $648,062.

This is obviously very concerning and certainly opens our eyes to the reality that malicious activity is occurring right in front of us.

The challenge with insider threats is that this type of activity can somewhat fly under the radar as normal day-to-day activity, making it extremely challenging to identify. Some of the ways that you can reduce risk in this area are to ensure the correct access controls are in place with your users such that the least privilege is in place and being enforced, ensure user activity is being monitored, provide awareness around insider threats and encourage any suspicious activity to be reported confidentially, conduct regular audits of user access, look at implementing insider threat security capabilities, and look at technologies like **Data Loss Prevention** (**DLP**) and information protection.

Physical Security

One important item to consider as part of your identity and access management solution is that of physical access to your offices and data centers. Ensuring that an intruder can't easily walk onto your site is not only critical to ensure the protection of your data but also the safety of your employees. Nothing is more important than human safety! Within the **National Institute of Standards and Technology 2.0** (**NIST**) Cybersecurity Framework, there is a specific category, subcategory, implementation examples, and informative references to ensure you implement the minimum recommended guidelines for physical security, as shown below. You can review more details on the informative references by navigating the following link: https://www.nist.gov/informative-references and clicking **Download** (**XLSX**) within the **Directly download all the Informative References for CSF 2.0** section.

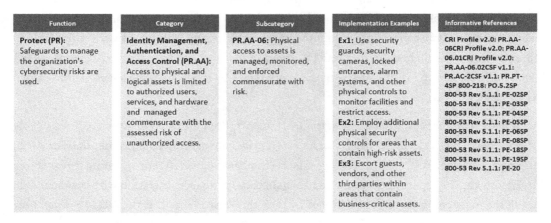

Function	Category	Subcategory	Implementation Examples	Informative References
Protect (PR): Safeguards to manage the organization's cybersecurity risks are used.	**Identity Management, Authentication, and Access Control (PR.AA):** Access to physical and logical assets is limited to authorized users, services, and hardware and managed commensurate with the assessed risk of unauthorized access.	**PR.AA-06:** Physical access to assets is managed, monitored, and enforced commensurate with risk.	**Ex1:** Use security guards, security cameras, locked entrances, alarm systems, and other physical controls to monitor facilities and restrict access. **Ex2:** Employ additional physical security controls for areas that contain high-risk assets. **Ex3:** Escort guests, vendors, and other third parties within areas that contain business-critical assets.	CRI Profile v2.0: PR.AA-06CRI Profile v2.0: PR.AA-06.01CRI Profile v2.0: PR.AA-06.02CSF v1.1: PR.AC-2CSF v1.1: PR.PT-4SP 800-218: PO.5.2SP 800-53 Rev 5.1.1: PE-02SP 800-53 Rev 5.1.1: PE-03SP 800-53 Rev 5.1.1: PE-04SP 800-53 Rev 5.1.1: PE-05SP 800-53 Rev 5.1.1: PE-06SP 800-53 Rev 5.1.1: PE-08SP 800-53 Rev 5.1.1: PE-18SP 800-53 Rev 5.1.1: PE-19SP 800-53 Rev 5.1.1: PE-20

Figure 6.8: The NIST 2.0 Cybersecurity Framework Physical Access to Assets Guidelines

 The following link gives you access to *NIST Special Publication 800-53*, referenced in the preceding figure: https://csrc.nist.gov/pubs/sp/800/53/r5/upd1/final.

Don't forget to integrate the physical access system (badge system) with your HR termination process to ensure badges that have access to facilities and data centers (if access has been granted) are terminated at the same time a user is terminated. If the physical access system is independent, badges may not get terminated promptly, which will allow a disgruntled employee to still have access to your facilities.

Now that we have completed our review of securing your identities, let's shift over to some of the more enhanced capabilities available to protect your identities.

Enhanced Identity Security

The last section of this chapter will cover some of the more enhanced capabilities you should be reviewing and deploying to best protect your identities. This includes providing biometric capabilities, moving to both phish-resistant MFA and passwordless, deploying a risk-based type of protection that can leverage AI capabilities, and ensuring your identity-based activity is being sent to your SIEM for 24/7 monitoring and review.

Biometrics

If you recall the *Authentication* section from earlier in the chapter, biometrics was referenced as a Type 3 authentication method to confirm your identity. Type 3 authentication is something that you are. Because of this, biometrics provides a much more secure method to authenticate as your biometrics are unique to you as a person and they can't easily be replicated, making it very difficult to compromise. In addition, biometrics provides a much-improved user experience as you don't need to remember anything to authenticate, other than the correct biometric method. Some of the more common methods you'll be familiar with include:

- Facial recognition
- Fingerprints
- Palm recognition
- Voice recognition
- Iris recognition or retina scan

Biometrics can be used as a second factor for authentication, or it can even be used alone, which falls within the passwordless model of authentication. which we will cover later in the section. Fortunately, biometrics has become widely available and is available with many of the familiar technologies we already use today. With Windows, biometrics can be leveraged using Windows Hello to log into your Windows device.

Options to authenticate include facial recognition, fingerprint, or iris recognition. With mobile devices, Apple provides the capability to authenticate using facial recognition or fingerprint using Face ID or Touch ID. Android has similar capabilities to allow facial recognition or fingerprints to be used. If you are using a **Mobile Device Management** (**MDM**) platform within your environment, you will be able to enforce the use of biometrics on your users' devices.

Like anything, biometrics doesn't come risk-free, and as biometrics becomes more adopted, the risk will also increase as threat actors continue to look for ways to abuse this data. In 2023, the US **Federal Trade Commission** (**FTC**) released a policy statement to provide some protection for consumers on how biometrics data is handled and used: `https://www.ftc.gov/news-events/news/press-releases/2023/05/ftc-warns-about-misuses-biometric-information-harm-consumers`. You will also need to be aware of whether there are any privacy laws or acts that may impact you and what they mean. For example, the state of Illinois in the US has enacted the **Biometric Information Privacy Act** (**BIPA**) to aid in protecting users' biometric data.

If biometric authentication is an option with any technologies you use, it is highly recommended to enable the capabilities for users to adopt. With increased security and convenience, you'll be reducing the risk of compromise along with providing your users a much better experience.

Phishing-Resistant MFA

Because MFA continues to be compromised by threat actors, we need to find stronger methods of authentication to better protect us. Phishing-resistant MFA is the next generation of MFA to provide improved protection against MFA being compromised. Referencing CISA, they have a great guide to better understand phishing-resistant MFA and how to implement it. You can access the CISA *Implementing Phishing-Resistant MFA* guide here: `https://www.cisa.gov/sites/default/files/publications/fact-sheet-implementing-phishing-resistant-mfa-508c.pdf`.

According to CISA, the following methods have been used by threat actors to compromise MFA:

- **Phishing** by tricking the customer to enter their login information including the code from their authenticator app.
- **Push bombing**, or you may be familiar with the term **push fatigue**, where a threat actor continuously sends the user notification in the hope they will finally click **Accept**.
- **SS7 protocol vulnerabilities** that allow threat actors to exploit the communications network to read texts and listen to phone calls to obtain MFA codes.
- **SIM swap** where a threat actor can transfer a user's phone number on a SIM card they have access to allowing them to take control of the number.

To combat these threats, CISA recommends using phishing-resistant MFA in the form of:

- **Fast IDentity Online (FIDO)/WebAuthn Authentication** using tokens or security keys.
- **Public Key Infrastructure (PKI)-based MFA** using smart cards as an example. This is a popular method used by the US government for their users to authenticate.

The least resistant and easiest method to adopt is that of FIDO/WebAuthn authentication, leveraging widely available security keys and gaining popularity.

 If you aren't familiar, FIDO is an alliance that works toward improving today's authentication challenges with passwords. They are looking to provide simpler and more secure authentication methods using open standards. You can view additional information about this at `https://fidoalliance.org/`. FIDO will be covered in more detail in the *Passwordless* section.

Deploying phishing-resistant MFA won't be easy and something with which you may not get 100% deployment anytime soon. Those of you who have already deployed MFA will be well aware of the challenges faced by users and the adoption of MFA. Phishing-resistant MFA is going to require additional hardware to support it, which means additional costs, a lot of logistics around the distribution of hardware, and ongoing support and operations. In addition, not all applications will support phishing-resistant MFA. For this reason, it is recommended you focus on high-risk areas with phishing-resistant MFA to begin. This will include those high-risk targets within the organization such as executive leadership and anyone with privileged access. Anyone with the highest level of access such as Global Administrators should have this enabled already.

Passwordless

As clearly stated several times, password-only authentication is becoming obsolete and not something we can rely on anymore. Because of this, we need to look at more modern and secure methods that will provide us with a world in which passwords no longer exist. Welcome to a new era of authentication with passwordless. The good news is that passwordless is already available and accessible today. Similar to phishing-resistant MFA, this method isn't going to happen overnight and the journey to a full passwordless world is far away. The first challenge will be getting users to change their ways of working and adopt passwordless. To make this happen, it will be critical that the passwordless experience is streamlined and easy to adopt and use. The bigger challenge will be getting applications up to date to support passwordless capabilities. With some applications still unable to support MFA, we clearly have a long way to go.

By eliminating the use of passwords, authentication is significantly improved by using something you have, such as a security key or phone, and/or something you are, such as biometrics, as previously discussed. Examples of current passwordless authentication methods include:

- Windows Hello on Windows devices leveraging facial recognition, fingerprint, iris recognition, or even a PIN.

- Microsoft mobile authenticator app that provides a number matching screen; either select the number on the screen or enter the number presented followed by biometric or PIN confirmation.

- Apple and Android devices using facial recognition or fingerprint capabilities.

- Both Apple and Google are adopting the use of passkeys that allow passwordless authentication to websites and applications.

- FIDO 2 security keys with supported IdPs, such as Microsoft Entra ID.

 Passkeys are the next generation of authentication that will allow us to eliminate passwords and provide a much simplified and secure user experience when logging into applications or websites. Passkeys essentially allow applications or websites to pass through authentication to a passwordless mechanism, such as biometrics. You can learn more about passkeys here: `https://fidoalliance.org/passkeys/`.

Ideally, we would like to get all our users leveraging passwordless capabilities as soon as possible. The reality is this isn't going to happen overnight, but as leaders, we need to keep pushing hard to make this happen as soon as possible. FIDO 2 security keys are the most secure method, but this comes at a cost and with a lot of logistics needed with physical key management and distribution. As stated in the *Phishing-Resistant MFA* section, you should deploy FIDO 2 biometric keys to your highest-risk users. First and foremost, the most privileged users, such as Global Administrators, then other privileged users throughout your environment, especially those who manage the environment, and then executive leadership, who are most prone to targeted phishing attempts.

Here is an example of a FIDO 2 biometric key that can be utilized by your users if your IdP supports FIDO. This key comes in both USB C and A:

Figure 6.9: The Yubico USB C FIDO 2 biometric key
source: https://www.yubico.com/product/yubikey-bio-series/yubikey-c-bio/

Although not considered as secure as a security key, the next best passwordless method that you can begin to enable for your users today is through the use of passkeys or an authenticator app that supports passwordless. If you already have an authenticator application deployed, and support is available for passwordless, it simply becomes a configuration change followed by change management and communication for your users. Although your users may not adopt the capability right away, at least you have the capabilities enabled to allow the adoption to occur without the need to deploy and manage physical security keys. An example of this would be using Microsoft Entra ID with the Microsoft Authenticator app. In the following screenshot, you can see that the user is being prompted to enter a number that will have been provided by the app.

An additional optional feature is the ability to provide a location of where the authentication is coming from so you can visually confirm the login attempt is from yourself.

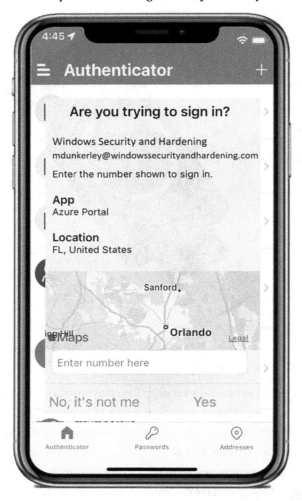

Figure 6.10: A passwordless option using Microsoft Authenticator using additional context

Next, we will move into conditional-based policies to better protect your user's identity.

Conditional-Based Policies

Conditional-based policies are a very powerful tool to protect your identities and reduce risk with potential account compromise. You'll need to check with your IdP on conditional-based policies and whether they are available.

To help provide a better understanding of what conditional-based policies are and how they will benefit you, let's reference Microsoft Entra Conditional Access capabilities. Microsoft Entra Conditional Access is a cloud-based policy tool that enforces compliance conditions for your users. These policies fall in line with ZTA.

Conditional Access policies make decisions based on signals along with enforcing configured policies. Signals include specific conditions such as user and sign-in risk (based on identity-based signals), device platform or type, sign-in location, specific applications, and device-level attributes. Based on these conditions, you can manage access controls by granting or blocking access, or by applying session-based controls such as specifying the sign-in frequency needed for an application. For example, if a user is not on a compliant managed device or they are not accessing from a trusted IP (company office, for example) and they are trying to access their email via a personal device, they can be blocked or you can require a strong form of MFA or passwordless to be satisfied before access is granted, then require them to re-authenticate every hour. This is a simple example, but the possibilities of this scenario are extensive and allow for much greater fine-grained security controls for your user identities. In today's world, conditional-based policies are a necessity, especially with the transformation to the cloud. If you aren't adopting conditional-based policies, you need to review and enable them immediately.

Some use cases that can be set up with Conditional Access include:

- Require phishing-resistant MFA for your admins (privileged roles).
- Require mobile devices to be compliant before allowing access to company resources.
- Enforce MFA for all guest users.
- Block all legacy authentication protocols.
- Allow users on trusted company devices to access resources without MFA.
- Block any access outside a specific region – for example, a US-based company can block any connection from outside the country.
- Block applications from being accessed on untrusted devices.

 Microsoft Entra Conditional Access is being shared as a reference to provide a better understanding of how conditional access works. Use this information to understand what capabilities your IdP has available. You can learn more on Microsoft Entra Conditional Access here: `https://learn.microsoft.com/en-us/entra/identity/ conditional-access/overview`.

Let's review another enhanced identity protection capability using risk-based protection.

Risk-Based Protection

Risk-based login, or behavior-type login, is an advanced protection mechanism to help protect your identities based on what is considered normal behavior versus abnormal behavior. These are capabilities that you need to be adopting immediately if not already being adopted. Your IdP determines the capabilities available for you to enable. I'd imagine whichever IdP you are with, there'll be an upcharge for these types of capabilities. Even with the upcharge, the feature will be well worth the investment.

Again, I'm going to reference the Microsoft capabilities with Microsoft Entra ID Protection as a baseline for what you should be thinking about within your environments. Microsoft Entra ID Protection is an identity-based security tool similar to Conditional Access. The protection works by analyzing signals from user activity to generate both a user risk level and identify any risky sign-ins in real time. A mind-staggering statistic from Microsoft states that they analyze 65 trillion (as of this writing) signals per day. These signals are used to identify the risk of a user and then provide automated remediation based on configuration. Some of the actions that are used to determine risk level and potentially generate an alert include:

- Atypical or impossible travel
- Anonymous IP addresses
- Malware-linked IP addresses
- Identified leaked credentials
- Unfamiliar or new locations
- Suspicious activity-linked IP addresses
- Suspicious inbox forwarding
- Password spray
- Admin confirmed compromise

The user risk detection mechanism determines whether an account has been compromised and is categorized as low, medium, or high. The same categories apply to the sign-in risk but instead, the determination is based on the legitimacy of the authentication request and whether is it from the real owner or not. Based on these determinations, policies can be configured to block access of the user, or you can get more granular by using the risk as a condition within Conditional Access (as discussed previously) to grant access based on additional controls, requiring phishing-resistant MFA as an example.

Again, this is used only as a reference to provide a better understanding of the concepts using Microsoft Entra ID Protection. Use this information to understand what capabilities your IdP has available. You can learn more about Microsoft Entra ID Protection here: https://learn.microsoft.com/en-us/entra/id-protection/overview-identity-protection.

Let's move on to the final section to briefly discuss SIEM integration and that importance of ensuring the identity-based events are all sent for review.

SIEM Integration

We will finish the chapter by explaining the importance of ensuring the integration of your identity and access management events are fed into your SIEM or security monitoring platform. We won't go into detail on SIEM here as it will be covered in more detail in *Chapter 7, Cybersecurity Operations*.

Obviously, there will be a need to ensure all events are being monitored within the organization. However, ensuring identity-based events are being sent and analyzed correctly is critical. As stated previously, identity is the new perimeter of our organization, and the quicker we can identify suspicious activity of these accounts, the quicker we can action and remediate. The first recommendation is to automate as much as you can based on your conditional-based policies and risk-based protections. At the same time, you need to ensure any actions from your automation are sent to your SIEM solution for review as an investigation and follow-up will be needed.

As an example, let's take a business user who is working their normal day from the office. Suddenly, your risk-based detection policy identifies a login attempt for this user coming from a different city, in which it would be impossible for the user to log in based on their current location and login time. If set up correctly, your risk-based identity policy should have blocked the new attempt if the signal determined this was a high risk. Once detected, this event should have been instantly sent to the SIEM where an incident would be triggered for the **Security Operations Center (SOC)** to investigate. This will require reaching out to the user to determine whether the abnormal activity was indeed them. Maybe they remoted in to a different location for the first time to execute some work. If so, the incident can be closed, and the user account can be unblocked so they can continue. If the user has no idea of the attempted login, it will then be determined as malicious. From here, you will need to follow the process to reset the user's password, scan their device for any malicious activity, and then once it is confirmed that it is clean, re-initiate access for the user.

Beyond this, you will want to ensure additional monitoring of the user account for a specified duration to ensure no more abnormal activity has occurred. In this instance, there is a high possibility the user's password has been compromised via a possible phishing attempt, the use of a common password, a reused password.

As you can see, the importance of efficient integration into the SIEM will allow for a much quicker response to reduce risk as quickly as possible. The reality is that your users' identities will get compromised at some point, and no one is immune to threat actors attempting to access our environment. Make sure you are providing as much visibility into your SIEM from your identity and access management activity to ensure quicker response for reduced risk.

Summary

This chapter shows how important the identity and access management function is within the cybersecurity program. Identity and access management is not just simply providing an ID to your user so they can accomplish their responsibilities, although, to them, it may look that easy. Under the covers, there is an extraordinarily large amount of processes and complexity involved. With complexity comes increased risk and the need to ensure you have the correct resources in place, and continuing to simplify and modernize your identity and access management program is critical. Every IdP will have different capabilities so it will be important to work closely with them to ensure they are providing the most advanced identity protection capabilities, as reviewed in this chapter.

As we journeyed through this chapter, we first completed a review of identity access and management and covered IAAA in detail, which provided the foundation for your identity and access management program. Next, we reviewed the reasons and need to modernize your identity architecture along with many statistics to prove the need for moving away from a legacy-based architecture. This was followed by going into detail with account and access management by covering the identity life cycle process, group management and RBAC, service and external accounts, privileged access, and finishing with governance, reporting, and auditing.

In the final couple of sections, we focused on securing your identity. In the first section, *Securing Your Identities*, we went into detail about protecting directory services, password management and vaulting, MFA, non-user account best practices, SSO, the different privileged account considerations, and physical security. We then moved into the last section, which covered *Enhanced Identity Security*. This provided a focus on protection using biometrics, phishing-resistant MFA, passwordless, conditional-based policies, and risk-based protection, and finished off with the importance of integrating your identity-based events with a SIEM solution.

In the next chapter, we will be reviewing cybersecurity operations. First, we will look at an overview of cybersecurity operations, including the management of your cybersecurity operations. We will follow this with a review of threat detection and prevention along with a detailed review of the SOC and everything involved with a SOC. Next, we will spend some time reviewing incident response followed by finishing off the chapter with a very important overview of **Business Continuity Planning (BCP)**.

Join our community on Discord!

Read this book alongside other users, Cybersecurity experts, and the author himself. Ask questions, provide solutions to other readers, chat with the author via Ask Me Anything sessions, and much more.

Scan the QR code or visit the link to join the community.

https://packt.link/SecNet

7

Cybersecurity Operations

Your most active function will be cybersecurity operations. If set up correctly, this function will be operating 24/7/365. If it isn't, you'll be incurring increased risk with the possibility of threat actors infiltrating your environment without anyone monitoring activity. Trends have shown that increased activity typically occurs during off-hours, weekends, and holidays. Also, this function will most likely be your largest staffed area within the cybersecurity team, more specifically, the **Security Operations Center (SOC)**. This team's primary responsibility is to detect and respond to cybersecurity incidents within the organization. This function serves as a critical component within your program and it's important you invest the time needed to ensure your cybersecurity operations are running as efficiently as possible.

We will begin this chapter with an overview of cybersecurity operations, and everything involved in running an efficient cybersecurity operations program for the organization. We will also briefly discuss AI and its impact from both an offensive and defensive perspective. This will lead to a detailed review of SOC operations and everything involved in running an efficient SOC. This will include a review of the different SOC models to consider; the SOC organization structure; log collection, analysis, and automation; SOPs and processes; SLAs; and key metrics, before finishing off the section by touching upon governance. We will then shift our focus to threat detection and what needs to be considered, including asset management and visibility, digital asset monitoring, threat intelligence, and threat hunting.

Next, we will transition into incident response, and everything involved in handling incidents, including the severity of an incident, incident investigation, **Root Cause Analysis (RCA)**, and **Digital Forensics Incident Response (DFIR)**, along with a brief review of SOC analyst tools.

We will then finish off the chapter by reviewing **Business Continuity Planning (BCP)** and the importance of BCP across the organization, in which crisis management, **Disaster Recovery (DR)**, and the **Cybersecurity Incident Response Plan (CIRP)** all play a critical role when undergoing a major incident within the organization. The following will be covered in this chapter:

- **An overview of cybersecurity operations**
- **Security Operations Center (SOC)**
- Threat detection
- Incident management and response
- **Business Continuity Planning (BCP)**

An Overview of Cybersecurity Operations

First, let's take a high-level look at all the sub-functions that should be addressed as part of cybersecurity operations. The following image captures much of what the cybersecurity operations function entails.

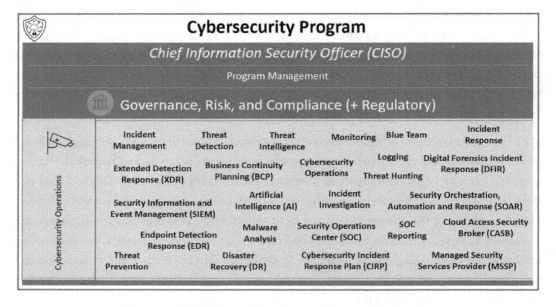

Figure 7.1: Sub-functions of the cybersecurity operations function

Similar to your IT operations, which overlook the day-to-day IT activities of your organization, cybersecurity operations overlook the day-to-day operation of your cybersecurity program for the organization.

One of the primary responsibilities is managing cybersecurity incidents submitted through the ticketing system, whether they originate from automated alerts or users reporting them. As part of your cybersecurity operations, the focus of the structure will be around three primary areas: the SOC, threat detection, and incident management and response. In addition, with the recent advancement of **Artificial Intelligence (AI)** becoming more accessible, we will look into its impact and usage later in this section.

Security Operations Center (SOC)

The first area we will cover in detail is the SOC, also known as the blue team, or the defenders of your organization. The primary responsibility of the SOC is to protect the organization from threat actors trying to compromise the organization, or at a minimum, detect them. This team will typically consist of a variety of different roles and will be required to operate around the clock to ensure malicious activity is being monitored 24/7. For this to be an effective operation, a well-defined organization structure for the SOC should be put in place along with firm processes. At a high level, the following are some of the areas and tasks for consideration that provide for more efficient operations:

- **Strategic and administrative***:*

    - **Cybersecurity strategy development**: Aligning SOC goals with the organization's cybersecurity strategy and business objectives

    - **Budget management**: Planning and managing the SOC budget to ensure optimal use of resources

    - **Policy development and review**: Establishing and regularly updating cybersecurity policies and procedures

    - **Compliance and audit management**: Ensuring SOC operations comply with relevant laws, regulations, and standards (e.g., GDPR and HIPAA)

    - **Cybersecurity operations model: Security Operations Center (SOC), Security Operations Center as a Service (SOCaaS), Managed Detection and Response (MDR), Managed Security Services Provider (MSSP)**, and hybrid

    - **Architecture model**: Selecting an appropriate model to meet your organization's strategy, such as leveraging an on-premises or cloud deployment

- **Operations and incident management***:*

    - **Standard Operating Procedures (SOPs) and processes**: Implementing well-defined SOPs and processes for your SOC to follow

- **Continuous monitoring**: Implementing continuous monitoring strategies to identify and mitigate risks in real time

- **Incident triage and prioritization**: Efficiently triaging and prioritizing incidents based on severity and impact

- **Post-incident analysis**: Conducting RCA and developing lessons learned reports after incidents

- **Incident communication protocols**: Establishing and maintaining communication protocols for internal and external stakeholders during incidents

- **Detection and response**:

  - **Log collection, analysis, and automation**: Ensuring you are collecting all relevant cybersecurity logs throughout the organization for analysis and applying automation where applicable

  - **Advanced threat hunting**: Proactively searching for cyber threats that evade existing security solutions

  - **Forensic analysis**: Conducting detailed forensic investigations to understand and mitigate breaches

  - **Vulnerability management**: Regularly scanning and addressing vulnerabilities within the organization's infrastructure

  - **Red-team/blue-team exercises**: Conducting simulated attacks (red team) and defense exercises (blue team) to test and improve the SOC's capabilities

- **Technology and tools**:

  - **Platform foundation**: *Ensu*ring the appropriate foundation is in place such as **Security Information and Event Management (SIEM)** and **Security Orchestration, Automation, and Response (SOAR)**.

  - **Tool integration and management**: Ensuring seamless integration and management of various SOC tools (e.g., SIEM, SOAR, and EDR)

  - **Ticketing system**: Ensuring there is a platform in place to efficiently manage tickets generated from any systems, your SIEM, and users

  - **Threat intelligence integration**: Incorporating threat intelligence feeds to enhance detection and response capabilities

  - **Automation and orchestration**: Leveraging automation to streamline repetitive tasks and enhance efficiency

- **Staffing and training***:*

    - **Organization structure***: Roles and res*ponsibilities of your staff, structure of your team, and hours of operation

    - **Staffing model**: Determining whether to outsource (MSSP), use in-house-only resources, or a combination of both (hybrid)

    - **Staff recruitment and retention**: Recruiting, training, and retaining skilled SOC personnel

    - **Continuous training and certification**: Providing ongoing training and certification opportunities to keep the team up to date with the latest cybersecurity trends and technologies

    - **Role-specific trainin***g*: Tailoring training programs to specific SOC roles (e.g., analysts, incident responders, and threat hunters)

- **Governance and metrics***:*

    - **Metrics and reporting**: Developing and tracking **Key Performance Indicators** (**KPIs**) and metrics to measure SOC performance and **Service-Level Agreements** (**SLAs**) to ensure a quality service is being provided

    - **Regular reportin***g*: Providing regular reports to leadership and stakeholders on SOC activities and security posture

    - **Risk managemen***t*: Identifying, assessing, and managing risks related to SOC operations

    - **Retention managemen***t*: Ensuring the retention of all cybersecurity logs and that data is retained per company policy

- **Communication and collaboration***:*

    - **Collaboration with IT and other departments**: Ensuring effective collaboration with IT, legal, HR, and other departments

    - **Third-party management**: Managing relationships with third-party vendors and service providers

    - **Public relations management**: Handling public relations and communications during and after security incidents

- **Innovation and improvement***:*

    - **Research and development**: Continuously researching new technologies and methodologies to enhance SOC capabilities

- **Process improvement**: Regularly reviewing and improving SOC processes and procedures
- **Benchmarking and best practices**: Benchmarking SOC performance against industry standards and incorporating best practices

We will cover the SOC operations in more detail within the *Security Operations Center (SOC)* section later in the chapter.

Threat Detection

Threat detection is one of the primary roles of the SOC. Because of the importance of this function, we will be covering it in depth to ensure the full scope of threat detection is understood as part of your broader program. In its simplest terms, the focus of threat detection and response is to first identify any cybersecurity-related threats that may be impacting or could have an impact on the organization. Once these threats have been identified, you'll need to determine how to prevent the impact from becoming a bigger issue if malicious activity has already occurred or you'll need to determine steps to reduce risk in the event no malicious activity has occurred yet. The following are the threat detection responsibilities that you need to consider:

- **Continuous monitoring**: Ongoing monitoring of all assets.
- **Threat intelligence integration**: Incorporate threat feeds and enrich alerts.
- **Log collection and analysis**: Centralized log management and anomaly detection.
- **Asset management and visibility**: Visibility into all assets and ensure security alerts are configured for all assets.
- **Alert generation and management**: Rule-based alerts and behavioral alerts.
- **Incident detection and response**: Early incident detection and containment.
- **Behavioral analysis: User and Entity Behavior Analytics (UEBA)**.
- **Threat hunting**: Proactive threat search and hypothesis testing.
- **Correlation and contextualization**: Event correlation and context enrichment.
- **Machine learning and AI integration**: Anomaly detection and predictive analysis.
- **Use case development**: Custom detection rules and use case updates.
- **Reporting and documentation**: Threat reports and trend analysis.
- **Collaboration and communication**: Incident handover and threat briefings.
- **Continuous improvement**: Process reviews and tool evaluations.
- **Compliance and audit support**: Compliance checks and audit trails.

We will cover threat detection in more detail within the *Threat Detection* section later in the chapter.

Incident Management and Response

Another primary role of the SOC is incident management and response. Like threat detection, incident management and response is a critical component of the broader SOC function that needs to be covered in more detail. Here, everything as it relates to any cybersecurity incident falls within scope, which is no easy task. Depending on the size of your organization, this can easily be thousands of incidents on a weekly basis that need to be reviewed. Once a true/positive cybersecurity incident has been determined, a process will need to be followed on how to best respond to the incident, including assigning the correct severity. The following should all be considered as part of your incident management and response responsibilities:

- **Incident handling and severity**: Incident triage and severity assessment.
- **Processes and procedures**: Incident response procedures and escalation processes.
- **Incident investigation and false positive reduction**: Detailed incident investigation and reducing false positives.
- **Root Cause Analysis (RCA)**: Identifying root causes of incidents.
- **Business Continuity Planning (BCP), Disaster Recovery (DR), and crisis management**: Ensuring continuity and DR planning.
- **Cybersecurity Incident Response Plan (CIRP)**: Developing and maintaining a CIRP and being familiar with the plan.
- **Backup and recovery**: Data backup strategies and recovery plans.
- **Playbooks (e.g., ransomware)**: Incident response playbooks and ransomware response.
- **Digital Forensics Incident Response (DFIR)**: Forensic investigations and evidence collection.
- **Malware analysis and malicious code review**: Analyzing malware and reverse engineering.
- **SOC analyst tools**: Various analyst tools, network analysis, and endpoint analysis.
- **Communication and coordination**: Coordination with stakeholders and communication during incidents.
- **Post-incident review and reporting**: Post-incident analysis and lessons learned.
- **Threat containment and eradication**: Containing threats and removing malicious artifacts.
- **Incident simulation and training**: Simulated incident response exercises and staff training.
- **Compliance and legal coordination**: Ensuring legal and regulatory compliance during incidents.

- **Continuous improvement**: Regularly updating **Incident Response (IR)** plans and improving response capabilities
- **Integration with third-party services**: Collaborating with external IR services and threat intelligence providers

We will cover incident management and response in more detail within the *Incident Management and Response* section later in the chapter.

Now that we have completed a brief review of the SOC, we are going to shift our focus over to **Artificial Intelligence (AI)**. We will be covering AI throughout the book because of its importance, but wanted to ensure a comprehensive review of AI is provided as part of cybersecurity operations because of its relevance in this area.

Artificial Intelligence (AI)

With the recent advancement of **Artificial Intelligence (AI)**, more specifically, Generative AI, or GenAI and the use of **Large Language Models (LLMs)**, it is important we cover this important topic in more detail. As we briefly touched upon in *Chapter 1, Current State*, the current *buzzword* and trend as of 2023/2024 is GenAI and for good reason. AI is being categorized as a game-changing technology, and many believe we are observing an *"iPhone moment"* with AI as the next big thing and what is predicted to be a turning point in technology. I guess time will tell how much of an *"iPhone moment"* we will see with AI, but in the meantime, we have new challenges ahead of us from a cybersecurity perspective. AI in cybersecurity will provide significant advancement for our defense against threat actors. Unfortunately, the opposite is already happening where threat actors have taken advantage of AI at lightning speed, and we are already challenged with new and more advanced techniques being used by threat actors using AI capabilities.

First, let's look at some recent statistics on AI to get a better understanding of the threat landscape. *Deep Instinct*, a company with a prevention platform built on a deep learning cybersecurity framework (https://www.deepinstinct.com/), released its fourth edition of the *Voice of SecOps 2023* report, *Generative AI and Cybersecurity: Bright Future or Business Battleground?*, which was conducted by Sapio Research. The report includes responses from 652 cybersecurity professionals working for organizations in the US with over 1,000 employees. The report found:

- Privacy concerns were reported by 39%.
- The volume and velocity of attacks increased by 33%.
- 37% of attacks were undetectable.

- A 75% increase in cyberattacks in the previous 12 months with 85% from Generative AI.

- Over the next 12 months, 51% of respondents may leave their jobs because of stress and threats related to Generative AI.

- 11% are uncertain of their future because of AI.

Source: `https://www.deepinstinct.com/voice-of-secops-reports`

One of the more concerning areas relating to AI is the use of email and significantly increased phishing and social engineering techniques. Statistics from a 2023 *Darktrace* report, *Generative AI: Impact on Email Cyber-Attacks*, which included 6,711 employees across multiple countries, concluded:

- Within the 6 months previous to the survey, 70% of the respondents noticed an increase in scam emails and texts.

- The use of Generative AI to create malicious emails that are like genuine emails was a concern for 82% of the respondents.

- 87% of respondents shared concerns about their personal information being used in malicious emails (phishing or other scams) because of the availability of it online.

Source: `https://darktrace.com/resources/generative-ai-impact-on-email-cyber-attacks`

Another observation was made across thousands of active Darktrace email customers from January to February 2023, stating there was a 135% increase in novel social engineering attacks. This timeframe corresponds with the increased adoption of ChatGPT.

As statistics are already showing, GenAI is helping threat actors to advance at a rapid pace. As we continue to understand more about AI and how this technology is being used against us, some of the ways we are understanding AI to support threat actors include:

- Enhanced social engineering, specifically with phishing and legitimate-looking emails.

- The ability to write zero-day malicious code with minimum effort.

- Collecting information more efficiently (reconnaissance).

- Enhanced automated attacks.

- The advancement of deepfake capabilities with sound, images, and videos.

- Threat actors can learn and attack much quicker with tools like ChatGPT.

On the flip side, we are seeing advanced capabilities becoming more accessible with vendors and current tools being enhanced to provide additional intelligence with AI. However, let's not forget that AI has been used by vendors for some time and this technology is not new. For example, the identity protection capabilities discussed in *Chapter 6, Identity and Access Management*, leverages AI capabilities as part of the risk-based protection provided by Microsoft. The difference now is that these capabilities are available to anyone to use at their disposal with the advancement of cloud compute. Some of the benefits we are currently seeing with GenAI from a defensive approach include:

- Improved cybersecurity for identity protection, network analysis, email malicious activity (phishing, BEC, etc.), vulnerability management, and security awareness
- Behavioral analytics and continuous learning and improvement over time
- The ability to enhance automation throughout the program
- Faster detection and remediation of incidents, allowing for reduced false/positives, less human error, and improved response times
- The ability to handle and manage greater volumes of data for improved analysis, increased visibility, and to allow for better-informed outcomes

I'm sure most of you can relate: as GenAI capabilities with ChatGPT, Google Gemini (formerly Bard), and others were released, we were instantly challenged with how to quickly tackle both privacy and security concerns. As you most likely observed, we were quickly seeing issues with employees using such tools for work-related activities. And, even more concerning, there were reports of employees uploading sensitive information with examples such as source code being uploaded into these tools. As we should all be aware by now, anything uploaded into these tools becomes public domain knowledge and is available for others using the model to access. Because of this, several companies announced that they were banning GenAI tools within their environment. Others were looking to quickly implement an AI policy to better inform users on how to use the technology.

The reality is, users are going to use these tools, and preventing them from using them will only be a temporary measure, especially since everything is now available via mobile devices and other channels can easily be used to access these tools. As leaders, we need to recognize this and determine what the best approach and path forward is to handle these types of situations. I believe in this situation, the most efficient action would be to focus on policy and user awareness as a first phase. First, create a new policy on AI that provides guidance and highlights risks for users to be aware of. Once published and available, communicate to your users so they are aware and understand what the new policy means moving forward.

You could also issue the policy to your users requiring them to review and acknowledge if you have tools in place to allow acknowledgment to occur. Beyond this, build in GenAI training that your employees are required to complete to ensure a safer work environment by allowing the use of GenAI. As a second phase, begin reviewing options for GenAI that can be deployed or used within your organization that isn't part of the public domain. For example, providing capabilities from Microsoft or OpenAI that contains your company data within your environment, but at the same time, being able to take advantage of the **Large Language Models (LLM)** available as part of public ChatGPT. We will be reviewing more details on both user awareness and policy within *Chapter 9, Cybersecurity Awareness, Training and Testing,* and *Chapter 13, Governance Oversight.*

Here are a couple of valuable resources on engaging with AI and secure AI system development that are worth reviewing:

- *Engaging with Artificial Intelligence (AI)* released by the **Australian Signals Directorate's Australian Cyber Security Centre (ASD's ACSC)** in collaboration with many other international partners: `https://www.cyber.gov.au/resources-business-and-government/governance-and-user-education/governance/engaging-with-artificial-intelligence`
- *Guidelines for secure AI system* development published by the UK **National Cyber Security Centre (NCSC)**, the US **Cybersecurity and Infrastructure Security Agency (CISA),** and many other international partners: `https://www.ncsc.gov.uk/collection/guidelines-secure-ai-system-development`

There is clearly a long way to go with GenAI and how we best protect our users' privacy and manage all the risks that GenAI brings with it. We are still in the early days of GenAI, but time will tell how well this technology progresses in the future. We will continue to touch upon AI throughout the book as it applies to the overall cybersecurity program.

As already mentioned, AI is moving at an extraordinarily fast pace and we can expect laws, regulations, and acts to be developed. We are already seeing movement such as examples with the AI Act within the European Union (`https://digital-strategy.ec.europa.eu/en/policies/regulatory-framework-ai`) and an Executive Order from the US: `https://www.whitehouse.gov/briefing-room/statements-releases/2023/10/30/fact-sheet-president-biden-issues-executive-order-on-safe-secure-and-trustworthy-artificial-intelligence/`.

Let's shift our focus on to the next section, which will cover more details on the **Security Operations Center (SOC)**.

Security Operations Center (SOC)

As you will be familiar, IT operations within an organization are very standard and mature operations. The IT operations function is core to the ongoing success of IT systems, users, and applications to ensure efficiency and availability for your business operations. If there is an outage or an issue, the operations teams typically follow very strict **Service-Level Agreements (SLAs)** to return the service back to normal operations. The same concept applies in the cybersecurity world today. The concept of cybersecurity operations and the SOC has grown exponentially and is becoming a requirement for organizations to maintain normal business operations.

As briefly touched up in the *An Overview of Cybersecurity Operations* section, the primary responsibility of cybersecurity operations is to run the day-to-day operations of your cybersecurity program and to protect the organization from threat actors trying to compromise the organization. This includes ongoing protection against cybersecurity-related activity, which means a function that doesn't rest, with a requirement to operate 24/7/365 and to not let your guard down. This is not an easy feat and there's a lot of complexity and resources needed to ensure an efficient operation model for your cybersecurity program.

 If you are building a SOC for the first time, there are a lot of good resources available to reference. One good example is that from the **National Cyber Security Centre (NCSC)**, *Building a Security Operations Centre (SOC)*: https://www.ncsc.gov.uk/collection/building-a-security-operations-centre.

First, let's review some of the different cybersecurity operational models being used today.

Cybersecurity Operations Model

The model you adopt will be determined by many different factors like your company size, budget availability, the type of industry you are in, regulatory reasons, skill set challenges, and resource availability. The following are some of the more common models.

Traditional Security Operations Center (SOC)

The most common and traditional model is a standard SOC. This is an in-house model that is staffed and operated by employees or contractors you onboard as part of your organization. The SOC will typically manage a SIEM and SOAR that is used to primarily monitor events to detect abnormal activity.

You will primarily see a traditional SOC within larger enterprises where they have the budget to staff a 24/7/365 SOC and they are able to attract the needed talent to efficiently operate.

Security Operations Center as a Service (SOCaaS)

A SOCaaS model is one where your SOC is hosted, managed, and operated by a third-party vendor who takes on the full responsibility of running all SOC-based operations. You will typically see all the same services being provided, or at least available, as in an internal SOC. As already mentioned, there are a variety of factors that will determine the model you implement. SOCaaS will most likely be adopted by medium to smaller-sized organizations that don't have the budget to staff their own in-house SOC and are not able to attract the talent needed to run an efficient SOC. When outsourcing your SOC, you will typically contract with a **Managed Security Services Provider** (**MSSP**) that specializes in cybersecurity services.

Managed Detection and Response (MDR)

While the SOC focuses more on the monitoring of events from SIEM, which can be looked at as more traditional, MDR is a more modern approach that focuses on enhanced threat detection capabilities. This includes using technologies like **Extended Detection and Response (XDR)** in addition to the SOC capabilities. MDR is also considered an outsourced service, which may make sense for medium to smaller-sized organizations.

Managed Security Services Provider (MSSP)

An MSSP, on the other hand, is more generalized in that it provides a broad portfolio of services for organizations beyond SOC types of services. An MSSP would be the vendor you use to run SOCaaS or MDR. In addition, an MSSP will provide a catalog of many services that can be contracted such as vulnerability management, testing services, vendor risk management, tabletop exercises, identity management services, and much more.

 You may be familiar with the concept of a **Managed Service Provider (MSP)**. An MSP is essentially the provider of outsourced IT services for an organization versus the cybersecurity-specific services provided by an MSSP. An MSP has been around for a while and has become a common model within IT.

Hybrid Model

The final model to share is a hybrid model. This essentially means mixing any of the mentioned services above that make sense for your organization. For example, you may opt to deploy your own in-house SOC, but you need to onboard a third-party MSSP to manage and operate the SOC.

As reviewed in *Chapter 1, Current State*, there are two distinct challenges specifically within cybersecurity operations. The first is a shortage of expertise. This is an issue across the broader cybersecurity program, but I would argue probably more so in the cybersecurity operations space for the simple reason that this has become a much more technical type of role and a lot of the roles within cybersecurity operations require a higher level of technical expertise. This, unfortunately, doesn't come easy. The second is well-being and burnout (we covered some thoughts on how to best address this in the *Prioritizing Well-Being* section of *Chapter 1*; make sure you review this section again if needed), and cybersecurity operations are likely the most impacted by this. It becomes exhausting having to deal with 24/7/365 operations and overseeing non-stop events and incidents. This specific function barely gets any air to breathe, and this can quickly lead to an environment with high turnover if not managed efficiently. Both of these challenges need to be taken into consideration when determining the best cybersecurity operations model for your organization.

SOC Organization Structure

Within the cybersecurity organization, your SOC function will most likely have the largest group of employees to allow efficient operations. Even as an outsourced SOC, you will find a large team behind the scenes to ensure 24/7/365 operations. If you are not staffed efficiently in this function, you will most likely be challenged with high turnover and potential burnout from employees over time. Within the SOC, there will essentially be a mini organization defined with multiple different roles to make up the team. The roles (including a brief description of responsibilities) you can expect within the SOC include:

- **SOC manager/director**: The SOC manager/director is responsible for overlooking the operations of the SOC to ensure processes are being followed and SLAs are being met and to ensure any escalations of incidents are being handled correctly. They will typically report directly to the CISO.

- **SOC lead**: The SOC lead serves more as a technical role that overlooks and serves as an escalation for the cybersecurity analysts. They will be the most knowledgeable technical resource who can provide the needed direction and guidance to the analysts. The SOC lead will typically report directly to the SOC manager/director.

- **Cybersecurity analyst 2**: The Cybersecurity analyst 2 will serve as a more senior analyst role on the SOC. They will be a lot more knowledgeable when it comes to incident management and response and will serve as an escalation to the Cybersecurity analyst 1. They will take on the more complex incidents for review and investigation and will typically have more access to the environment to allow a more thorough investigation. They may also leverage more advanced tools that allow for deeper analysis of security incidents. The Cybersecurity analyst 2 will typically report to the SOC lead.

- **Cybersecurity analyst 1**: The Cybersecurity analyst 1 will typically be an entry-level role within the SOC. They will primarily handle all the incidents coming into the SOC and determine what can be closed as false/positives versus what needs further investigation along with any security incidents that will need to be escalated to the Cybersecurity analyst 2. They will complete a basic investigation of incidents and will typically contain enough access throughout the environment to complete an initial review of an incident. The Cybersecurity analyst 1 will typically report to the SOC lead.

- **Incident response lead**: An Incident response lead will oversee and manage all activities related to incident response. For the most part, the majority of the incidents will be handled by the SOC Cybersecurity analysts per standard processes and procedures. Although the Incident response lead will review all incidents, a primary function of their role is to overlook escalated incidents and major incidents that will require a lot of coordination and collaboration across other teams. They will also oversee the **Cybersecurity Incident Response Plan** (**CIRP**) and be the point lead when the plan needs to be invoked. The Incident response lead will typically report to the SOC manager/director.

- **Other relevant roles**: Depending on the size of your organization, some other roles you may see within cybersecurity operations or the SOC include a *Threat hunter, Penetration tester, Forensic investigator, Vulnerability analyst,* and *Cybersecurity architect/engineer.*

The resource model you decide to implement may change slightly depending on the size of your organization and the current roles in place as multiple responsibilities may fall to a couple of key employees. The important point to take away is to understand all the responsibilities and expertise involved with a SOC to ensure the program runs effectively. If you move forward with a hybrid model, it's highly recommended you employ some key resources in-house, such as the manager/director-level resources along with an incident response lead. At a minimum, having your manager/director in-house will be needed for more efficient operations, especially to build better synergy across the broader cybersecurity program and organization. In addition, you'll also want to consider an in-house incident response lead, especially as major incidents need to be managed and coordinated beyond the cybersecurity teams with the business, executive leadership team, and possibly the board.

Every organization is different and there will not be a single model that fits all. But it is important you build an organization structure for the SOC because of the complexity involved in running day-to-day operations. This also ensures everyone has full visibility and transparency into how cybersecurity operations are run.

At a very high and basic level, your SOC organization structure may look something like the following:

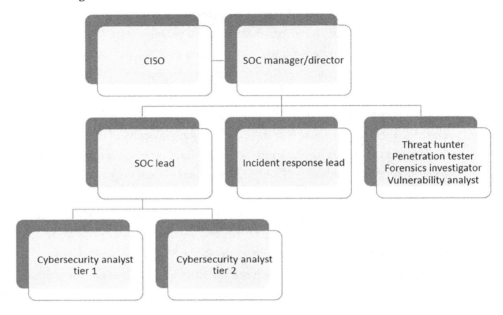

Figure 7.2: Example of a SOC organization team structure

Let's look at some of the more operational items with log collection, analysis, and automation.

Log Collection, Analysis, and Automation

One of the key components of your SOC is the collection and visibility into all logging throughout the environment, specifically, security logging to allow ongoing analysis and determining if there is any malicious activity occurring, or any risk to be aware of. Part of this ongoing log collection will be the continuous analysis of all the collected data, which will include large amounts of data. Doing this manually isn't possible so using advanced tools to help parse the logs, look for anomalies, and provide automation is a must. This is where your architecture and engineering function will come into play as you determine the best architecture and tools needed to run and operate your SOC.

Without even looking at specific vendors, there are a few primary components that you will need to understand to determine the best deployment for your environment. Your cybersecurity operations model (in-house, outsourced, etc.) will also determine how your architecture model will be deployed.

For example, opting for an on-premises SOC will possibly include different components to that of an MDR. The following components will need to be considered as part of your core SOC technology stack.

Security Information and Event Management (SIEM)

SIEM, is the traditional method that most of you will be aware of and most likely the solution you have deployed today if you have a SOC in place. SIEM, in simple terms, provides the capability to collect all logs throughout the environment. At the same time, the SIEM tool will analyze all the collected data to determine if there are any potential cybersecurity incidents that will need to be reviewed further and investigated by the SOC. Traditional SIEM models typically consist of a traditional server deployment within a traditional infrastructure, whether it be in an on-premises data center or IaaS within the cloud. Modern SIEM models are considered cloud-native reducing the need to deploy any traditional infrastructure. Some of the more popular SIEM tools you may be familiar with include IBM QRadar, Splunk, LogRhythm, and Microsoft Sentinel. If you manage the SOC in-house, you will need to ensure you have the correct resources or staff to manage the environment, which will include ongoing maintenance with updates and upgrades. These environments can become very complex, so ensure they are well documented for everyone to understand the architecture and configuration. The following are events and logs that will need to be collected at a minimum:

- **Servers**: Windows, Linux, and Unix
- **Network**: Firewalls, switches, routers, IDP, DNS, and WAFs
- **End-user devices**: Windows, Linux, Mac, Chrome, iOS, and Android
- **Collaboration**: Email, IM, voice, and intranets
- **Identity sources**: AD and cloud IdP
- **Cloud environment**: Azure, AWS, and GCP
- **Security tools**: DLP and advanced threat protection
- **Application and databases**: On-premises and SaaS apps, SQL, Oracle, containers, DevOps tools, and CI/CD pipelines
- **Other areas to consider**: IoT/OT/ICS technology, backup systems, virtualization platforms, physical security systems, and third-party service providers and integrations
- Any other security-related logs within your environment

Next, we will review SOAR and the benefits it adds in addition to SIEM.

Security Orchestration, Automation, and Response (SOAR)

The SOAR solution supplements SIEM. Because of the volume of data that needs to be handled within SIEM, additional tools are needed to be more efficient, and SOAR is one of these tools. SOAR is a solution that provides capabilities to build automation against the data that is received by SIEM. With SOAR, use cases or playbooks can be set up to orchestrate and automate the collection of specific data and then follow a specified process to improve overall threat and detection response times. Some examples where use cases or playbooks will benefit the team include ransomware response, phishing or **Business Email Compromise (BEC)** investigations, vulnerability detection and patch management activities, privilege escalation, and lateral movement investigation and containment. SOAR is another solution that is typically deployed on a server within the traditional infrastructure where the orchestration and automation are set up. For the most part, you can expect SOAR capabilities to be included with your SIEM, although they may come at an extra cost. Overall, the implementation of SOAR must be part of your SOC architecture because the reality is, we as humans cannot efficiently review the volume of data being collected by a SIEM tool efficiently. We need the help of tools that can speed up threat detection and response, reduce false/positives, and support our staff to prevent them becoming burnt out.

Extended Detection and Response (XDR)

XDR can somewhat be considered more of a modernized capability compared to SIEM. XDR is a much newer concept that extends the capabilities of detection and response beyond one technology to include all the detection and response capabilities throughout the environment. A couple of examples include **Endpoint Detection and Response (EDR)** for your end-user devices and **Network Detection and Response (NDR)** for your network equipment. With XDR, you can expect comprehensive coverage across your entire environment, including areas such as email, servers, cloud infrastructure, identity and access management, network, and applications. One of the primary benefits of implementing XDR capabilities is the ability for it to consolidate all data into a centralized view for improved efficiency, improved correlation of events, and increased response time to alerts and incidents. XDR doesn't necessarily replace a SIEM or SOAR deployment. It complements it by adding intelligence and automation and providing advanced analytics with data to correlate events and allow focus on higher-priority incidents. XDR is now available from many vendors, and it's highly recommended you review and understand XDR and the benefits it can provide to your organization. We will be covering other detection and response capabilities in the *Threat Detection* section that follows.

Processes and SOPs

A component of the SOC that provides for much greater success is having a well-defined process in place, along with well-documented **Standard Operating Procedures (SOPs)**. To achieve efficient operations, everyone within the SOC needs to work coherently with full transparency. Collaboration is an absolute requirement in the SOC, and the more efficient the team works together as one, the more effective they will be at their job. To make this successful, well-defined processes need to be documented and everyone needs to be fully aware of them. In short, a process is a step of defined tasks to achieve a specific outcome. Because you are running 24/7/365 operations, shifts will change frequently and there will never be one individual overlooking everything. As incidents come into the SOC and investigations occur, there will be a need for hand-offs to happen between different analysts. Ensuring a clean hand-off and allowing investigations to continue efficiently will require well-defined processes to be in place, with great collaboration. This type of efficiency doesn't occur overnight and will constantly need to be reviewed with ongoing process improvement. In addition, there will be turnover in which having a well-documented process will provide a smoother onboarding experience to allow new staff to get trained much faster. Your managers/directors and leads will be the ones to ensure these processes are in place and being followed. When looking at process management, some items that you'll need to consider include:

- How to best document your processes and where will they be hosted with versioning control.

- The need for a standardized template that is consistent and easy to follow. Make sure purpose, scope, roles and responsibilities, step-by-step procedures, and references sections are included.

- Ensure easy access to processes but ensure that only those who need access have access.

- Efficiently train your staff on the processes in place, including a comprehensive onboarding program in addition to ongoing training.

- Assign a process owner and backup resource to keep processes up to date. Ensure processes are reviewed regularly to ensure they are up to date.

- Ensure continuous process improvement through feedback channels, audits, and reviews of processes. Use KPIs where applicable to measure processes and identify areas of opportunity.

- Implement processes for incident detection, response, and recovery.

- Make sure shift handover procedures are in place to ensure seamless transitions. Implement checklists and communication protocols for smoother transitions.

- Implement clear communication protocols and escalation paths. Set up regular team meetings and briefings to keep everyone up to date.

- Ensure processes comply with any regulations and industry standards. Ensure they remain up to date with regulations and industry standards.

- Implement a documentation quality control process that includes reviews and approvals.

- Ensure technology integration and usage processes are in place and aligned with the tools and technologies used in the SOC.

- Make sure **Disaster Recovery Planning (DRP)** and **Business Continuity Planning (BCP)** processes are in place to maintain SOC operations during a major event.

- Ensure processes are in place for performance monitoring and generating regular reports.

- Establish processes for improved collaboration with other teams throughout the organization.

In addition to your processes are SOPs. In simple terms, these are step-by-step instructions for your staff to follow to complete a specific task. For example, if you have a defined process in place, which will be a series of tasks, you may have several SOPs defined to complete each of those specific tasks within your process. Having an SOP in place allows for consistency across your program. When an SOP is created, everyone now has the instructions needed to complete a specific task, allowing for improved efficiency. This will allow users to complete tasks a lot quicker with the hope of reducing mistakes. When well-documented SOPs are in place, you will have greater success with onboarding new employees, allowing them to become more efficient quicker. This is important in a very fast-paced and highly demanding function. The following image demonstrates a high-level view of a process with an SOP embedded into one of the tasks as part of the broader process. This example only shows one SOP but there could be several SOPs embedded within this process.

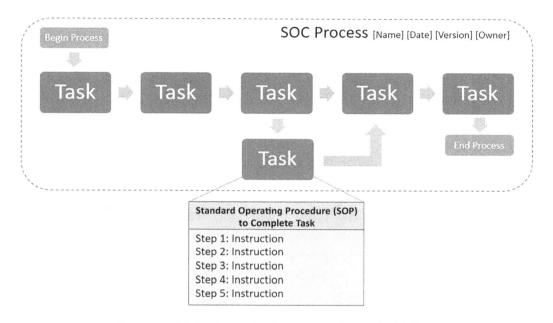

Figure 7.3: High-level example of a process with an embedded SOP

To provide a little more clarity on a process and an SOP, let's refer to the example shared earlier of a SOC analyst changing shifts. The documented process will essentially provide a set of steps that analysts must take to successfully change the shift from one cybersecurity analyst to another. At a high level, the process may look like the following:

- **Task**: Cybersecurity analyst starting the shift to clock in and log in to PC.
- **Task**: Cybersecurity analyst starting the shift to open and access all applications.
- **Task**: Cybersecurity analyst starting the shift to determine if any incidents need to be handed off:
    - **Decision**: If yes, work with the cybersecurity analyst ending the shift to transfer active incidents to them.
    - **Decision**: If no, begin the shift by taking the next incident in the queue for investigation.
- **Task**: Cybersecurity analyst ending the shift to log out of all applications once the transfer of any active incidents has been complete to the cybersecurity analyst starting their shift.
- **Task**: Cybersecurity analyst ending the shift to log out of PC and clock out of shift.

From an SOP standpoint, let's look at the step within the process that determines if any incidents need to be handed off. If the answer is yes, an SOP will need to be created that the cybersecurity analyst can follow so an efficient and clean transfer of the incident occurs. This will allow the investigation to continue smoothly. Within the SOP, there will be a set of instructions that both cybersecurity analysts will need to follow to ensure the cybersecurity analyst starting the shift is fully prepped and ready to proceed with the incident to continue the investigation until closure.

SLAs and Key Metrics

One important consideration for your SOC is the **Service-Level Agreement (SLA)**. An SLA defines the expected service to be delivered, backed by metrics for measurement. This helps to ensure the services are being delivered optimally. For the most part, you will see SLAs between an organization and a service provider, and they are typically defined within the contract with the service provider. But, regardless of whether you outsource your SOC or not, you should be defining SLAs for the broader organization. Since the SOC is dealing with incidents and working within a ticketing system, there could be an impact on the organization from an incident. Because of this, transparency needs to be provided to the organization when they can expect operations to resume if there has been a declared incident impacting any business operations. There may be SLAs for different types of services being provided, but for a SOC operation, the primary SLA will be for incident response and resolution. For the declared incidents with the SOC, the severity of your incident will equate to more aggressive SLAs, as shown in the following example of defined SLAs:

- **Incident declared as Critical**: 15-minute response time with 12-hour resolution
- **Incident declared as High**: 30-minute response time with 24-hour resolution
- **Incident declared as Medium**: 2-hour response time with 72-hour resolution
- **Incident declared as Low**: 8-hour response time with 120-hour resolution

 It is important you continuously monitor your SLAs to ensure you remain in compliance with your incidents. There may also be a need to adjust your SLAs to become more aggressive or less aggressive depending on the volume of incidents and any potential staffing concerns.

Another important area to efficiently track, especially as a leader, is key metrics to measure your SOCs' performance. You may also see the term **Key Performance Indicators (KPIs)** used to provide metrics on your SOC. Tracking metrics of your SOC is a very important activity that is ongoing.

You cannot measure the performance of your SOC without any metrics. If there are no measurements, how will we know how effectively the SOC is operating? Having metrics in place allows us to better determine how well (or not) the SOC is performing on a day-to-day basis. They will also provide a baseline that will allow for process improvement activities to drive the SOC to perform better. There are many metrics that can be monitored; some of the more relevant and important metrics you will typically see within a SOC operation include:

- Any defined SLA metrics
- Number of alerts/events
- Number of incidents (also broken into categories, for example, email, user, network, etc.)
- False/positive and true/positive rate
- **Mean Time to Detection (MTTD)**
- **Mean Time to Recover (MTTR)**
- Other metrics may include time to mitigate, time to respond, and time to resolution

Once you can produce all this information, you need to determine the frequency of this data being generated. On-demand will be ideal, but your environment may not support on-demand, so weekly, every other week, or monthly will be the next best option. Once determined, the data will need to be shared in a presentable form or within a report as part of ongoing meetings with your SOC to review. These activities will typically fall under governance, which will be covered next.

Governance

The last area we will cover is governance of your SOC. Although we do cover governance more broadly in *Chapter 13, Governance Oversight*, as a leader, you need to ensure the correct governance for your SOC operations is in place. This allows full transparency to be provided on how the program is operating on a frequent cadence. From a leadership perspective, some of the governance you will want to ensure is in place include:

- Frequent reports of key metrics and KPIs (weekly, monthly, quarterly, and yearly)
- Ongoing scheduled meetings to review key metrics and discuss challenges
- Reviews of high-priority incidents
- Ensuring continuous improvement
- Ongoing reviews of processes and SOPs
- Data handling and retention
- Regulation requirements

As already mentioned in other functions, governance is a requirement, and as a leader, you must commit the time to ensure each of the functions is running as planned. Next, we will review threat detection.

Threat Detection

One of the primary reasons we need to have a SOC in place is to detect threats within the environment. It is the SOC's responsibility 24/7/365 to do all they can to detect any activity that may be a threat to the organization. This requires the ongoing scanning and analyzing of all activity within the environment to identify any anomalies. This is not an easy task and there is a lot involved to ensure efficient threat detection is in place. Throughout this section, we will cover everything you should consider as part of threat detection within your cybersecurity operations.

Asset Management and Visibility

Before we go into any more technical areas, one area that cannot be ignored is your assets. If you do not know what is within your environment, how will you ever be able to detect if there are any threats within those assets? Because of this, it is critical that you dedicate the time to ensure you have full visibility into all your assets and that there is an ongoing process to identify any new assets that may have been introduced without going through the correct process. One of the more traditional methods used to inventory all your digital assets is the use of a **Configuration Management Database (CMDB)**. Implementing and managing a CMDB is no easy task and requires a lot of effort, and I'm confident a lot of organizations don't even have a CMDB. If they do, there's a high probability it is incomplete. A more recent and modern approach to asset management is **Cybersecurity Asset Management (CSAM)**. The idea behind CSAM is that it can continuously scan your environment to ensure all digital assets are identified to understand what risk exposure you may have from your digital assets. The following are some of the digital assets you can expect to be inventoried:

- Endpoints
- Servers
- Network equipment

- Cloud resources
- Software and applications including SaaS
- Users and other accounts
- IoT, OT, and ICS

Another concept becoming more widely adopted is **External Attack Surface Management (EASM)**. This focuses primarily on your external-facing footprint for any vulnerabilities that can be found on the internet vs your entire portfolio. You are essentially focusing on your higher-risk entry points for any vulnerabilities. We will review EASM in more detail in *Chapter 8, Vulnerability Management*.

Digital Asset Monitoring

With all assets identified, you need to ensure you can monitor and scan all digital assets within your environment. More specifically, this refers to the ability to monitor and scan for any cybersecurity-related events that may be malicious across your entire portfolio. Once we are able to monitor all digital assets, capabilities will be needed to collect and analyze the cybersecurity-related events and logs to determine if there is any malicious activity occurring. One of the methods to accomplish this is via SIEM, which we covered in detail within the *Security Operations Center (SOC)* section. It is important you can feed all events and logs that are relevant for any analysis into the SIEM tool to ensure there are no gaps in the ability to monitor for any malicious activity. As covered previously, the SIEM model will be constantly monitored by the SOC for any malicious-related activity.

Extended Detection and Response (XDR)

Another method that will provide the capabilities for monitoring is through XDR, which we also covered within the *Security Operations Center (SOC)* section. With XDR, it's going to be critical that the solution being used provides capabilities across your entire digital footprint. As we covered, there are many standalone detection and response capabilities, like EDR. With XDR, we need to ensure detection and response capabilities provide comprehensive coverage across all areas. To keep consistency, we'll reference back to the ZTA model, which covers all relevant areas that must be monitored.

The following diagram represents everything that should fall within the detection and response capabilities:

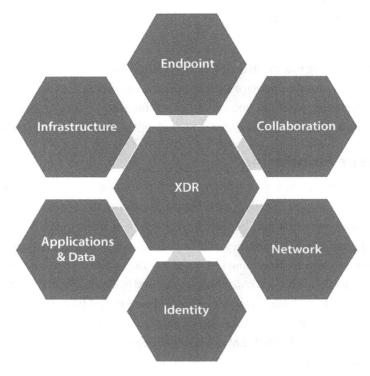

Figure 7.4: XDR threat and detection capabilities

One area that is of importance if applicable is **Operational Technology (OT)** and the **Internet of Things (IoT)**-related technology. It's essential to monitor and collect data from assets equipped with these technologies within your environment for review and analysis.

It's important to discuss the need for a unified platform for the collection and monitoring of all events and logs. It is critical that the correlation between events within your environment is established. This will not be possible without the ability to bring all logging into one unified platform. Ensure you complete reviews of your digital asset inventory to ensure all events and logs are being fed into your centralized logging platform, such as your SIEM.

Let's shift our focus onto the **Cloud Access Security Broker (CASB)** to better understand the capabilities to provide additional protection for your SaaS environments.

Cloud Access Security Broker (CASB)

With the ongoing shift to SaaS-based applications, the need for improved visibility and protection against these environments is required. Most vendors are now adopting SaaS-based models, which significantly increase the footprint and risk profile of the organization. As the business continues to onboard more and more SaaS types of services, we need to ensure we have visibility into these applications to provide improved security controls. In short, a CASB is a solution that is deployed between your users and a SaaS (or other cloud)-based application, allowing for increased visibility and security. The following are some of the primary features you can expect from a CASB:

- Advanced threat detection and prevention for SaaS apps
- **Data Loss Prevention (DLP)**
- Shadow IT visibility
- Governance of SaaS apps, including usage
- Compliance enforcement
- Risk profile
- Encryption capabilities
- Auditing

Once you have a CASB in place (which should be a requirement if there are SaaS applications deployed in your organization), you will need to ensure integration is configured into your SOC, or more specifically, your SIEM. The SOC will need to receive alerts so they can review and investigate any identified risk with your SaaS-based applications.

 An emerging technology to be aware of is **Secure Access Service Edge (SASE)**, which brings together some of the more modern cloud security capabilities with CASB, including the **Secure Web Gateway (SWG)**, **Zero Trust Network Access (ZTNA)**, and more: https://www.gartner.com/en/information-technology/glossary/secure-access-service-edge-sase.

Now that we have reviewed the CASB and understood how to gain better visibility into our SaaS environments, let's shift our focus to threat intelligence and the importance of ensuring the SOC is receiving threat intelligence data from as many sources as possible for a more thorough review and analysis.

Threat Intelligence

Threat intelligence, or **Cyber Threat Intelligence (CTI)**, is data or knowledge that allows us to be more proactive with our prevention measures. Threat intelligence is information gathered from the analyses of threat actors and groups along with already-known compromises. Information provided as part of CTI can include the following with some examples of data sources:

- **Indicators of Compromise (IoCs)**: Malicious IPs, domains, file hashes, and email addresses
- **Dark web activity**: User accounts, passwords, PII, financial data, and dark web forums
- **Vulnerabilities**: Active exploits, zero days, security updates, and known vulnerabilities
- **Active malware intel**: Ransomware, viruses, spyware, and botnets
- **Tactics, Techniques, and Procedures (TTPs)**: Threat actor intel, social engineering, **Advanced Persistent Threat (APT)** actors, and common attack vectors
- **Network information: Next-Generation Firewall (NGFW)**, intrusion detection and prevention system, anomalous network traffic patterns, and data exfiltration indicators
- **Public sources or Open-Source Intelligence (OSINT)**: News, forums, blogs, social media, and publicly disclosed breaches
- **Other sources**: Threat actor profiles, threat feeds, geopolitical information, phishing intelligence, threat-hunting reports, collaboration and sharing platforms, behavioral analytics, and machine learning and AI analysis

This information can then be used to add additional controls to your environment to prevent the risk of any known threats occurring. CTI can be obtained via multiple different sources in the form of a feed. You will be able to obtain CTI via a paid service from a vendor; many cybersecurity vendors offer this as a service. In addition to paid services, you can also use **Open-Source Intelligence (OSINT)** feeds, which are free.

 The OSINT Framework is a great resource for cybersecurity-related resources: https://osintframework.com/

Once you have identified any feeds that you would like to retrieve intel from, you'll need to ensure they are configured and integrated into your SIEM to allow correlation between all your events and logs to identify if there is any malicious activity. Don't overlook the power of threat intel for your program. This data can help provide a much more proactive approach to your defense strategy to prevent an attack from being successful.

MITRE ATT&CK

One of the more popular tools used to detect specific attack scenarios is the **MITRE ATT&CK** matrix. **ATT&CK** stands for **Adversarial Tactics, Techniques, and Common Knowledge**. MITRE ATT&CK is a knowledge base based on real-world scenarios used by threat actors. The model allows cybersecurity experts to better understand how threat actors are infiltrating environments so they can assess and identify any gaps within their own environments. You can learn more about MITRE ATT&CK here: `https://attack.mitre.org/`.

The following are the most common MITRE ATT&CK use cases:

- Detection and analytics
- Threat intelligence
- Adversary emulation and red teaming
- Assessment and engineering

The following is a snippet of the MITRE ATT&CK matrix taken from `https://attack.mitre.org/`:

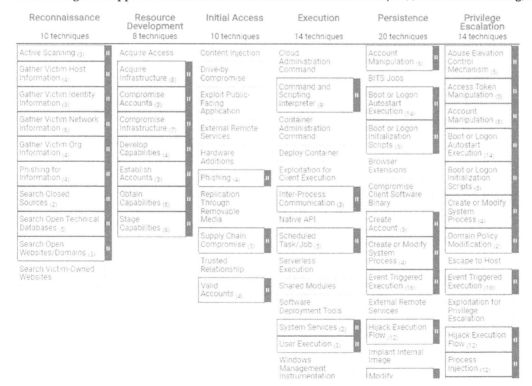

Figure 7.5: Snippet of the MITRE ATT&CK matrix

MITRE has a short e-book that guides you on how to use MITRE ATT&CK: *GETTING STARTED WITH ATT&CK* (`https://www.mitre.org/sites/default/files/2021-11/getting-started-with-attack-october-2019.pdf`).

Next, let's look at one way to proactively detect if there is any malicious activity occurring in your environment using threat-hunting techniques.

Threat Hunting

Threat hunting is a technique used to determine if there is any active malicious activity occurring within your environment. More specifically, CTI data is used to detect any current exploitation within your environment. Although threat hunting is considered more of a proactive approach to detect any current malicious activity in your environment, this exercise is typically run by your SOC or blue team. Threat-hunting activities will most likely follow one of these methods:

- **Structured**: This type of activity focuses primarily on TTPs and **Indicators of Attack (IoAs)** when hunting.
- **Unstructured**: Here, a threat hunter will focus primarily on IoCs as part of their activity.
- **Situational**: These types of hunts will typically use data from assessments executed against your environment.

Running ongoing threat-hunting activities is the ideal scenario but this may not be feasible for many. At a minimum, make sure you are running these activities as frequently as possible to detect any possible advanced threats against your organization. If you don't have the skill sets internally as part of your SOC, or with your outsourced SOC, you'll want to ensure you are working with an outsourced partner who can provide threat-hunting activities within your environment.

 In addition to threat hunts, you will want to ensure you are executing penetration and application testing activities. These will be covered in more detail in *Chapter 11, Proactive Services*.

Now that we have covered threat detection activities and what is involved, next, we will cover incident management and response and what is involved once an incident has been detected.

Incident Management and Response

Incident management and response involves everything that is needed to investigate any identified incident within your environment. Once an incident has been determined, the incident management and response protocols will need to be invoked.

This process can become complex depending on the severity, impact, and magnitude of the incident. The key to more efficient incident management and response is having a well-documented repository of all processes and procedures to ensure the identified incident can be resolved as quickly as possible.

Incident Handling and Severity

An important component of your SOC operations is the ability to track all the incidents and manage them efficiently from beginning to end. To accomplish this, you are going to need a ticketing system to handle all your incidents efficiently. The least-resistant path to enabling this functionality will most likely be through your current ticketing system within your IT function. There's a high probability there is already a ticketing system in place with IT for all users in the organization. Leveraging this system will provide easier enablement since all users will already be set up within this system and they will be familiar with its functionality. If a separate system must be used for your cybersecurity-specific incidents for reasons like outsourcing, ensure you make it a requirement to e-bond the cybersecurity ticketing system with the primary ticketing system used by your users to provide a simplified user experience.

Incident Reporting Methods

Once you have a ticketing system in place, you will need to determine how incidents will be opened within the ticketing system. Typical contact methods are via a telephone number or an email address that will open an incident within the ticketing system. You may look at some more modern methods such as chat functionality for your users if it makes sense for your environment. From a systematic perspective, you'll need to determine how incidents are opened based on any identified risk from your events and logging. Maybe there are email-generated high alerts that will need to immediately open an incident within the ticketing system. Or there may be a need to set up automation using SOAR that identifies incidents based on all the events and logs within the SIEM model. Regardless of the approach, you will need to ensure time is spent to ensure all scenarios for opening an incident are thought through and well documented. In addition, you'll need to ensure your users are aware of the communication options for them to open a cybersecurity-related incident.

Incident Categorization

Another important factor with your incidents is the categorization of your tickets. When an incident has been opened, how do you classify the incident category? For example, is the incident a network-related incident or a phishing-related incident? You'll need to ensure the categories are predefined and agreed upon before going live with the ticketing system.

The following are some of the more common categories you'll most likely want to use:

- Data loss/breach
- Policy violation
- Device loss or theft
- Social engineering/phishing
- Malware/ransomware
- Account compromise
- Network event/denial of service
- Insider threat
- Application/website compromise
- Zero-day vulnerability
- Other

Severity Assessment

Once an incident has been identified, a severity will need to be assigned. Determining the severity will depend on several factors for the incident, including impact, urgency, number of users, type of data, business outage, etc. Once all these factors have been determined, a severity will need to be assigned. A traditional severity model will consist of Critical, High, Medium, and Low, or Priority 1, 2, 3, and 4. Your priority matrix can be customized to meet your needs, but a very simple high-level model that can be used for reference is the **Information Technology Infrastructure Library (ITIL)** incident priority matrix. The following is a very basic example using impact vs urgency:

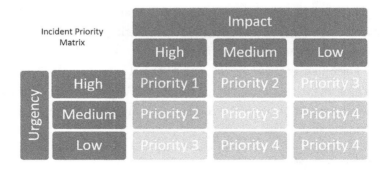

Figure 7.6: An example of an incident priority matrix

The severity will determine which path to take for incident investigation. You should also have SLAs in place against each of the severities, which we covered earlier in the chapter. For example, a critical-priority incident will need attention immediately versus a lower-priority incident. Once the investigation of an incident has begun, it will also need to be determined if the **Cybersecurity Incident Response Plan** (**CIRP**) will need to be invoked at some point. If so, this will kick off a whole separate process that will need to be followed. The CIRP will be covered in more detail within the *Business Continuity Planning (BCP)* section.

To reference a more detailed guide for incident handling, NIST has a great reference and publication available with details specifically focused on cybersecurity-related incidents. The *NIST Special Publication 800-61, Revision 2, Computer Security Incident Handling Guide* uses a four-step process for the incident response life cycle:

1. Preparation
2. Detection and analysis
3. Containment, eradication, and recovery
4. Post-incident activity

To learn more about the NIST incident response life cycle, visit the NIST *Computer Security and Incident Handling Guide* at https://nvlpubs.nist.gov/nistpubs/ SpecialPublications/NIST.SP.800-61r2.pdf.

Incident Investigation

As users open incidents and your SIEM continues to send alerts to your ticketing system, you are going to be challenged with ongoing investigations into each. One of the biggest challenges is the need to deal with false/positives versus true/positives. Ideally, you will want to ensure your false/positive rate is kept as low as possible. For this to occur, you'll need ongoing reviews as to why you are receiving false/positives and to fine-tune them as best as possible. At the same time, you need to be careful that you aren't allowing false/negatives to occur as you continue to fine-tune your false/positives.

The following are brief descriptions of a true/positive, false/positive, true/negative, and false/negative:

- **True/positive:** An alert or incident is triggered that is an actual threat.
- **False/positive:** An alert or incident is triggered that is not an actual threat.
- **True/negative:** No alert or incident is triggered and there is no actual threat.
- **False/negative:** No alert or incident is triggered when there is an actual threat.

The hope is that you will be able to orchestrate and automate your alerts as much as possible, but there will always be a need for analysts to investigate incidents as they come into the ticketing system. Referring back to the NIST incident response life cycle shared above, the analyst will need to quickly move into detection and analysis of an incident as they begin their investigation of the incident. The nature of the incident and the current threat being posed will determine how the incident will need to be investigated. Hopefully, there are well-documented processes and procedures the analysts can follow that will allow them to quicky analyze the incident to determine the next steps. These types of investigations will require a dynamic mindset as the majority of incidents will be unique and, at times, will need out-of-the-box thinking. Depending on the incident type, escalation or bringing in additional resources may be needed. For example, will there be a need for detailed malware analysis or malicious code review that is beyond the capabilities of the analyst who is currently investigating the incident? Is the incident turning out to be a much broader issue in which more senior analysts will need to be engaged? Or will there even be a need to invoke the major incident response plan? Once the analysis is completed, the next steps will need to be determined. If the incident is a false/positive, then it can be closed and noted as a false/positive. If it is a true/positive, you will need to move into the containment, eradication, and recovery phase as noted in the NIST incident response life cycle. Again, having a well-defined process and documented procedures will be critical in how you contain and eradicate an incident based on the type of incident. There may also be a need to escalate or engage more senior staff as these activities occur to ensure that the incident is contained as quickly as possible.

Root Cause Analysis (RCA)

After an incident has been resolved, it is important to ensure that an RCA is complete to understand how the incident occurred. This is a very important process and is often overlooked once an incident has been resolved. An RCA is a process that identifies the root cause of the incident, essentially identifying how the incident was able to occur.

This involves the collection, review, and detailed analysis of all evidence from all assets affected by the incident. It will be critical that all events and logs are accessible to conduct a thorough RCA in addition to ensuring everything is documented thoroughly with all details captured to provide clear reporting. Understanding or knowing the root cause will help prevent the same incident from occurring again. It will also help validate that the vulnerability that caused the incident has been remediated. Depending on the severity of the incident and the impact caused, you may need to provide the RCA of an incident to your leadership team and the board. In addition, there may be a need to provide additional details to your cybersecurity insurance, an external party if you are regulated, or even as evidence in court if any legal action is being taken. Because of this, the RCA should clearly define the following:

- Title of incident
- Date that incident occurred
- Description of incident
- Timeline of events including containment and restoration
- Root cause of incident
- Impact of incident
- Damage caused by incident
- Remediation or mitigation steps
- Future preventative measures
- Lessons learned

Next, let's look at **Digital Forensics Incident Response (DFIR)** and why it may be needed.

Digital Forensics Incident Response (DFIR)

DFIR is a capability and/or service that you will need to leverage in the event of a major cybersecurity incident. This is a very specialized service that will provide a detailed analysis of a cybersecurity incident. With DFIR, the focus will evolve around evidence collection, preservation of evidence, and specialized analysis of an incident. The output of this analysis will help with the RCA, better understanding what damage has been caused (any data exfiltration) and who may have initiated the attack, and confirming that the threat is still not active. This type of investigation requires a very specialized skill set, most likely one you won't have in-house. Even if you do have these skill sets in-house, you would need to bring in an external resource as a segregation of duty for any litigation issues that may occur. In today's world, with the ongoing increase of threats we face, it is important that you can quickly engage a company to provide DFIR services.

If you have a cybersecurity insurance policy (if you don't, you should consider purchasing one), you'll need to check with them about approved vendors that you can engage when DFIR services are needed. You'll need to use a vendor approved by your cybersecurity insurance company or they may not cover the costs of the DFIR services. Once you are aware of the approved DFIR vendors from your cybersecurity insurance, it is recommended to onboard one (or a couple in the event of any conflict of interest) and have a contract in place to allow immediate engagement when needed. This will also allow you to define SLAs, pricing, and any other requirements needed. Like any vendor, you'll need to work through your due diligence with vendor onboarding, which will be covered in *Chapter 10, Vendor Risk Management*. If you don't onboard a DFIR, you'll need to work with a vendor assigned by your cybersecurity insurance, which will require additional contract work to be completed before any work can start. We will cover cybersecurity insurance in more detail in *Chapter 14, Managing Risk*.

SOC Analyst Tools

One final item to cover is the tools that you can expect your analysts to use as part of an ongoing investigation. There may be various tools needed for each investigation and every investigation may need different tools depending on the incident type. It will be important that your SOC team is enabled with the correct tools so they can be more efficient in determining the root cause of an incident and to ensure a thorough evaluation has occurred.

For example, one very common tool is *VirusTotal*: https://www.virustotal.com/. This is an open-source tool that allows you to scan for malware by uploading files or scanning domains, IPs, and URLs. There are many tools available and too many to list within this chapter. Some may require a license or subscription, but you should be able to reference a great deal of open-source resources as there are many available. A great resource to reference for resources or tools is the *OSINT Framework*, which we briefly mentioned earlier in the chapter. If you visit the OSINT Framework site (https://osintframework.com/), then click on a topic for which you would like resources or tools, you will be presented with more details or a final reference that will direct you to access the resources or tools listed. As an example, if you were looking for threat intelligence resources or tools, you'd open the link above and click on the **Threat Intelligence** option, and you'd be presented with resources or tools as shown in the following image:

OSINT Framework

Figure 7.7: Threat intelligence within the OSINT Framework

Now that we have completed our review of incident management and response, we will begin our review of **Business Continuity Planning (BCP)**. With BCP being such a critical area of focus, we have dedicated a section to reviewing it in more detail along with the components included as part of a comprehensive BCP program. Let's take a deeper look into BCP and the importance of this critical program.

Business Continuity Planning (BCP)

To finish this chapter, we will cover details on the importance of having a business continuity plan in addition to a couple of the sub-plans as it relates to cybersecurity and your infrastructure. The two additional plans are the **Disaster Recovery Plan (DRP)** and the **Cybersecurity Incident Response Plan (CIRP)**. When we look at the **Business Continuity Plan (BCP)**, it is considered a business-wide plan that focuses on the business as a whole to ensure continued operations in the wake of a major disruption. Within the BCP, multiple plans come together that comprise the overall plan, and there are many frameworks available to assist with the BCP program. The *SP 800-34 Rev. 1, Contingency Planning Guide for Federal Information Systems* NIST framework publication is a great resource for your BCP and can be found at `https://csrc.nist.gov/publications/detail/sp/800-34/rev-1/final`. The following image from NIST *SP 800-34 Rev. 1* represents all the plans covered within the publication.

Figure 7.8: NIST SP 800-34 Rev. 1 individual plans that make up the BCP
Source: Figure 2-1 in the following document `https://nvlpubs.nist.gov/nistpubs/Legacy/SP/nistspecialpublication800-34r1.pdf`

Although an older visual, I am more partial toward the visual provided in the older version of the document NIST SP 800-34, which I tend to use as my reference point for how a BCP program should look. The image can be found in the older NIST contingency planning publication *NIST SP 800-34, Contingency Planning Guide for Information Technology Systems (Figure 2-2):* `https://nvlpubs.nist.gov/nistpubs/Legacy/SP/nistspecialpublication800-34.pdf`.

Let's take a brief look at each of the components that comprise the BCP:

- **Business continuity planning**: The broader plan covering the entire business that documents the process and procedures to ensure ongoing operations in the event of a major disruption.

- **Crisis communications plan**: This plan covers all communication plans (internal and external) for your BCP.

- **Continuity of Operations Plan (COOP)**: This plan covers the details and procedures for ensuring critical operations can resume at a secondary site for up to 30 days.

- **Occupant Emergency Plan (OEP)**: The OEP is used in response to a physical threat. This plan covers reducing the impact on people and any potential damage to property.

- **Disaster Recovery Plan (DRP)**: This is the technical-specific recovery plan and will be covered in more detail later in the chapter.

- **Information System Contingency Plan (ISCP)**: Tied closely with the DRP, the ISCP focuses on the procedures for the restoration of a specific system or service.

- **Cybersecurity Incident Response Plan (CIRP)**: This is the cybersecurity-specific response plan and will be covered in more detail later in the chapter.

- **Critical Infrastructure Protection (CIP) plan**: The CIP will not apply to everyone and focuses on ensuring the protection and restoration of critical infrastructure.

- **Business Recovery (or Resumption) Plan (BRP)**: This plan is not covered in the updated version of the NIST contingency planning guide, but it is important to reference as you may come across it. This plan focuses specifically on the recovery steps needed to bring your business operations back to a normal operating state.

Not all these components may be necessary to make up the broader BCP for your organization. But make sure you are familiar with them as they all serve a unique purpose. At a minimum, make sure you have the higher-level BCP in place with a well-documented CIRP, DRP, and crisis communication plan.

The BCP is not a simple plan to create; it requires a lot of time and resources to make it successful. In addition to building a well-documented plan, it is just as important to ensure that everyone is familiar with the plan and that it has been tested in some way. When it comes to executing the BCP in a real-world scenario, you don't want to be doing so for the first time without at least being familiar with the process and steps involved. To be better prepared, different exercises can be completed to become more familiar with the BCP, and one of the more common BCP exercises is in the form of a tabletop exercise. In a tabletop exercise, you bring together the key personnel in your organization and walk them through a crisis that may occur. When walking them through the situation, you request a response on how to handle the situation to identify any gaps in a real crisis. Tabletop exercises will be covered in more detail in *Chapter 11, Proactive Services*.

Some examples that may trigger the execution of the BCP include natural disasters, such as hurricanes, earthquakes, and floods. Depending on the severity, fires or power outages are other examples that may cause the BCP to execute. Today, more common threats that will trigger the BCP include cyberattacks, which can easily bring a business to a halt very quickly. To create a BCP for your organization, you may want to bring in an external resource that specializes in creating a BCP. Or if you are up to the task, there are many examples available on the internet to give you a start. Either way, make sure you are using the resources at your disposal as you create your plan. One example is that provided by the **Federal Emergency Management Agency (FEMA)**, *FEMA Continuity Plan Template and Instructions for Non-Federal Governments*: `https://www.cisa.gov/resources-tools/resources/fema-continuity-plan-template-and-instructions-non-federal-governments`.

FEMA Continuity Plan Template and Instructions for Non-Federal Governments includes the following major sections to include in your BCP:

- Promulgation Statement
- Confidentiality Statement
- Essential Functions
- Essential Records and IT Functions
- Human Resources
- Communications
- Alternate Locations and Telework
- Reconstitution
- Devolution
- Budget and Acquisition

- Multi-Year Strategic Planning
- **Training**, **Testing**, and **Exercising** (TT&E)
- Supporting Appendixes

As part of your BCP, it will be important to ensure you have references to other plans you have in place for the organization, like crisis communication, CIRP, DRP, etc. In addition, you will find that there will most likely be conflicting sections across your plans. To allow for easier governance and updating of the documents, you'll want to centralize specific sections so you aren't maintaining several copies of the same information that will need to be updated throughout several documents. For example, centralizing all your contact information into the BCP may make the most sense, then simply reference each of your other plans back to the BCP contact so you only have to maintain one contact list.

One important item to mention is **Business Impact Analysis (BIA)** and the importance of completing one for every function within the organization to understand the impact of an outage on the business. The BIA will also allow you to prioritize restoration activities for the more critically identified business functions to allow the business to get up and running as fast and efficiently as possible. Referring back to *SP 800-34 Rev. 1, Contingency Planning Guide for Federal Information Systems, Appendix B* provides a sample BIA along with a BIA template.

Disaster Recovery Planning (DRP)

DRP is primarily technical in nature and focuses on the recovery of IT infrastructure and systems. DRP falls within a larger BCP plan for an entire organization, and the execution of DRP will be unique, depending on the situation, and it may not impact the entire business. Events that trigger BCP will typically have some form of impact on your IT systems and may require the execution of DRP at the same time. Some examples include a natural disaster, such as a hurricane or fire destroying a data center.

As stated already, the entire DRP may not need to be fully executed in certain situations. For example, if a business-critical application becomes corrupted and is not highly available, you may only need to execute the DRP for that specific application to ensure ongoing operations. Many different instances can cause the DRP to be activated, so it's critical to account for each of these scenarios and make sure you can accommodate recovery for individual systems or an entire data center. A key component of DRP is understanding the impact of a service or function and the **Maximum Tolerable Downtime (MTD)** that a service or function can withstand before your business is negatively impacted. Two important factors that also need to be understood are the **Recovery Time Objective (RTO)** and **Recovery Point Objective (RPO)**.

- The RTO is the maximum amount of time a system can be unavailable before negatively impacting a service or function within the business and preventing it from being able to operate normally.

- The RPO is the point in time at which the service or function can suffer data loss without being negatively impacted.

As you build your DRP and understand the expected restoration of each system based on the RTO and RPO, you are going to need to ensure that you have the process and infrastructure in place to meet those requirements. A few examples to consider include the following:

- Having **High-Availability (HA)** configurations for systems may help restoration after any local issues within a data center but will come at a cost.

- Understand your backup strategy as it relates to full backups, incremental backups, and differential backups, and also consider how often each backup needs to be taken and retained.

- Consider what type of failover is needed for a complete restoration of a data center, should you have a cold, warm, or hot site available.

All these considerations will depend on the business requirements and needs, along with the expense. Having a hot site on standby will come at a much greater cost than a warm or cold site. More importantly, as you implement your DRP and backup strategies, the role security plays in each should be well understood. Ensuring that your data is backed up securely, off-site storage of data is secure, and your standby data center maintains the same level of security as your primary data center should all be taken into consideration. One last point to mention is testing. Like your BCP, it is critical that you test your DRP to ensure that everyone is familiar with the restoration procedures and to confirm that the DRP works. This is typically validated by executing the DRP, restoring part of or the entire infrastructure in a secondary location, and validating that services and applications are functioning. This should be an annual event at a minimum.

If you search on the web, you will find many examples and templates at your disposal. These will help provide you with a baseline to get started when needing to create a DRP for your organization. An example for reference is IBM's *Example: Disaster Recovery Plan*: https://www.ibm.com/docs/en/i/7.5?topic=system-example-disaster-recovery-plan. This plan includes 13 major sections to include in your DR plan:

- Major goals of a disaster recovery plan
- Personnel
- Application profile

- Inventory profile
- Information services backup procedures
- Disaster recovery procedures
- Recovery plan for mobile site
- Recovery plan for hot site
- Restoring the entire system
- Rebuilding process
- Testing the disaster recovery plan
- Disaster site rebuilding
- Record of plan changes

One important consideration that mustn't be overlooked is your backups as part of your DRP. With the advancement of ransomware and threat actors using tactics to take ownership of your backups or destroy them, it is critical that they have extra protection measures in place. The following is a list of modern-day best practices for your backups:

- Maintain multiple copies of your backups. An example is the 3-2-1 backup method, which requires three copies of the backed-up data on two different types of media, with one being off-site.
- Implement an airgap with your backups. This is done by maintaining a copy of your backups offline and ensuring they cannot be accessed by any network with internet connectivity.
- Ensure all backups are encrypted and protect all private keys used to encrypt your backups.
- Make sure you utilize immutable backups to prevent them from being altered.
- Ensure application backups are being executed. You may need to coordinate with the business to document backup procedures for these applications.
- Test your backups regularly.
- Ensure you have well-documented backup procedures, review them frequently, and keep them up to date.
- Ensure you have a multi-layered access model to access your backups, which includes phishing-resistant MFA, PIM, and/or PAM to ensure accounts are protected.
- Implement monitoring and alerts on all backup activity and access.
- Ensure auditing is in place for all backup activity.

Cybersecurity Incident Response Plan (CIRP)

As you are aware, one of the more advanced and disruptive threats is ransomware, which is known to severely impact business operations. Many companies aren't prepared for these types of attacks, which can take days, weeks, or even months to recover from and get operations back to normal. Depending on the size of the business and the severity of the breach, the impact could be as damaging as a complete business shutdown. With that, it is critical that your organization fully understands the impact of a potential breach and how to best be prepared to handle one. For optimal preparation, a CIRP can help organizations navigate through a breach when it occurs. Your CIRP should include elements such as identifying the roles and responsibilities of the incident response team and updating the contact information of everyone involved, including vendors, incident response procedures, communication procedures, and playbooks.

Once you have a production-ready CIRP, all parties involved in a cybersecurity incident should become familiar with the CIRP. You will also want to test the response of both your technical and executive teams with a tabletop exercise to address any gaps in your CIRP and better prepare the team for what decisions may need to be made in such a situation.

 We will cover the CIRP in more detail in *Chapter 11, Proactive Services,* including going into detail with what is required to build a CIRP in addition to covering different playbooks and their importance.

For any plans that have been created to support the business, including the BCP, DRP, and CIRP, it will be critical that they are reviewed and updated on an annual basis at a minimum. Within these plans, you are going to have contacts, key personnel, vendors, and processes that will change throughout the year and it will be critical that the most current information is listed in each of the plans. As you review and update your plans, ensure you version control and add the date of modification. There should also be an approval process for when plans are reviewed and updated. A thorough BCP, DRP, and CIRP will require a lot of thought and collaboration between the IT team, cybersecurity team, broader business stakeholders, and executive team. The success of these plans will require the support and involvement of leadership to ensure that the business is best prepared for when an incident does occur.

Summary

As you have seen from this chapter, cybersecurity operations is a very busy and active component of the overall cybersecurity program. Ensuring this function is operating as efficiently as possible will allow for quicker response and resolution as incidents occur within the organization. Ensuring you have some form of 24/7/365 operations in place with your cybersecurity program is no longer a choice these days. Obviously, this comes at a cost, and accomplishing this internally may not be realistic. But there are options such as outsourcing and engaging MSSPs who specialize in 24/7/365 operations. Because of the ongoing demand on your employees within this function, ensure you have their wellness top of mind and that they are not being overworked and burnt out.

The first part of the chapter provided an overview of cybersecurity operations and the main components involved in completing the cybersecurity operations program. This included the SOC, threat detection, and incident management and response. In addition, we covered AI in more detail and the impact it is having from a cybersecurity perspective, both defensive and offensive. We then moved into a detailed review of the SOC and everything involved in operating and running an efficient SOC. This included reviewing the different cybersecurity operational models like a traditional SOC, SOCaaS, MDR, MSSP, and a hybrid model. We then reviewed the SOC organization structure before reviewing log collection, analysis, and automation for your SOC. We then finished off the section by looking at the importance of processes and SOPs, along with a review of what SLAs and key metrics are and why they are important, and finally, the importance of governance.

In the next section, we covered threat detection and the different components involved. This included asset management and visibility, digital asset monitoring, threat intelligence, and threat hunting. This led into the next section, which provided details of incident management and response in which we covered incident handling and severity, incident investigation, RCA, and DFIR, before finishing off the section with an overview of the importance of the tools needed for your SOC. In the final section, we reviewed the importance of a BCP and what exactly is involved within a BCP. We then focused our attention on two important topics, DR and CIRP.

In the next chapter, we will be reviewing vulnerability management. This is where we will be covering everything related to identifying, tracking, and mitigating vulnerabilities within your environment. We will begin the chapter with an overview of the broader vulnerability management program along with managing it. This will be followed by vulnerability discovery and alerting within your environment before moving on to tracking and remediating any identified vulnerabilities. Next, we will go into detail on update management and email protection for your organization. We will then finish the chapter by reviewing other vulnerability considerations.

Join our community on Discord!

Read this book alongside other users, Cybersecurity experts, and the author himself. Ask questions, provide solutions to other readers, chat with the author via Ask Me Anything sessions, and much more.

Scan the QR code or visit the link to join the community.

`https://packt.link/SecNet`

8
Vulnerability Management

In this chapter, we will be reviewing everything you need to consider as part of your vulnerability management function. This function, to some extent, will be an extension of your cybersecurity operations, or SecOps, function. However, vulnerability management needs to be tracked separately because of the vast amount of effort needed to run this function efficiently. If you are managing vulnerabilities correctly, you will find this function to be very active, with your vulnerability analysts/administrators working constantly to remediate vulnerabilities. Like the cybersecurity operations function, you will need to ensure you have the dedication needed to run this function efficiently, along with ensuring you aren't overworking those responsible for vulnerability management activities.

We will begin the chapter with an overview of managing your vulnerability management program. First, we will look at some building blocks needed to create your program. We will then investigate program management and governance before finishing the section with more details on asset management. Although we covered asset management in the previous chapter, it is important we continue to stress the importance of asset management, especially as it relates to vulnerability management. If we don't know it exists, we can't remediate any vulnerabilities. In the next section, we will take a deeper look into vulnerability discovery and alerting, where we provide an overview of vulnerabilities before moving on to the different places you can expect to receive vulnerability alerts and the different ways to scan your assets for vulnerabilities.

Following, we will review vulnerability tracking and remediation. Here, we will look at different options to track your vulnerabilities along with gaining a better understanding of scoring your vulnerabilities based on criticality. This will lead to the topic of vulnerability remediation, where we will look at how to best track your vulnerability remediation activities.

We will finish the section with a look into some ways to modernize your vulnerability management program. Next, we will turn our focus to two areas that require the most effort for vulnerability management: first, the ongoing update management lifecycle, followed by a deeper review of how to best protect your email service from ongoing exploitation. We will finish off the chapter with a look into other vulnerability management considerations, which includes hardware vulnerabilities, your virtualization infrastructure, network infrastructure, cybersecurity testing activities, and auditing and assessments. The following will be covered in this chapter:

- Managing your vulnerability program
- Vulnerability discovery and alerting
- Vulnerability tracking and remediation
- Update management and email protection
- Other vulnerability management considerations

Managing Your Vulnerability Program

First, let's take a high-level look at all the sub-functions that should be addressed as part of vulnerability management. The following image captures much of what the vulnerability management function entails.

Figure 8.1: Sub-functions of the vulnerability management function

As a reminder, vulnerability management is the process of identifying and remediating vulnerabilities – more specifically, identifying and remediating vulnerabilities as they relate to and impact your organization. The following image provides a more detailed visual of the steps involved in vulnerability management.

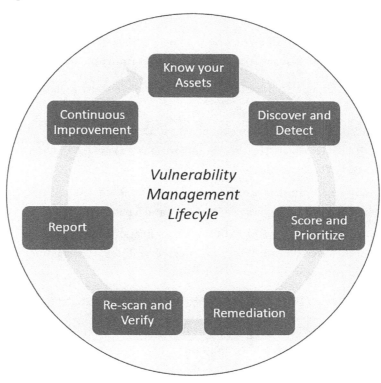

Figure 8.2: The vulnerability management lifecycle

Vulnerability management is yet another function within the cybersecurity program that requires never-ending commitment, unfortunately. This chapter comes right after the cybersecurity operations chapter for a reason. In some respects, your vulnerability management activities can be considered part of your cybersecurity operations as there will be overlap in some of the work, along with synergies needed between the two functions. But because vulnerability management requires such a large effort to manage and operate efficiently, it needs to be managed separately as its own function. In addition, dedicated resources are going to be needed to bring efficiency in this area. Similar to cybersecurity operations, your staff in this area will need to be managed carefully in terms of being overworked and burning out.

As already mentioned, vulnerabilities will always exist, and they are only becoming more pervasive. The more technology we continue to introduce, the more vulnerabilities are detected. There are thousands and thousands of vulnerabilities being documented monthly, which requires evaluation to understand what impact they may pose to our organizations. Our challenge is remediating or mitigating the ongoing vulnerabilities in a timely manner.

Before we go into more detail on how to manage and oversee your vulnerability management program, let's take a look at some high-level statistics from the 2023 Qualys TruRisk Research

- 2.3 billion anonymized vulnerabilities were detected by the Qualys **Threat Research Unit (TRU)** in 2022.

- The average noted time to weaponize a vulnerability is 19.5 days, while the same vulnerabilities have a **Mean Time to Remediate (MTTR)** of 30.6 days and are only patched on average 57.7% of the time.

- Windows and Chrome weaponized vulnerabilities are patched much faster than all other applications with an average of 17.4 days and a patch rate of 82.9%.

- In 2022, Qualys scanned 370,000 web applications in which more than 25 million vulnerabilities were found. 33% of these vulnerabilities were classified as OWASP Category A05, which represents misconfiguration.

Report Source: `https://www.qualys.com/forms/tru-research-report/`

Now that we have better insight into the challenge ahead of us, let's take a look at the building blocks needed for your vulnerability program.

Building Blocks for Your Program

Like with any of your functions within the cybersecurity program, it is important to define how each individual program will run. You can't expect vulnerabilities to be resolved on an ad hoc basis. A formalized program needs to be defined that allows the team to efficiently work through remediating or mitigating identified vulnerabilities. A policy will need to be defined and processes will need to be documented for your staff to follow. If thousands of vulnerabilities are identified on a monthly basis, how does the team know what to prioritize? How does the team identify which vulnerabilities apply to the organization? How does the team know which systems to update? How does the team know who to work with from the business? There are many more questions. All these items will be addressed as part of building and defining the vulnerability management program for your organization.

As we continue to mention throughout the book, it is important to use resources like the **National Institute of Standards and Technology (NIST)** guidelines, CIS Controls, and others that are at our disposal. Governments and organizations around the world provide a lot of great material to better support organizations of all kinds to reduce risk and build stronger cyber resilience. The same applies to vulnerability management. There is a lot of great material available to support the building of your vulnerability management program. A recent example is from the **National Cyber Security Centre (NCSC)**, which has recently made a great reference guide available on vulnerability management, which can be found here: `https://www.ncsc.gov.uk/collection/vulnerability-management`. The guide provides five distinct areas to cover as part of vulnerability management:

- Put in place a policy to update by default.
- Identify your assets.
- Carry out assessments by triaging and prioritizing.
- The organization must own the risks of not updating.
- Verify and regularly review your vulnerability management process.

These items will all be covered throughout the chapter as we journey through each section.

Program Management and Governance

As you formalize and begin to mature your vulnerability management program, there will be several governance-related items that you want to ensure are in place as the vulnerability management team works through each identified vulnerability:

- **Risk management:** This is obviously at the core of the vulnerability management program. As stated, managing risk is our primary role, and this couldn't be more relevant with managing vulnerabilities. It is important that we understand what the risk of each vulnerability poses to the organization. Understanding the risk will determine how we prioritize our work to ensure those vulnerabilities that come with greater risk receive the needed attention before lower-risk vulnerabilities. This essentially will be documented as part of your vulnerability policy and procedures, which we cover next. It is also important to remember that risk is something that will ultimately need to be accepted by the business, specifically at the executive leadership and board level. You as a leader will need to determine when any vulnerabilities that can't be remediated or mitigated will need to go to the leadership level for awareness and acceptance. Make sure the risk acceptance is not happening at the vulnerability-team- or CISO-only level.

- **Policy and procedures:** As briefly mentioned, it is critical that your cybersecurity program includes well-documented policies and procedures for each function within the program and is signed off by the executive leadership team. Your vulnerability management policy will need to cover items such as the purpose, scope, prioritization of vulnerabilities, remediation, exceptions, etc. The CIS Critical Security Controls Version 8 includes several policy templates, including one on vulnerability management, which can be found here: `https://www.cisecurity.org/insights/white-papers/vulnerability-management-policy-template-for-cis-control-7`. The same applies to procedures; well-defined, step-by-step instructions must be in place for the vulnerability analysts to follow. We will cover policies and procedures in more detail in *Chapter 13, Governance Oversight*.

- **Roles and responsibilities:** Within your vulnerability management team, you will most likely have roles such as a vulnerability manager, a vulnerability administrator/engineer, and vulnerability analysts or general cybersecurity analysts. Smaller organizations may have a joint role for these responsibilities. It's important to understand that the roles and responsibilities don't stop within the vulnerability management team. The extent of vulnerabilities will impact many different areas within the organization. Anything technology-related will have vulnerabilities identified at some point. This means responsibilities will need to include the broader IT team to cover infrastructure (and other areas), business application owners who manage and administrate applications within their functions, outside vendors who manage applications within your organization, etc. Because of this, it will be critical that a well-defined roles and responsibilities matrix is in place for everyone responsible for everything related to the vulnerability management program. This includes having up-to-date contact information as the vulnerability management team will need to constantly engage asset owners in a timely manner.

- **Vulnerability tracking:** As you receive notifications about vulnerabilities that impact your organization, you will need to ensure that you are efficiently tracking them to resolution. As you track your vulnerabilities, it will be important that all relevant information is included, such as the vulnerability owner, risk identified, initial date identified, systems/applications impacted, etc. How you track these vulnerabilities will be the difficult part. For the most part, you may have a primary tool that is used to identify vulnerabilities and allows you to track vulnerabilities, but there will be other sources you will need to track. Maybe you build a digital tracking tool that allows you to track everything, or to get started, use a spreadsheet that will allow you to track everything in one place for review.

- **Meetings and status calls**: Once you have determined how you will track your vulnerabilities, it will be important to book some time in your calendar at a frequent cadence to review all vulnerabilities and discuss the current status. At a minimum, your vulnerability management team should be connecting multiple times a week as they work through vulnerabilities. A minimum of a weekly status call with management should be in place to review any outstanding vulnerabilities and the potential for any escalations that may be needed.

- **SLAs, metrics, and reporting**: Like everything within the cybersecurity program, you are going to want to ensure metrics are being produced to ensure the program is running efficiently and any SLAs that are in place are being met. This will also tie back to the policy with any defined timeframes of when vulnerabilities should be remediated based on criticality. A couple of examples include the average time it takes to patch your systems/applications since a patch was made available, or the time taken to patch your systems/applications since a vulnerability was identified. Once you have your metrics, you'll want to ensure reports are compiled and available regularly so you can track progress over time and share the reports with the executive leadership team and the board.

- **Vulnerability tools**: Finally, it will be important that the vulnerability management team is enabled with the right capabilities to ensure they are able to efficiently scan the entire environment for vulnerabilities. There may be multiple tools needed to efficiently complete a thorough vulnerability assessment with ongoing scans throughout the environment, but it's important that the right tools are made available and utilized fully to ensure risk is being reduced as much as possible. Some of the more popular vulnerability management tools will be referenced in the *Vulnerability Scanning* section.

 Program management and governance of the vulnerability program will in turn tie back to the broader **Governance, Risk, and Compliance (GRC)** for the cybersecurity program. Everything from managing risks with vulnerabilities, building policy, and tracking vulnerabilities to building metrics, creating reports, etc. will need to be tracked as part of the broader GRC program. We will cover GRC in more detail in *Part 3, Bringing It Together*, of the book.

It is important to note that this function will require ongoing collaboration beyond the IT teams with the broader organization. Ensure the vulnerability administrators and/or analysts have good collaboration and communication skills as they will need to reach out and work with a much broader group of users.

Asset Management

With asset management being so important for your cybersecurity program as a whole, it will be referenced several times throughout the book. If you recall, we covered asset management and visibility briefly within *Chapter 7, Cybersecurity Operations*. As part of your cybersecurity operations function, not knowing all the assets within your organization translates into not being able to monitor for any malicious activity occurring. The same applies to vulnerability management. If we don't have a complete inventory of all assets within the environment, then we can't efficiently protect the organization from known vulnerabilities. Clearly, if there is a gap within our asset management processes, the risk of a compromise increases significantly.

As reviewed in *Chapter 7, Cybersecurity Operations*, tracking all assets is not easy and requires a lot of invested time to ensure everything is being tracked. One of the more traditional methods reviewed is a **Configuration Management Database (CMDB)**, which will typically fall within your **Information Technology Services Management (ITSM)** function. Having a CMDB in place today is a great start and should capture the majority of your assets. With the advancement of technology in general, and more specifically cloud technology, efficiently tracking all your assets becomes extremely challenging, especially as resources are continuously spun up and decommissioned. Leveraging more modern capabilities is going to be required to provide better visibility into all our assets. One new approach we also discussed in *Chapter 7* is **Cybersecurity Asset Management (CASM)**. This concept essentially allows for a more dynamic and up-to-date asset inventory of your environment. Instead of a more static approach like your CMDB, a CSAM solution will continuously scan your environment for new assets to ensure no gaps. In addition, you'll be able to configure multiple sources to allow ingestion from different areas. Examples are integration with your CMDB for a baseline, configuring integration with your cloud environments to ensure all those assets are being accounted for, bringing in your vulnerability scanning tools, and the ability to integrate with your endpoint management solution like Microsoft Defender or CrowdStrike Falcon. As you begin to collect all these different data points, you'll have a much clearer picture of all assets with the goal of ensuring there are no gaps within your asset management program.

Obviously, getting to this level of asset management isn't going to happen overnight, but it needs to be the goal as all it takes is one rogue device that hasn't been updated to lead to a major breach within your environment. As hard as this may be to accept, it is the reality we face today, and it has happened many times (and will continue to happen). If you don't have an asset inventory today, at least start with the basics by using a spreadsheet. While not ideal, it will allow you to better understand and track everything in your environment.

When you are working through your vulnerabilities, you need at least something to reference and track as you need to remediate vulnerabilities. Having nothing will provide no structure or efficiency as you work through remediating or mitigating your identified vulnerabilities.

As a reminder, you will need to ensure all digital assets are being tracked as part of your asset management. This will include, at a minimum, all:

- Endpoints
- Servers
- Network equipment
- Cloud resources
- Software and applications, including SaaS
- Users and other accounts
- OT & IoT

 Something else to consider is the **Software Bill of Materials (SBOM).** This is essentially a detailed inventory of all components and dependencies associated with an application. This helps provide better visibility into all components of an application, allowing for increased security along with being able to more efficiently manage risk within the software supply chain. This in turn can allow your vulnerability management team to track any vulnerabilities with the components used with an application.

One final consideration to track is asset owners, more specifically, application owners, and up-to-date contact information for each. Whether on a spreadsheet as your first revision of asset management or within a system where you're tracking your assets, it will be critical that you have the most current owner and contact information available. As vulnerabilities are identified against assets, you will need to quickly identify who owns the asset and who is responsible for maintaining the asset so they can quickly remediate any identified vulnerabilities. Time is of the essence and the faster the vulnerability management team can identify the asset owners to remediate a vulnerability, the safer the organization will be. This again falls back to having vulnerability management staff who are efficient and collaborative at what they do to work through vulnerabilities as quickly as possible. The tracking of owners and contact information will be discussed further in *Chapter 13, Governance Oversight*, where this type of information becomes critical to track as part of the broader governance program for your cybersecurity program.

Now that we have completed an overview of managing your vulnerability management program, let's shift our focus onto the important aspect of discovering and being alerted of vulnerabilities.

Vulnerability Discovery and Alerting

Now that you have a little more insight into some of the ways you should be overseeing your vulnerability management program, the next step is to begin execution to better understand what vulnerabilities exist within your environment and how to be well informed of the most current vulnerabilities. First, let's make sure we fully understand how vulnerabilities are tracked and managed.

Vulnerability Overview

You may have noticed that as vulnerabilities are published, they each have a **Unique Identifier (UID)** to reference the vulnerability, beginning with CVE. **CVE** stands for **Common Vulnerabilities and Exposures** and is the standard for vulnerability management, allowing one source to catalog and uniquely identify vulnerabilities. CVE is essentially a list of disclosed vulnerabilities discovered by someone or an organization and made available for the public to review. CVE is operated by the MITRE Corporation and funded by the US **Department of Homeland Security (DHS)** and CISA. You can learn more about CVE here: https://www.cve.org/About/Overview.

The following are the high-level steps in the lifecycle of a CVE record:

1. **Discover:** A new vulnerability is discovered.
2. **Report:** The vulnerability discovered is reported to CVE.
3. **Request:** A unique **CVE Identifier (CVE ID)** is requested.
4. **Reserve:** A CVE record is initiated with a reserved ID.
5. **Submit:** Details of the vulnerability are submitted to the CVE.
6. **Publish:** Once minimum requirements are met, the CVE is made available for the public to view.

Information on the CVE record lifecycle process can be accessed here: https://www.cve.org/About/Process. To access individual CVEs or download the entire catalog, access the CVE home page (https://www.cve.org/) and enter the known CVE or follow the instructions to download the entire catalog. As of September 2024, there are 240,830 officially documented CVEs:

CVE® Program Mission

Identify, define, and catalog publicly disclosed cybersecurity vulnerabilities.
Currently, there are **240,830** CVE Records accessible via Download or Search ⬀

Figure 8.3: The CVE home page with the total number of vulnerabilities

Now that we know where vulnerabilities are formally issued and tracked, let's review NVD. **NVD** is the **National Vulnerability Database** (https://nvd.nist.gov/), which is an additional resource for vulnerability management provided by NIST. NVD is synchronized with CVE to ensure the latest updates appear within its repository. NVD provides additional analysis of the vulnerabilities listed in the CVE dictionary by using the following:

- **Common Vulnerability Scoring System (CVSS)** is used to measure the severity of a vulnerability using a numeric value from 0 to 10. The higher the number, the more critical the vulnerability.

- **Common Weakness Enumeration (CWE)** is a categorization for identified software and hardware vulnerabilities. For example, a vulnerability categorized as CWE-284 would translate to improper access control.

- **Common Platform Enumeration (CPE)** is a structured naming standard used to identify hardware and software. An example of a CPE for Microsoft Windows 10 version 1507 looks like *cpe:2.3:o:microsoft:windows_10_1507:-:*:*:*:*:*:x86:**.

If you search for a vulnerability within NVD, you will be provided with additional details on an identified vulnerability. The most important detail is the calculated CVSS, which helps us to better prioritize. To search for a vulnerability, browse to https://nvd.nist.gov/vuln/ then type in your criteria to search. I simply searched VMware and was presented with VMware-specific vulnerabilities.

In the following screenshot, you will see a column referencing CVSS severity.

Figure 8.4: NVD showing vulnerabilities with CVSS severity

You can view additional details on the vulnerability by clicking on the CVE number, or if you click on the CVSS severity number, you will be provided with more details on how the vulnerability was scored:

⊞ Common Vulnerability Scoring System Calculator CVE-2023-34045

Source: NIST

This page shows the components of a CVSS assessment and allows you to refine the resulting CVSS score with additional or different metric values. Please read the CVSS standards guide to fully understand how to assess vulnerabilities using CVSS and to interpret the resulting scores. The scores are computed in sequence such that the Base Score is used to calculate the Temporal Score and the Temporal Score is used to calculate the Environmental Score.

Figure 8.5: CVSS v3.1 calculator scoring of a CVE

A notable change to be aware of is the CVSS severity. You may be familiar with the scoring system for vulnerabilities of low (0.0-3.9), medium (4.0-6.9), and high (7.0-10.0), which is reflected on older vulnerabilities. This was known as the published v2.0 of CVSS ratings. NVD is now on its fifth version (previous versions: v1.0, v2.0, v.3.0, v3.1) and using the following CVSS v4.0 ratings:

- **Low**: 0.1-3.9
- **Medium**: 4.0-6.9
- **High**: 7.0-8.9
- **Critical**: 9.0-10.0

As CVSS has matured through the years with its versioning, improvements have been made to the scoring system. For example, within the Base Metrics group, some new metrics were added in v3.0 when moved from v2.0. These new metrics included a **Privileges Required (PR)** and **User Interaction** metric in addition to some other changes. CVSS v4.0 was released on November 1, 2023, in which some of the benefits included increased granularity for the Base Metrics group in addition to a completely new group named Supplemental Metrics. The following shows the first two sections of the CVSS v4.0 calculator, which can be found here: `https://nvd.nist.gov/vuln-metrics/cvss/v4-calculator`.

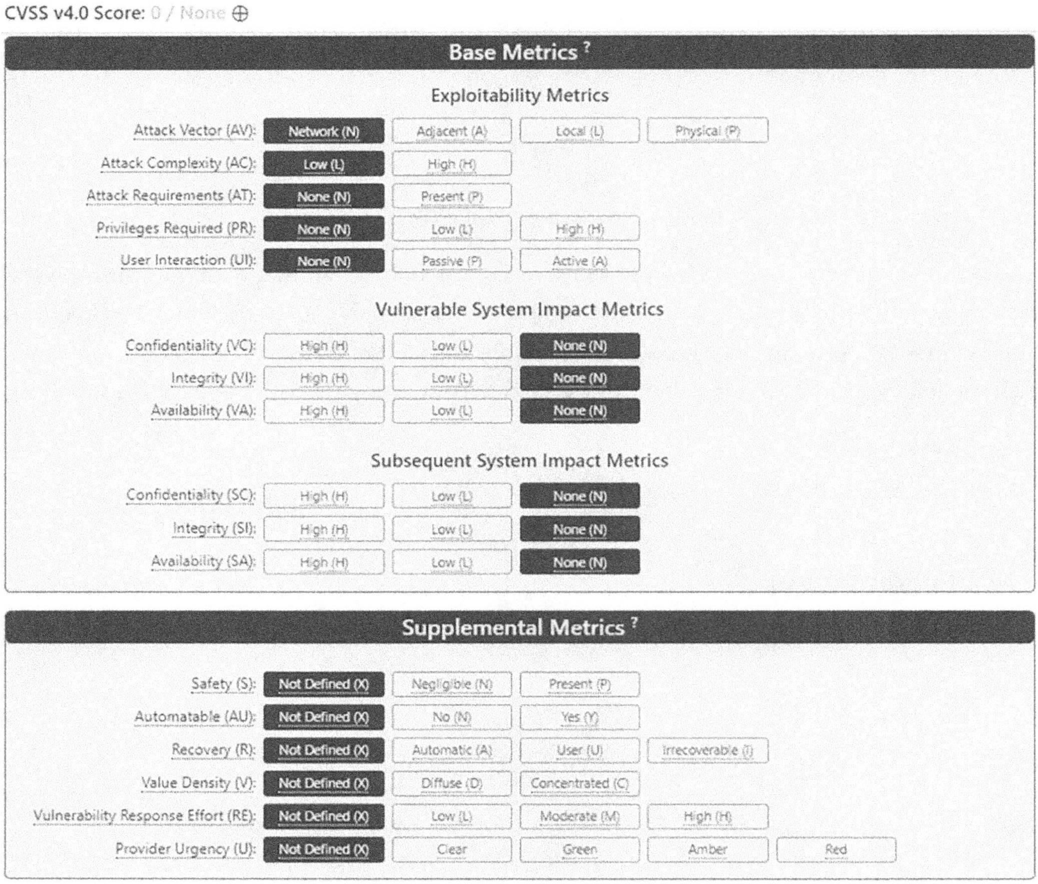

Figure 8.6: The Base Metrics and Supplemental Metrics of the CVSS v4.0 calculator

You may not see the v4.0 score appear for vulnerabilities right away. It may take time for them to begin to appear as vulnerabilities are added. The VMware CVE example above only had scoring available with v3.1, which is why the image shows the v3.1 calculator.

To finish the section, I think it's important to share a representation of how challenging the vulnerability landscape has become. Again, this should be another visual and metric your executive leadership and board are made aware of to better understand how important the investments being made are to continue to strengthen the cybersecurity program. The following image represents all documented vulnerabilities from 1988 through to July 2024.

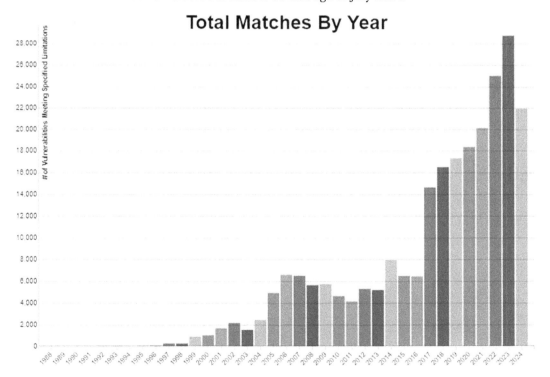

Figure 8.7: The total number of vulnerabilities in the NVD by year

Source of chart: Select **Statistics** in **Results Type** then click **Search** within this link: https://nvd.nist.gov/vuln/search.

As you can see, there was a significant increase from 6,447 vulnerabilities in 2016 to 14,643 vulnerabilities in 2017. Every year onward has shown a significant increase with a massive 28,830 vulnerabilities in 2023, the biggest to date. The scary part is that, as of July 2024, there are already 21,953 documented vulnerabilities. This will clearly surpass 2023 by the end of the year. So, as we continue to digitize more and more, and the more we continue to adopt technology as it becomes more widely available, vulnerabilities will only continue to increase. These numbers clearly coincide with the data and statistics shared in *Chapter 1, Current State*, where we continue to observe increased activity from threat actors. Essentially, the increase in technology equates to increased vulnerabilities, which in turn means more compromises to vulnerable technology. Because of this increase, it is critical we do everything we can to identify vulnerabilities within our environments. So, we will move our focus over to vulnerability scanning in the next section.

It is important to be aware that in February 2024, NVD announced it was struggling with the increase in vulnerabilities submitted and will only be focusing on significant vulnerabilities until it determines longer-term solutions. NVD has stated that they are committed to the continued support of NVD, but we don't know what that means at this time. The latest update to the announcement can be found here: `https://nvd.nist.gov/general/news/nvd-program-transition-announcement`.

Vulnerability Scanning

One of the more efficient ways to understand what vulnerabilities exist within your environment is through vulnerability scanning or assessments. For the most part, you'll be able to gain visibility into most of your environment if you have the right vulnerability scanning tool(s) in place.

When reviewing the market for a tool, it is important that you have clearly defined your requirements and understand what capabilities are needed to cover as much of your environment as possible. Like with any review of tools, you'll need to do your research and complete a **Request for Proposal (RFP)** to ensure you are making the best decision for the organization.

Vulnerability scanning tools will provide additional insight into your environment by identifying known exploits and weaknesses and raising awareness about how vulnerable your systems are against new and emerging threats. Your vulnerability management team should receive instant notifications as vulnerabilities are identified, allowing them to quickly react based on criticality. Ideally, your vulnerability scans will be set up on an automated schedule to look for and identify known vulnerabilities within your environment or systems on an ongoing basis.

For example, your vulnerability scan might detect that an application or operating system version on the network has a known vulnerability. Depending on the configuration, you may receive an instant notification if the vulnerability is critical, or you may simply receive an overview report that highlights all the vulnerabilities and actions that need to be remediated. Your vulnerability scanning tool will most likely provide dashboards and on-demand reports that can be viewed for further analysis. The following is an example of the Tenable vulnerability management overview dashboard provided on their website: `https://www.tenable.com/products/vulnerability-management`.

Figure 8.8: The Tenable vulnerability management overview dashboard

Similar to your asset inventory, you are going to want to ensure scanning capabilities are available for all digital assets within your environment. The following is a list of some of the more common types of vulnerability scanning tools you'll need to consider:

- Network/wireless scanning or assessments
- Web application scanning or assessments
- Application scanning or assessments
- Database scanning or assessments
- Host-based scanning or assessments
- Cloud infrastructure scanning or assessments
- OT & IoT scanning or assessments

Ideally, you'll want to strategize on a platform that can scan across multiple areas. This may not be possible when it comes to coverage across OT & IoT but it's important that you are able to retrieve as much vulnerability data on all your digital assets as possible.

There are many vulnerability scanning tools available today and if you don't have one deployed, you should be looking to deploy these capabilities. Like with everything that needs to be deployed, these tools will not turn on and run themselves; you need to ensure you have the resources available to manage and maintain them. Typically, your vulnerability management team will oversee your vulnerability management platform. Within a smaller organization, this may be a general cybersecurity administrator or engineer who oversees multiple platforms. Either way, it's important to note that in order to get the best value out of these types of capabilities, you need the committed resources to ensure they are set up and running per best practices. Some of the more common tools that you may be familiar with include the following:

- **Qualys**: https://www.qualys.com/
- **Tenable (Nessus)**: https://www.tenable.com/
- **Rapid7**: https://www.rapid7.com/products/insightvm/

You'll notice the above tools are all services that will require some form of subscription. There are a lot of open-source tools available that can accomplish different types of vulnerability scanning, but they will be more involved versus opting for a platform that will cover most of your environment and provide the needed support. Whichever tools you look to utilize, it will be important that your staff receives training on the capabilities in addition to receiving continued training to ensure you are taking advantage of all functionality as it is released.

In addition, you'll need to perform reviews of the tools you are using to ensure they still meet your requirements and confirm they are providing the latest capabilities. If not, you may need to revisit the market to review other options.

External Attack Surface Monitoring

The foundation of your vulnerability management program is to identify any vulnerabilities within your organization, whether external or internal. A new concept that is gaining popularity is **External Attack Surface Monitoring (EASM)**. EASM essentially focuses primarily on your external Internet-facing footprint for any vulnerabilities that can be found on the Internet versus your entire portfolio. This provides the ability to focus on remediating high-priority vulnerabilities since they are accessible from the Internet, which means threat actors are most likely already trying to exploit them if they haven't already. As vulnerabilities continue to increase and make the challenge to remediate more difficult because of the volume, we need to continue to identify ways to improve on prioritization. EASM is one of those areas that allows us to prioritize more efficiently to ensure our external attack surface has reduced vulnerabilities. We will cover prioritization in more detail later in the chapter. As EASM becomes more widely adopted, it will be important that these capabilities are built into your traditional vulnerability management platform. Ideally, work with your current vulnerability management platform to see if these capabilities are available as an add-on or will become available in the future. If you have a separate platform for EASM, you will need to ensure there is integration with your current vulnerability management capabilities to efficiently track these vulnerabilities with all other vulnerabilities.

We'll now transition away from vulnerability scanning to review the need for vulnerability alerting in more detail.

Vulnerability Alerting

You are going to find that the vulnerability management team will quickly become bombarded with many different vulnerability alerts from many different sources. Although more information can be good, it can also be challenging to handle. It will be important that alerts don't get lost, especially those deemed high-priority and critical. In addition, it will be important to ensure the vulnerability management team and the the **Security Operations Center** (**SOC**) team are collaborating closely as vulnerability alerts will need to be available for review by both teams in some instances. These teams will need to have a strong relationship. You can expect to receive alerts from the following sources.

Vulnerability Management System

The first area you can expect alerting to come from is your vulnerability management system, as we referred to in the previous section. The majority of your alerts will come from this system, so it will be important that the team is efficiently working through the vulnerabilities that are identified based on priority. It will also be important to ensure that alerting is set up appropriately in the vulnerability management system. The last thing you need is to receive alerts that aren't relevant. For example, informational findings will provide little value and should only be reviewed when reviewing the console or extracting reports for review. The same should apply to low and medium findings; leave these vulnerabilities for review as part of your frequent meetings, then ensure only high and critical findings are sending alerts to the vulnerability management team for review and remediation. This will help reduce a lot of the noise.

SOC

You will find your vulnerability management team will constantly receive vulnerability alerts from the SOC on an ongoing basis. Depending on your organization structure, you can expect ongoing collaboration between the SOC and vulnerability management team if the right processes and synergies are in place. As the SOC receives incidents from the SIEM, there will potentially be a need to apply some sort of remediation or mitigation for the identified vulnerability that is being investigated. This will require the vulnerability management team to review and work through the recommended remediation or mitigation steps. Ensuring the correct processes are in place to support this ongoing collaboration will be critical.

Threat Intel

As discussed in the previous chapter, **Cyber Threat Intelligence (CTI)** will provide a significant amount of information. This, for the most part, will be integrated into the SOC if set up correctly, and in turn, the SOC will reach out as intel is received with identified vulnerabilities within your environment.

In addition, your cybersecurity vendors may be sending you intel. For example, there may be daily intel emails being sent from your cybersecurity vendors.

As these feeds come into the team, they will also need to be reviewed and potentially addressed. Each one of these intel feeds, whether going directly to the SOC or coming into the vulnerability management team, may contain different information that will help you more efficiently protect the organization. As difficult as it may be, it is important that all intel information is reviewed to allow any identified vulnerabilities to be actioned within your organization.

 The more efficiently you can integrate threat intel with the vulnerability management processes, the quicker you will be able to address vulnerabilities within the organization. Look for ways to automate these processes that will allow you to strive for a real-time vulnerability management model.

External Sources

I want to make sure you are aware of a couple of extremely valuable external resources that you should be signed up for. If you don't have any threat intel services or you are just in the early phases of building a vulnerability management program, these two resources will provide you with a great start. Although there will be some overlap with the notifications, they will both provide intel to benefit your vulnerability management team.

The first resource is the **Multi-State Information Sharing and Analysis Center (MS-ISAC)**, which is supported by both CISA and CIS. When you sign up for subscriptions or alerts from MS-ISAC, you will receive newsletters and notifications on known vulnerabilities.

To subscribe to alerts with MS-ISAC, browse to https://www.cisecurity.org/cybersecurity-threats, click on **Subscribe to Advisories**, and follow the instructions to subscribe.

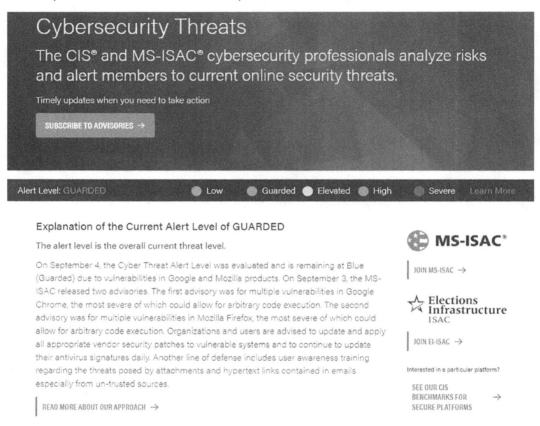

Figure 8.9: The CIS security subscribe to advisory webpage

The next important resource is available through CISA and they have many alerts you should consider. They have an option for an RSS feed or you can opt to receive emails for the team to review. The most important alerts from CISA to closely monitor are **Known Exploited Vulnerabilities (KEV)**. This is a catalog of all known vulnerabilities that CISA has identified that are currently being exploited. This catalog should be your first and utmost priority for review and remediation if you have any products in this catalog within your environment. You can also access the live catalog here: https://www.cisa.gov/known-exploited-vulnerabilities-catalog. To sign up for CISA KEV email alerts, access the following link, https://www.cisa.gov/about/contact-us/subscribe-updates-cisa, and within the **Subscribe to Email Updates** section, click on **SUBSCRIBE**. Once clicked, enter an email address to which you would like to receive **alerts** and follow the remaining steps.

When you get to the alerts section, select all the alerts you would like to receive within the **Cybersecurity** section.

Welcome **mdunkerley@coke-bsna.com** (Sign Out)

Quick Subscribe for mdunkerley@coke-bsna.com

Cybersecurity and Infrastructure :   ow.
Subscribe by checking the boxes, unsubscribe by unchecking the boxes.
Access your subscriber preferences to update your subscriptions or modify your password
or email address without adding subscriptions.

Subscription Topics

☐ CISA Careers

☐ Cybersecurity

   ☑ FedVTE

   ☑ General Cyber Training

   ☑ Incident Response

   ☑ Training Advocates at Organizations

   ☑ Continuous Diagnostics and Mitigation (CDM)

   ☑ Known Exploited Vulnerabilities Catalog

   ☐ Federal Cyber Defense Skilling Academy

   ☑ Cybersecurity Advisories

   ☑ Vulnerability Bulletins

   ☑ Industrial Control Systems (ICS) Cybersecurity Advisories

   ☑ ICS Medical Advisories

   ☐ CDM Dashboard Training Program

   ☐ Incident Response Awareness Webinar Training

   ☐ Incident Response Cyber Range Training

☐ Critical Infrastructure

Figure 8.10: The CISA subscribe to notifications options

Once you have selected your preferred alerts, click **Submit** and then **Finished**. Next, let's review the expected alerting from your vendors.

Vendors

Another area you'll want to focus on is ensuring you are receiving cybersecurity or vulnerability alerts from your vendors. There's a chance you will have tens to hundreds (thousands for some) of vendors within your portfolio. And there's a good possibility some of your vendors may not even be notifying CVE of any vulnerabilities or making them public knowledge. Because of this, it will be important to ensure you are receiving alerts from them on any applications you are using within your environment. If there are no alerts from the vendor, you will need to work with them to ensure this is part of their process moving forward. Managing your vendors is an entirely separate function that will need addressing as part of the onboarding process along with any ongoing cybersecurity reviews needed for the vendor. Collaboration with the vendor risk management team will be very important to the vulnerability management team similar to that of the SOC team. We will cover vendor risk management in more detail in *Chapter 10, Vendor Risk Management*.

Other

If you have CTI subscriptions already in place, you may already be receiving some of this information through that channel. If not, some additional areas for consideration as part of your vulnerability alerting can include:

- *Password monitoring*: Ideally, dark web monitoring services will cover this. But a free service you can take advantage of is Have I Been Pwnd. Have I Been Pwnd will provide you with notifications of any compromised accounts within its database. You can sign up your business domain to receive notifications: `https://haveibeenpwned.com/`.

- *Typosquatting*: Typosquatting is when threat actors leverage like domains in an attempt to trick users through phishing or other malicious types of attacks. They may also use like domains to set up websites that look like your organization's to lure users. An example of a standalone service to monitor this type of activity is dnstwister: `https://dnstwister.report/`.

- *Enhanced email services*: With threat actors continuously trying to abuse your brand and domain, one vulnerability you'll be challenged with is threat actors spoofing your company domain to trick users into thinking an email is coming from your domain, when in fact it is not. Services like MxToolbox can provide additional insight into this type of activity with deeper insights into SPK, DKIM, and DMARC usage. In addition, monitoring blacklists where your domain may be getting abused is another benefit.

We will cover these items in more detail in the *Email Protection* section later in the chapter. You can access MxToolbox here to learn more: `https://mxtoolbox.com/`. Another option to review is Valimail: `https://www.valimail.com/`.

There are many more standalone services that you can subscribe to for vulnerability alerting. Make sure you are also targeting alerts that are specific to your organization and the industry in which you operate.

Vulnerability Tracking and Remediation

In the past, vulnerability management has taken more of a passive approach. When updates are released, they are typically pushed into a non-production environment and then finally pushed into the production environment once validation has been complete and sign-off occurred by the business. This process could be weeks to months depending on the organization. The reason for this has traditionally been to ensure no impact on business applications. Today, we must weigh up the risk of possibly impacting a business application against the chance of a major cybersecurity incident occurring because of an unpatched system. Because of this, our strategy must change to get updates pushed out to our devices and applications as quickly as possible. The quicker the better, and your strategy should even consider immediate updating as updates are released, especially for **Known Exploited Vulnerabilities (KEV)**.

 You'll also need to ensure your vulnerability management strategy aligns with any certifications, compliance, or regulatory requirements like the **General Data Protection Regulation (GDPR)**, **Health Insurance Portability and Accountability Act (HIPAA)**, and **Payment Card Industry Data Security Standard (PCI DSS)**. It will be important you are aware of any requirements for your organization. We cover regulatory matters in more detail in *Chapter 15, Regulatory and Compliance*.

Tracking Your Vulnerabilities

We briefly discussed tracking vulnerabilities earlier within the *Program Management and Governance* section. As we stated, you may already have a single tool that manages most of your vulnerabilities. For example, *Qualys*, *Tenable*, and *Rapid7*, which we shared earlier in the chapter, are very popular vulnerability platforms for vulnerability management. The challenge with these tools is that you will only be able to manage the vulnerabilities being identified within the tools – that I'm aware of anyway. You will be receiving alerts from multiple different sources for which there will be a need to centrally manage all sources of vulnerabilities for review and remediation. Unfortunately, there is no easy answer to efficiently manage your ongoing vulnerabilities.

Some suggestions that you can adopt include:

- **Spreadsheet:** This may not be the most efficient option, but it will quickly allow you to begin tracking all your vulnerabilities in one centralized location to review the status.

- **Digitized list:** This could be a customized list within an intranet such as SharePoint. You'll be able to create a customized digital list that can track all vulnerabilities that need to be reviewed and remediated.

- **GRC tool:** Maybe your GRC tool can be customized to track all vulnerabilities, which can then tie back to your assets and applications for more efficient tracking and reporting. We will cover the GRC tool in more detail in *Chater 13, Governance Oversight.*

- **In-house built app:** You could work with a development team to build a custom app that will allow you to efficiently track all your vulnerabilities more efficiently than a spreadsheet.

- **Off-the-shelf app:** You will need to do some additional research here if this is the direction you would like to go to see what is available to meet your needs. There are applications available to allow you to efficiently track your vulnerabilities, but they will come at a cost and the need for your vulnerability management team to learn and operate a new application.

As mentioned, the least resistant method will be to set up a spreadsheet to track vulnerabilities as you work on a longer-term, more efficient digitized option. If setting up a spreadsheet to get started, some of the data you should be tracking includes:

- Internal incident number
- Unique ID (CVE)
- Date identified
- Date modified
- Name of vulnerability
- Description of vulnerability
- Publication date of vulnerability
- Impact
- Severity

- CVSS score

- Impacted products

- Application owner

- Additional information

- Remediation steps

- Date remediated

- Status

Scoring

When tracking your vulnerabilities, it is important you score each one. The scoring of each vulnerability essentially indicates the criticality of your vulnerabilities. Referring back to NVD, which we reviewed earlier, it uses the CVSS framework scoring system to publish a score that translates to critical, high, medium, and low. With a score assigned to your vulnerabilities, you can now prioritize which vulnerabilities need attention first. For the most part, your vulnerabilities will be scored once they have been made public and entered as a CVE. NVD will then take the CVE and assign a score using the CVSS framework. This makes prioritization easy when vulnerabilities have scores assigned for us.

The challenge is scoring vulnerabilities that haven't been published, are identified internally based on a cybersecurity incident, or someone external notifies you of a vulnerability. When this occurs, you will need to assess the vulnerability and apply a score so you know how to best prioritize remediating or mitigating the identified vulnerability. There are a few options you can use to do this. First, you could use the same scoring method as that published by CVSS by leveraging the latest published calculator: `https://nvd.nist.gov/vuln-metrics/cvss/v4-calculator`. Another option could be to use the *OWASP Risk Rating Methodology* found here: `https://owasp.org/www-community/OWASP_Risk_Rating_Methodology`.

There's also a template provided based on the OWASP scoring, which can be accessed here: `https://wiki.owasp.org/index.php/File:OWASP_Risk_Rating_Template_Example.xlsx`. It looks like the following.

Risk: Full database theft from datacenter

Likelihood								
Threat agent factors					Vulnerability factors			
Skill level	Motive	Opportunity	Size		Ease of discovery	Ease of exploit	Awareness	Intrusion detection
4 - Advanced computer user	1 - Low or no reward	access or resources required	5 - Partners		3 - Difficult	3 - Difficult	4 - Hidden	3 - Logged and reviewed
			Overall likelihood:	3.375	MEDIUM			

Technical Impact					Business Impact			
Loss of confidentiality	Loss of integrity	Loss of availability	Loss of accountability		Financial damage	Reputation damage	Non-compliance	Privacy violation
2 - Minimal non-sensitive data disclosed	0 -	0 -	9 - Completely anonymous		1 - Less than the cost to fix the vulnerability	1 - Minimal damage	0 -	5 - Hundreds of people
	Overall technical impact:	2.750	LOW			Overall business impact:	1.750	LOW
			Overall impact:	2.250	LOW			

Overall Risk Severity = Likelihood x Impact					Likelihood and Impact Levels	
Impact	HIGH	Medium	High	Critical	0 to <3	LOW
	MEDIUM	Low	Medium	High	3 to <6	MEDIUM
	LOW	Note	Low	Medium	6 to 9	HIGH
		LOW	MEDIUM	HIGH		
			Likelihood			

Figure 8.11: The OWASP risk calculator template

If you wanted to keep the scoring as simple as possible, you could use your judgment by using only the risk assessment matrix shown in the bottom left of the image above. Simply determine what is believed to be the impact along with the likelihood to give you a quick idea of the vulnerability's priority.

Vulnerability Remediation

Once a vulnerability has been determined and a score has been assigned, you will need to remediate or mitigate the vulnerabilities based on priority. Remediation is not an easy task. Each identified vulnerability will potentially possess unique characteristics on how they will need to be remediated. For the most part, remediation will come in the form of an update or patch in which a process set out by the vendor will need to be followed. Other vulnerabilities may need a developer to remediate, or you may need an administrator or engineer to make a configuration within a cloud environment or network infrastructure to remediate a vulnerability. A remediation for a vulnerability may be easy, complex, uneventful, impactful, and so forth.

It is important that the vulnerability management team understands exactly what is required when remediating a vulnerability and any potential impact it may cause on the business. Depending on the impact, communication may be needed to inform users of any potential downtime when vulnerabilities are being remediated.

Prioritization

As we have mentioned multiple times already, it is important that prioritization occurs as vulnerabilities are remediated. The vulnerability management team is going to be inundated with hundreds if not thousands of vulnerabilities on a monthly basis and priority is going to be key. This is why every vulnerability must be assigned a consistent score so priority can be determined. Based on the scoring system and some of the discussion points mentioned above, I'd recommend you focus your remediation methods in the following order:

- **KEV**: KEV should be of the highest and utmost priority on your list. Here, you will need to immediately remediate any technology in your environment that has been identified as being exploited. I recommend you review the entire list immediately to identify if you have any technologies within your organization listed in the KEV catalog. We covered how to receive notifications on KEV earlier in the chapter.

- **Critical**: Anything that has been flagged as critical by NVD or internally from your teams should also be of the highest priority after KEV. I would consider all zero-day venerabilities critical; they may not necessarily be critical, but I would closely assess them and treat them as critical until proven otherwise. Since zero-days will typically have no update or patch available, you'll need to look at temporary mitigation efforts.

- **High**: Although a level below critical, I would consider KEV, critical, and high to all be ongoing priority items. I would ensure these three categories continue to have the utmost attention and are addressed before working on medium or below.

- **Medium**: Address these after KEV, critical, and high are remediated.

- **Low**: Address these after KEV, critical, high, and medium are remediated.

- **Informative**: This is most likely not worth the effort but certainly review it to ensure there are no concerns from anything provided as informative.

As discussed earlier in the chapter, you will also want to take into consideration EASM priorities. For example, a critical vulnerability identified by EASM will take priority over a critical vulnerability that is only accessible on the internal network.

Make sure you define specific **Key Performance Indicators (KPIs)** to track the performance of your vulnerability management program. Some high-level KPIs you should be measuring against include the average time to detect vulnerabilities, average time to remediate vulnerabilities, number of open and closed vulnerabilities, number of open high and critical vulnerabilities, overall exposure score, etc. Your vulnerability management tool should have these built in as part of its reporting.

Let's finish off the section with some thoughts on the modernization of your vulnerability management program.

Modernizing Your Program

As vulnerabilities continue to grow every year, we need to look at ways to modernize our vulnerability management programs. This isn't easy, especially when teams are always heads down putting fires out on a day-to-day basis. But long term, looking at ways to modernize will benefit not only the team but the organization as a whole, as you strive to reduce the time it takes to remediate vulnerabilities. Let's look at some areas that can help with the modernization of your vulnerability management program:

- **Immediately install updates**: One mindset we need to change is that of following a mundane process of updates being released > updating the test environment > waiting 30 days for any impact > updating the production environment. We need to change our mindset to have updates pushed ASAP. Look at opportunities to configure your centralized updating services to push updates as soon as possible, configure applications to auto-update as updates are released, etc.

- **Automation**: As mentioned several times, managing hundreds to thousands of identified vulnerabilities monthly has become impossible for your staff to manually manage. You need to look at ways to automate as much as possible, especially higher-priority vulnerabilities where applicable. Look at your current tools and take advantage of any automation opportunities available. This applies to both identifying vulnerabilities as well as deploying updates for identified vulnerabilities. Ensure you are streamlining and looking to automate your vulnerability management lifecycle from beginning to end.

- **Continuous scanning**: Traditionally, you may have scheduled scans every 24 hours or weekly on a schedule. Although you need to balance any potential impact or performance with continuous scanning, it is important that you look at opportunities to enable continuous scanning for vulnerabilities.

At a minimum, you should be continuously scanning your external surface-facing footprint. The quicker you identify any external-facing vulnerabilities, the quicker you can patch any risk that can be exploited.

- **Broaden the program**: Continue to look at ways to broaden the vulnerability management program. Look beyond vulnerability scanning at other types of activities like pen testing (apps, network, laptops, etc.), bug bounty programs, attack surface monitoring, etc. We will be covering all these in more detail later in the chapter and throughout the book.

- **Process improvement**: This goes without saying for everything you do as a leader. Always ensure reviews of current processes are occurring and look for opportunities to improve, improve, and improve. New capabilities continue to be released and we must stay current. This is where you need to continue to instill an innovative mindset in your employees. You will find process improvement will occur within the team if you provide an environment that allows the team to innovate and bring ideas to the table.

- **Integration with DevSecOps practices**: Look to work with your internal development teams to integrate vulnerability management into the **Software Development Lifecycle (SDLC)** as early as possible. Essentially, this means pushing for a shift-left model where testing occurs early in the project timeline with development to ensure vulnerabilities are addressed more efficiently.

- **Emerging technologies**: Make sure you educate yourself and are aware of emerging technologies that will continue to evolve to enhance your vulnerability management program. Examples include **Machine Learning (ML)** for predictive vulnerability management and **risk-based vulnerability management (RBVM)**.

 RBVM expands on traditional vulnerability management by identifying, prioritizing, and mitigating vulnerabilities that pose the biggest risk to your organization. This approach helps focus on the most impactful and critical vulnerabilities identified within an organization. You can learn more about RBVM here: `https://www.tenable.com/source/risk-based-vulnerability-management`.

- **Artificial Intelligence (AI)**: Although considered an emerging technology, AI is growing fast with a lot of unknowns. I expect AI to significantly impact the vulnerability management space and it's important we keep a close eye on the market to understand the benefits it can provide. Make sure you are discussing this with your vulnerability management vendors and continue to push them to do better with AI opportunities.

As you can see, there is a lot of opportunity to allow the modernization of your vulnerability management program. This is a very important area to ensure efficiency is being maintained and all opportunities are being explored to reduce the time it takes to remediate vulnerabilities.

Update Management and Email Protection

Although there are going to be many different areas that you'll need to address as part of your vulnerability management program, I believe it's important we spend some time focusing on two of the more challenging and biggest areas as it relates to vulnerability management. Your update management processes, specifically as it relates to your end-user devices and traditional infrastructure, will be the largest component of your vulnerability management program. The next area we will look at is your communication systems and, more specifically, email as it continues to be abused from a vulnerability perspective.

Update Deployment

Before we go into more detail on the update process, let's look at some data and statistics to give us a better idea of what we face when it comes to the end-user device footprint we need to keep updated. For these statistics, we will refer to an online service called *StatCounter Global Stats*: https://gs.statcounter.com/. This dataset is not all-inclusive, but there is a very large sampling of data used to give us a good idea of worldwide usage. As of the time of publication, StatCounter Global Stats collects its data through web analytics via a tracking code on over 1.5 million websites globally. The aggregation of this data equates to more than 5 billion page views per month. What does this mean? Essentially, all these devices have a public presence based on the way this data has been collected. This translates to any of these devices having an increased risk of becoming vulnerable as they are accessible on the **World Wide Web (WWW)**. First, let's take a look at the global **Operating System (OS)** market share to understand global usage: https://gs.statcounter.com/os-market-share.

Figure 8.12: StatCounter OS market share worldwide

As you can see, Android, Windows, and Apple hold the majority of the market share, although we know that the majority of these devices will be mobile and personal-based devices. If we break down the data further, we will see the difference between the mobile and desktop (including laptops) market share: `https://gs.statcounter.com/platform-market-share/desktop-mobile-tablet`.

Figure 8.13: StatCounter mobile vs desktop market share worldwide

Within the business world, the majority of devices within an organization will be categorized as desktop (including laptops), and the majority of those will most likely be Windows within your environments: `https://gs.statcounter.com/os-market-share/desktop/worldwide`.

Figure 8.14: StatCounter desktop (including laptops) OS market share worldwide

Now that we have reviewed some statistics and have a better idea of the different footprints that will need to have their vulnerabilities remediated, let's start to focus on Windows.

Windows

Based on the data presented by StatCounter, there's a high possibility many of the **OSes** in your organization are Windows for your end users. To make vulnerability management more challenging, Windows typically always has a large subset of identified vulnerabilities. With this, patching/updating these systems has morphed into a full-time and very specialized role over the years. To put things into perspective, a quick search on the NVD website for Windows vulnerabilities from its inception shows 11,605 vulnerabilities as of July 2024: `https://nvd.nist.gov/vuln/search`.

Simply enter `windows` in the Keyword Search field and click Search.

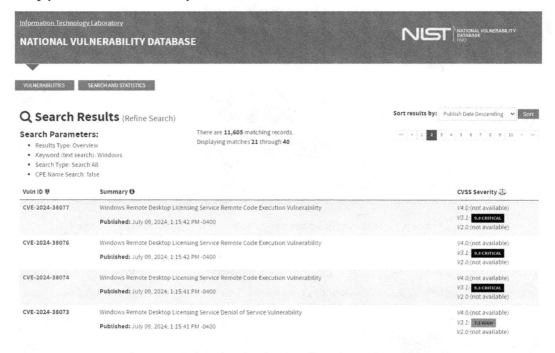

Figure 8.15: NVD with a search to list all Windows vulnerabilities

One term you may be familiar with as part of vulnerability management with Microsoft is the famous *Patch Tuesday* (also referred to as *Update Tuesday*). Patch Tuesday occurs on the second Tuesday of every month and is the day that Microsoft will release its monthly patches for Windows and other Microsoft products. Microsoft provides its own authoritative source regarding security updates, which can be found here: `https://msrc.microsoft.com/update-guide/en-us`.

 Although these patches are released on the second Tuesday of every month, updates will not be installed until determined by the policy or configuration specified within an organization. As noted already and with all updates, you need to be as aggressive as possible and deploy updates ASAP.

Obviously, you are going to need an efficient way to update your Windows devices, both end-user devices and the server environment. If you have a large Windows footprint, you may want to strategize on the Microsoft updating technologies, which include **Windows Server Update Services (WSUS)**, Azure Update Manager, or **Windows Update for Business (WuFB)**. Make sure you spend time on your patch management strategy as it can become complex to manage.

For example, WuFB for Windows 11 has a servicing model that covers four update groups: feature updates, quality updates, driver updates, and Microsoft product updates. From a security perspective, you'll need to focus on the quality update group deployment times as it contains critical updates and security patches.

Other OSes

In addition to Windows, you are guaranteed to have a mixture of other OSes within your environment. And all of them are going to need ongoing updates. Your OSes can be categorized into the following categories:

- *End user*: Windows, macOS, and ChromeOS
- *Mobile*: iOS and Android
- *Infrastructure*: Windows, Linux, and Unix

 OT and IoT will also fall within this category, but they will be covered in more detail in *Chapter 12, Operational Technology (OT) & the Internet of Things (IoT)*.

It is important that all OSes receive updates as they are released. You may opt for a centralized third-party tool that can manage the majority of the OSes mentioned above, or you may be challenged with needing to use individual systems to update your devices and infrastructure. Centrally managing everything would be ideal, but the reality is there may not be a tool to accomplish this, especially as you may have different systems to manage your mobile devices. For your infrastructure, the Microsoft tools will also manage Linux updates, which can help simplify your server updates. For end-user devices, you'll need to review your options further. Obviously, Microsoft will allow updating of Windows devices, and macOS if you enroll them in Intune. For ChromeOS, there is an enterprise connector into Intune that you will be able to configure, but you may need to manage updates from the Google Admin console. Other options are to deploy a reliable centralized update management solution if it supports all end-user devices. But again, more management consoles equate to more cost, resources, and increased risk. So, make sure you carefully weigh out your options.

Finally, your mobile devices will potentially have another management console. A couple of the more common management tools for mobile devices that you may be familiar with are Microsoft Intune and Omnissa Workspace ONE. Depending on how you manage your mobile devices, whether you use **Mobile Device Management (MDM)** or **Mobile Application Management (MAM)** will also determine how updates are managed on these devices.

With MAM, you won't necessarily have the ability to push updates, so you may need to use restriction methods to your corporate data based on OS versioning. Again, it's important you understand all the different challenges to ensure the most efficient methods are used to update all OSes in your organization. Most importantly, make sure you have a policy that clearly defines how updates for your OSes will be handled.

Browsers

Browsers continue to be a challenge with vulnerability management. The first challenge you'll find is the use of multiple different types of browsers within your environment. Ideally, you'll want to strategize on one or two at a minimum, but there will most likely be a need for additional browsers based on legacy apps, the need to test applications on multiple browsers for compatibility, vendor requirements, etc. To get an idea of the usage of browsers used worldwide, let's refer back to StatCounter Global Stats to look at the browser market share: `https://gs.statcounter.com/browser-market-share`.

Figure 8.16: StatCounter browser global market share

As you'd expect, most leverage the more common browsers we are familiar with, such as Chrome, Safari, Edge, and Firefox. Based on the frequency of browser-based vulnerabilities, the only option is to configure all your browsers to auto-update. This will require the need for your end-user device/system administrators to configure policies for your browsers or modify your device settings to ensure your browsers auto-update as soon as updates are available. In addition, it is highly recommended to ensure your browsers are hardened based on best practices using the CIS Benchmarks. The top four browsers listed in the above image all have CIS Benchmarks available: `https://www.cisecurity.org/cis-benchmarks`.

Applications

The last area we will touch upon is application updates. This will get very complex as you will find there are tens if not hundreds of applications being used in your environment. Applications continuously need to be updated, and the same concept applies here.

If possible, auto-update where applicable. Although this may not be feasible for all applications, you will need to work closely with your end-user device/system administrators as vulnerabilities are identified as they may need to re-package new versions of applications to remediate any identified vulnerabilities. Make sure your vulnerability scanning tool is scanning applications on end-user devices and servers to ensure any vulnerable applications are identified as soon as possible.

A couple of high-priority applications to focus on are remote access tools and web conferencing tools. These tools can bring some higher risks because they can allow direct access to your end-user devices or servers if you have any of these tools installed. For web conferencing tools, think of Zoom, Cisco Webex, and Microsoft Teams. Again, make sure these are set to auto-update if possible or have processes to ensure you can quickly remediate any vulnerable versions. A set of tools I would say carries greater risk is remote-based access tools. Think of TeamViewer, AnyDesk, LogMeIn, etc. New vulnerabilities are continuously identified with these tools and it's critical they are updated right away when any vulnerabilities have been identified. Also, make sure you are aware of any **Managed Service Providers (MSPs)** you may be using or vendors that leverage these types of tools to access your network. This could allow for a back door into your environment if you aren't managing the usage of these tools closely.

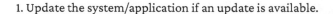

It is important to be aware that there may not always be an update available for your system and/or application. Because of this, it is important you have a strategy in place to address the vulnerability at hand. The following is a high-level process that can be followed in the event an update is not available yet:

1. Update the system/application if an update is available.

2. If no update is available, can a configuration change be made to mitigate the vulnerability?

3. If no update or configuration change can be made, can you apply defense in depth (additional security measures)?

4. If none of the above are applicable, you'll need to receive acceptance of the risk and document it in the risk register.

Email Vulnerability Management

We are going to finish off this section with a review of vulnerability management as it relates to your email systems. If you have a legacy type of infrastructure like Microsoft Exchange Server, which will be deployed on a traditional server-based architecture, you'll need to continuously keep these systems updated as vulnerabilities continue to be identified. Ideally, moving to a cloud-based SaaS email model is recommended to eliminate the need for having to continuously update Microsoft Exchange and the underlying infrastructure such as the server OS. There will always be risk with every model, but the more you can reduce risk, especially with legacy infrastructure, the better. Although it is important to ensure your legacy infrastructure is free from vulnerabilities, the premise of this section is to cover vulnerability management from the perspective of threat actors exploiting email as a delivery mechanism to penetrate your environment. With email being one of the more common attack vectors, everything must be done to prevent vulnerabilities from coming into the environment.

Here is a list of some recommendations to help reduce the number of vulnerabilities coming into your environment:

- Ensure spam filtering policies are configured with strong protections.
- Use allow/block lists.
- Ensure anti-spam, anti-phishing, and anti-malware configurations are set up with strong protection.
- Enable quarantining functionality.
- Enable any advanced threat protection capabilities, such as attachment, URL scanning, and QR code scanning.
- Look at preventing auto-forwarding.
- Review and lower thresholds for the volume of outbound email.
- Prevent attachments that are known to be used to contain malware.
- Ensure modern authentication has been enabled and disable all legacy protocols.
- Enforce MFA for your users and require strong unique passwords.
- Enable advanced email protection features, such as **Secure Email Gateway** (**SEG**).
- Take advantage of any spoofed domain capabilities.
- Ensure efficient management of transport rules.
- Enforce DLP capabilities.

- Ensure encryption is enabled and enforced as needed.

- Ensure TLS capabilities are being enforced between trusted partners.

- Use **Sender Policy Framework (SPF)**, **DomainKeys Identified Mail (DKIM)**, and **Domain-Based Message Authentication, Reporting, and Conformance (DMARC)**

- Implement Mailbox Intelligence protection if applicable.

- Investigate tools to enhance capabilities with AI type of technology.

- Ensure you have good monitoring and integration with your SIEM.

- Make sure you implement the CIS Benchmarks, which are available for both Microsoft and Google: `https://www.cisecurity.org/cis-benchmarks`.

 One of the more beneficial methods of reducing risk with email vulnerability management is through user awareness, training, and testing. We will cover this in more detail in *Chapter 9, User Awareness, Training, and Testing*.

Managing vulnerabilities within your email systems is a very challenging task. As threat actors come up with new ways to bypass your filtering settings, you are challenged with vulnerable emails sitting in your users' mailboxes. It is important you continue to fine-tune and evolve your email security to help reduce vulnerabilities reaching your users as much as possible.

Other Vulnerability Management Considerations

Let's finish off the chapter with some other considerations for your vulnerability management program. Here, we will look at some other areas of importance relating to ensuring vulnerabilities are remediated as soon as possible. We will also review other activities that can be included to improve your vulnerability management program.

Hardware Vulnerabilities

One area that can be overlooked is vulnerabilities with any hardware within your organization. Vulnerabilities such as Meltdown and Spectre are prime examples of this (you can learn more about Meltdown and Spectre here: `https://www.cisa.gov/news-events/alerts/2018/01/04/meltdown-and-spectre-side-channel-vulnerability-guidance`). Your hardware is not just limited to your end-user devices but any infrastructure that you have in place, such as network equipment, servers, IoT, etc. Over the years, we have invested heavily in the software side of security, but recently, the impact of hardware vulnerabilities has shown the criticality of ensuring your hardware is protected from exploits, thus requiring more attention in this area to address risks.

Some hardware-specific vulnerabilities to be aware of include:

- Rootkits embedded in the BIOS and UEFI
- Side-channel attacks toward CPUs
- Kernel-level exploits
- Firmware attacks
- Memory vulnerabilities such as buffer overflows

Another area of concern regarding vulnerabilities with your hardware is the supply chain process. NIST has some great material on this, which can be found here: *NIST Cybersecurity Supply Chain Risk Management*: `https://csrc.nist.gov/Projects/Supply-Chain-Risk-Management`. The following are some of the supply chain risks to be aware of as noted by NIST:

- Any insertion of counterfeit items as part of the supply chain process
- The production of items that have not been approved
- Tampering with any items within the process
- Theft of any items
- The insertion of malicious software and hardware at any time during the process
- Manufacturing and development practices not maintaining expected standards during the supply chain process

Another good reference to be aware of is the **Common Weakness Enumeration (CWE)** *Most Important Hardware Weaknesses* list released by MITRE and CISA. You can view the list here: `https://cwe.mitre.org/scoring/lists/2021_CWE_MIHW.html`. Make sure you are paying attention to the certification of hardware. Ensuring your hardware is certified is a critical process of the overall cybersecurity program. As you purchase new servers, PCs, storage, and peripherals, it is critical you validate that the hardware is compatible with your deployed systems. Using non-compliant hardware could make your hardware vulnerable to a compromise, or the additional hardware components could even have a compromise already embedded in them.

An example would be allowing the use of **Universal Serial Bus (USB)** drives on your devices. Users receiving a free USB drive don't realize that the drive itself could be infected and that, once inserted into a company device, it could compromise the entire organization. Because of this, it is critical you only allow pre-certified USB drives that are encrypted and provided by the organization to be used by employees. Any data that is copied from a company device to a USB drive must require encryption.

Ensuring the vendor has certified the hardware significantly reduces risk the hardware could be pre-infected with vulnerabilities. This doesn't necessarily mean it will be 100% guaranteed, but your risk is reduced significantly.

Here are some considerations with your hardware to best protect it from vulnerabilities:

- Only purchase hardware that has been through a proper hardware certification program. Examples include:

  - The Windows Hardware Compatibility Program certification for Windows devices: `https://partner.microsoft.com/en-us/dashboard/hardware/search/cpl` and `https://www.windowsservercatalog.com/`

  - Red Hat certified hardware: `https://catalog.redhat.com/hardware`

  - Ubuntu Certified Hardware: `https://ubuntu.com/certified`

  - ChromeOS Flex certified models list: `https://support.google.com/chromeosflex/answer/11513094?hl=en`

- Keep your hardware up to date. Just as with software, hardware continues to evolve to become more secure. This is not the easiest of tasks and is known to be disruptive if a hardware update goes wrong. You'll need to determine the likelihood and impact versus the overall effort when reviewing these vulnerabilities.

- Have an effective and secure process for upgrading firmware/BIOS and ensure the proper protections are enabled to ensure only approved sources can update them. Like the previous bullet, this can be a difficult task with severe implications if not completed successfully.

- Other considerations include **Trusted Platform Module (TPM)** 2.0, **Dynamic Root of Trust for Measurement (DRTM)**, **System Management Mode (SMM)**, Secure Boot, **Direct Memory Access (DMA)** Protection, memory encryption with AMD's **Secure Memory Encryption (SME)** and **Intel's Total Memory Encryption (Intel TME)**, hardware-based isolation of application code in memory (**Trusted Execution Environment (TEE)** with Intel **Software Guard Extensions (SGX)**), and **virtualization-based security (VBS)**.

 You can learn more about some of the referenced technologies and other considerations on the Windows Hardware Security webpage: `https://learn.microsoft.com/en-us/windows/security/hardware-security/` and the Intel® Software Guard Extensions (Intel® SGX) webpage: `https://www.intel.com/content/www/us/en/products/docs/accelerator-engines/software-guard-extensions.html`.

Now that we have a better understanding of managing vulnerabilities with hardware, let's take a brief look at some virtualization concerns.

Virtualization Infrastructure

Although there has been a big shift to the cloud, where the virtualization plane becomes less of a responsibility for organizations, it is still important to be aware of these concerns as many continue to have traditional virtualization infrastructure deployments. We continue to see KEV with common virtualization technologies such as VMware and Citrix, so it's important the vulnerability management team is fully aware of any virtual infrastructure within the organization so they can quickly address any vulnerabilities. With virtualization, the capability to centralize hundreds and thousands of standalone workloads on less hardware significantly increases the risk profile for the organization, and it is critical that we ensure our virtualized infrastructure is protected correctly from vulnerabilities. A breach of a virtualized host could mean a compromise to hundreds of servers, instead of a single physical server in past models.

Here are some risks of virtualization that must be addressed:

- Everything listed in the previous section on hardware vulnerabilities. Your virtualized infrastructure will be running on the same hardware.
- Hypervisor threats.
- VM escape or the ability to interact with the physical host OS or hypervisor directly from a VM.
- Non-segregation of resources, network, and data.
- VM sprawl.
- Non-encrypted storage, physical drives, virtual disk files, and network traffic.
- The potential for corruption of the physical host OS and/or VMs.

Although there are security tools and configurations that help ensure the security of your VMs and the data within them, you must consider the physical separation of specific functions within your virtualized infrastructure if it hasn't been implemented already. For example, the management plane, the production environment, the **Demilitarized Zone (DMZ)**, and highly confidential applications and databases should be separated. Ensure the separation of these functions includes the physical host, network, and storage, which will help safeguard and reduce risk. In addition, it will be critical that your virtualization infrastructure including all VMs is backed up per best practices, as covered in *Chapter 7, Cybersecurity Operations*.

Another area of consideration that you will need to manage with vulnerabilities is containerization. With the increased use of containerization, you will need to ensure you are familiar with this technology and the tools used to create, manage, run, and orchestrate like Docker and Kubernetes. A couple of resources of value include the OWASP Kubernetes Security Cheat Sheet: https://cheatsheetseries.owasp.org/cheatsheets/Kubernetes_Security_Cheat_Sheet.html and the Docker Security Cheat Sheet: https://cheatsheetseries.owasp.org/cheatsheets/Docker_Security_Cheat_Sheet.html.

Network Infrastructure

The network architecture within your organization is most likely extremely complex. Many of these devices are Internet-facing and bring a lot of risk if they aren't managed and secured correctly, including the need to remediate vulnerabilities in a timely manner. In addition to the ongoing vulnerability management of these devices, some other challenges we face include ensuring the data we transmit is secure and no perpetrators are accessing our network who shouldn't be, preventing traffic that isn't welcome, and ensuring confidential data is isolated. There are many components involved within a network architecture, which can make vulnerability management very challenging. The following technologies are considered more critical for your enterprise deployment as they relate to your network security and should be implemented to protect your organization:

- Routers and switches using VLANs
- Next-generation-type firewalls
- A VPN to encrypt connections
- **Intrusion Detection Systems (IDSs)/Intrusion Prevention Systems (IPSs)** to proactively detect and prevent threats
- Wi-Fi with a minimum of WPA2-Enterprise security
- **Network Access Control (NAC)** to better manage endpoint access to your network
- Proxy/web content filters to prevent malicious websites
- **Distributed-Denial-of-Service (DDoS)** protection
- **Data Loss Prevention (DLP)** to prevent the loss of sensitive data
- DNSSEC to protect your DNS services
- **Public Key Infrastructure (PKI)** to provide digital certificates for encryption

Like all other areas discussed, there is ongoing KEV with networking equipment whether it be with client VPNs, firewalls, or anything else related. Your team will need good collaboration with the network team as vulnerabilities are identified so they can work through remediation as soon as possible, especially with critical and highly categorized vulnerabilities.

Here are some important considerations for your networking equipment that will help reduce risk:

- Ensure you keep the software of your network devices current.
- Enable auditing on devices.
- Integrate authentication using LDAP.
- Use a **Privileged Access Management (PAM)** solution.
- Disable or prevent local account access and change default usernames and passwords.
- Ensure the management of devices is encrypted (SSH).
- Isolate the management network.
- Don't allow the management of network devices from the Internet.
- Make sure you implement the CIS Benchmarks that are applicable to your networking equipment: `https://www.cisecurity.org/cis-benchmarks`.

As stated, networking is a very complex area to run and, in most cases, requires specialized skill sets to manage efficiently. There are a lot of considerations to be aware of as it relates to networking. Others we haven't mentioned include vulnerabilities around Bluetooth and voice-related technologies, which bring their own level of risk. It is important that your vulnerability management team is well versed in networking capabilities so they can quickly respond to anything related to networking.

Cybersecurity Testing

As part of your cybersecurity program, you are going to need to run testing for various reasons. Some of these tests may not fall within the vulnerability management scope, but there will most likely be findings from the tests that will need remediation. These findings will translate into some form of vulnerability that has been identified and, in turn, will need to be remediated. The findings should be listed by criticality, which will help prioritize remediating them. The following are some of the more common cybersecurity testing activities you should be aware of:

- Penetration testing (external and internal)
- Application testing (standard and mobile)
- Web application/API testing

- Network testing (routers, wireless, etc.)
- Infrastructure testing (cloud, AD, etc.)

In addition to those listed above, end-user testing will also be part of your cybersecurity testing. End-user testing won't provide vulnerabilities for your vulnerability management team to remediate, but it will provide a better idea of where to focus training for your users. This will be covered in more detail in *Chapter 9, Cybersecurity Awareness, Training, and Testing*. In addition, the above testing activities will primarily be covered in *Chapter 11, Proactive Services*.

Auditing and Assessments

Your company is also most likely running internal audits and assessments, or you may even need to run audits from an external auditor depending on any potential regulations that need to be complied with. Similar to the testing activities, there will probably be findings or recommendations from the audits and assessments that identify vulnerabilities and areas for improvement. These findings and recommendations should come based on criticality, which will allow you to prioritize what to remediate first. Some of the audits and assessments you can expect include:

- **International Organization for Standardization/International Electrotechnical Commission (ISO/IEC)** 27001
- System and Organization Controls (SOC) 1 and 2
- NIST Cybersecurity Framework
- **Health Information Trust Alliance Common Security Framework (HITRUST CSF)**

Another important area is regulatory compliance, which is becoming more of a requirement around the world today. Where you are located and what industry you are in will determine if you have any regulations enforced on your organization. The following are some you should be familiar with, and all have very stringent requirements as part of their regulations or standards. This in turn will require continuous remediation of any findings or vulnerabilities to ensure you keep within compliance:

- **Sarbanes-Oxley (SOX)** Act
- **Health Insurance Portability and Accountability Act (HIPAA)**
- **Federal Information Security Management Act (FISMA)**
- **Payment Card Industry Data Security Standard (PCI DSS)**
- **General Data Protection Regulation (GDPR)**
- **California Consumer Privacy Act (CCPA)**
- **Gramm-Leach-Bliley Act (GLBA)**

There will be other assessments and exercises that will continuously be run providing feedback and potential identified vulnerabilities. Another example is tabletop exercises, which may not necessarily identify vulnerabilities, but there may be areas of improvement that need to be addressed. Regardless of the exercise, assessment, audit, or regulation, you'll need to ensure everything is being tracked and prioritized for remediation. We will cover tabletop exercises in more detail in *Chapter 11, Proactive Services*.

OT & IoT

We haven't touched a whole lot on OT & IoT in this chapter, and for a reason. Since these technologies are a lot more specialized, we will cover them in more detail in *Chapter 12, OT & IoT*. Within this chapter, we will essentially cover everything that comprises the broader cybersecurity program for just the OT & IoT technologies. With the specialized skill sets required, you'll need to run this function somewhat like a miniature version of your broader cybersecurity program to ensure there are no gaps with these technologies. You will want to ensure unification and synergies with certain areas like your cybersecurity operations and vulnerability management functions. It will be important that monitoring of these systems is filtered into your centralized SOC as well as tracking vulnerabilities along with all other vulnerabilities. Also, you'll need to ensure everything follows the same ARB process, of course. You'll just need to ensure well-documented processes and procedures are in place to engage and collaborate with the teams that manage this technology.

Other Activities

We'll finish the chapter with a review of some other activities that are relevant to the vulnerability program. It may not be feasible for you to implement all these activities, but it is important you understand them and determine what you need to execute based on your priorities for the cybersecurity program. Like everything else we have discussed throughout the chapter, these activities will have a direct impact on the vulnerability management team as vulnerabilities are identified and need to be addressed:

- Bug bounty program
- Threat modeling activities
- Vulnerability disclosure program
- Compliance reviews
- Configuration management reviews
- Insider threat reviews

- Breach and attack simulation activities

 Insider threats pose a significant risk to organizations and are another vulnerability that must be accounted for. If you don't have protocols in place to help detect insider risk, you'll want to look at what options you have to provide notifications of any suspicious activity with your users. We reviewed insider risk in detail in *Chapter 6, Identity and Access Management.*

One additional activity that you will want to incorporate into your vulnerability management program is extending the notifications of vulnerabilities to your end users. For example, as you receive high-rated vulnerabilities that impact iOS or Android devices, extend the notification to your users as they will be impacted. This may not be a direct responsibility of your vulnerability management team. But this should be information that is filtered over to the cybersecurity awareness, training, and testing function, which will have the appropriate channels to share this type of information that will benefit your end users. We will cover cybersecurity awareness, training, and testing in more detail in *Chapter 9*.

Summary

As you have observed, vulnerability management is a large endeavor to deploy and manage efficiently. We have covered a lot of different areas, and the reality is, we could cover a whole lot more within this area. Efficiently managing ongoing vulnerabilities requires a strong and collaborative team. As hundreds of vulnerabilities continue to be identified, the team must work fast to determine what the impact on the organization is to assign a priority. Once this has been done, the team will need to work with many different areas throughout the business to ensure vulnerabilities are remediated or mitigated as quickly as possible. Make sure you understand the workload for this team, and that you staff appropriately to support the ongoing demand within this area. Constantly look for continuous improvement, specifically around automation opportunities, to allow for greater efficiency with vulnerability remediation.

In the first part of the chapter, we covered how to best manage your vulnerability management program. We reviewed the building blocks for your vulnerability management program along with the program management and governance needed to oversee the program. We then discussed the importance of asset management. This is one of the most important components of your vulnerability management program.

If you don't know what and where your assets are, how can you protect them? Next, we began to review some of the core functionality of your vulnerability management program with vulnerability discovery and alerting. We first provided an overview of vulnerabilities before going into detail about scanning your assets, especially the important area of external attack surface monitoring. We then finished off the section with a detailed review of vulnerability alerting from multiple sources, such as a vulnerability management system, your SOC, threat intel, external sources, and vendors.

The next section covered vulnerability tracking and remediation activities in which we reviewed scoring in more detail. Within the vulnerability remediation activities, we reviewed the prioritization of vulnerabilities and what metrics you should be tracking, before finishing off the section with some thoughts on modernizing your cybersecurity program. The following section covered update management and email protection. I would consider these two of the most active areas within the vulnerability management program that will require more of your time over some of the other activities to reduce risk. Finally, we finished off the chapter by looking into some other areas of consideration for your vulnerability management program. This included hardware, virtualization, and networking, to name a few.

In the next chapter, we will be reviewing cybersecurity awareness, training, and testing. This, in my mind, is the most important chapter for your cybersecurity program. Here, we will cover why the human element is the most important factor within the cybersecurity program. This will transition into the building of a current and more modern cybersecurity program to meet the demands of today's threats. We will then go into detail about user awareness and what you should be considering regarding user awareness for your users. This will lead on to a section on user training and testing, which will provide more details on everything that should be executed to train and test your users. The chapter will finish off with a look into other items you should be considering as you look to evolve your cybersecurity awareness, training, and testing program.

Join our community on Discord!

Read this book alongside other users, Cybersecurity experts, and the author himself. Ask questions, provide solutions to other readers, chat with the author via Ask Me Anything sessions, and much more.

Scan the QR code or visit the link to join the community.

`https://packt.link/SecNet`

9

User Awareness, Training, and Testing

In this chapter, we will be moving on to what I consider the most important functions within your cybersecurity program: user awareness, training, and testing. The human element is critical to the success and ongoing operation of any organization, but they are the most vulnerable as it relates to cybersecurity. Our cybersecurity programs need to evolve to put our users first and not treat the user awareness, training, and testing program as a check box to ensure we meet compliance or regulations. Traditionally, an annual cybersecurity training requirement along with an annual testing exercise is most likely what is currently being executed. This provides very minimum benefit to our users. The user awareness, training, and testing program needs dedication to provide users the attention they need to be better informed and prepared for today's current cybersecurity risks. Throughout the chapter, we will provide you with details on what should be considered as part of a modernized user awareness, training, and testing program.

We will begin the chapter with an overview of why the human element is the most important aspect of the cybersecurity program. Having a good understanding of why the human element is so important will allow you to better understand why we must do better with our user awareness, training, and testing programs. This section will primarily be driven by statistics that show the challenges we face with our users. We will then cover what should be included in your user awareness, training, and testing program. Here, we will provide an overview of an annual schedule for your program with all the components that make up a more robust program. At a high level, we will briefly review each of the components involved that comprise your program in its entirety.

Next, we will see user awareness in more detail. As part of your user awareness activities, making sure you are constantly communicating and initiating outbound awareness is a must. You'll need to consider how you efficiently communicate outbound to your users, whether you use email, newsletters, company town halls, or so on. Also, providing a location for your users to access cybersecurity-related information is important, like an intranet or knowledge base of some sort. Following the awareness section will be user training and testing. Here, we will review in detail what should be included as part of your ongoing user training and awareness to continuously provide your users with current and relevant up-to-date training. In addition, continuous testing with your users is not an option these days, you need to ensure you are testing your users the same way they are being targeted in real life. Finally, we will look at additional areas beyond the traditional methods of awareness, training, and testing that you should consider implementing to best prepare your users against today's threats. The following will be covered in this chapter:

- Why the human element is the most important
- Building a user awareness, training, and testing program
- User awareness
- User training and testing
- Expanding beyond the traditional channels of awareness

Why the Human Element is the Most Important

First, let's take a high-level look at all sub-functions that should be addressed as part of the user awareness, training, and testing program. The following image captures much of what the user awareness, training, and testing program entails.

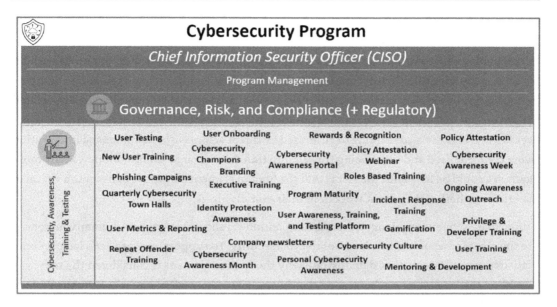

Figure 9.1: Sub-functions of the user awareness, training, and testing program

Our organizations are dependent on both technology and users to be successful in today's world. There is a dependency on both to allow for success. Because we live in a world where everything is digitized, and with the availability of the internet, the risk profile of our data has increased substantially. The challenge is that we can put all the technical controls in place to protect our data, but all it takes is for a single user with the right access to become compromised, and the rest is history. In addition, it's our users, or some users somewhere, who are designing and creating the software we use. This equates to vulnerabilities within the applications we use because of human errors whilst creating these applications.

Outside of the organization, our users continue to become victims of cybercrime in their personal lives. This is occurring in many different ways. The first with their data being trusted in the hands of companies who continue to leak data. This is happening on multiple different levels with basic **Personally Identifiable Information** (PII) such as email addresses and telephone numbers being compromised to catastrophic information such as sensitive PII including health-related PII, government IDs, and even data such as biometric and genetic information. Another distinct area is fraud and users continuing to be abused by threat actors with their financial information being compromised and users being tricked into transferring large sums of money. Many more examples could be listed but the fact is that the human element continues to become a victim of cybercrime whether they are at work or in their everyday lives.

Unfortunately, our users, along with statistics, continue to show that "we," the human component, are the most vulnerable when it comes to cybersecurity. To support our case, let's look at some statistics over the years that demonstrate why the human element is considered the most vulnerable in this challenging cybersecurity landscape. First, let's review statistics from the Verizon 2023 DBIR report (which we have already referenced a couple of times) that are both directly and indirectly tied to the human element. From the Verizon report, it is extremely concerning that we continue to observe an extremely high percentage where the human element is involved with a breach. The report clearly shows we have much improvement needed with training our users, more specifically around stolen credentials and social engineering:

- Within the social engineering category, **business email compromise (BEC)** represents more than 50% of incidents.
- The human element is included in 74% of breaches. This is through either human error, privilege misuse, stolen credentials, or social engineering. In the 2022 report, it was reported that the human element was included in 82% of breaches.
- Stolen credentials, phishing, and exploitation of vulnerabilities are the primary entry points for attackers.

Source: https://www.verizon.com/business/resources/reports/dbir/

Other great reports with eye-opening statistics include those from Proofpoint's **State of the Phish** reports as shown in the following list. This is just a snapshot of some of the data provided by the report, but as you can observe, we are still challenged with some very rudimentary cybersecurity awareness. What is more concerning is the percentage of these questions coming back is high; we clearly have our work cut out to reduce risk with our users for our organization.

Here are the details from the 2022 report:

- 42% reported taking a dangerous action, whether it was exposing personal data or login credentials, clicking on a malicious link, or downloading malware.

- Friends and family were provided access to employer-issued devices (to shop, stream, and play games) by 56% of those surveyed.

- User awareness of key terminology is still a challenge, with only 36% recognizing the definition of ransomware, 24% vishing (voice calling), 53% phishing, 63% malware, and 23% smishing (SMS text messages).

- More than one personal device is used for work purposes by over 80% of workers in the US.

- A risky action was taken by 55% of US workers.

Source: https://www.proofpoint.com/us/newsroom/press-releases/proofpoints-2022-state-phish-report-reveals-email-based-attacks-dominated

Here are the details from the 2023 report:

- When an email contains familiar branding, 44% believe it to be safe.

- It was reported by 33% of employees that cybersecurity is not a top priority within the workplace.

- Password re-use for multiple work accounts was reported by 28%.

- Work devices are used for personal use by 78% and personal devices are used for work by 72%.

- The default network name on work Wi-Fi routers is unchanged by 71% and 80% reported not changing the default admin password.

- Users who left a job and took data with them was reported by 44%.

- It was reported by 34% of users that they did something that put their organization or themselves at risk.

Source: https://www.proofpoint.com/us/resources/threat-reports/state-of-phish

Some more eye-opening statistics are from Tessian (acquired by Proofpoint) with their 2nd edition of the *Psychology of Human Error* report, as shown in the following list. Similar to the previous statistics, we are seeing a very high response of actions taken from employees that have the potential to evolve into a major cybersecurity incident and breach.

The statistics clearly show how vulnerable the human element is with a need to focus primarily on basic cybersecurity hygiene and social engineering type of awareness:

- 52% of employees reported falling for a phishing email which was an impersonation of a senior executive.
- In the past year, 26% of employees reported clicking on a work-related phishing email.
- It was reported by 40% of employees within the US and UK that they sent an email to the wrong person within a 12-month timeframe.
- 32% of the 56% of employees who received a scam text message responded and complied with the request from the scam message.
- In the last 12 months, 36% of employees reported making a mistake they believe compromised security.

Source: https://www.tessian.com/resources/psychology-of-human-error-2022/

Looking at the *Global Risks Report 2022* published by the World Economic Forum, it was shared that 95% of issues related to cybersecurity were traced back to human error: https://www.weforum.org/publications/global-risks-report-2022/.

Before we look at some real-life examples of actual cybersecurity incidents caused by human error, let's review the password problem we continue to be challenged with. The challenge is one we need to prioritize and continue to educate as users don't understand the impact they are causing by not using strong passwords. Regardless, the following image speaks volumes about the current issue at hand. The following figure shows the top five most commonly used passwords reported by NordPass with their 2023 insights research, *Top 200 Most Common Passwords*. More concerning is the count represented for each of these passwords. You can access the dataset here: https://nordpass.com/most-common-passwords-list/.

RANK	PASSWORD	TIME TO CRACK IT	COUNT
1	123456	< 1 Second	4,524,867
2	admin	< 1 Second	4,008,850
3	12345678	< 1 Second	1,371,152
4	123456789	< 1 Second	1,213,047
5	1234	< 1 Second	969,811

Figure 9.2: The top five most common passwords reported by NordPass

As you can see, the data available is endless and shows the challenge we have ahead of us as we continue to educate and make our users more aware of the ongoing cybersecurity risks we face today. When looking at real-world examples of breaches caused by human error, the list quickly becomes extremely lengthy. As an example, let's refer back to the **Identity Theft Resource Center (ITRC)** reviewed in *Chapter 1, Current State*. If you access `https://www.idtheftcenter.org/notified`, scroll to the **Custom Breach Search** section, change the **Timeframe to Last 30 Days**, click **Filter by Attack Vector**, select **System and Human Error**, and then click **Search**:

Figure 9.3: System and human error breaches from ITRC

This search returned 25 items related to system and human error for the previous 30 days at the time searched. Some of these are very small compromises, but they have still been reported and some of them were because of human error. Change this to 1 year or more and this number increases substantially. This highlights the critical need for robust measures to address both system and human errors effectively. That said, let's move to the next section, where we will discuss how to build a user awareness, training, and testing program.

Building a User Awareness, Training, and Testing Program

For those of you who don't have a user awareness, training, and testing program in place today, I would suggest you make this one of your highest priorities. This program is core to the success of your broader cybersecurity program. Executing one-time onboarding training, a one-time annual phishing campaign, and a one-time annual training event does not constitute having a user awareness, training, and testing program in place. This shows you are simply meeting a requirement enforced and checking a box to show you are in compliance. Unfortunately, this is no longer acceptable, and we need to evolve our user awareness, training, and testing programs to a much higher standard.

For those who do have user awareness, training, and testing programs in place today, there is a need to continue to review and assess your programs to ensure they continue to evolve. The reality is this program needs resources either dedicated or at least committed to ensure greater success. If you are a larger organization, dedicated resources shouldn't be a question. If you are running this program as needed, a lot of time and commitment will be required to execute it efficiently. There is a lot involved in this program if you are looking to show positive results, such as continuous training, interactive training modules, phishing simulations, comprehensive metrics, feedback mechanisms, and tailored content.

Your user awareness, training, and testing program needs to focus on the primary goal of building a culture around cybersecurity for your users, and the focus of this culture should not be scoped to your organization, it needs to relate to the users' personal lives also. Not only will you find greater engagement from your users, but they will also typically gain a better understanding when something is related to their personal lives and real-world examples are provided. It is important that the program runs continuously throughout the year, and you need to ensure it creates an engaging and purposeful experience. In addition, ensure the program is dynamic, threats are changing daily, and your program needs to account for these ongoing changes to keep your users up to date with the current threat landscape.

Let's look more into security culture and maturity before we take a deeper dive into the user awareness, training, and testing program.

Security Culture and Maturity

With the ongoing evolving threats that we continue to face as individuals and within our organizations in today's world, we need to shift our mindset with our user awareness, training, and testing programs that allow for a security culture to take place. This is a shift from the past where cybersecurity didn't necessarily need the attention it does today. Today, everyone within the organization needs to understand cybersecurity and it needs to be part of their everyday lives, similar to the shift from using pen and paper to a computer. Today, using a computer comes naturally. The same applies to cybersecurity; this should be natural as part of our everyday work and personal lives. Every user should have some natural awareness about cybersecurity as part of their job and everyday lives.

As part of this culture, there also needs to be a shared sense of responsibility as it relates to cybersecurity. The business must become more responsible and accountable for cybersecurity throughout the organization. As our users become more aware, they have a responsibility to identify and bring awareness around bad cybersecurity practices within the organization. As they manage and execute their daily processes, they should be able to identify when something isn't secure versus knowing what they are doing is secure and meets compliance. To ensure the success of a shared responsibility model, it will be important that leadership supports the need for the cybersecurity program to be more engaging. Encouraging and allowing cross-collaboration across the organization will provide a significant benefit and it will be important that time is made available for users to focus on cybersecurity awareness, training, and testing outside of their daily tasks without expecting them to overwork.

As you continue to provide awareness and demonstrate the shared sense of responsibility, the idea is that our users share this sense of shared responsibility throughout the organization with their peers. This in turn will create champions throughout the organization to help bring more awareness at a much quicker rate. In addition, this sense of shared responsibility will naturally become part of your users' everyday personal lives, which will equip them to support family and friends to become more conscious and secure with everything they do in life. This is essentially the ultimate goal, to spread a cybersecurity mindset and culture throughout the world to allow everyone an opportunity to best protect themselves against ongoing cybercrime.

Ensuring we continue to mature our programs is a must with our user awareness, training, and testing programs. We can't just set up the program one time and expect it to meet our needs in the long term. To mature the program, we must be able to assess and measure the current state so we can better understand what is needed to continue to mature the program. This isn't a one-time event; this needs to be continuous to meet the fast pace and dynamics of the threat actors we are challenged with. Like anything in life, if we don't have any measurements to work against, how can we show improvement?

With this sense of culture and maturity, we need a foundation and model to follow. Fortunately, there's an excellent model readily available to support your security culture journey with a maturity model to follow. KnowBe4, a leader in cybersecurity user awareness and training has created a five-level security culture model that begins with level 1 for basic compliance through to level 5 for achieving a sustainable security culture. The five models represented here can be found at this link: `https://www.knowbe4.com/security-culture-maturity-model`.

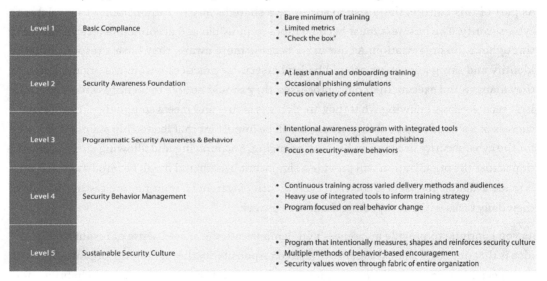

Figure 9.4: KnowBe4 five maturity levels for the security culture maturity model

In addition to these five levels, KnowBe4 provides a great visual of the correlation between achieving a level 5 and the reduced likelihood of a breach based on the increased awareness of your users and a culture being built around cybersecurity at the same URL referenced above.

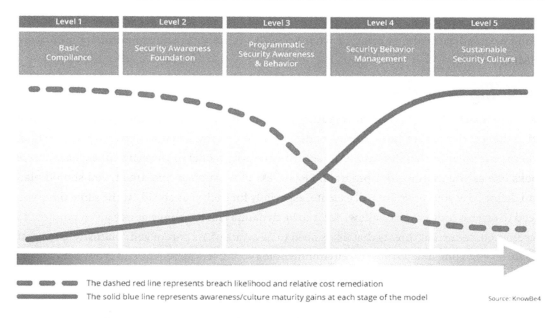

The dashed red line represents breach likelihood and relative cost remediation

The solid blue line represents awareness/culture maturity gains at each stage of the model

Source: KnowBe4

Figure 9.5: KnowBe4 research security culture maturity model visualization

To access a detailed copy of the *Security Culture Maturity Model* framework, browse to `https://www.knowbe4.com/security-culture-maturity-model` and click on the **Get This Resource** button, as shown in the figure:

Figure 9.6: KnowBe4 Security Culture Maturity Model

Once opened, you will have access to a lot more details on how to best achieve a level 5 maturity along with indicators for success, or what KnowBe4 refers to as **culture maturity indicators** (**CMIs**).

Defining Your Program

As you look to build your program or mature your current program, you must take into consideration the security culture and maturity model just reviewed. You'll first want to better understand where your organization sits within the security maturity model so you can define what success looks like as you mature your program. Regardless of what stage you are at, you should plan and define how your program will execute, especially for each year ahead. At the same time, you need to ensure your program allows for it to be dynamic and to change at any given moment to accommodate current threats that users need to be aware of. As part of your maturity, you must create an engaging and purposeful environment for your users to benefit and grow from.

When defining your program, there are three core components to consider in a robust user-centric cybersecurity program:

- The first is *testing*, commonly executed through the likes of phishing campaigns as an example.
- The second is providing a robust and well-rounded *training* program, which includes providing and keeping records of employees who have taken and passed assigned cybersecurity training.
- Lastly, you need to provide constant *awareness* to your users as deemed appropriate.

Keeping security at the forefront by sending weekly communications with tips, tricks, and recommendations, or providing somewhere for users to learn more about security-related topics, such as the company intranet, will help strengthen your users' mindsets. As already stated, these three core components shouldn't be one-time events, but rather ongoing events, to align with attack methodologies that are constantly changing.

As you build your program, you are going to need tools to enhance and mature your user awareness, training, and testing programs. There will only be so much you can do with in-house capabilities. As we reviewed in *Chapter 2, Solidifying Your Strategy*, you are going to want to strategize around a core foundational platform that will provide the capabilities you need. You may not get all the capabilities, but you should be looking for a platform that provides the majority of what you need to deliver your program. There are many vendors and tools available to comprise your program.

But as we've already stated, adding more vendors only brings more complexity, more risk, and an inconsistent user experience, and creates more challenges for your staff to manage as they will need to learn and maintain multiple platforms. Make sure you do your research, and you are familiar with the cybersecurity user awareness, training, and testing platforms available. A few platforms worth noting include the following:

- **KnowBe4** (`https://www.knowbe4.com/`), which is known for its extensive library of training materials and robust phishing simulation capabilities.

- **Proofpoint**: (`https://www.proofpoint.com/us/products/security-awareness-training`) provides a broader portfolio of services such as advanced threat detection but also provides a library of training materials and phishing simulation capabilities. These capabilities were acquired by Proofpoint with their acquisition of Wombat Security Technologies, Inc. in 2018.

- **SANS** (`https://www.sans.org/security-awareness-training/`) provides in-depth training courses and certifications, renowned for their technical rigor and practical focus.

As we have discussed with other functions within your program, ensuring the resources and time for the success of your user awareness, training, and testing program is critical. This is a program that will not run itself; it needs dedication and commitment to ensure success. With a smaller organization, it may not be possible to dedicate resources to this program. If that is the case, make sure someone is assigned responsibility over the program to ensure the success long term. If you are a larger organization, you are going to need a dedicated resource at minimum to enable the maturity of the program. There is a lot involved to make this a successful program and without commitment, which needs to start at the leadership level, the program will not mature.

Resources in this program don't need to be highly technical, but there should be some fundamental understanding of what is needed to best educate your users, and the need to keep current. A cybersecurity analyst/administrator, or if possible, a training administrator type of resource would go a long way. Someone who has a background in presenting or teaching would be an added benefit to help with end-user engagement. To make this program more successful, you are going to need people who will be able to present to broader audiences, work directly with users across the business, create quality material that can be consumed by the end users, and much more. The dedicated people must hold some quality in these areas. It will also be important that this team collaborates with the broader cybersecurity functions. As a leader, you must enable cross-collaboration. The more you can bring in data from other functions and allow other functions to share and present content to the end users, the more successful the program will become.

Next, let's shift our focus onto how you should be planning your user awareness, training, and testing program for your organization. To be best prepared, you should look to have your program planned for the year ahead. There's no reason why you can't pre-build a footprint that can be followed throughout the year. Although areas of interest may change, the scheduling and events will all remain the same. This also allows transparency to be built around how the program is being run and everything involved in running the program for your end users. It also brings visibility to your executive leadership team, the board, and the broader organization on what to expect throughout the year. This allows for buy-in and sign-off at the highest level to ensure success.

The following figure provides a high-level example of how you should look to define your program. Everything listed within the figure will be covered in more detail within each of the core sections previously mentioned.

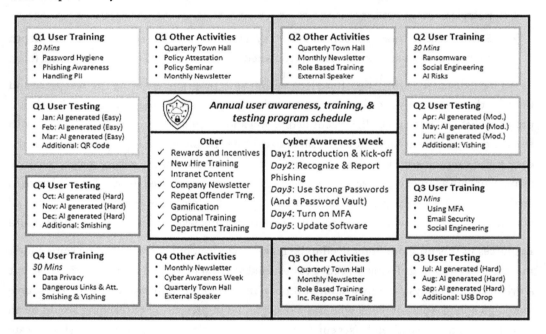

Figure 9.7: Cybersecurity program plan for the year ahead

It is important to remember that as you continue to build and mature your user awareness, training, and testing program, you need the ability to provide intelligence, dynamic learning, and automation. The ability for a solution to understand your users' behaviors and risks to allow content to be delivered that provides more value is key. For example, if a user fails a simulation or is compromised through a real incident, we need to make them aware of those failures and provide them with the relevant material to learn and continue to improve.

Ideally, this needs to be automated. Remember, humans are our greatest asset, but they are our biggest threat vector when it comes to cybersecurity. We as cybersecurity professionals can change this!

Ongoing Program Management

Oftentimes, we deploy new technologies or roll out a new program and don't take into consideration the ongoing operation of the program. Once we have enabled something new in the organization, we must ensure a support model is in place to support the ongoing success. Too often I observe projects being completed to meet a specific need, only to move into production without taking into consideration ongoing support and operations. This, in turn, creates more burden and work across the support teams who need to pick up additional responsibilities and creates frustration for end users who can't receive the expected support. The same applies to the user awareness, training, and testing program. This program doesn't run itself once a plan has been made for the year ahead. Many tasks need ongoing support along with regular maintenance to keep the program relevant and running effectively.

Some of the ongoing operational items that you can expect as part of the program include the need to support users across a variety of areas, including supporting phishing campaign issues and training assignment challenges. You will also need to follow up with users who have become victims of a real-life security incident, such as phishing or some other form of social engineering activity. It will be the responsibility of the team to work with and educate users as they fall victim to these events. You'll also need to assign relevant training based on the incident that impacted them. Other ongoing tasks will include the need to build presentations and collect relevant information that can be used to better educate the users. It will be important that you keep the content relevant for your end users so they can better relate to current trends and threats. You will also need to work closely with the cybersecurity operations team and **Security Operations Center (SOC)** to ensure they are keeping you informed of end-user activity and any threats they deem relevant for users to be more aware of.

Another important area that will be a requirement for the team is to ensure all users within the organization are receiving user awareness, training, and testing. As the program is (or was) being built, all users will need to be onboarded and enrolled into the program, whether you have a single platform or multiple platforms, you will need to ensure all users are in scope. When referencing all users, don't just include your full-time users, include all contractors and other third-party entities that access your systems. The cybersecurity program must be all-inclusive of anyone responsible for managing and operating your systems, specifically anyone with privileged access.

In addition, ensure there is a process for newly onboarded users being accounted for. This will ensure they receive all training and testing moving forward. This is not an easy task and will require ongoing reconciliation on a frequent basis to ensure compliance with all users in the organization.

One final area we will touch upon is the need to be dynamic and keep the program relevant for your users. You must stay current with the latest threats and trends so you can quickly educate your users as much as possible. For example, you have a quarterly training scheduled around password management, but suddenly you are seeing a significant increase in QR code-based phishing attempts. To better educate your users immediately, you'll want to switch the training for password management to something that will provide awareness to the users around QR code phishing, or quishing. This would be in addition to other communication channels you will take advantage of to keep your users current.

There are many more ongoing operational tasks needed to keep the user awareness, training, and testing running efficiently. We have only reference on a few items to ensure you fully understand the importance of ongoing operations. As leaders, we can influence and ensure the program receives the attention it needs to be successful.

Program Management and Governance

As we have touched upon in every other function, the program management and governance of each function mustn't be taken lightly. As you operate each of the functions within your cybersecurity program, you need to ensure everything is being executed as planned and agreed upon to ensure the program is delivering as expected.

To begin, ensure policies and procedures are in place with your user awareness, training, and testing program. Your broader policy should include language that sets expectations for your users with the program and what they must complete as part of their responsibility whilst working for the organization. It is also important that users are made aware of your policies around the program, which we will cover in more detail within the *User Awareness* section next. You'll also want to ensure well-defined processes are in place. For example, what is the process when an end user becomes compromised by clicking on a malicious link or opening a malicious URL? If the same user repeats the same mistake multiple times, how do you handle the situation? This will need to be addressed within a documented process, which should be agreed upon by others throughout the organization, for example, the HR and legal teams will most likely need to review and approve.

You will also need to ensure processes are in place throughout other areas of the program as you work with the cybersecurity operations team to receive updates on users who require additional awareness, the need to handle situations where end users are not completing training or failing phishing simulations, and so on. All your policies, procedures, and processes must be well defined and signed off by your executive leadership team to ensure enforcement, otherwise, the program will not get the success it needs.

It will be important you understand the roles and responsibilities of the user awareness, training, and testing program. Internally, your roles and responsibilities will be fairly simple as your team overlooks the program. The complexities of the roles and responsibilities will be as you look to involve the legal and HR teams as you need to ensure policies and procedures are in place and require their support as any personnel issues continue to arise. You will also want to define the end users' roles and responsibilities as it relates to cybersecurity and what is expected from them as part of the program. It will be important that they know what is expected of them to remain compliant with the program. Other areas of responsibility will include those from the broader cybersecurity organization that need to support the user awareness, training, and testing program. For example, cybersecurity operations will play an important role as information will need to be shared to ensure the team can take corrective actions with end users and ensure the content being delivered is relevant.

You will also want to ensure you have frequent meetings to ensure the program is running as expected and review any outstanding issues that need attention. The more you run training campaigns and user testing, you will want to review training completion rates to escalate if users aren't taking training and review failure rates to better understand any trends that may need more attention. In addition, you will want to ensure you keep any awareness material current as you look to publish content for users to access and as you share information with the users via different communication channels.

Finally, you will need to ensure you are capturing relevant metrics to allow measurement of the program maturity over time. More importantly, you'll need to make sure you are capturing all relevant metrics that measure your end users. A couple of common metrics will include items such as user failure rate with testing from phishing simulations and user completion rate with training campaigns that are being assigned. You will need to continuously track these metrics over time to analyze whether you are seeing improvements or not with the overall program.

Depending on progress (or not), the program can be adjusted to support the identified metrics if improvement is needed. Once you have all the agreed-upon metrics, you'll need to ensure reports are created and made available as part of your broader reporting strategy for the organization, including the executive leadership team and the board. It will also be good to share this data with your end users, so they are aware of how well (hopefully) they are doing with the overall program.

It's important you don't overlook the program management and governance aspects of your functions. These become critical in ensuring a successful delivery of the program for your organization as well as providing full transparency on what is involved to make it a success.

User Awareness

The first of the three core components we will cover is user awareness. Of the three core components, this tends to get overlooked and not provided with the attention it needs for your users. Typically, as a user awareness, training, and testing program evolves, there is an emphasis on the training activities and testing type of simulations for the users. Because of this, the user awareness component doesn't get the focus it needs to be successful. At a high level, user awareness provides users with the information they need to be more aware of the environment around them as it relates to cybersecurity. Although training and testing do bring awareness, there is a focus area around user awareness that needs to be addressed as part of your broader program.

We must provide our users with awareness so they can keep current with the latest threats, which allows them to make better-informed decisions when dealing with potential threats. If a user isn't aware of the correct actions they should be taking with potential threats, the risk of becoming compromised increases significantly. The more we provide awareness of how the end user can better protect themselves and the organization, the likelihood of a compromise will be reduced, although it will never be eliminated. It is also important that users are aware of expectations and any cybersecurity policies they must follow. Without providing the correct awareness, users will not know how to comply with what is expected of them.

It is also important to ensure that the awareness component is not only relevant but is an ongoing effort. Sending out an annual email reminder does not comprise a mature awareness outreach program. This component needs ongoing outreach using various channels to reach your users. It also will require engagement and interaction in certain use cases to ensure awareness is being delivered effectively. Ensuring a comprehensive outreach program for awareness will require ongoing effort and collaboration across the business for success. The following components are areas you should be focusing on to support your awareness program.

Awareness Channels

As you look to provide increased awareness around cybersecurity, you'll need to understand what channels you have available to push information to your users. Unfortunately, the use of one channel will not suffice these days and you'll need to leverage multiple different channels to broaden your reach as much as possible. Thinking outside the box will significantly benefit awareness as you look for more creative and effective ways to communicate awareness to your users. You'll also need to partner closely with your marketing and/or communications team to follow any standards or procedures they have in place. They will already have channels in place that you will be able to take advantage of as you look to integrate awareness across the broader organization. You'll also want to build a brand around your cybersecurity program and build a standardized style with a logo that provides consistency for your program and allows users to better identify when anything cybersecurity-related is being communicated. Some channels to take advantage of include those outlined in the following subsections.

Portals

A couple of thoughts come to mind with portals. First, there's a chance you already have some form of internal documentation repository or intranet available for your users. If so, it would be advantageous to leverage what is already available today. It would be more beneficial if you build a cybersecurity-specific section that becomes a known place for your users to navigate when accessing the latest and most current cybersecurity information. The second thought that comes to mind with portals within your company is any company-wide type of portal that all users access for up-to-date company information. Is there a possibility you can get some real estate (even if it is small) to share any current or relevant information that will reach a broader audience, even if it's a notification that directs the user to more detailed information. This portal will be managed by a different team in the organization so you'll need to collaborate and show the benefit it will create for the broader organization. At a minimum, make sure the portal covers the following:

- Latest threats and news
- Policies and procedures
- Training resources
- FAQs and tips
- Incident reporting

Email

The obvious and most common channel is your email communications system. Email still being one of the most commonly used communication channels will allow information to reach a large portion of your users for more visibility around cybersecurity awareness. Sending emails to all your users will most likely require your marketing and/or communications team, who will most likely own this process for the organization. You'll want to be strategic with this method as you won't want to overwhelm the users with constant communication, or they will begin to ignore it. Ensure the content via this channel provides purpose and is relevant. Too much information through this channel will simply be lost. Also, check whether there is a broader frequent outbound email communication in which you can provide some quick high-level awareness to all your users, again, even if it is just information on where to get more information. Some guidelines you can follow to ensure you aren't overwhelming your users include:

- Monthly e-newsletters to users with recent updates.
- Urgent alerts as immediate notifications for critical threats.
- Quarterly reports, which will summarize key metrics and trends.
- Event-driven updates, that send emails based on specific incidents or new policies.

Newsletters

Newsletters can be a great place to include cybersecurity information and I'd imagine newsletters are already being distributed throughout the organization in some fashion. Ideally, your organization is using e-newsletters and everyone has moved beyond physical copies. Look for opportunities to embed cybersecurity information into already published e-newsletters and see whether you can get some permanent real estate within them moving forward to provide ongoing awareness. If you are ambitious, you could look to create a cybersecurity-only newsletter that can be distributed monthly or quarterly to your users. This will require a lot of effort, especially to edit and produce in quality. Just make sure the information is current and relevant.

You can find an example of a newsletter from SANS. SANS publishes a monthly newsletter named *OUCH!*, which is geared toward everyone. You can view all their newsletters here: https://www.sans.org/newsletters/ouch/. The example figure is from the July 2024 publication.

Browse to the following link `https://www.sans.org/newsletters/ouch/text-messaging-attacks-smishing-saga/` and click on **Download** to download a PDF version of the newsletter.

Mark was perplexed by the text message, a package delivery notification from Amazon - "Delivery attempt missed! Click the link now to reschedule or your package will be returned." Mark could not remember ordering anything online recently, but to be honest, he ordered so many things online it was easy to forget. Not wanting to miss any packages, he clicked the link, and a page loaded asking for his contact information "to ensure proper rescheduling." The message seemed a bit odd, but Mark figured better safe than sorry. He entered his home address details and was then asked for additional information, including his credit card information. Trusting the company, he entered everything it asked to ensure delivery. The page then said his package should be delivered soon. Then, within fifteen minutes Mark received a phone call from his credit card company notifying him that his card was being used to make numerous online charges from all over the world. Mark froze as he realized that there was no package and that the text message had been a scam to trick him out of all his information, including his credit card.

What Are Messaging Attacks (Smishing)

Messaging attacks, also called Smishing (a combination of the words SMS and Phishing), occur when cyber attackers use SMS, texting, or similar messaging technologies to trick you into taking an action you should not take, such as giving up your credit card or bank account password or installing a fake mobile app. Just like in email phishing attacks, cyber criminals often play on your emotions, such as creating a sense of urgency or curiosity. However, what makes messaging attacks so dangerous is that there is far less information and fewer clues in a text than there is in an email, making it much harder for you to detect that something is wrong.

Sometimes cyber criminals will even combine phone calls with messaging attacks. For example, you may get an urgent text message from your bank asking if you authorized an odd payment. The message then asks you to reply YES or NO to the message. If you respond, the cyber criminal now knows you will engage with the message and will then call you on your phone pretending to be the bank's fraud department. They can then try and talk you out of your financial and credit card information, or even your bank account's login and password.

SECURITY
AWARENESS
© SANS Institute 2023   www.sans.org/security-awareness

Figure 9.8: SANS July 2024 OUCH! monthly newsletter (Page 1 of 2)

User Awareness, Training, and Testing Platform

As we discussed in the *Defining your Program* section, to be more successful with this function, you are going to need a foundation in place to help with issuing training and testing activities. With this platform, take advantage of any awareness materials or built-in functionality your users can benefit from. For example, there may be toolkits available for you to take advantage of to provide additional awareness to your users. In addition, there may be an opportunity for your users to access additional content within the platform for continued learning. For example, is there a catalog of additional content or videos that users can access in their own time to learn more? One excellent example is *The Inside Man* from KnowBe4. The Inside Man is an ongoing award-winning series that provides current and relevant security awareness to the user as part of each episode. You can review more on the series here: `https://insideman.knowbe4.com/`.

Figure 9.9: KnowBe4 The Inside Man

Other Awareness Channels

The sky is the limit with your communication channels for awareness. Make sure you take advantage of all possible channels and every opportunity to provide awareness to your users. Some other ideas that come to mind include the use of poster boards around your offices, if applicable. Be creative with leaflets that can be handed out around the office, or business cards that include how users can reach out for support. Maybe you can create laptop stickers with your cybersecurity brand on them and information on how to reach out for support. You could create standup cards that are placed on end-user desks with information tips on them. Another opportunity is for your cybersecurity team to include information within their signatures that provides users with where to go for additional information and support.

Does your organization have an internal social media platform that can be leveraged to share information? If so, see how you can incorporate awareness into the platform for your users. A more relevant opportunity could be around the usage of AI with chatbot type of functionality. Is there a way you can provide your users a channel to interact with a bot that can provide awareness from any user guides or documentation you have available for them? This will help provide a cleaner and simplified user experience if you can set up efficiently with access to all end-user cybersecurity information. One final channel that needs to be noted is that of out-of-band communication, better known as an **Emergency Mass Notification System** (**EMNS**). This is not a tool that should be used for standard awareness, but a tool that should be in place in the event any of your other awareness channels are unavailable due to some form of ransomware or major cybersecurity event. There may be a need to quickly communicate to your users to inform them of an attack and to follow some immediate instructions. For example, shut down your laptop immediately if an active ransomware event is occurring. To accomplish this, you will need an out-of-band channel to provide awareness to your users.

 Make sure you constantly remind your users of the correct process to reach out for cybersecurity support or issues. You should have a separate channel that users can contact, whether by email, phone, ticketing system, or so on. The quicker they reach out with a cybersecurity issue, the quicker the SOC can review and remediate.

Again, your imagination is the limit with awareness, and anything is possible. Make sure you are collaborating with your marketing and/or communications teams who are the subject matter experts within this field and will certainly bring more value to your awareness efforts. Now that we have reviewed the various user awareness channels, let's shift our focus to the month dedicated to cybersecurity awareness.

Cybersecurity Awareness Month

One event that you should all already be aware of is Cybersecurity Awareness Month. If you are not aware, the month of October has traditionally been dedicated as a month for cybersecurity awareness. This initiative was launched by the **Cybersecurity and Infrastructure Security Agency (CISA)** and the **National Cybersecurity Alliance (NCA)** back in 2004. The month provides continuous awareness to the public and businesses on how to best protect themselves and their data from being compromised. The great benefit from this month is that there is a lot of material available from various government entities and vendors to support the additional awareness for your users. Some of the materials I've referenced in the past include content from:

- **KnowBe4**: https://www.knowbe4.com/cybersecurity-awareness-month-resource-kit
- **Proofpoint**: https://www.proofpoint.com/us/resources/awareness-materials/cybersecurity-awareness-kit
- **CISA**: https://www.cisa.gov/cybersecurity-awareness-month
- **NCA**: https://staysafeonline.org/programs/cybersecurity-awareness-month/

To expand on CISA's 2023 campaign, a new theme was released for the program named *Secure Our World*. This theme will be used indefinitely for all future cybersecurity awareness months. For the 2023 theme, there was a focus on four actions that you should take to provide a more secure workplace, home, or school environment. The four actions are:

- Recognize and report phishing
- Use strong passwords
- Turn on MFA
- Update software

As mentioned, CISA provided a lot of great material to support the theme released. This is designed to support your awareness activities, so you don't need to spend time creating your own material; it is available at your disposal, so take advantage of it. To access the toolkit for the 2023 Cybersecurity Awareness Month, go to https://www.cisa.gov/cybersecurity-awareness-month, scroll down the page, and click on **DOWNLOAD TOOLKIT (ENGLISH)**. Here, you will be able to download and use all the great materials provided by CISA. An example of a tip sheet looks like the following.

Figure 9.10: The CISA Secure Our World tip sheet

As you can see, you must take advantage of the resources available to us for our users. It is also important to consider that although Cybersecurity Awareness Month is a great initiative and we must take advantage of it for the month of October, it essentially doesn't stop with one month of awareness. We need to continue raising awareness for our users on an ongoing basis. The new theme we need to keep in mind is *Cybersecurity Awareness Forever*. Now that we have reviewed more on Cybersecurity Awareness Month and the opportunities it provides, let's shift our focus to cybersecurity policy awareness.

Policy Awareness

One area you may not be taking into consideration is ensuring that your users are aware of and acknowledge that you have a cybersecurity policy (or policies) in place. If you don't have a cybersecurity policy in place, you will want to address this immediately. We will be covering policies in more detail in *Chapter 13, Governance Oversight*. You may have one unified policy, or multiple cybersecurity policies in place for your organization. There's a possibility these policies are sitting somewhere on an intranet, or they've been sent to your users via email for review. Although this is a good start for your policy awareness, more effort must be put into ensuring your users are aware that there is a cybersecurity policy in place and more importantly, what is contained within the policy that they need to be aware of.

A couple of actions to think about as it relates to awareness around your cybersecurity policy or policies are as follows:

- **Attestation**: It will be important that you require your users to view the current policy in place and acknowledge that they have read the policy. This becomes important for several reasons. First, you should review your policy annually and make any necessary changes. When this happens, you will need to ensure your users have reviewed and acknowledged changes to the policy. This provides evidence that the user is aware of the policy that they must follow. Second, as you look to implement a framework and audit against that framework, there'll most likely be a control that validates if your users are communicated and made aware of your cybersecurity policy on an annual basis. Third, when a user doesn't comply with your policy or puts the organization at risk because they weren't following policy, you'll have evidence that the user was aware of the policy even though they didn't follow it. This, of course, becomes an HR and potential legal issue as they arise. With this, you are going to need a tool that allows you to share the policy with your users and requires them to attest via e-signature or some form of assignment completion. Tools like Adobe and Docusign provide these types of capabilities. If you have a current **learning management system** (**LMS**) in place, you could leverage it as part of an assignment to ensure users review and complete the assignment. The ideal solution would be to deliver the policy through your user awareness, training, and testing platform where you deliver all your other cybersecurity-related content.

- **Policy review and/or training**: In addition to attestation of your cybersecurity policies, it is important you provide an opportunity for your users to receive a review or some form of training as it relates to the policy.

Policies can become long and complex, and users may not easily understand what is expected of them, especially if legal language is used throughout. Because of this, it will be important to provide sessions or channels for your users to learn more about the cybersecurity policy and what is expected from them as an employee or contractor after attesting to the policy. You may want to consider a shortened and simplified version that highlights the key points your users should be aware of, so it doesn't overwhelm them. This event also needs to be annual as there will be a need to provide updates to users as the policy is updated and modified annually.

Managing your policy and providing awareness is not a simple task and will require time and effort to ensure this requirement is met, especially as you look to receive 100% compliance for policy attestation. There may be a need to work with HR and legal as you may have some users who won't attest because they have concerns with the language within the policy. These will need to be addressed on a case-by-case basis.

User Awareness Content

The most important aspect of awareness for your users is that of the content you share. Your content must be engaging for the users. It also needs to be relevant to current threats and trends. To ensure you are getting the attention you need from them, make sure you are including data that relates to your users on a personal level. There is nothing more important than users realizing how cybersecurity can impact their personal lives. This in turn begins to create the mindset that cybersecurity is cybersecurity no matter whether it is within the organization you work for or your personal life. Once this begins to happen, a shift to a security culture and shared responsibility within cybersecurity becomes more of a reality amongst your users. The following are some suggestions for you to share relevant and engaging content with your users:

- **Statistics**: This one is big for users, ensure you leverage any statistics available that your users can reference. I have used statistics many times throughout the book and many of them are very relatable to your users so ensure you use them to bring more awareness. In addition to external statistics from case studies, also make sure you bring any internal statistics from your cybersecurity program. It is always good to bring transparency to the users based on any statistics that are measuring the cybersecurity program, or more importantly, statistics that measure your users.

- **Personal impact**: This one is extremely powerful and should be used all the time. Ensure you bring examples of users who have been personally impacted in their everyday lives. A quick search will reveal endless examples of articles and news feeds sharing financial fraud, cybercrime, and so on with users in their everyday lives.

- **Free content:** Leverage as much of the free content as possible, as long as it's good quality. For example, we shared resources made available by KnowBe4, Proofpoint, CISA, and NCA earlier in the chapter. All these resources were provided as part of Cybersecurity Awareness Month but can be used at any time to provide awareness for your users. These materials can save you a lot of time and provide some quick and easy awareness for your users.

- **Platform:** Ensure you are taking advantage of all the content made available from your user awareness, training, and testing platform. Although the focus will be on your initial training and testing, the platform will most likely provide a lot of additional content and material that will be available for your users. Make sure you are enabling access to this content for your users to provide additional awareness. A good example is the *The Inside Man* series provided by KnowBe4.

- **Actual data:** Work with your SOC to provide examples of cybersecurity incidents that have come into the organization. Use these examples to provide better guidance to your users and help them better understand the signs of something being legitimate or fake. For example, if there is a phishing email that a user became compromised by, use that example (anonymously of course) to physically show the email and the signs that users should be aware of when similar types of email come into the organization in the future. Also, provide data on how many cybersecurity incidents the SOC handles, the types of incident, and so on. Users will have a better appreciation for the work that goes on behind the scenes.

- **Training and testing:** Ensure you are sharing the results from your ongoing user training and testing efforts. It will be important that your users are aware of how many are completing the training required along with how well users are identifying phishing types of emails, with a hope of seeing gradual improvement over time.

- **Threat briefs:** Although threat briefs can get very deep and technical, think of opportunities to work with your cybersecurity vendors to create an end-user type of threat brief; something that your end users can relate to with actual data coming from a third party on the current state and what they should be more aware of within the organization and their lives personally. We will be covering threat briefs in more detail in *Chapter 11, Proactive Services*.

- **Applications and tools:** Make sure you are bringing awareness around any of the cybersecurity-related tools that your users may have at their disposal. It will be important that users take advantage and use any cybersecurity-related functionality or tools within the organization to better protect them. The same applies to them personally – are there applications or tools you can make them aware of to better protect them in their everyday lives?

For example, providing them more awareness around the use of password safes for personal use, making them aware of tools like Have I Been Pwned (`https://haveibeenpwned.com/`) to better understand if they have any compromised accounts, and so on.

 The following link offers a great example from CISA with many great resources for reference: `https://www.cisa.gov/resources-tools/resources/cisa-cybersecurity-awareness-program-toolkit`.

This list is not comprehensive but should provide you with a better idea of the content and resources you should be considering as you provide more awareness for your users.

Current Trends

It is important that you are always bringing current trends in front of your users so they can be better prepared for the latest threats. Make sure collaboration is occurring with the SOC and vulnerability management teams as they will have the latest intel from what is being observed in the wild. In addition, as you follow the latest news (breaches, scams, etc.) and keep up with the most current research, you will be able to take this information and provide more awareness of the latest threats to your users. Trends will continue to change, but some of the more familiar trends related to your end users in the current state as of July 2024 include:

- **Social engineering**: You may already be aware that social engineering continues to be a significant challenge for our users. Our users continue to become compromised via social engineering techniques and it's not easing up. For some great examples of how easily one can be hacked via social engineering, take a look at `https://www.socialproofsecurity.com/` and some of the videos available on their website.

- **Email**: As we've continued to demonstrate throughout the book with statistics, email continues to be a challenge for our users. Traditional methods of phishing and **BEC** continue to exist and now we are seeing more advanced vulnerabilities via email through quishing and the use of AI. Outside of email, threats continue to rise with both vishing and smishing types of attacks.

- **Ransomware**: This one isn't slowing down. As threat actors continue to make big profits from ransomware attacks, we are going to continue to see an increase in this area. Some companies are paying the ransom, which encourages threat actors to continue to breach as many organizations as they can. Make sure your users are aware of the damage that ransomware can do to an organization.

- **Passwords:** Data on passwords continues to show how far we have to go with better managing our password hygiene. Statistics are readily available in this area, and we have covered a lot within this book, including earlier in the chapter with the NordPass (`https://nordpass.com/most-common-passwords-list/`) example of the most commonly used passwords and the time it takes to brute force a password (`https://hivesystems.com/blog/are-your-passwords-in-the-green`). Clearly, this shows how bad hygiene with passwords is. In addition, passwords continue to be stolen as we have shared via `https://haveibeenpwned.com/`. Make sure you continue to share the information provided in the *Securing Your Identities* section from *Chapter 6, Identity and Access Management.*

- **AI:** AI is probably one of the latest trends that is impacting both the organization and our users. We are observing everything from advanced email threats to the ability to write malicious code, to more extreme instances of advanced deepfake capabilities using sounds images, and videos. A very unfortunate example is that of a deepfake video call a finance worker had with the supposedly **Chief Financial Officer** (**CFO**) and paid out $25 million (`https://www.cnn.com/2024/02/04/asia/deepfake-cfo-scam-hong-kong-intl-hnk/index.html`). We touched upon AI in more detail in *Chapter 7, Cybersecurity Operations.*

- **Data Privacy and PII:** An item that will always be a trend is that of data privacy and PII. One of the primary drivers for threat actors is to obtain our data to profit from it. More and more companies are becoming compromised with our data and more concerning, companies continue to sell our data for profit, which only increases the footprint of where our data is stored. We need to continue to bring these concerns to our users' attention and make sure they are more conscious when sharing data and understanding what companies are doing with our data. Regulations also continue to increase, which will be good for your users to be aware of so they understand their rights.

Again, these are just a selection of some of the many trends we continue to be challenged with. These trends can be impactful at both a professional and personal level. It will be important to relate these trends to the users at a personal level, so they better understand the impact and current risk of each of these trends.

User Training and Testing

In this section, we will cover two of the more common and executed components of your program. Traditionally, there has most likely been some required cybersecurity training, whether it be when you were onboarded as a new employee or on an annual basis. For some, there may be nothing in place.

Moving forward, more focus and effort are needed around employee training activities, which we will cover in more detail throughout this section. The same applies to user testing; there may have been some phishing type of simulation sent to your users on an ad hoc basis, or maybe never at all. Unfortunately, this will not suffice, and the user testing activities need significant improvement from what may have been executed in the past.

User Training

First, let's review the user training components for your overall user awareness, training, and testing program. Training is a critical component of everything we do, not just in life, but more importantly within the organizations where we work. We have grown up constantly being trained how to navigate through life and (hopefully) do things right. The same applies within an organization. Starting a new job without any training will only set you up for failure. Even if formal training isn't provided, to be successful, you will need to self-train to accomplish the role you have been assigned. Training provides us with the foundation of what is expected of us and provides us with the information needed to complete specific tasks, and our job as a whole. The same applies to cybersecurity; if you are not trained on what to be aware of, how do you know that you are taking the incorrect action? This is where training comes into play, which serves as a very important role to better prepare your users against cyber-related activities within your organization. Your training program must be built thoughtfully to provide the intended value to your end users.

Before we move into the schedule section, let's look at some concerning numbers as they relate to training. The *Oh Behave! The Annual Cybersecurity Attitudes and Behaviors Report 2023* by the **NCA** has a sample size of over 6,000 people from 6 global countries and identified the following:

- 64% of the users from the sample reported having no access to training, 10% with access but didn't use it, and 26% with access and used it.
- The dataset showed training was primarily accessible for those employed at 47% or studying at 49%. Although, 53% of those employed still didn't have access to training.
- Of those who had access to training, 84% found it useful and 78% found it engaging.
- It was reported that 79% of the respondents were able to take the advice learned from training and put it into action.
- After taking training, 34% of users responded that they began using MFA, 50% were able to better recognize and report phishing emails, and 37% began using strong and unique passwords.

Source: https://staysafeonline.org/online-safety-privacy-basics/oh-behave/

As the data shows, users are struggling to gain access to training, but when they do, the positive impact is being reported. This is why we all need to make a bigger effort to make cybersecurity training accessible to as many people as possible. Let's shift our focus over to the training schedule component of your user training program.

Training Schedule

It's going to be important that you schedule your training in advance. Although the content may change based on current threats, ensuring you have buy-in and alignment from your executive leadership team is critical. Without alignment from them, you will struggle with compliance throughout the organization. You are going to be requesting time from your users, which will take them away from their primary role. This can also be challenging with hourly users who are staffed to cover a specific schedule. Because of this, it will be important to ensure you have an agreed-upon allocation of time that can be used on an annual basis for all required cybersecurity training. This will also require collaborating and working with business function leaders to ensure you receive their buy-in and for them to set expectations for their respective staff.

For your training schedule, you can look at different approaches. Depending on the platform used to deliver training and the strategy you have in place, you may opt for shorter-style monthly training videos versus a longer quarterly assignment. At a minimum, your program should require a quarterly 30-minute allotment for all users within your organization. This may not sound like a lot, but you can get a lot covered in 30 minutes as part of your requirements. Within each of these 30-minute allotments, you'll be able to run multiple training assignments depending on the length. As part of each quarterly requirement, allow the users flexibility to complete the training as time permits on their schedule, but with a deadline of completing by the end of each quarter. If there are multiple training assignments, the user can complete one at a time without feeling rushed to complete everything at once, unless you have one assignment that takes the entire 30 minutes, which I wouldn't recommend.

Outside of the quarterly standard requirements for training, you can also provide additional optional training for those who are interested in learning more. You may also need to schedule targeted training for the HR group handling sensitive PII, the finance group handling financial payments, and so on. Having all of this scheduled ahead of time will allow greater success as you look to deliver training throughout the organization.

Per the high-level schedule shared here, the following training will all need to be scheduled as part of your program:

- Quarterly 30-minute training with multiple videos

- Department-specific training (i.e., HR and finance)

- Role-based training (i.e., executives, development, privileged users)

- Incident response training for SOC, service desk, and so on.

- New hire onboarding training

- Optional training modules

- Failure training for failed phishing (or other) tests

- Failure training for real-life events (true/positive incidents)

Once your schedule is defined, you will need to ensure you have the correct assignment groups in place so you can support the deployment of training to the correct groups of users.

Training Assignment Groups

It will be important you spend some time setting up your assignment groups ahead of time so they can be applied correctly to each training assignment efficiently. For example, as you build out more specific assignments like role-based, you will have different groups of users that will need to take each of the role-based training. As you build out your groups, some things to think about include:

- **Automation**: Wherever you can, build automation or auto-assignment into your groups. This will allow more of a hands-off approach and will save a lot of time. There will be some dependencies to make this successful.

- **Leverage attributes**: Use attributes on your user accounts to allow for easier group assignment. Using attributes to identify which departments your users are in, which division they belong to, who's the manager, and so on, will allow for auto-assignment to become much more efficient. Depending on the setup, this may be something you have to complete within your enterprise user directory and not the user's properties within your training platform.

- **Auto-provisioning and de-provisioning**: Take advantage of user provisioning and de-provisioning capabilities. As users are onboarded, ideally, they should be auto-enrolled into the training platform and assigned training through the use of assignment groups.

- **Single sign-on (SSO) and System for Cross-Domain Identity Management (SCIM)**: SSO is more of an added secure and convenient feature for users allowing them to use their company credentials to authenticate instead of needing a separate account. This should be part of your broader application strategy to begin with. SCIM, on the other hand, is a provisioning component that will provision your users from your enterprise directory into the awareness platform.

This allows your users to be auto-provisioned, then with the use of attributes within your user accounts, auto-assignment of your users into groups will occur based on pre-defined assignment groups. The same will apply when user accounts are disabled; they should be removed from the training platform automatically.

Another important area to consider is auditing of your assignment groups. It will be important to confirm that each of your users is receiving the correct training and that no one misses any training. The more time you spend on setting up and configuring your assignment groups upfront, the more efficiently your program will run long-term.

Training Content

To provide engaging and impactful training as part of your program, you are going to need to ensure you have quality content for your users. You could look to build your own training content, but I'd highly recommend against this, other than building presentations and providing documentation type of material. The content that will be most challenging to create is video-related and interactive-based training. This content needs to constantly evolve and it would be in your best interest to leverage content readily available by strategizing on a vendor who specializes in providing this type of content, such as KnowBe4 and Proofpoint, as referenced earlier in the chapter.

One area of consideration with your content is where it will be housed and how will it be delivered to your users. Strategically, you'll want to justify using your unified user awareness, training, and testing platform, where you can provide direct access to training and allow better insights into your user's risk. This will also allow the auto-assignment of training based on failed phishing simulations, and in general, will provide a better user experience based on all the data being collected into one environment. On the flip side, your organization may have a separate centralized LMS from which all training needs to be delivered. Although there are benefits to this from an organization perspective with other non-cybersecurity training assignments being delivered centrally, you'll lose many of the benefits and maturity opportunities available with a platform dedicated to cybersecurity training and testing. If you must use an LMS because of a company policy, you'll need to coordinate with your HR and/or learning management team to provide them with the cybersecurity content needed in a format, most likely a **Sharable Content Object Reference Model (SCORM)** file, that can be uploaded into the LMS.

Once you know where you are delivering your content from, you'll need to determine the content type and select the training that will be most appropriate to your users. Pre-determining your schedule ahead of time will make this task a lot easier. If you do have a user awareness, training, and testing platform, there will most likely be a lot of content available at your disposal.

You'll just need to determine which content to select from. The major platforms also support exporting the content into the SCORM format for your LMS system if required. If you don't have a user awareness, training, and testing platform, there is ample content available on the internet you can reference and use for your users. The following are some examples of training to be aware of in the event you are in the infancy of your program and need to get some quick wins for your users:

- NIST has compiled a catalog of (free and low-cost) training: `https://www.nist.gov/itl/applied-cybersecurity/nice/resources/online-learning-content`

- Some free courses for the public from the **Federal Virtual Training Environment (Fed-VTE)**: `https://fedvte.usalearning.gov/public_fedvte.php`

- The **National Cyber Security Centre (NCSR)** has provided an e-learning package for use: `https://www.ncsc.gov.uk/blog-post/ncsc-cyber-security-training-for-staff-now-available`

- Amazon has also built cybersecurity training available for anyone to use: `https://learnsecurity.amazon.com/en/index.html`

Figure 9.11: Amazon Cybersecurity Awareness training portal

As already stated, you will need to keep the training current for your users. You'll have a limited timeframe to train your users and you'll need to take advantage of it.

This means keeping the content extremely relevant and ensuring it is providing learning based on current threats and trends that are being observed in the organization and within their everyday lives. For example, some emerging threats to provide training on as of July 2024 include QR code phishing, or quishing, and AI-based threats such as deepfakes.

New Hires

In addition to your ongoing training schedule, you are going to need to set up a separate process to capture any new hires within the organization, whether it be an internal employee or a contractor. A lot of companies will typically have some form of onboarding program in place for new users within the organization. This will most likely be within the HR group, so you'll want to collaborate with the HR team to see how you can incorporate cybersecurity training as part of the onboarding experience. It will be important that new employees are provided training as soon as they join the organization so they understand what is expected of them as it relates to protecting the organization and its data.

This process will need to be separate from the standard training material and all new users should be required to complete this process. At a minimum, I would recommend a couple of more generic and basic training courses that will get the new user acquainted with the expectations of cybersecurity within the organization. In addition, you'll need to have the user review and attest to the cybersecurity policy (or policies) that are currently in place. You may even want to do a live training session that focuses on cybersecurity-related policies to ensure they are fully aware of what is expected and if there are any questions or concerns, they can be addressed immediately.

One last item as it relates to new hires is ensuring they have the information and guidance needed to ensure they know how to access and use cybersecurity-related tools. If there is an onboarding package, you'll want to ensure you embed any cybersecurity-related documentation and how-to guides for the new users, especially the process of how to contact the SOC for any cybersecurity-related issues that arise.

Compromised User

It is important that you incorporate compromised or failed user training into your program. There are a couple of areas where this will apply, first from a user taking an incorrect action from a testing or phishing type of simulation. When a user takes an incorrect action, they need to be made aware of what they did wrong so they can learn for the next time. Ideally, you will want to have your user awareness, training, and testing platform auto-assign training that is relevant to the exercise that they took the incorrect action.

If this is not automated, the logistics around training users on how they took the incorrect action will become a lot more challenging, especially if you are using dynamic simulation tests for your users. Regardless of how you are testing your users, they must be informed of what incorrect actions they took so they can address the mistake in the future. Without training them on their mistakes, they will continue to take incorrect actions.

The second and more important activity for compromised users is follow-up training for those who became compromised in an actual real security incident. It will be important that the user undergoes training that will help them understand what to be aware of next time a similar incident occurs. In addition, you will want to spend some one-on-one time with the users so you can walk through the events that occurred with them to become compromised. If you fail to educate and train the users on where they became compromised, they will continue to repeat the same mistake, which significantly increases the risk for the organization.

Role-Based

As your program continues to evolve and mature, you are going to want to include role-based training for your users. This can create a more complex program to manage but it will be important you provide relevant training, especially to specific roles and areas of the business with higher risk. You will also need to determine how to best assign role-based training. Will it be OK to assign additional training in addition to the agreed-upon standard schedule? Or, are you going to need to assign training using the already agreed-upon schedule, which will require you to spend more time on the schedule and assignments. The following are some of the role-based trainings you will want to consider as part of your training schedule. At a minimum, you should ensure your privilege management training and executive leadership training are the highest priority:

- **Privileged users**: It will be important for your privileged users to receive training focused on how important the level of access they have is and to enforce the due care they must take with their access. A privileged role can be anybody with elevated access within your environment, specifically the group of users who manage and operate the infrastructure and have access, which, if compromised, could be catastrophic to the organization. Don't forget your support desk, which may have the ability to change passwords, MFA, and so on.
- **Executive users**: Since executive users may have access to highly confidential information, or, in some instances, trade secret type of information, they can be a primary target for threat actors. In addition, since executives typically have the authority to make higher-level decisions, their accounts are targets for threat actors to use against other employees to take some form of action, typically with finances.

Because of this, executives must be aware of the different types of attacks that could directly target them like BEC, phishing attempts, or social engineering type of tactics.

- **Developers:** It is well known that a lot of vulnerabilities find their way into applications because of bad development practices. Because of this, we need to ensure that anyone within the organization who develops is following best practices as they develop their applications. In addition, developers tend to have a lot more access than standard users, more specifically to a broader set of data that may at times be confidential or contain PII. Because of this, they must be provided development-specific training that provides them with better knowledge of how to develop with security in mind first.

- **Department-focused:** Finally, you may want to look at department-specific training where it makes sense. A few areas that come to mind are HR, finance, and the support center:

  - HR handles sensitive PII daily so it will be critical that they fully understand their role and the importance of protecting the data they handle.

  - Finance is another area threat actors continue to target organizations to trick them into changing financial information so they can receive payments. These are primarily happening in email with phishing and BEC types of tactics.

  - The support contact center is another important area to consider. They are constantly targeted with vishing and social engineering types of tactics to spoof users and have passwords and/or MFA reset to allow the threat actor to gain access.

Ensuring targeted training for each of these groups will make them more aware of the different types of threats they need to be better prepared for.

There's a lot involved in building a successful training program for your organization, which requires committed time and dedication. The more you provide relevant training for your users, the more it will pay off and reduce risk in the long term for your organization.

User Testing

The last core component of your user awareness, training, and testing program is the testing capabilities. In addition to your awareness and training components, you are going to need to test your users. Think of testing as a way for your users to better recognize threats so they become more aware of what is safe and what is not within their inbox. Like everything in life, the more you practice, the better you become. This is the whole purpose of testing your users. This isn't a one-time event but needs to be a continuous event to meet the ongoing change of threats that our users face daily.

Unfortunately, the only efficient way to execute testing is with a platform that specializes in testing activities for your users. We covered tools that fall in this category earlier in the chapter, which includes KnowBe4, Proofpoint, and SANS. There are other tools available, but you'll need to do some additional research on capabilities. If you don't have a platform today, it is highly recommended you implement a user awareness, training, and testing platform for your users because the overall benefit is well worth the investment. Additionally, these platforms are very reasonable in pricing.

Another option to get you up and running a lot quicker (if applicable) if you don't have a user awareness, training, and testing platform, is to look and see whether there is one with the vendors you already subscribe to. For example, if you are a Microsoft customer, Microsoft has a tool available to test your users, named *attack simulation training* (included with specific licenses). With attack simulation training, like the other platforms, you can measure your user's awareness with testing by choosing between different real-world scenarios commonly used in social engineering attacks, such as phishing emails.

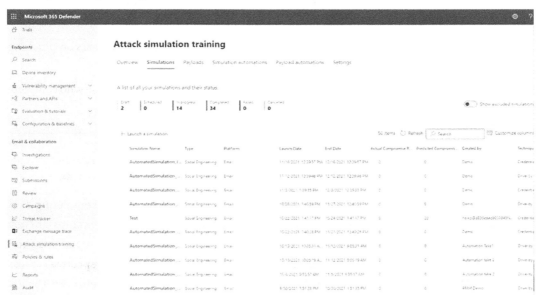

Figure 9.12: The Microsoft Attack simulation training portal
Image source: `https://www.microsoft.com/en-us/security/business/threat-protection/attack-simulation-training`

As you can see from the preceding figure, Microsoft also has training capabilities built in as part of their offering if you are licensed for these capabilities. Again, this may meet your needs depending on your organization, but you may be missing some capabilities with a platform built specifically for user awareness, training, and testing.

Testing Schedule

As we already mentioned with the other components, executing a testing or phishing exercise once in a while is not going to suffice. You need to be strategic with your testing schedule and build a schedule for the year ahead. Based on research and recommendations from others in the industry, a minimum of a monthly user test should be executed. To back up the recommendation, KnowBe4 has some incredible research on why you should be looking to execute at minimum monthly user tests. A KnowBe4 white paper named *Data Confirms Value of Security Awareness Training and Simulated Phishing*, released in October of 2023, analyzed data from over 60,000 worldwide customers, which consisted of over 32 million end users, who took over 493 million **phishing security tests** (**PSTs**) and took at least one training course within a year. Some of the highlights from this white paper are:

- The more PSTs that were completed, the better the performance from users on phishing tests.
- Performance increased in detecting simulated phishing campaigns for those who complete frequent PSTs compared to those who did not complete frequent PSTs.
- Performance increased on simulated phishing tests the longer users trained.

Source: `https://www.knowbe4.com/press/knowbe4-analysis-finds-security-awareness-training-and-simulated-phishing-effective-in-reducing-cybersecurity-risk`

The same white paper also found that those completing weekly PSTs versus those completing PSTs less than quarterly were 2.74 times more effective at reducing risk. Ideally, the more campaigns you can run, the better, but you'll need to find a reasonable balance that works for your organization. Running weekly may become overwhelming from an oversight and end-user perspective (unless fully automated), so working with a monthly schedule may be more feasible to begin with. Clearly, the long-term results show progress with more frequent user tests. To visually demonstrate the outcome, the following graph provided in the same white paper clearly shows a decrease over time, with a significant improvement from end users with frequent PSTs.

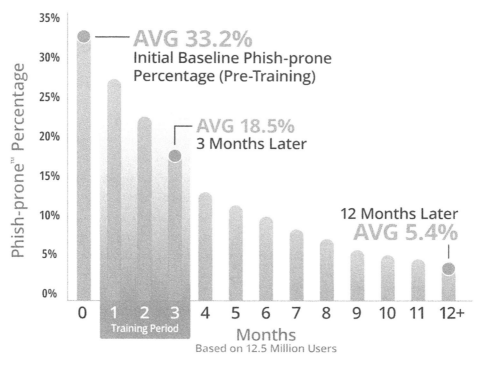

Figure 9.13: A KnowBe4 graph showing a constant decrease in failure after frequent training and PSTs

Source: https://www.knowbe4.com/hubfs/Data-Confirms-Value-of-SAT-WP_EN-us.pdf?hsLang=en-us

It is important to remember that, to get improved results from your users, training must be considered for your program. As we covered in the previous section, the recommendation of quarterly training was made for your users. This is completely independent of your user testing activities. We also recommended, in addition to standard training, the need for remediation training when users fail a PST (and a real-life compromise); more specifically for PST, automated training that provides the user training around the theme of the PST they failed, which is assigned immediately. This in turn will help the user understand what they need to be aware of next time, which in turn should reduce the PST failure rate over time. Again, this is exactly what the white paper and the preceding figure demonstrate. This visual alone should be enough for you to justify the need for a minimum of monthly PSTs if you do not have them in place today.

User training must be sent to the users dynamically (unique campaigns) and sent randomly to the users. Traditional methods of phishing exercises would send the same campaign to all users at the same time. This provides a lot less value as users would quickly notify each other of the campaign and provide a false sense of awareness. Modernized platforms support dynamic and randomly sent emails.

Not to sound like a broken record, but if you don't have a unified user awareness, training, and testing platform that is purpose-built, you will lack a lot of the advanced capabilities that will allow you to schedule your testing campaigns automatically and assign auto-remediation training when users take an incorrect action with a test. You need to make good use of your time and spending unnecessary time running reports and manually setting up remedial training if the capabilities are not available is not a good use of anyone's time. In addition, the user experience will not be seamless, which is not what we need when trying to get users to comply.

Testing Assignment Groups

Similar to training, you are going to want to ensure that you have assignment groups set up and in place as soon as possible. Ideally, this would be a task you would spend time on when the platform is set up so you can leverage the defined groups moving forward. If you don't have any defined assignment groups set up, it will be worth the investment, especially as you look to provide more target-based testing. There's no need to go into detail again, but as mentioned previously, ensure auto-provisioning and de-provisioning are in place, leverage attributes where applicable, set up SSO for a more secure and improved user experience, set up SCIM if it's supported, and automate as much as possible to minimize the administrative overhead, which will allow you to work on more value-add activities for the program.

Testing Types

We will finish off the section with a look into some of the different types of user testing simulations and/or activities that you should be considering as part of your ongoing user testing. You need to focus on testing that is relevant based on current threats and recent malicious activity attempts within your environment. The most common user test you will execute will be email-based since this is where the user primarily becomes most compromised. With your email-based tests, the following are some of the more common types you'll most likely execute against your users:

- Traditional phishing emails.
- Malicious attachments within emails.
- Malicious URLs within emails.

- Emails with call-back details to submit information.

- Emails containing QR codes with malicious URLs.

- Credential harvester links within emails.

 Make sure you assign relevant training to users who fail a simulation. For example, if a user fails a phishing email simulation, ensure training on how to identify phishing emails is assigned to the user. Ideally, this should be an auto-assignment as part of your training platform.

In addition to your email-based testing activities, some other end-user testing activities to consider include those described in the following list. You should also check with your user awareness, training, and testing platform to see whether they provide any of these or additional testing capabilities:

- **Vishing**: Attempt to call your users or service desk to see whether you can extract any type of personal information or gain access to an account.

- **Smishing**: You can attempt to text your users to see whether they are willing to click on a link or share any confidential information.

- **QR code**: In addition to sending QR codes via email, you could leverage this as part of a smishing campaign by texting your users to see whether they access the QR code. Or, you could print off QR code posters/leaflets and leave them around the office or hang them on poster boards to see whether users access the QR code.

- **USB drive**: Create USB drives (and leave them around the office) that contain a harmless payload that warns them of the incorrect action they took and notifies the cybersecurity team for a follow-up.

Like what we mentioned in the training section, you'll also want to consider role-based testing as an advanced step for your users. The same role-based groups will apply as referenced in the training section. Using these groups to target your specific roles based on current threats will provide them with more awareness and help them better understand what to expect with actual malicious attempts.

A final consideration that will help take your program to the next level is the use of AI-based testing. If the user awareness, training, and testing platform you use incorporates AI-based testing, this may negate the need to set up and manage role-based types of training. The AI functionality should take into consideration the user's role and historical data from training and testing to provide more realistic testing campaigns targeted toward each user.

In reality, if your platform doesn't provide these capabilities, you'll want to push your vendor to provide them immediately or look for a platform that does provide these capabilities. Unfortunately, threat actors are already taking advantage of AI capabilities, so our users need to be best prepared against these more advanced types of capabilities. This is only possible with a testing platform that can provide the next generation of end-user testing by using AI. An example of AI-induced user testing is provided by KnowBe4 as part of their platform known as **Artificial Intelligence Driven Agent (AIDA)**: `https://www.knowbe4.com/what-is-aida`. AIDA also provides relevant training for the users based on their activity.

Expanding beyond the Traditional Channels of Awareness

In the final section, we will take a closer look at some of the activities you may want to consider beyond the traditional awareness, training, and testing components we have already reviewed. As you continue to evolve your user awareness, training, and testing program, it is important you look at new ways to educate your users on an ongoing basis. Bringing diversity with alternative types of events and activities will help keep the program current and engaging. Let's review some ways to bring new opportunities into your overall program.

Personal Awareness

As briefly mentioned earlier in the chapter, bringing the personal aspect of cybersecurity into your awareness program is very powerful. The more you relate cybersecurity to one's personal life and how it can impact them, a family member, or a friend, the sooner you'll get their attention. This in turn will bring more appreciation for cybersecurity within the workplace. A few areas to focus on include using real-life example use cases, sharing consumer-based statistics, and providing guidance on how to best protect themselves.

Real-Life Examples

This one is impactful and should be used to demonstrate how cybercrime is impacting everyday users' lives when the money they have worked hard for is suddenly gone through some form of fraud. It is happening every day, and the more we can raise awareness about it, the better. Simply searching for news articles about users becoming victims of cybercrime will help end users realize how important this topic has become. Take every opportunity to share examples of real-life personal data/monitory theft with your users so they can better understand the real challenges we face and allow them to learn from others' unfortunate mistakes.

Consumer Statistics

As we continue to show throughout the book, statistics speak for themselves. They show the real issue we have in front of us with actual data. The same applies to the consumer and everyday life. A couple of great references from the US are:

- **Internet Crime Complaint Center (IC3)**: `https://www.ic3.gov/`
- **Federal Trade Commission (FTC)**: `https://consumer.ftc.gov/`

There's a lot of great information for users to be aware of in their everyday lives that you can share with them. In addition, the FTC has a *Consumer Sentinel Network Data Book 2022* they publish every year with some good insights that can be shared with your users: `https://www.ftc.gov/reports/consumer-sentinel-network-data-book-2022`.

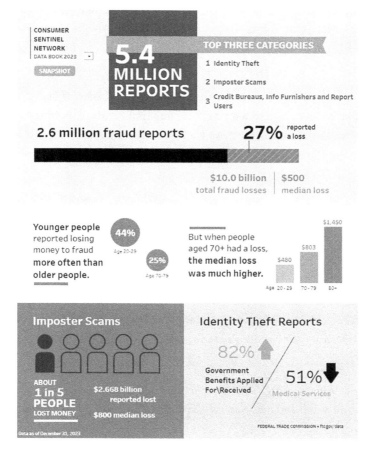

Figure 9.14: The FTC Consumer Sentinel Network Data Book 2022 data dashboard

You'll need to do some additional research in the country where you reside to see whether your government has similar statistics available for your users.

There's a lot of data available that will help the end user understand why they need to take cybersecurity seriously within their professional and personal lives. Make sure you are taking advantage of the data to bring more visibility to them.

Personal Protection

The last section within personal awareness is to guide your users on how they can better protect themselves in their everyday personal lives, for example, encouraging them to use MFA on all personal accounts, consider using a password vault to store all their passwords, use unique and long passwords for all their accounts, and so on. In addition, provide them with some guidance on how to be more proactive with their identity and ways they can better protect themselves. I often share the following best practices for personal proactive measures:

- **Check whether your email or phone has been in a data breach**: https://haveibeenpwned.com/ is a place you can go to see whether your email address or phone number has been compromised. There are currently 14,115,116,543 (yes, that is billions) accounts within the database as of the August 22nd, 2004. You can also sign up to be notified of any future breaches and it is highly recommended you do this.

- **Obtain ID protection for you and your family**: There are several available, LifeLock and Identity Guard being a couple of the more common. Don't forget about your children; they are issued a social security number (or some form of national ID within your country) when they are born and they can just as easily be used by someone else.

- **Freeze your credit with the three credit bureaus**: The three credit bureaus are Experian, Equifax, and TransUnion. It only takes less than five minutes to freeze your credit through each bureau, and it could save you years of financial struggle and potential debt. Also, consider freezing your children's credit.

You'll need to research similar capabilities within your country to freeze your national identification and freeze any credit that is similar to the US system.

External Speakers

Bring some diversity to your awareness program with external speakers who can provide a different or fresh perspective on cybersecurity risk and the current challenges faced from threat actors. Also, don't just make this a one-time event; look for ongoing opportunities to bring in external speakers. Some options include:

- Reaching out to other leaders such as a CISO from another organization.
- Bringing in cybersecurity experts from within the industry.
- Looking to execute threat briefs from external partners.
- Leveraging your current cybersecurity vendors for expertise.
- Reaching out to your local FBI office to get an agent to address your users.
- Bringing in advocates within the industry. For example, KnowBe4 has a great program in which you can schedule one of their advocates who can present on many different topics: `https://www.knowbe4.com/security-awareness-training-advocates`.

Rewards and Recognition

Recognition will go a long way if you are consistent and make it a permanent part of your program. Look at thoughtful ways to recognize your users throughout the year as you execute against your user awareness, training, and testing program. Some ways to recognize your users include:

- Recognizing those who complete their training first (or top xx)
- Recognizing those who've reported the most true-positive phishing emails within a month.
- Recognizing the top users who have completed additional training.
- Recognizing those who have made a positive impact by being proactive with cybersecurity and helping raise awareness.

Once you build measurements for recognition, you can use them as a template to repeat every month or quarter. As part of recognizing your end users, you could also give awards to those who have been recognized. This could be in the form of certificates, incentives, prizes, company gear, and so on. When you add rewards to the picture, you tend to get a lot more interaction and involvement from the users, which will help with the promotion and engagement of the user awareness, training, and testing program.

Gamification

This in some regards will tie back into rewards and recognition. Gamifying the program for the users will bring some fun competition, which should entice the users to engage more with the program. We are naturally competitive, and gamification can bring a lot of fun to the program if executed correctly. You can also use the ideas shared in the rewards and recognition section to bring some fun competition to each department or function level within the organization. Recognize those departments that finished their training first and those that received the lowest PST scores. You can also implement organization-level gamification that tracks the overall training completion rate along with the PST failure rate and set a specific goal for the organization to meet. For example, set a target of 95% or greater completion of all cybersecurity training and a PST failure rate of 5% or less. An example of gamification includes KnowBe4's Learner Experience with capabilities like Group Leaderboards, as shown below. You can learn more about KnowBe4's Learner Experience and gamification capabilities here: `https://support.knowbe4.com/hc/en-us/articles/360014116634-Learner-Experience-LX-Guide`.

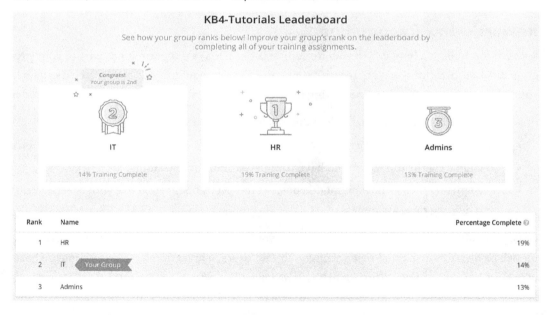

Figure 9.15: KnowBe4's Learner Experience Group leaderboard

In addition, throw in prizes and rewards for your users to encourage more engagement. Who doesn't like a fun prize?

Quarterly Cybersecurity Town Halls

Another great opportunity is to create a quarterly cybersecurity town hall like your regular town halls. You don't necessarily have to make them required, but they should give the users another avenue to learn more about the broader cybersecurity program and be a place where you can share relevant information with them. From experience, if you implement a cybersecurity town hall that is relevant, you will get an amazing turnout and response from your end users. From what I've observed, end users are looking to learn more once they have been engaged in a meaningful way. The following are some examples of agenda items you can share:

- In the first town hall, share the purpose.
- Introduce the cybersecurity team.
- Provide some current state news, maybe an insight into a recent breach, some recent trends, or provide an overview of some relevant recent statistics.
- Provide the latest training and testing (phishing) results.
- Issue rewards and recognition.
- Review statistics from users, leaderboards, and so on.
- Review actual incidents from the SOC.
- Review cybersecurity tools available for users within the organization.
- Provide examples of personal stories where users' lives have been impacted in some way.
- Provide the contact information for the SOC.
- Bring in a guest speaker.
- Provide any company goals and current results.

You will want to engage your marketing and/or communications team to support this activity and help coordinate the communication and invitation for your users. This will help formalize the event and help incentivize more users to attend. Also, having executive leadership to support and use their channels will help. For example, using company-wide town halls to advertise the cybersecurity town hall will help a lot. Once you execute your cybersecurity town hall, ensure you are sending post-event surveys to receive feedback on the content being presented and allow users to provide input on the type of content they would like to see. This will help with continuous improvement and ensure the town hall is effective and relevant.

Cybersecurity Awareness Week

Another effective awareness campaign to consider is executing a cybersecurity awareness week for your users. A week where you provide daily sessions, outbound communications, prizes, bring in guest speakers, and so on. Although a lot of effort is needed to make one of these events successful, the end users will get a lot of value if executed effectively. If you do implement a cybersecurity week, I'd highly recommend coordinating it with Cybersecurity Awareness Month. By doing this, you can follow the same theme and a lot of newly created content will be available at your disposal for the week. Based on the most recent Cybersecurity Awareness Month (2023), the following could be used as a high-level agenda for your cybersecurity awareness week:

- **Day 1**: Introduction and kick-off
- **Day 2**: Recognize and report phishing
- **Day 3**: Use strong passwords (and a password vault)
- **Day 4**: Turn on MFA
- **Day 5**: Update software

The more beneficial events you provide to the users, the more engagement you will receive, and in turn, you will begin to observe a more positive security culture within the organization. Similar to your quarterly town hall, ensure you are sending post-event surveys to measure the success of the event and to understand what can be done better or differently with the next events. It will be important that these events are kept relevant and users are engaging.

Cybersecurity Champions

Look to implement cybersecurity champions within your organization. You may be thinking what this means exactly. This role is essentially someone (or multiple people) who shows initiative within the business to learn more and help others learn more. They show passion for the topic, and they are willing to share their knowledge with others to help them learn and grow. Within your business, look for those who understand cybersecurity more than the standard user and those who are willing to learn. Every organization has a handful of employees who will go above and beyond and bring that extra level of energy around the workplace. Create a program with a positive culture that will encourage these users to want to be part of it. Provide them with a pre-defined list of responsibilities that will allow them to be successful, ensure training materials and resources are made available to them, and ensure you are supporting them with the promotion of a cybersecurity culture within the organization.

Provide branded clothing and other unique gifts that show their role as a cybersecurity champion within the organization. Make sure you are connecting with your champions periodically to maintain motivation and provide ongoing support.

 Make sure leadership is engaged and encourages users to become cybersecurity champions for greater success. It will also be important to ensure cybersecurity champions are provided with the time and resources needed to be successful.

Branding

Another way to better promote the cybersecurity program in general is to create a brand for the program. Work with your marketing and/or communications team to build a theme for the cybersecurity program. Some items to support the theme are:

- A cybersecurity logo for the team
- Cybersecurity template presentation slides and Word documents
- Cybersecurity branded website/knowledge base
- Cybersecurity branded email signature
- Branded T-shirts to wear at work

This theme can be used as you create and deliver presentations, send outbound communications, and for any guides users will need to reference. This will help users know when something is cybersecurity-related and will help give the program that extra touch.

Mentoring and Development

Finally, one of the more important areas I like to focus on is mentoring and allowing for development within your team and others throughout the organization who would like to learn more, including those who may have a desire to move into cybersecurity. As we covered in *Chapter 1, Current State*, we are dealing with a skillset shortage within cybersecurity. The more we can mentor and share knowledge with those around us, the more successful we can continue to grow the cybersecurity industry. In addition, you must be continuously providing development opportunities whenever possible. This field is constantly evolving and changing on a daily basis. It is important that we allow our users to keep current and have the opportunity to take courses and certifications. It is also important to keep growing your staff to allow them continued growth and success.

In *Chapter 1, Current State*, we covered some areas that will support the development of your users within the *Methods of Staying Current* section. For example, we provide some details about some common certification providers.

A lot of organizations have mentoring and development programs in place today, so make sure you are aware of any channels you can leverage within your organization to help with mentoring and development. If you need to build a mentoring and development program, make sure you set clear goals and milestones, and allow feedback mechanisms for continuous improvement. Remember, mentoring and development are all about the employee and the opportunity to learn and grow.

Summary

This chapter has covered a lot and has hopefully demonstrated the need for a very comprehensive user awareness, training, and testing program for your cybersecurity program and broader organization. We cannot take this function lightly by requiring only a single annual training event and a single user test. This will not improve a user's ability to reduce risk for the organization. We must provide ongoing awareness, training, and testing to all users. Data also backs this up: the more we provide awareness, require training, and execute testing, the more aware our users become and the less vulnerable and prone to threats they become. Clearly, running an effective program is going to take dedicated resources; this program will not run efficiently with limited resources. It needs dedicated time and commitment to ensure longer-term success.

To begin the chapter, we covered why the human element is the most important. Here, we reviewed multiple sets of statistics to show the challenges we face with the human element as it relates to cybersecurity. We then moved into the next section, which covered what is needed to build a user awareness, training, and testing program. In this section, we discussed the importance of building a security culture and what maturity for your program looks like by referencing the KnowBe4 Security Culture Maturity Model. We then looked deeper into defining the program, reviewing the importance of ongoing program management, and then finishing off the section with some insight into program management and governance.

The next section covered one of the three core components of your program: user awareness. Here, we covered everything that you should consider through awareness channels. This includes items like portals, newsletters, and a user awareness, training, and testing platform. We then touched upon Cybersecurity Awareness Month and ensuring you are taking advantage of this opportunity every October.

We then reviewed the importance of providing awareness on policies for the cybersecurity program before looking at different sources of content and then finishing off with a review of some current trends.

In the section following, we reviewed both user training and testing. For training, we covered scheduling, assignment groups, content, new hires, compromised users, and role-based. This transitioned into the testing component, which covered scheduling, assignment groups, and types of training. We then finished off the chapter with a look at some areas to consider beyond the traditional methods. This covered items such as personal awareness, external speakers, rewards and recognition, gamification, quarterly town halls, cybersecurity awareness week, cybersecurity champions, branding, mentoring, and development.

In the next chapter, we will be reviewing vendor risk management. Currently, this is an extremely important function because of all the ongoing compromises we are observing with our vendors. To begin the chapter, we will provide an overview and a better understanding of vendor risk management and what it entails. This will transition into developing a vendor risk management program for your cybersecurity organization. We will then review the importance of integrating the process across the broader business before going into the details of contract management and what is involved. We will then finish the chapter with more details on managing your vendors and how you should be monitoring your vendors on an ongoing basis.

Join our community on Discord!

Read this book alongside other users, Cybersecurity experts, and the author himself. Ask questions, provide solutions to other readers, chat with the author via Ask Me Anything sessions, and much more.

Scan the QR code or visit the link to join the community.

`https://packt.link/SecNet`

10

Vendor Risk Management

Today, Vendor Risk Management should be considered one of your top priorities. We are continuously seeing vendors of all types become compromised on a day-to-day basis. As cybersecurity professionals, this doesn't make our lives any easier. In addition to having to protect our own organizations, we also need to manage and challenge our vendors to do better with their own cybersecurity programs. In the past, with all our systems being hosted in a traditional data center, we knew where our data resided, and protecting it was a much easier task than the distributed architecture we have today with our data. In today's modern world, with the adoption of third-party cloud services and most vendors providing SaaS services, our data has become significantly distributed. Vendors now have access to more data than they have had in the past, and we must ensure they are doing everything to best protect the data they are storing on our behalf. This is one of the primary reasons we must build a mature Vendor Risk Management function within the cybersecurity program to hold our vendors more accountable.

We will begin the chapter with an introduction to Vendor Risk Management to provide a better understanding of what exactly is entailed within this function for the broader cybersecurity program. Like most other chapters, we will look at the current state of Vendor Risk Management and review some statistics to provide a better understanding of the challenges we face within this function. These statistics will also help show why the prioritization of this function is so important. In the following section, we will review what is needed to develop and mature a cybersecurity Vendor Risk Management function for your organization. We will look into all the components required to build your Vendor Risk Management program, along with understanding who must be involved with the broader process for Vendor Risk Management. For example, this program will not be successful without the involvement of your procurement and legal teams.

The next section will cover the details around processes and ensuring the broader organization is aware of your processes. Since vendors will be onboarded and used by functions across the organization, it is critical a centralized process is in place and being followed to reduce risk. We cannot allow vendors to be onboarded by the business without the correct due diligence being complete and without understanding the risk profile for a known vendor. We will move on to contract management in the next section, which will cover the different types of contracts you need to be aware of, along with the importance of understanding and reviewing contracts from a cybersecurity perspective. You will need to be involved in ensuring the right contract type is being used for each of your vendors, and each contract will be different. We will finish the chapter by covering the ongoing management and monitoring of your vendors in more detail. Once a vendor is onboarded, the work doesn't stop. You will need to continue to manage and monitor your vendors for any identified risk that may impact them in the future. The following will be covered in this chapter:

- Understanding Vendor Risk Management
- Developing a Cybersecurity Vendor Risk Management Program
- Integrating a Process Across the Business
- Contract Management
- Managing Your Vendors and Ongoing Monitoring

Understanding Vendor Risk Management

First, let's take a high-level look at all the sub-functions that should be addressed as part of Vendor Risk Management. The following diagram captures much of what the Vendor Risk Management function entails.

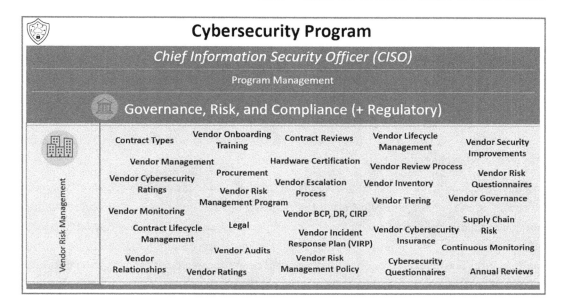

Figure 10.1: Sub-functions of the Vendor Risk Management function

As more and more vendors continue to become compromised, we need to push them to do better with their cybersecurity practices. And the unfortunate reality is, there doesn't seem to be any light at the end of the tunnel; it is only going to get worse before it gets any better. To make this more challenging, organizations continue to onboard vendor after vendor, constantly increasing the size of vendor portfolios. Unless it is a requirement from a regulation standpoint or there is some form of certification required for your organization, there's a high possibility Vendor Risk Management is not being considered for your organization. Today, we don't have an option but to set up a Vendor Risk Management program if one doesn't exist.

It is important that you understand who all your vendors are and ensure they are centrally documented for quick reference. In addition, it will be critical that you know exactly how a vendor is connected to your network (if applicable) and what data they are accessing, storing, and/or processing on your behalf. You will also need to ensure you have up-to-date contacts for each of your vendors in the event you need to quickly engage them. Without good processes in place, this will become a very challenging task and you will quickly find there will be vendors that have been onboarded without going through the correct processes. This realization will usually come when an incident has occurred, unfortunately.

In *Chapter 4, Solidifying Your Strategy*, we discussed the need to strategize with your cybersecurity product and vendor portfolio. Part of that strategy focused on the need to keep your vendor portfolio to a minimum because of the increased footprint and risk we inherit by adding more and more vendors. The same strategy should be applied at the broader organization level. This may not be as feasible, however, because there will be unique needs for each function within an organization. But there should be a process that checks new vendors to make sure there is an actual need to onboard them, and that they are not onboarded because it is nice to have them. For this strategy to work, you will need the support of your executive leadership team. Here's a reminder of some of the reasons for consolidating your vendor portfolio:

- Reduced attack surface risk across the organization
- Reduced overall costs (savings on infrastructure, licensing, personnel costs, etc.)
- Less complexity and reduced integration points
- The ability to build stronger relationships with key vendors
- Streamlined operations and support
- The ability for your staff to focus on fewer technologies to become more efficient
- Improved security monitoring with a smaller footprint

In addition, by allowing a smaller vendor footprint, your Vendor Risk Management program will become much more efficient because you need to manage fewer vendors. Again, this will need support from the very top to enforce. With the current state of Vendor Risk Management, it shouldn't be hard to gain alignment.

Types of Risk

When we look at broad Vendor Risk Management within the organization, risk is much bigger than just cybersecurity. Risk comes in many forms, and a mature Vendor Risk Management program will assess more than just cybersecurity risk.

Obviously, we will be focusing the chapter on cybersecurity Vendor Risk Management, but it is important to understand that risk goes way beyond cybersecurity. As mentioned, there may already be some form of Vendor Risk Management program in place today assessing specific risks while onboarding vendors. If there is, you'll want to understand what is in place, specifically as it relates to cybersecurity (if cybersecurity risk is being monitored) so you can partner with the relevant teams. If a program is in place today but cybersecurity is not being monitored for risk, you'll want to collaborate with the team to enable cybersecurity risk as part of the broader Vendor Risk Management process. If there is a general Vendor Risk Management program in place, it will most likely be operating under the procurement team, risk management team, or possibly the legal team. It will be important that you have a close relationship with each of these teams (if they exist) as they all will play a critical part in the overall Vendor Risk Management process.

The following figure is not a comprehensive list of all risks that may need to be reviewed and understood, but a lot of the risks listed will need to be considered for your organization's broader risk program.

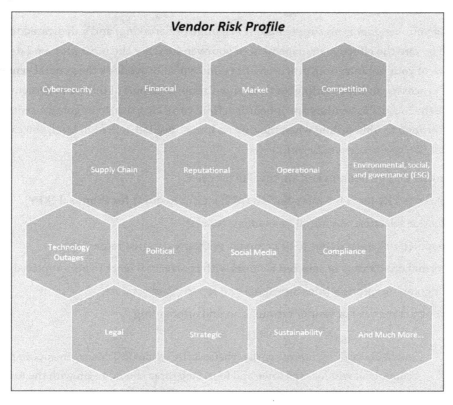

Figure 10.2: Potential risk to be aware of for vendors

In order to manage risk efficiently for your vendors, a platform dedicated to risk management will prove to be advantageous. You could manage the process manually, but this will prove to be a very large undertaking and will most likely contain gaps. There are many vendors available that provide broader Vendor Risk Management capabilities. Some third-party Vendor Risk Management tools that come to mind are OneTrust, Archer, Prevalent, and ServiceNow.

One thing to point out is that you may see different terms used for "vendor." For example, depending on the organization and industry, you may see the term "supplier." Although there are differences between them, they mean the same from a cybersecurity perspective. Another common term you may see in place of vendor is "third party," a more generic term that may entail non-vendors as part of risk management. Within this book, the terms vendor, supplier, and third party can be considered interchangeable, and the process presented will apply to all. An additional term that I consider more detailed is supply chain risk management. Since this is unique, we will briefly touch upon this later in the section.

Vendor Lifecycle Management

Managing your vendors is no simple task. It is a huge undertaking, and a dedicated team is required to ensure the effective management of your vendors. For the most part, and depending on the size of your organization, procurement teams typically overlook the general vendor management processes and onboarding for their organizations. As part of this management, there is a lifecycle that occurs as vendors are onboarded, from reviewing risk, to ongoing monitoring, all the way through to offboarding when they are no longer needed. At a high level, you can expect the vendor lifecycle process to be as follows:

1. Initial idea generation and approval to proceed.
2. Execute a **Request for Information (RFI)** and/or **Request for Proposal (RFP)**.
3. Vendor selection and initial assessment request.
4. Detailed risk assessment to be complete with any required remediation.
5. Formal onboarding of selected vendors with contractual work to be completed.
6. Ongoing monitoring (measure SLAs) and reporting of vendors.
7. Contract renewal or vendor termination and offboarding.

 In the above lifecycle process, the cybersecurity Vendor Risk Management component should fall within steps three and four. Step three should begin with the **Architecture Review Board (ARB)** process, which was covered in *Chapter 5, Cybersecurity Architecture*.

You may come across other variations of a vendor lifecycle process; the one above simply provides an understanding of how the vendor lifecycle process should work. Make sure you are familiar with the defined vendor lifecycle process in place for your organization. If one doesn't exist, you will need to work with the appropriate teams to determine how one can be implemented for your organization.

Current Landscape

Data continues to show the challenge we face with our vendors today. And the reality is, it is only going to get worse before it gets any better. We continue to hear news of vendor breaches on a frequent basis, and a simple search will provide many examples of vendor breaches. There is also a very high possibility that one of your vendors has already suffered a breach in the past. According to a report released by *SecurityScorecard* and the *Cyentia Institute* named *Close Encounters of the Third (and Fourth) Party Kind*, within the previous 2 years, over 98% of organizations have had at least one of their third-party relationships suffer from a breach (source: `https://securityscorecard.com/research/cyentia-close-encounters-of-the-third-and-fourth-party-kind/`).

Another report, *The State of Cybersecurity and Third-Party Remote Access Risk,* released by *SecureLink* (an *Imprivata company*) and *Ponemon* in 2022 found:

- 56% of organizations have experienced a data breach in the past caused by a third party that resulted in the misuse of its sensitive or confidential information, whether directly or indirectly.

- 49% of organizations have experienced a data breach or cyberattack caused by a third party in the past 12 months, whether directly or indirectly.

- Of the organizations who have experienced a third-party data breach or cyberattack, 70% resulted from giving too much privileged access.

- Organizations that allow third-party access to their networks, and do not have a proper understanding of the inventory of their vendors was reported by 48%.

Source: `https://security.imprivata.com/wp-state-of-cybersecurity-third-party-remote-access-register.html`

Another report, released by *KPMG International*, named *Third-Party Risk Management Outlook 2022*, found that 73% of its respondents, within the 3 years prior to the survey, experienced at least one significant disruption that was caused by a third party. You can find more details on the report here: `https://kpmg.com/xx/en/home/insights/2022/01/third-party-risk-management-outlook-2022.html`.

One final report I'd like to share is the 2023 Prevalent *Third-Party Risk Management Study*, which covers broader third-party risk management. It found the following, among many great statistics:

- For 71% of the respondents, the top concern for third-party risk management is a data breach or security incident because of bad security practices. The remaining concerns reported were an audit finding (8%), supply chain disruption from a third-party failure (10%), and legal, reputational, or financial issues (11%).

- Within the previous 12 months of the report, a data breach or other security incident had a tangible impact on 41% of the respondents.

- For risk management tracking of third parties, a staggering 48% are still using spreadsheets.

- A concerning statistic shows that tracking or remediating risk was not being done by approximately 20% of companies.

Source: `https://www.prevalent.net/content-library/2023-third-party-risk-management-study-infographic/`

Although some of the reports provided are a little old, the reality is our vendors are continuing to become compromised, and it isn't slowing down. The current landscape isn't good and Vendor Risk Management must be a priority for all organizations in the current state.

Cybersecurity Risk

As you have observed, cybersecurity risk has become the number one risk concern for organizations today, and for good reason. There are many areas to be concerned about with our vendors becoming compromised. One prominent concern is data loss from a vendor. Vendors are storing, managing, and accessing more and more of our data than ever before, especially with the prevalence of cloud and SaaS environments. Once a vendor becomes compromised, any data they have stored or have access to within our environment may potentially become compromised. If this is sensitive PII or highly confidential data, the impact could be extremely damaging for the organization. Another prominent concern is the potential disruption a vendor breach could have on our operations. Think of a company that has a big dependency on a manufacturing supplier that is needed to create a product. If the manufacturing supplier is hit by a ransomware attack, they could become inoperable for days, weeks, and possibly months. If this does happen, how are you going to source materials for your product? In this situation, you will have no control over how quickly the supplier will be able to become operable again, so you will need to invoke your internal **Business Continuity Plan** (BCP) and any backup plans to procure an alternative supplier to source the needed materials. This may not be easy, but this is the level of planning we need to think about with the current state of cybersecurity.

As you look to embed cybersecurity risk into your broader Vendor Risk Management program, you are going to need a formal cybersecurity Vendor Risk Management program that falls within the broader risk management program. Managing cybersecurity risk for your vendors is not a small feat. Doing this efficiently will require dedicated time and resources. This task is not as simple as reaching out to your vendor with a few questions to be answered. Understanding cybersecurity risk for your vendors is very involved, and the program around this needs to be comprehensive for it to be effective. Without going into too much detail (as we will be covering this throughout the chapter), the following are some of the items that will need to be included as part of managing your vendors' cybersecurity risk:

- Checking if the vendor has a cybersecurity program in place and how mature it is through meetings and with the use of questionnaires.
- Understanding what policies, procedures, and other documents are in place to support the maturity of the cybersecurity program.
- Reviewing if a framework or any type of certifications/audits, like SOC2 Type 2, are in place.
- Understanding if there have been any breaches with the vendor in the past.
- Reviewing a cybersecurity risk score to understand any external-facing risk with the vendor, essentially understanding the external attack surface.
- Checking if the vendor executes any type of testing, like penetration testing, and reviewing the outcome.
- Confirming if the vendor has BCP in place, including crisis management, disaster recovery, and cybersecurity incident response plans.

Again, this program has become critical in recent years and, as you can see from the previous list, a lot of effort and coordination is going to be needed to ensure the cybersecurity Vendor Risk Management program matures in the right direction. If a cybersecurity Vendor Risk Management program is not in place, you'll need to prioritize the creation of one. If there is no general Vendor Risk Management program in place, start by building the cybersecurity Vendor Risk Management program.

Supply Chain Risk

It's important to understand the depth of supply chain risk compared to the standard Vendor Risk Management process. As stated earlier in the chapter, Vendor Risk Management is the program that reviews and assesses the risk of your direct vendors or third-party vendors. Essentially, this will be everyone you have direct contracts with.

Supply chain risk, on the other hand, goes a lot deeper; it is essentially understanding the risk of your fourth-party vendors and beyond. Ultimately, you need to understand the risk of vendors and/or suppliers involved in the end-to-end process of anything you manufacture, purchase, use as a service, etc. This level of review expands significantly beyond traditional Vendor Risk Management, and your organization and the industry you are in will affect how important it is to implement a full supply-chain risk review for your organization.

Although we won't be going into supply chain risk in great detail, it is important we briefly cover it so you fully understand the risk and any potential impact it could have on your organization. There are also capabilities with Vendor Risk Management platforms that can help you better understand your broader portfolio as it relates to the supply chain. The challenge with supply chain risk management is the fact that you won't necessarily have contracts with any of the fourth party or beyond vendors. This can make it extremely challenging if there is an event further down the supply chain that impacts your organization. Although supply chain risk management should be a consideration for all organizations, industries that are dependent on manufactured goods, specifically any type of electronic goods, will need a very clear understanding of their supply chain risk from beginning to end.

In addition to the potential effect of anyone in the supply chain being compromised and disrupting operations or any manufacturing processes, there are additional cybersecurity concerns we need to be aware of within the supply chain. In the **National Institute of Standards and Technology (NIST)** Cybersecurity Framework, *Cybersecurity Supply Chain Risk Management* was added in version 1.1 and is still listed as a category in the most recent version, 2.0. NIST references the following cybersecurity supply chain risks to be aware of, which we also mentioned in *Chapter 8, Vulnerability Management*:

- Any insertion of counterfeit items as part of the supply-chain process
- The production of items that have not been approved
- Tampering with any items within the process
- Theft of any items

- The insertion of malicious software and hardware at any time during the process
- Manufacturing and development practices not maintaining expected standards during the supply chain process

As a reminder, you can view additional information about the NIST *Cybersecurity Supply Chain Risk Management* project and the risks at this link: `https://csrc.nist.gov/Projects/Supply-Chain-Risk-Management`

Hardware Compatibility and Certification

As we look at the concerns listed within supply chain risk, one area we should be paying more attention to is the purchasing of hardware. As you purchase new servers, PCs, storage, and peripherals, it is critical you validate that the hardware is compatible with your deployed systems. Using non-compliant hardware could make your hardware vulnerable to a compromise, or the additional hardware components could even have a compromise already embedded in them. Ensuring the vendor has certified the hardware will help to reduce some of this risk. This doesn't necessarily mean it will be 100% guaranteed, but your risk is reduced significantly.

Let's look at an example of a certification process. Since many of us probably have Windows devices within our organization, you will want to ensure you are using hardware that has been certified and approved by Microsoft to run Windows. Microsoft has a very well-defined Windows Hardware Compatibility Program for vendors to follow to ensure they are maintaining the highest standards. Although a primary focus of these programs is on compatibility and interoperability, Microsoft also specifies a focus on security and reliability, as stated in their specifications (`https://learn.microsoft.com/en-us/windows-hardware/design/compatibility/whcp-specifications-policies`). Using any hardware outside of this compatibility list could render your Windows OS unstable and, even more importantly, create security gaps within your systems. To view the *Windows Compatible Products List*, browse to `https://partner.microsoft.com/en-us/dashboard/hardware/search/cpl`, type in a product name, company name, or operating system, and click **Search**.

You will be provided with a list of compatible hardware based on your search. The following screenshot shows the HP EliteBook products that are compatible with Windows 11:

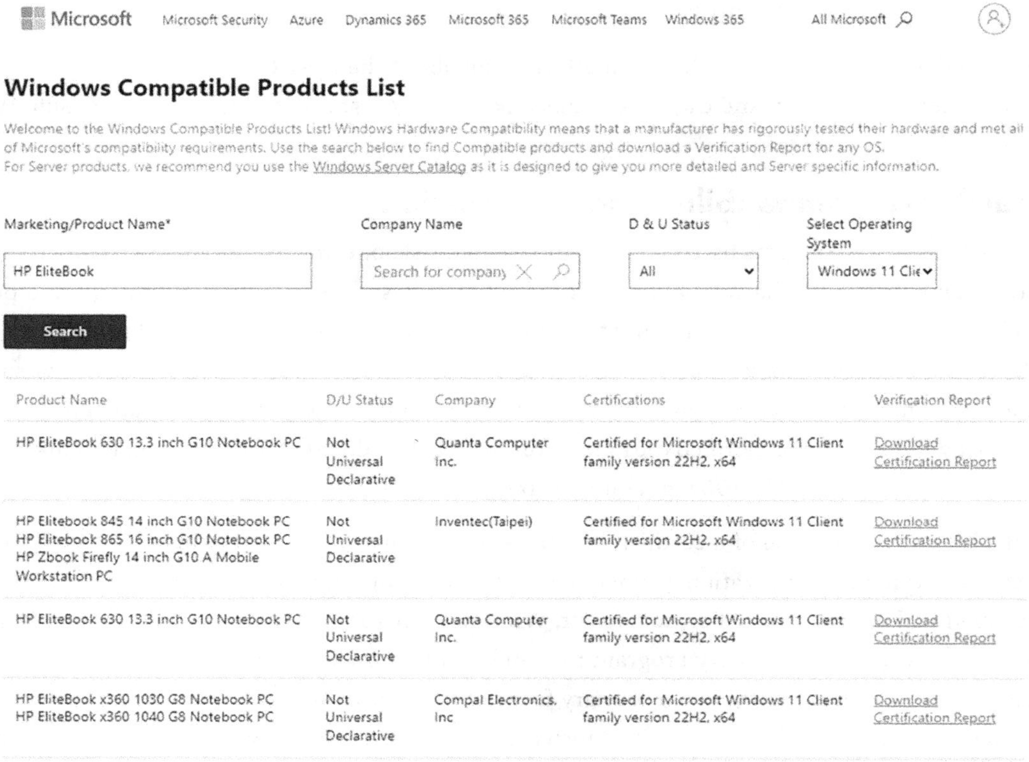

Figure 10.3: Windows Compatible Products List for hardware compatibility

There is an additional portal where you can view certified products specific to Windows Server within the Windows Server Catalog. To view supported hardware for Windows Server, browse to `https://www.windowsservercatalog.com/hardware`. We are simply using Microsoft as an example of the due diligence needed as it relates to your hardware. If you are using a reputable vendor for any hardware-related products, they should be able to show certification of the products you are purchasing.

You should also be careful about going cheap on your hardware purchases. There is always a drive to bring costs down, but opting for hardware that is cheaper than certified hardware could be a costly mistake. It may prove more cost-effective to enter into a contract with a vendor, standardize certified hardware, and purchase a warranty program. There's a good saying in life: you get what you pay for!

Hopefully, you understand the importance of cybersecurity risk within the supply chain. There is a lot involved with managing the deeper aspects of supply chain risk beyond the traditional process of managing risk for your direct vendors. It is important that you understand the difference and ensure you are reviewing cybersecurity risk within the supply chain wherever applicable to your organization. Now that we have reviewed Vendor Risk Management, let's move on to developing a program for it.

Developing a Cybersecurity Vendor Risk Management Program

As touched upon, the job of managing vendors can be quite a challenge for organizations. Onboarding vendors can be a thorough and lengthy process involving different departments, such as the business, risk, legal, and procurement, to help ensure risk is assessed and contracts are written and executed correctly to reduce any liability. As more services shift to third-party cloud-based vendors that host our user and customer data, the onboarding process must be more rigorous than ever before. This becomes even more challenging as you need to deal with both current and new privacy requirements. As part of the onboarding process, the right personnel must be included in the process.

It is also important to remember that Vendor Risk Management is not just a one-time exercise but one that needs a life cycle attached to it. At a minimum, annual reviews should occur as audits and certifications expire. Because of this, it is important that a formalized program is built to support the ongoing efforts of your vendor's risk management concerns. You may already have a Vendor Risk Management program in place today. If you do, this is a great start, but it's recommended you have ongoing reviews of the program to ensure process improvement is occurring and to meet the ongoing challenges we face with our vendors. If you don't have a program in place, or it's not formalized, it is highly recommended you implement a formalized Vendor Risk Management program today.

Policy and Procedures

Let's start by reviewing policies and procedures as part of the Vendor Risk Management program. As we've discussed in other functions, policies are critical for supporting the formalization and efforts of your cybersecurity program. This is no different for the Vendor Risk Management program. Your policy should also have sign-off from your executive leadership team, enforcing the need for your full time associates to follow the policy. You may already have a broader cybersecurity policy in place that references Vendor Risk Management, or you may want to consider a separate policy specifically for Vendor Risk Management if it makes sense.

Some of the Vendor Risk Management items you will need to address as part of your policy are:

- Onboarding expectations
- Roles and responsibilities
- Tiering vendors and minimum requirements
- Review expectations including information collection
- Data classification with business impact
- Contract expectations
- Risk assessment
- Centralized management of all data including contacts
- Incident response and business continuity
- Escalation requirements
- Training requirements
- Ongoing review and assessments

We will be reviewing these in more detail as we progress through the chapter. We will be covering policies in more detail in *Chapter 13, Governance Oversight*.

Once you have your policy in place, the vendor risk program will then require processes defined and in place to ensure cybersecurity is part of the broader vendor onboarding process. You may need to collaborate with various parts of the business to understand how the beginning-to-end process works for onboarding vendors. You'll also want to understand if there are any current processes in place (such as a project intake process, an architecture review process, a contract review process, etc.) so you can have an approval step inserted for the cybersecurity team to ensure the due diligence of that vendor has been completed.

Ideally, all vendors will go through a standard process of onboarding, hopefully through the procurement team, which should help ensure the process needed for the cybersecurity review are efficiently embedded into the overall process. This will allow a streamlined approach. Of course, you will find vendors being onboarded outside of this process, which will create additional challenges. For example, you may have business units that have the ability to directly purchase services or subscriptions using a company credit card, or you may not have an efficient way of tracking when vendors are being onboarded. Whatever the challenge, you will need to ensure you are doing all you can to bring all vendors in scope for a cybersecurity review. Your policy, along with your process, should help force the need for the broader organization to maintain compliance.

Roles and Responsibilities

With the cross-collaboration required as part of your Vendor Risk Management program, it will be important you have the roles and responsibilities defined for everyone involved in the onboarding and risk management of vendors. For this program to be successful, it will be important that all involved groups are fully aware of each other's responsibilities to allow an efficient onboarding process. If you have been involved in onboarding vendors, you will be aware that this is not a quick process, especially if you are running through the correct due diligence as part of the onboarding process. Because of this, you may need to establish SLAs for each of the involved groups, including the cybersecurity team, to set expectations on when reviews will be turned around and completed.

The size of your organization will determine how many teams are involved with the Vendor Risk Management process. Your organization may not necessarily have all the departments listed below, but the responsibility of each of them will essentially sit with someone or a group within the organization:

- **Procurement**: The team that is responsible for the relationship with the vendor and the overall lifecycle of the vendor, from onboarding, assessing general risk, relationship management, etc.
- **Finance**: The team that overlooks the financial process for the vendors, including issuing **Purchase Orders (POs)**, making payments, etc.
- **Cybersecurity**: The team responsible for all risks related to cybersecurity for a vendor.
- **Information Technology**: This team's involvement will be related to any technical requirements for a vendor, including supporting the architecture review.
- **Risk Management**: This team will be involved in any identified risk that needs to be addressed further, reviewed, and potentially approved.
- **Legal**: Anything contract-related, including language reviews and approval, will be the responsibility of the legal team.
- **Business Function**: Each function will typically be the sponsor for the vendors and, depending on the size of your organization, may be responsible for some of the onboarding activities and may hold the primary relationship with the vendor.
- **Audit**: This team will ensure policies are in place, that they are being followed, and that compliance from the vendor is occurring by confirming all due diligence has been complete and ongoing reviews, monitoring, etc. are in place.

- **Executive Leadership/Board**: Executive leadership and the board are the ultimate sponsors of the Vendor Risk Management program and hold accountability at the highest level. Any critical or high risks will need to be reviewed and a direction determined.
- **Vendor**: The vendor is responsible for ensuring all requirements are met for the onboarding organization and ensuring risk is reduced as much as possible within their own organization. They will typically work very closely with the procurement team through the vendor lifecycle process.

Clearly, the onboarding process for your vendors is not simple, which is why the importance of roles and responsibilities must be defined with expected SLAs.

Vendor Management

To efficiently operate the broader Vendor Risk Management program, you are going to need to implement a platform and/or tools to track your vendors, as briefly mentioned in the previous section. As stated, there are platforms available for your broader Vendor Risk Management needs. It will be important that you partner closely with the team that manages this platform, if it exists, as there may already be some cybersecurity capabilities in place, or the ability to integrate cybersecurity risk management. Depending on the capabilities (or lack thereof) of your Vendor Risk Management platform, you may need to focus on adopting something specific for cybersecurity risk. If this is the case, you'll want to focus on how you integrate the cybersecurity risk management process with the broader Vendor Risk Management program. In addition, you'll need to think about integration with your **Governance, Risk, and Compliance (GRC)** platform, if one exists. It will be important that you are somehow tying your vendors back to your applications, which in turn ties back to your risk register. We will be covering this in more detail in *Chapter 13, Governance Oversight*, and *Chapter 14, Managing Risk*. This can become complex with all the different components involved, but keep simplicity in mind as much as possible as you look to integrate the cybersecurity Vendor Risk Management component with both the broader Vendor Risk Management process and your GRC platform.

Cybersecurity Vendor Risk Scoring

As we just discussed, there are third-party services that provide additional insights into your vendor's security posture by means of a cybersecurity score or a rating system. The score is built around scanning publicly available vendor domains, IPs, and any associated assets reachable over the internet to identify any known vulnerabilities.

This can also be considered **External Attack Surface Management (EASM)**, as we briefly discussed in *Chapter 8, Vulnerability Management,* only it would be EASM for your vendors versus your own environment. A few examples of cybersecurity risk scoring platforms are:

- **SecurityScorecard**: `https://securityscorecard.com/`
- **BitSight**: `https://www.bitsight.com/`
- **RiskRecon**: `https://www.riskrecon.com/`

These platforms are designed to help assess and monitor your vendors for cybersecurity risk. There is a lot of information provided by these platforms to help with the review and assessment of your vendors. They can be very powerful tools as you look to have constructive conversations with your vendors about their security state and the potential need to improve. When using a vendor to provide a score of your vendors, use the tool from the perspective that if they can see the vulnerabilities being presented, then in all probability they are already being exploited. In turn, if a vendor has a low score because of many identified vulnerabilities, this potentially poses a significant risk to your organization.

These platforms work by assessing our vendors' external digital footprints to build a risk profile based on what threat actors can see externally. It is important to note, though, that some vendors provide services to customers who operate from the vendor's own environment, such as Salesforce, Microsoft, Google, and AWS. These vendors' scores may include services that other organizations are subscribing to, thus providing a lower score. Because of this, you'll need to ensure you are reviewing the vendor from a fair perspective, defining the scope of their business environment and removing any scope that pertains to their customers. The same applies when completing any initial reviews of a vendor with these platforms. Ensure the defined footprint for the vendor is correct as there may be domains or IPs that are no longer (or have never been) associated with a vendor.

Bear in mind that these platforms are not a one-stop shop for your vendor cybersecurity reviews and assessments. They may become more prominent at some point, but you need to reference multiple data points as part of your overall risk review from a cybersecurity perspective. Although they are very powerful tools as capabilities continue to improve, these platforms are simply another tool in your toolbox to provide you with an overall risk assessment of your vendors.

Questionnaires

Questionnaires are a very important component of not only the cybersecurity risk management program but also the broader Vendor Risk Management program. As you look to collect information from your vendors, there will be quite a lot to complete through questionnaires.

For the broader Vendor Risk Management program, there'll most likely be several questionnaires already being distributed for your vendors to better understand all other risks listed earlier in the chapter. To ensure an efficient process with questionnaires, you must ensure the correct and up-to-date contact information is in place for the vendor. You'll need to check frequently as turnover will be common among your vendors, and it will be important that you ensure the questionnaires are received. This becomes more important when you are looking to receive a quick response regarding any cybersecurity-related questionnaires for identified zero days, incidents, breaches, etc.

Although you could send questionnaires via email, and there is nothing wrong if this is your only option, it is recommended to use a platform that supports these types of capabilities. Ideally, you'll want to use the Vendor Risk Management platform or, if supported, the cybersecurity vendor risk rating platform if it provides these capabilities. By using a platform designed to manage questionnaires, you'll be able to pre-build all your questionnaires and easily distribute them to all your vendors, as needed. As it relates to cybersecurity, some of the questionnaires you'll need to consider for both onboarding and the ongoing management and monitoring of your vendor are:

- A generic onboarding cybersecurity questionnaire to collect some initial information to understand risk.

- Your detailed cybersecurity questionnaires using **Standardized Information Gathering (SIG)** or the **Cloud Controls Matrix (CCM)**, which includes the traditional **Consensus Assessments Initiative Questionnaire (CAIQ)**.

- Annual review questionnaires.

- Determining if the vendor is categorized as **Important** or **Standard** (which we will discuss in the next section).

- Understand the risk of a vendor when **Known Exploited Vulnerabilities (KEV)** or high or critical **Common Vulnerabilities and Exposures (CVE)** are detected within their environment.

- To receive more details on a cybersecurity incident or breach that you have been informed or made aware of.

- If there is a known breach with a fourth-party vendor that your vendor uses to understand if there is any impact on your organization or data.

One additional process that will be addressed as part of the onboarding process that directly supports the cybersecurity Vendor Risk Management components is the architecture review process.

Any new vendors coming into your organization will also need to go through the defined ARB process that we covered in *Chapter 5, Cybersecurity Architecture*. Going through the ARB process will determine the data classification, data type, any mapping of data between you and the vendor, and a whole lot more to help determine the business impact if the vendor is subjected to a major cybersecurity incident.

Tiering Vendors

A very important consideration for your cybersecurity Vendor Risk Management program is the tiering of your vendors to determine what level of review is needed for a particular vendor. The reality is that you are not going to be able to complete a detailed review of all your vendors, and for the most part, it won't even make sense to complete a detailed review of some of them. There are many ways in which you can tier your vendors, and it will all depend on how large your organization is and how complex the tiering system becomes. For example, you could put your vendors into **High, Medium, and Low** tiers, in which you will have a pre-determined matrix of what determines the risk a vendor will fall within. For example, a vendor storing the PII of your users that is integrated into your environment would be considered a High risk. A vendor that isn't storing any of your data and has no integration would be considered a Low risk. And anything in the middle would be considered Medium risk. For example, a vendor storing information that is non-sensitive or non-confidential and with no integration. The following is an example of how you may determine which tier your vendors fall within and the requirements for each of those tiers:

Risk Rating:	Low	Medium	High
Criteria:	• No impact to business continuity in the event of a breach. • Doesn't have access to or isn't storing any data. • No integration or network requirements with environment. • Vendors grade or score is above "TBD."	• Possible impact to business continuity in the event of a breach. • Access to information that is not considered PII and/or confidential information. • If any infrastructure is required including IaaS, PaaS, and SaaS. • Requires any data feeds or interfaces set up in environment. • Vendor has disclosed a breach. • Vendors grade or score is between "TBD."	• Guaranteed impact to business continuity in the event of a breach. • Access to or storing PII and/or confidential information. • Requires any network requirements or integration into environment. • Vendors grade or score is below "TBD."
Requirements:	• Add to cybersecurity vendor risk rating platform for ongoing monitoring. • Cybersecurity exhibit with minimum requirements.	• Add to cybersecurity vendor risk rating platform for ongoing monitoring. • Cybersecurity exhibit with full requirements. • Additional contracts as required such as a Software as a Service (SaaS) agreement. • Standard cybersecurity questionnaire • BCP, CIRP, & DR Plans. • Limited documentation requirements depending on solution such as audit reports, application testing results, etc.	• Add to cybersecurity vendor risk rating platform for ongoing monitoring. • Cybersecurity exhibit with full requirements. • Additional contracts as required such as a Data Protection Agreement (DPA). • Detailed cybersecurity questionnaire • BCP, CIRP, & DR plans. • Full documentation requirements such as audit reports, penetration testing results, cybersecurity program documents, etc.

Figure 10.4: An example of a three-tier ranking system for your vendors

You could simplify the model by using a two-tiered system. As you've observed throughout the book, I like to keep simplicity at the top of my mind as it helps streamline processes and operations. I'm not suggesting this is the only way or the right way, but to keep the concept simple, we will use a two-tiered tiering system as an example throughout the chapter. To support this model, we will simply categorize the vendors as either Important or Standard. By simplifying the tiering model, we can also streamline the requirements needed to determine what tiering the vendor falls within. This will also help your procurement and business teams determine where a vendor falls based on the requirements. Although the cybersecurity team will be involved, the more we can enable the business to become self-sufficient and move more efficiently, the better.

Requirements

As you look at the requirements to determine what tier your vendor falls within, you are essentially assessing the risk that vendor poses to your organization. This can be as complicated as you make it, or you can keep simplicity at the top of your mind to enable a much more efficient program. As we just stated, we will focus our requirements on a two-tier system: Important and Standard. This allows a simple risk assessment to determine how the vendor will be reviewed. To make the determination, the following questionnaire can be used to classify whether a vendor will fall within the Important or Standard tiering structure. Only one of these requirements will trigger the need for a vendor to fall within the Important category:

Are any of the following applicable with the vendor	Place a checkmark in any that are applicable
Is there is any infrastructure, Infrastructure as a service (IaaS), Platform as a service (PaaS), or Software as a service (SaaS) being provided?	☐
Is any information or data is being stored or accessed that falls within the data classification schedule? For example, confidential or PII data.	☐
Are any data feeds or interfaces needed with company systems? For example, APIs, Secure File Transfer Protocol (SFTP), etc.	☐
Is there are any integration requirements into the company infrastructure? For example, Single Sign-On (SSO), System for Cross-Domain Identity Management (SCIM), Simple Mail Transfer Protocol (SMTP), etc.	☑
Is there any Network requirements? For example, VPN tunnels, RDP access, client access, etc.	☐
Has the vendor ever had a disclosed breach?	☐
Does the vendor have a [specific grade or score "TBD"] or below on a vendor risk-rating platform.	☐

Figure 10.5: Checklist requirement to determine if a vendor is Important or Standard

The questionnaire to collect these requirements can be distributed through the broader Vendor Risk Management platform, if one exists. If one doesn't exist, you could look at your cybersecurity vendor risk rating platform to see if it is able to distribute the questionnaire. As a last resort, you could use basic systems such as Microsoft Forms or Microsoft Word/Excel to get you started.

Now that we have determined how the vendors will be tiered, we can differentiate between the level of review needed and the contract requirements for that vendor. Essentially, we will categorize the review into two separate tracks:

- **Important Vendor:** If a vendor is categorized as Important based on meeting at least one of the requirements in the previous checklist, they will undergo a detailed cybersecurity review and require specific cybersecurity requirements within the contract.
- **Standard Vendor:** If a vendor is categorized as Standard based on not meeting any of the requirements in the previous checklist, they will be added to the cybersecurity vendor risk rating platform for ongoing monitoring and will include minimal cybersecurity requirements within the contract.

Based on these requirements, we will now look at the information collection process to support the detailed cybersecurity review. We will cover contracts in more detail within the *Contract Management* section.

Information Collection

Now that we have determined whether a vendor is an Important or Standard vendor, we can begin the next step in the process, which is the collection of information needed to complete the review and determine what the risk profile is for the vendor. If the vendor falls within the Standard category, meaning there are no integrations, no data being accessed or sent to the vendor, no noted breaches, etc., we will not need to collect anything. You will simply need to ensure the vendor is added to your cybersecurity vendor risk rating platform for ongoing monitoring and work with the appropriate team to ensure the cybersecurity language is added to the contract.

For vendors that fall within the Important category, a much more thorough and detailed review will be needed. In addition to the detailed review, you will need to ensure the correct language within the contracts is being used. This will all be determined based on the type of vendor and data they will be handling and/or the type of service they will be providing. There will be more on contract management later in the chapter. For the detailed cybersecurity review, you are going to need to collect as much of the following information as possible.

Cybersecurity Questionnaire

Although you may have a generic cybersecurity intake questionnaire for initial onboarding, it is recommended you have a much more detailed questionnaire for your Important vendors. You could create your own, but this wouldn't be a good use of your time. Instead, you should look at using an industry-standard questionnaire that you will be able to keep updated with the current trends. A third-party cybersecurity questionnaire that is commonly used is the **Standardized Information Gathering (SIG)** questionnaire, provided by Shared Assessments (`https://sharedassessments.org/sig/`).

Standardized Information Gathering (SIG) Questionnaire

The SIG is a configurable solution enabling the scoping of diverse third-party risk assessments using a comprehensive set of questions used to assess third-party or vendor risk. The Shared Assessments SIG was created leveraging the collective intelligence and experience of our vast and diverse member base. It is updated every year in order to keep up with the ever-changing risk environment and priorities.

Direct Mappings:
Widely Accepted Regulations, Frameworks and Industry Guidance

The SIG aligns with the most updated domestic and international regulatory guidance and industry standards for risk management. Since its inception, the SIG has been regularly updated for emerging global risks, regulations, guidelines, and standards for a wide range of industries.

Figure 10.6: Shared Assessments website

Another is the **Consensus Assessment Initiative Questionnaire (CAIQ)**, which is part of the **Cloud Controls Matrix (CCM)** provided by the **Cloud Security Alliance (CSA)**: `https://cloudsecurityalliance.org/research/cloud-controls-matrix`

The CSA Cloud Controls Matrix (CCM) is a cybersecurity control framework for cloud computing.

It is composed of 197 control objectives that are structured in 17 domains covering all key aspects of cloud technology. It can be used as a tool for the systematic assessment of a cloud implementation, and provides guidance on which security controls should be implemented by which actor within the cloud supply chain. The controls framework is aligned to the CSA Security Guidance for Cloud Computing, and is considered a de-facto standard for cloud security assurance and compliance.

The CCM now includes the following:
- CCM v4 Controls
- Mappings
- CAIQ v4
- Implementation Guidelines
- Auditing Guidelines
- CCM Metrics
- CCM Machine Readable (JSON/YAML/OSCAL)

The download file also contains the following:
- STAR Level 1: Security Questionnaire (CAIQ v4)

Learn more about the transition to CCM v4 in this blog.

Figure 10.7: CSA website

These questionnaires can be lengthy, but they provide better insights into the overall cybersecurity program for your vendors. In some instances, your vendors may already have one of these pre-populated, which will allow them to easily provide it to you instead of needing to complete the survey.

Third-Party Audit Report

You'll need to request and review any third-party audit reports, compliance requirements, framework certifications, etc. For example, if you are looking to onboard a SaaS vendor, you'll want to ensure they have a SOC Type 2 report, and it is current. If a vendor will be dealing with credit card payment information, they will need to prove they are **Payment Card Industry Data Security Standard (PCI-DSS)** compliant. Regardless of the vendor, nowadays, you will want to ensure they are working towards standardization on a framework, whether it is the NIST or ISO 27001 framework. It will also be important that you continue to validate that the vendor maintains any compliance or certifications, as they tend to expire annually.

Depending on your vendor and how large and mature they are, there may already be some form of compliance portal at your disposal to access and download the latest certifications, regulations, etc.

These documents will not be available publicly, so you will need to work with your account representative to gain access to and download any relevant documents. A couple that I'm familiar with are Microsoft's *Service Trust Portal* (`https://servicetrust.microsoft.com/`) and Salesforce's Compliance portal: (`https://compliance.salesforce.com/en`). Below, you can see the Salesforce Compliance portal:

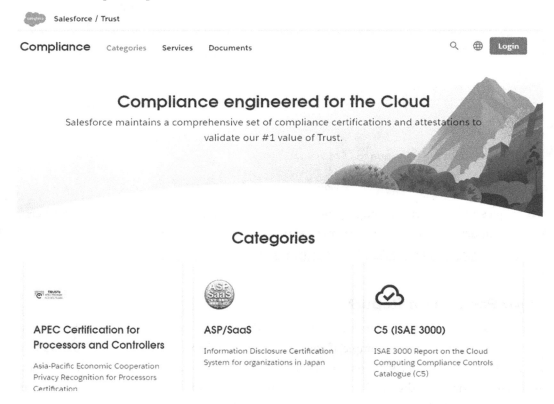

Figure 10.8: Salesforce Trust portal for compliance documents

Let's look at the next collection of documents as part of your detailed cybersecurity review.

Third-Party Testing Results

It will be good to understand if the vendor conducts any testing activities within their environment. Are there any external penetration testing activities occurring on an annual basis, or internal penetration testing activities? If you are onboarding a vendor that is providing an application, is there any application testing being complete? Whatever testing is being conducted, you should be reviewing the outcome of these tests.

Bear in mind that you will most likely be provided the summary findings and attestation from the third party who completed the testing as the detailed reports will be confidential. This is fine as long as you can see the outcome with the findings to ensure any Critical or High findings have been remediated. It is always preferred to have an external company complete these types of activities.

Information Security Management Program Documents

Although the detailed questionnaire and any compliance or certifications should cover this, it is always a good idea to ask the vendor to share any documentation that demonstrates how they run their cybersecurity program. This can be in the form of policies for the cybersecurity program if they can be shared, or even a high-level overview document that shows the broader cybersecurity program for the vendor. These documents can give you a much better idea of how mature (or immature) the cybersecurity program is for the vendor being onboarded.

Business Continuity Planning (BCP), Cybersecurity Incident Response Plan (CIRP), & Disaster Recovery (DR) Plans

Because of the ongoing compromises of vendors, it will be critical that you understand how prepared the vendor is for dealing with a major cybersecurity incident. To get a better sense of this, you will want to review whether the vendor has a BCP, DRP, and CIRP in place. Hopefully, they have all of them. If not, you will want to ensure they prioritize getting these plans in place so they are prepared for when a cybersecurity incident occurs and know how they can minimize the risk and impact on your organization. This may be another area where you will not get access to the documents in their entirety, but make sure you can confirm they have them in place and view the contents to get a better sense of how mature these plans are.

Other Supporting Audit, Risk, and Security Documentation

It is always good to ask for any additional supporting documentation that may not be covered above that will help you complete your detailed assessment. For example, are they conducting tabletop exercises to test their CIRP? Do they outsource any of their operations? Have they suffered a breach in the past? What were the remediation steps, lessons learned, and what was changed to prevent the incident from happening again? It is important the vendor provides full transparency, and this is the time to make the request for all relevant cybersecurity documentation that can be reviewed.

Cybersecurity Vendor Risk Rating Platform Detailed Report

This is where you will want to extract a detailed report from your cybersecurity vendor risk rating platform. All findings from these reports are publicly accessible, and you'll want to make the vendor aware of these issues so they can address them, especially if there are many critical and high-risk findings. This report can prove to be extremely valuable as you need to have conversations around improvement or get-well plans if needed.

 It is important to note that you will need to have a **Non-Disclosure Agreement** (**NDA**) in place to collect these documents unless a contract is already in place. We will cover what an NDA is in more detail later in the chapter.

As you go through the collection process and review everything, you are going to need to ensure you have a centralized place to store all the collected documents. An option could be the broader risk management platform, if one is in place and it can accommodate centrally storing all these documents. Or, are there capabilities within the GRC platform (if one exists) that allow you to track all your vendors, along with all supporting documentation as part of the detailed review? If you are building the program for the first time, you may need to use a secure collaboration site (like SharePoint) or traditional storage (like **Server Message Block** (**SMB**) file shares) to ensure the supporting documents are being retained for review later if needed. Regardless of where you store the vendor's documents, you will need to ensure the process is well documented to ensure consistency with the collection of the documents for review.

Risk Management

Once all documentation has been collected, you will need to complete a thorough review of everything to determine if there are any risks that need to be addressed. You should be able to create some default triggers that initiate the need for further action. For example, if there was a previous breach, you should document it as a risk and escalate for awareness. If there are high-risk findings in the vendor's external penetration test report that haven't been resolved, document them as risks and mention them to the business. Another risk could be the vendor has a poor score on the cybersecurity vendor risk rating platform, meaning an improvement plan will need to be put in place. On the other hand, there will be a lot of information that needs to be reviewed, and your team will need to use their best judgment to determine if any additional actions are needed or if any follow-up items are needed with the vendor.

For example, reviewing the detailed cybersecurity questionnaire that consists of hundreds of questions will need someone on your team to review and look for any red flags that will need to be discussed with the vendor. The same applies to any assessments or certifications that have any identified findings.

Once all the above is reviewed, you will need to, you will need to either approve the vendor or document any risk and invoke an escalation process, if needed. Either way, you will need to ensure the approval or not-recommended reason is documented appropriately with any notes to support the decision. For any non-recommendations, you will need to follow the process for the next steps, which may require the completion of formal documentation to send to the business owner or any committee in place to review and determine the next steps. Again, there may be different scenarios for the next steps, depending on the review and any actions needed. You will find most situations with each vendor will be unique and will require further collaboration to determine the next steps to ensure the vendor becomes compliant if any action is needed. As stated earlier in the book, any identified risk will need to be formally documented and presented to the business to either accept or determine the next steps before proceeding. The Vendor Risk Management team should not accept risk on behalf of the business.

If you need to formally score any identified risk, you can follow the process identified in *Chapter 8, Vulnerability Management*, with the provided risk matrix as an example. You will need to compare the potential impact and the likelihood of an event taking place to understand the risk level of the vendor. We will also cover this in more detail in *Chapter 12, Managing Risk*. Now that we have provided an overview of reviewing and identifying any risk with a vendor, we will look at governance and reporting, which includes the need to centrally manage and store all collected information, including a place to track any identified risks.

Governance and Reporting

As covered in other chapters, governance and reporting are standard across the entire program. Here, you will need to formally track the program to ensure compliance and ensure everything is being tracked with your vendors. You will need to link the Vendor Risk Management program to your broader GRC program, where you will centrally track all GRC activities in one place. As part of your broader governance activities, you are going to need to ensure everything is being documented and all policies and processes are being followed. This is why setting up the foundation for your program with a centralized location to track and store everything pertaining to your vendors is important. In addition, you will want to set up frequent meetings with your team to review the current status of reviews, discuss any escalations to make any necessary decisions, look at process improvements, and so on.

You will also want to ensure you are running reports for the overall program to track progress, provide visibility into the program, and provide insights to your executive leadership team. Some areas you will want to report on are the total number of vendors in your portfolio, the average score of your vendors, the total number of vendors that are compliant, the total number who are not compliant, how many vendors are currently being onboarded, how many have contract renewals in the next "TBD" months, have there been any breaches with your vendors, and are there any risk escalations in progress. You'll most likely need to provide high-level reports to your executive leadership team and the board on a quarterly basis. Automating your reporting will be ideal, but this may not be something that can easily be achieved overnight.

Integrating a Process Across the Business

The most important part of your program is ensuring a robust process is in place for your cybersecurity Vendor Risk Management program and that it has been shared across the broader organization. In addition to it being shared, it is important the broader organization is aware of what is expected from them when onboarding vendors. This may not be the easiest task to complete, but it is important that time is spent creating a process that guides the business through onboarding a vendor correctly. Ensuring success and enforcing the process to be followed will require support and enforcement from the executive leadership team.

Earlier in the chapter, we reviewed the roles and responsibilities of everyone who may be involved with your Vendor Risk Management program. Depending on how your organization manages vendors, whether it is each individual function or a centralized group like procurement, will determine how to best integrate the process across the business. Ideally, having a centralized process through the procurement team will provide the easiest path to integrate the process. If this is the case, you are most likely going to collaborate very closely with the procurement and legal teams when onboarding vendors. If the responsibility of onboarding vendors is decentralized, you will need to work on a formal change management approach to educate the broader business on the cybersecurity requirements. A decentralized approach will require close collaboration with all the business functions, in addition to any procurement resources and the legal team. No matter the model of onboarding vendors within your organization, you're going to need a process flow to help everyone with taking vendors through the cybersecurity requirements.

Review Process

You are going to need to build a visual process map for your users to understand how to navigate through the cybersecurity vendor risk process.

The process will require multiple steps to complete and may become quite involved, so it is important that the process is well documented and clear. Every organization will have its own variation of implementing processes, but the following high-level process map will serve as an example from which you can build your own process that can be used within your organization:

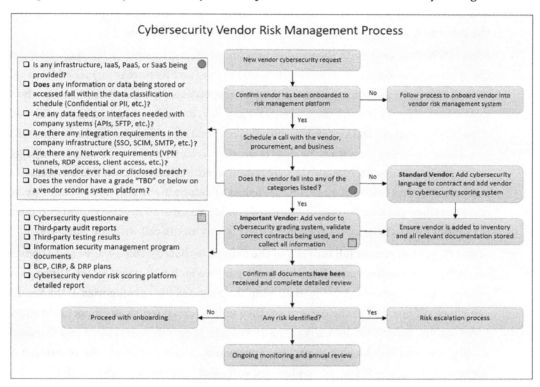

Figure 10.9: Cybersecurity Vendor Risk Management onboarding process

Now that we have a documented process that can be followed, let's briefly review each of the steps in the process:

- **Step 1 (New vendor cybersecurity request)**: This is where the initial request for the cybersecurity review will be initiated. To make this process efficient, you will need some form of intake process to initiate the request. Although email is a typical communication channel for this type of request, look at more modern options by using service requests within a ticketing system (such as some type of web app that allows a request to be submitted), or even see if there are any capabilities within the current Vendor Risk Management platform to initiate the request.

- **Step 2 (Confirm vendor has been onboarded to the risk management platform)**: Hopefully, this has already been completed because this should be one of the first steps of the onboarding process. If not, make sure the vendor is onboarded into the broader risk management platform before proceeding with the cybersecurity review. If your organization doesn't currently have a broader risk management platform, simply proceed and treat the cybersecurity Vendor Risk Management process as standalone.

- **Step 3 (Schedule a call with the vendor, procurement, and business)**: It is recommended the cybersecurity Vendor Risk Management team (and architecture team) schedules a call with the vendor (and procurement, if they are responsible for onboarding) and the business owner. To ensure you fully understand any risk, it will be good to discuss and walk through the solution and/or services with the vendor. I often find the vendor begins with no technical requirements from the sales team; then, once the technical teams are engaged, access to data is needed, integrations are required, etc. It is critical that this conversation happens and that the right questions are asked to determine whether the vendor needs to be classified as an Important or Standard vendor. You'll always want the procurement (or vendor onboarding team) and/or business teams on the call with the vendor.

- **Step 4 (Does the vendor fall into any of the categories listed)**: Once you've had a conversation with the vendor, you will be able to determine whether the vendor is classified as an Important or Standard vendor by reviewing the provided list of categories. If the vendor can check off just one of the categories, they will be classified as an Important vendor. If the vendor doesn't fall into any of the categories, they will be classified as a Standard vendor, and you will be able to provide approval to proceed through the remainder of the onboarding. As a Standard vendor, you will need to ensure that the vendor is added to the cybersecurity scoring platform for ongoing monitoring and ensure the minimal cybersecurity language is inserted into the contract for review. You will need to ensure the vendor is reviewed again once their contract comes up for renewal or if any requirements of their services change at any time. In addition, you will need to ensure that the vendor is added to any vendor inventories you have in place, including the GRC platform. If there do happen to be any supporting documents, make sure they are filed in the correct centralized location for any future reference.

- **Step 5 (Important Vendor: Add the vendor to the cybersecurity grading system, validate correct contracts being used, and collect all information)**: If it has been identified that the vendor is an Important vendor, you will need to initiate a detailed cybersecurity review.

The vendor should already be in the cybersecurity scoring platform as part of the initial review, but if they weren't saved, make sure they have been added and are being monitored in the correct group. You will need to validate that the correct contracts are being used with the vendor from a cybersecurity perspective. You will also need to work with the vendor onboarding team or the vendor directly to collect all information listed within the process for a detailed review. In addition, you will need to ensure the vendor is added to any vendor inventories you have in place, including the GRC platform, and make sure all documents collected are filed in the correct location. One final thought is to define an SLA for the business to be aware of. For example, they can expect the detailed review to be complete within five business days.

- **Step 6 (Confirm all documents have been received and complete detailed review):** Once everything has been provided by the vendor, you will need to complete a detailed review of all documents. This includes reviewing all questionnaires, certifications, audits, tests, etc. You'll need to ensure you review the detailed output from the cybersecurity scoring platform and any other relevant documents.

- **Step 7 (Any risk identified?):** The next step is to determine if there is any risk with the vendor after completing the detailed cybersecurity review. If everything is in order, you can approve the vendor to proceed with the onboarding process. At this point, you will most likely need to review any cybersecurity-specific language in the contract if any changes are being proposed by the vendor, which is normally the case. This includes confirming the cybersecurity insurance requirements are in place and that the vendor meets those requirements. If any risk has been identified that the team is not comfortable with, you will need to invoke the risk escalation process to document the risk and present the risk to the business and possibly executive leadership for review and to approve or deny proceeding. We will cover the escalation process in more detail later in this section.

- **Step 8 (Ongoing monitoring and annual review):** Once the vendor has been onboarded, you will need to ensure they are set up for ongoing monitoring. This will include the need to monitor for any identified breaches with the vendor, whether their score changes within the cybersecurity scoring platform, if any of the provided services and scope changes, etc. In addition, you will want to ensure an annual review is complete for your Important vendors to ensure any certifications or anything that expires is validated to be current.

This process will be essential to ensure the risk of your vendors is reviewed thoroughly. Although somewhat lengthy, it is important that everyone understands these requirements and that the correct due diligence is completed with all vendors. There should be no exceptions.

Escalation Process

As shown in the process map for onboarding, there is a risk escalation decision point that will need to be followed in the event that a vendor does not meet the expected requirements. There may be several different scenarios that trigger this process. From an onboarding perspective, the following are some examples that may trigger the risk escalation path for approval:

- The vendor has a low score on the cybersecurity scoring platform.
- There are several concerns with the completed questionnaire.
- The vendor doesn't have a well-defined cybersecurity program.
- The vendor has suffered a recent major cybersecurity incident or breach.
- There are high-risk and critical findings on testing reports that are not remediated.
- Certifications or regulation requirements are not up to date.
- Outside of the onboarding process, there may be triggers in an operational state that also trigger the need for the risk escalation process to be invoked. This may include items like:
  - The score on the cybersecurity scoring platform falls below a specific score or grade.
  - The vendor suffers a major cybersecurity incident or breach.
  - Known exploitable vulnerabilities have been identified with the vendor that are not being remediated.

It will be important that a process for any escalation is put in place and well documented. Like the overall cybersecurity Vendor Risk Management process, you will want to create a process map that can be easily followed. As part of the process, you are going to need to formerly document any risk on the risk register for tracking purposes. Depending on the type of risk or severity of the risk, the first step may be to implement a get-well plan or give the vendor the opportunity to review the identified risk and provide a response. Regardless, you will need to share the documented risk and inform the business and council or committee you have in place for review. If a decision is needed from the business and the council or committee, you will need to formally present the risk identified so they can make an informed decision on the next steps. There may also be a need to escalate to the executive leadership team and/or the board for a final decision depending on the severity of the risk being presented. Once a decision has been made, whether it is to accept the risk and proceed or not to proceed (possibly even terminate a current contract), you will need to formally document the risk with sign-off from the accountable party.

Cybersecurity Incident Process

If you haven't already dealt with a major cybersecurity incident or breach with a vendor, it is only a matter of time before you will need to go through an incident process with them. As you continue to mature your cybersecurity Vendor Risk Management program, it will be important that you have collected all the information (including architecture) during onboarding from the vendor to support you when you have to work through a cybersecurity incident with them. The last thing you need to be doing when dealing with a cybersecurity incident with a vendor is trying to understand how you are integrated with them, what type of data they handle on your behalf, who you should be contacting, etc. In an ideal world, all this information should have been collected from the initial onboarding phase. But the reality is, you probably don't have all this information for all your vendors, especially vendors that you have had on contract for many years. Because of this, you will want to backtrack and determine which vendors you need to collect the following information from:

- Current contact information for the vendor.
- What is the architecture between your organization and the vendor?
- What type of data is stored or handled by the vendor?
- Where is the data being stored within the vendor?
- What is the impact on the organization if the vendor is unable to operate?

 Not having this information in place for your vendor prior to handling an incident with them will become very challenging when dealing with an active incident, especially as you need to determine potential impact and any risk for your organization.

Once a vendor has declared a cybersecurity incident, there will be some immediate information you will need to request from the vendor as part of the process, although they may not be able to provide the details right away depending on how bad the incident is. Regardless, you will need to ensure the following details are provided:

- Collect all possible details about the incident. Understand the status and if it is ongoing.
- What is the impact of the incident on your organization?
- Is any of your organization's data impacted? If so, what type of data?
- Are there any integrations or services that directly impact your organization?

Depending on the severity of the incident, and especially if it's a ransomware event with the vendor, you will need to make some quick and potentially hard decisions about whether to terminate any connections or integrations with the vendor to prevent any possibility of lateral movement into your organization. Every situation will be unique, and you will need to act fast, depending on the situation at hand.

You will also need to be familiar with the current contract in place once a vendor has notified you of a cybersecurity incident. Hopefully, they notified you within a timely manner and in the timeframe agreed upon (if one exists) within the contract. You will need to understand any other obligations in place between you and the vendor as you work through the cybersecurity incident and if there will be any compromise of SLAs or a possible breach of contract.

Another important consideration is the relationships with your vendors, especially the Important vendors. You will want to ensure you build good, positive relationships with your Important vendors so you can pick up the phone and make a call to the right resource and get a response when needed. Don't let the contact information for your vendors become outdated, as they will very quickly. Make sure whoever is the point lead for each vendor is keeping in constant contact with them and updating the contact inventory frequently. When a cybersecurity incident occurs, the quicker you can get a response and information from the vendor, the more you can reduce risk and any impact on your organization.

Based on everything we have just reviewed as part of the cybersecurity incident process; it will be important to ensure you have everything formally documented so a process can be followed. From an internal perspective, you'll have a **Cybersecurity Incident Response Plan** (**CIRP**), which will typically be invoked for major cybersecurity incidents within your organization. We cover the CIRP in more detail in *Chapter 11, Proactive Services*. Unfortunately, the incident response plan for your vendors will be different, and a much lighter version. Because of this, it is recommended that you build a separate process for your vendors' cybersecurity incidents, or even add the plan to your broader CIRP as an attachment. It's possible that because of the increase in these incidents, we may need to formalize an official **Vendor Incident Response Plan** (**VIRP**) document. I just made that acronym up, but it may become a real thing. Once you have your response plan or process in place, make sure anyone who manages or has a direct relationship with your vendors is fully aware of the process to follow when an incident occurs with a vendor.

Training

To finish off this section, it would be remiss if we didn't discuss the training aspect of the Vendor Risk Management process being integrated across the business.

We need to remember that cybersecurity has traditionally not been a part of the Vendor Risk Management process for most. Because of this, we need to be empathetic that we are suddenly creating a lot more work for the business and/or procurement teams than they are traditionally familiar with. For them to truly become efficient and understand why we must run through and follow the process today, quality time and training will be needed with the appropriate groups. As you have observed, the addition of a cybersecurity Vendor Risk Management process is no easy feat, and it will be important that time is spent ensuring a thorough review is complete.

At a minimum, you will need to create material for the initial training of the users who will need to execute the cybersecurity Vendor Risk Management process, along with training material for newly onboarded users, or those who are new to their role who will be onboarding vendors. In addition, you will need to run a minimum of an annual refresher to remind the users and provide any updates and changes to the process throughout the year. You will want to ensure you cover everything we have covered in the chapter so they are familiar with what is expected of them, in addition to providing details about where any documentation can be accessed, including the training materials, and where to go for any needed support.

One final suggestion is to run a vendor/supply chain tabletop exercise that can help those who manage the vendor onboarding process to be better prepared to handle a situation when a vendor declares a major cybersecurity incident or breach. This will best prepare everyone for a cybersecurity incident by identifying any potential gaps that need to be addressed with your internal processes. We will cover tabletop exercises in more detail in *Chapter 11, Proactive Services*.

Contract Management

First, I want to provide a disclaimer that I'm not a legal expert and it is important you collaborate with your legal teams on any contracts or contractual language with your vendors. The information provided here is based on my experience; every organization will have different contract requirements or needs. This section will be somewhat high-level and mere guidance as part of a cybersecurity role and what can be expected as it relates to contracts with your vendors. Whether you like it or not, you will need to be involved with contracts to ensure the cybersecurity language is not overlooked and that any proposed changes from the vendor are acceptable. With that, you will need to understand the different types of contracts and have some basic knowledge of what should be included within the contracts, specifically as it relates to the technology and cybersecurity requirements.

As we covered as part of the cybersecurity Vendor Risk Management process, every contract moving forward should contain some form of cybersecurity language to provide better protection for your organization. This was not the case in the past, and you may not be doing this today. Whether the vendor provides technology-based services or not, it will be guaranteed that technology is a part of their organization in some way. This means they are vulnerable to a cybersecurity attack that could impact you. Because of this, ensuring there is some minimum cybersecurity language included in the contract between your organization and the vendors will help provide transparency in the event the vendor suffers a major cybersecurity incident or breach.

As you work through contract reviews, you will be in constant collaboration with the procurement team, the legal team, and possibly the business owner. Make sure you build positive relationships with each of these teams and any other team involved in the contract review process. This will allow a much more efficient process to occur, as contracts will typically go back and forth several times before final approval. Sometimes, it will be easier to jump on a quick call with all parties to gain alignment and allow the process to move faster.

Managing Your Contracts

An important part of your vendor onboarding is the management of the contracts with your vendors. The contracts with your vendors serve a very important role: a mechanism to ensure they are following best practices relating to cybersecurity, such as ensuring they are abiding by a standard cybersecurity framework, that they are completing testing activities, they have incident response and business continuity plans in place, etc. They also hold vendors accountable to ensure they are reporting cybersecurity incidents in a timely manner, and they allow a mechanism to provide protections for your organization (if written correctly) in the event of a major cybersecurity incident, such as requiring them to have insurance in place.

Managing your contracts can become very complex as you have multiple parties who need to review the contracts and you need to make and track changes throughout the document review. To manage the lifecycle of your contracts, a **contract lifecycle management** (**CLM**) platform should be considered to provide more efficiency with your contract management. There are many CLMs available to help with the lifecycle of your contracts, and a quick search online will provide you with many of the available CLMs. Without a CLM platform in place, the process will be very difficult to manage as the contract will need to go back and forth between multiple parties, and you will need to track several (most likely) rounds of reviews with edits within the contract. A CLM will allow you to track the contract through its lifecycle, with all reviews being tracked centrally in one place.

The platform will also provide auditing capabilities to track all actions taken by everyone to ensure completion. There are different variations, but at a high level, the CLM process will typically follow these steps with the contract:

1. Request
2. Creation
3. Negotiation
4. Review
5. Approval
6. Execution
7. Filing
8. Renewal or termination

Referencing back to the roles and responsibilities section, you will need to ensure everyone who is involved in the contract review is able to access and manage the contracts relevant to them. For the most part, I'd expect this to be legal, procurement (or the team managing the vendor), the business owner, the cybersecurity team, and the vendor all involved in the lifecycle of the contract. An additional consideration will be the approval and execution process. For approvals, you will most likely need to follow a **Chart of Authority (CoA)** to ensure approvals are complete by the correct personnel. This may include multiple levels of approvals (and should include the cybersecurity team) on the contract, and you'll ultimately need someone on the executive team to make the final approval. This should certainly be the case if the amount on the contract is over a specific amount. When ready for execution, the contract will need to be signed by both parties. The CLM platform will need to support e-signatures, whether natively or by using a third-party add-on.

Types of Contracts

There is an overwhelming number of contract types available. Your industry, your business type, whether you are government, whether it's for personal needs, signing up for services, business acquisitions, etc., will all determine what type of contract will be needed between an organization (or you personally) and whoever you are conducting business with. And possibly more than one contract will be needed, depending on the situation. Keeping the focus on the business perspective, some of the many contracts and/or agreements you can expect to come across from a business and/or IT/cybersecurity perspective include:

- **Master Service Agreement (MSA)**
- **Non-Disclosure Agreement (NDA)**

- **Statement of Work (SOW)**
- Non-Compete Agreement
- Managed Services Agreement
- Software License Agreements
- Professional Services Agreement
- Hardware and/or Lease Agreement
- Amendments

In addition to the above contracts, you'll find the following agreements/exhibits are specifically for cybersecurity requirements and protections. Some of them may be standalone or they may be embedded in broader contracts, depending on the vendor you are working with:

- **Cybersecurity Exhibit (Full)**: This will be a very comprehensive exhibit including all requirements expected from the vendor as part of their cybersecurity program and best practices.
- **Cybersecurity Exhibit (Minimum Requirements)**: This exhibit (or provision) will be a light version of the full exhibit that will be included in all contracts that don't require a detailed cybersecurity review for better protection.
- **Software as a Service (SaaS) Agreement**: This agreement will be in place for all your SaaS providers and will be a very comprehensive contract.
- **Data Protection Agreement (DPA)**: This agreement will be used when PII data is in scope. It will ensure the requirements are in place to provide protection for all personal data they are accessing, handling, or processing on your behalf.

Again, these are just a handful of the many contracts that can be used between you and the vendor. For the most part, the cybersecurity role will ensure there is appropriate cybersecurity language and will review any proposed changes by the vendor. In an ideal world, your legal team will have drafted standardized templates that already include the expected cybersecurity language that vendors need to use. If this is the case, your scope simply becomes reviewing any proposed language changes made by the vendor. If, unfortunately, you get contracts from vendors who insist on using their templates, you will require a deeper review to ensure the correct cybersecurity language is included.

Because of the potential of receiving contracts written by your vendor, you are going to need to ensure the cybersecurity requirements are very thorough. This includes ensuring everything you expect from a vendor to best protect your data is included.

These requirements will become very lengthy, especially for some of the Important vendors, depending on their services, so it is highly recommended that you work with the legal team to ensure there is a minimum cybersecurity requirement for all contracts. Some of the cybersecurity requirements that might be considered within a contract for your Important vendors (an example contract would be the Cybersecurity exhibit with full requirements) are:

- The requirement to meet standard cybersecurity protection measures to ensure all data is protected appropriately. This will include access controls, accounts management (SSO, MFA, password requirements), user awareness training, strict email protections, keeping systems up to date, etc.

- Ensure basic privacy language is included to protect your organization and your users' information.

- Cybersecurity incident notification requirements. For example, a vendor must disclose a major cybersecurity incident or breach within 48 hours of identifying it.

- The need for an inventory of all data they are storing on your behalf, including location, classification, security protections, etc., to be kept up to date. Make sure the language states that this data cannot be shared without any prior approval.

- The requirement for a dedicated cybersecurity program to be in place in addition to a framework like ISO 27001 or the NIST Cybersecurity Framework. You will also need to ensure the framework's maturity is shared.

- Depending on the services being provided, the need to provide (with attestation) evidence for anything that requires certification, like SOC 2 Type 2, PCI-DSS, etc.

- The requirement for a detailed cybersecurity questionnaire to be complete, like the SIG or CAIQ.

- Ensure language is included to ensure action is taken with any risks identified by questionnaires, audits, testing, frameworks, etc. This includes the right to implement a get-well plan for the vendors if needed.

- The right to receive details on a cybersecurity incident when one occurs, along with items like a final incident report, remediation plans, and action to prevent the incident from occurring again.

- Ensure data is being retained and log collection is occurring to meet any record retention policies in place. This will be important for any audit requests and for incident investigation purposes.

- Requirements for application and system testing to occur with any services being provided. For example, if a web app is being delivered, it must conform to OWASP testing standards.

- If there are development activities, ensure DevSecOps is in place and best practices are being followed with all development work.

- Any integrations between organizations must comply with security best practices, and all security controls must be in place, like data encryption at rest, **Transport Layer Security (TLS)**, VPN, etc. Make sure the requirement for architecture documentation is included for your vendor.

- Ensure a BCP, DR, and CIRP are all in place, current, and being tested at least annually.

- Depending on the services and data being stored, you may require a DPA (user protection with PII, etc.) or SaaS agreement in addition to the security exhibit, which will contain a lot more details and requirements for your vendor.

- To ensure any cybersecurity-related documentation can be requested and reviewed upon request, for example, testing results, continuity plans, etc.

- Cybersecurity insurance (this may already be part of the general insurance requirements).

 This is merely guidance as to what might be considered part of your detailed cybersecurity exhibit and does not necessarily include everything that may need to be in your contracts.

As we've mentioned earlier, all contracts should include some cybersecurity language for both Important and Standard vendors. Some of the cybersecurity requirements that might be considered within a contract for your Standard vendors (an example contract would be the Cybersecurity Exhibit with Minimum Requirements) are:

- Cybersecurity incident notification requirements.

- The requirement for a cybersecurity program or framework, like NIST, to be followed.

- To ensure any cybersecurity-related documentation can be requested and reviewed upon request, for example, testing results, continuity plans, etc.

- The right to receive details on a cybersecurity incident when one occurs, along with items like a final incident report, remediation plans, and action to prevent the incident from occurring again.

- Cybersecurity insurance (this may already be part of the general insurance requirements).

If you want to learn more about different contracts and the potential language to include in them, there are many online services available where you can view contract templates.

I'm not recommending you use these services to create your contracts but making you aware of them as a learning opportunity to better understand contracts and the language within them if you currently don't have any experience. A popular service you may already be familiar with is Rocket Lawyer, where you can view many different types of legal documents: `https://www.rocketlawyer.com/business-and-contracts`.

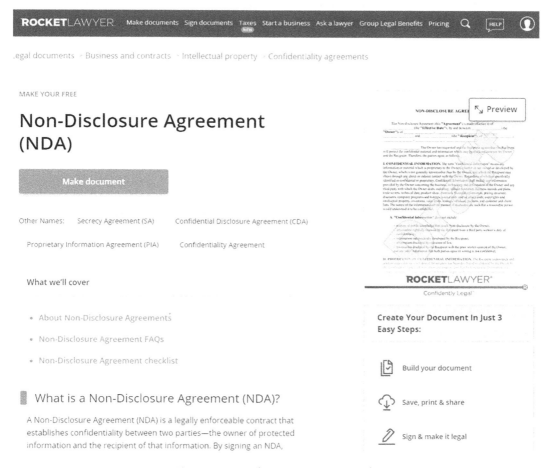

Figure 10.10: Rocket Lawyer NDA example

The above NDA can be accessed from Rocket Lawyer through the following URL: `https://www.rocketlawyer.com/business-and-contracts/intellectual-property/confidentiality-agreements/document/non-disclosure-agreement`. If you click **Preview** in the top-right corner of the page, you will be able to review the language within the template to get a better understanding of the language included.

Insurance Requirements

We'll finish off the section with a brief review of the insurance requirements within your contracts. At a high level, insurance is available for many different risks. Hopefully, you already have insurance requirements in your contracts today. I'm no insurance expert, but it is a good idea to have a general understanding of insurance requirements as you work with your legal and procurement teams through contract reviews. Some common insurance requirements you will see in contracts are:

- Commercial General Liability
- Workers Compensation
- Employers' Liability
- Automobile Liability
- Professional/Errors and Omissions Liability
- Property Insurance
- Umbrella/Excess Liability
- Directors and Officers Liability

The obvious requirement purposely left off the list is cybersecurity or cyber liability insurance. No contract should be agreed if there is no cyber liability insurance in place with the vendor you are doing business with. If you do not require cyber liability insurance, you will need to start requiring it ASAP. You will need to review all your old contracts to see where the requirement is missing, and you may need to amend your contracts to ensure the requirement is in place. You'll also want to review and determine what the maximum amount per claim needs to be with the vendor. This may be different for certain vendors, depending on what services and/or data they are storing and/or accessing on your behalf. In addition, you'll need to verify that the vendor has both first-party and third-party cyber liability insurance coverage. First-party insurance will protect the vendor from their losses, and third-party insurance will protect them from any claims you need to make against them for any losses. Make sure you understand the coverage being purchased as part of the cyber liability insurance plan for the vendor. The following are some examples of the coverage you can expect to see as part of a cyber liability insurance policy:

- Network Security Liability
- Privacy Liability
- Data Loss/Breach
- Business Interruption

- Media Liability
- Extortion
- Reputation or Brand Protection

As part of the due diligence of your onboarding, make sure the vendor provides the **Certificate of Insurance (COI)** for all coverage, including cyber liability insurance. We will be covering cyber-security insurance for your own organization in more detail in *Chapter 14, Managing Risk*.

Managing Your Vendors and Ongoing Monitoring

To finish off the chapter, we will cover the ongoing activities needed as part of your cybersecurity Vendor Risk Management program. We briefly reviewed this within the review process:

- **Step 8 (Ongoing monitoring and annual review)**: Once the vendor has been onboarded, you will need to ensure they are receiving the correct ongoing monitoring detections. This will include the need to monitor for any breaches with the vendor, whether their score changes within the cybersecurity grading system, if any of the provided services and scope changes, etc. In addition, you will want to ensure an annual review is carried out for your Important vendors to ensure any certifications or anything that expires is current.

It is important to note that even when a vendor is onboard, our job is far from complete. There must be continuous monitoring and ongoing review and check-ins with the vendors, especially the Important vendors. The great news is that there are tools and capabilities available today that allow us to manage and monitor our vendors on an ongoing basis more efficiently. It is so much better compared to five-plus years ago, when most activity for our vendors was primarily manual monitoring and activities.

Continuous Monitoring

Traditionally, Vendor Risk Management has been a point-in-time review. Although this is good to an extent, we need to enhance the program to be able to enable continuous monitoring, along with any automated opportunities. The more proactive we can be with our vendors, like the way we monitor our internal organization assets, the more quickly we can react to reduce risk when an incident does occur. To get your program to a place where you can allow for continuous monitoring, you are going to need a platform to provide these capabilities. As we have discussed throughout the chapter, you will need a cybersecurity vendor risk scoring platform to allow for continuous monitoring capabilities.

Depending on the platform you use and the capabilities available, some of the continuous monitoring capabilities I would suggest understanding and enabling are:

- Monitor and send a notification for a change in the vendor's risk grade or score if it falls below the threshold you specify. You may be able to get granular for specific categories versus the overall score if it makes sense for your program.

- Monitor and and send a notification if there are any breaches or major security incidents detected or reported with the vendor.

- Are capabilities available where you can monitor activity on the dark web for your vendors' IPs and/or domain names, or search intel feeds for activity? This may be more of a SOC capability, but it is something to consider.

- Can alerts be set up to notify the team if any ransomware or malware is detected with the vendor?

- Monitor and notify if the vendor has any publicly identified KEVs.

- Monitor and notify if there are any CVEs with a Critical-or High-rated CVSS.

- Is it possible to set up notifications for any other High or Critical issues identified with the vendor?

In addition, does the cybersecurity vendor risk-scoring platform allow setting up improvement or get-well plans for your vendors? If so, you'll want to monitor and track the progress of any improvements made for the vendor as they remediate identified items. Finally, look to automate all this as much as possible. Are you able to set up and configure alerts that will automatically monitor your vendors as they are onboarded to allow continuous monitoring?

Annual Reviews

Another consideration is that annual reviews should be part of your ongoing monitoring of your vendors. There will be certifications and other types of audits and tests that are most likely completed annually that you will need to validate have been updated and are current. If there are any noted issues, you will need to address them with the vendor. You'll also want to re-check the cybersecurity vendor risk-scoring platform to ensure the score has not decreased, although you should have been notified if it falls below your defined threshold. As you collect updated documentation and certificates, you'll want to ensure you update the document repository where you are storing all your vendor documents for any future reference and to remain in compliance. Additionally, you will want to confirm all contact information is current and up to date.

It would be good to ensure that your questionnaires are updated and current as well. This can be an automated task depending on the Vendor Risk Management platform you are using. For general risk management, I'd imagine there is already an annual notification being sent to vendors to request updated information.

If so, you'll want to ensure any cybersecurity questionnaires are also included in the annual review for the vendor. If not, make sure the questionnaire is being sent to the vendor, whether through your cybersecurity vendor risk-rating platform or manually if needed. Obviously, this will only apply to the Important vendors. Once you receive updated questionnaires, make sure a review is complete to ensure there are no major red flags or changes from the previous year.

You should also be tracking any changes with the vendor throughout the year, meaning are there any changes to the contracts, new services being added, new projects being undertaken, renewals, etc.? This could be through SOWs, **Change Orders (COs)**, amendments, etc. Any change of scope will need to be reviewed to ensure nothing changes with the vendor's current responsibilities and, most importantly, no architecture changes are occurring. This includes Standard vendors as well. Make sure Standard vendors don't become Important vendors as a result of any changes. If they do, a detailed review will need to be initiated for the vendor. Make sure processes are in place and that those managing the vendors are aware that if any changes occur before an end-of-year review, the cybersecurity team is engaged.

Business Continuity Planning (BCP)

We'll finish off the chapter by discussing BCP for your vendors. We did cover BCP in detail for the internal organization within *Chapter 7, Cybersecurity Operations*, but the situation with your vendors may present more unique scenarios to take into consideration. As part of your review, you will need to ensure your vendors have BCP in place, and you will need to validate these plans. You'll most likely not be provided the actual plans, but make sure you can get as much insight into the plans as possible to ensure maturity in this area. In the event a cybersecurity incident does occur, it will be important you understand how your vendors will be able to react and, more importantly, whether will they be able to provide services. If they undergo a critical ransomware incident and their operations are inoperable, do they have good enough continuity plans to support or provide services/products to your organization? The reality is you can't rely on your vendors with the impact cybersecurity incidents are having on organizations today.

Because of this risk, you are going to need to expand your BCP to take into consideration business continuity in the event one of your Important vendors is unable to provide services and/or products for months on end.

And this happens, so we have no choice but to prepare for the worst. A BCP for each of your vendors may not be realistic, but you will need to identify the most critical vendors that keep your business running. Once you have identified them, what are the BCP for each of these vendors? Do you need a backup or alternative vendor on contract to allow the business to continue? Will there be alternative options? Can other vendors support additional capacity? That's a lot of questions to ask and to be prepared for.

Ideally, you would have run a **Business Impact Analysis (BIA)** on all your vendors as they were onboarded, so understanding who the most critical vendors are shouldn't be difficult. If you haven't run a BIA or it's outdated, you'll want to look at running the analysis against all your Important vendors sooner rather than later.

As discussed earlier in the chapter, another consideration for your vendors as it relates to BCP is running tabletop exercises with the specific theme of your Important vendors being impacted. For example, you could run a supply chain-themed tabletop exercise that focuses on how you react to a major supply chain interruption because of a major cybersecurity incident or breach at one of your vendors. We will cover tabletop exercises in more detail in *Chapter 11, Proactive Services*.

One final comment to finish off the chapter is a reminder to ensure you have strong relationships with your vendors, specifically your Important vendors and those who are critical to the success of your operations. The stronger the relationships, the quicker you will be able to collaborate to understand the impact and allow action to be taken more efficiently and, in turn, reduce risk for your organization.

Summary

This chapter turned out to include a lot more content than I expected. This goes to show how important Vendor Risk Management has become recently, especially as it relates to cybersecurity. Unfortunately, cybersecurity Vendor Risk Management is not something any of us can ignore. We continue to be challenged by vendors falling victim to breaches and major cybersecurity incidents. As we have shown, the impact of these breaches can be catastrophic. Examples include your confidential data or user PII being exfiltrated from one of your vendors, your vendor allowing a back door into your organization, or your vendor suffering a breach that prevents them from providing services and/or products for your organizations. This is all realistic in today's world, and we must hold our vendors accountable to do better because it is only going to get worse before it gets any better.

To begin the chapter, we covered understanding Vendor Risk Management and what is involved within the program. This included reviewing details on the different types of risks an organization can expect to see, in addition to cybersecurity.

We then looked further into the vendor risk lifecycle process from beginning to end, which took us to looking at what challenges we face in the current landscape, with data and statistics from multiple sources. Next was a deep dive into cybersecurity risk, before finishing the section with supply chain risk and what exactly it entails. In the following section, we provided details on what should be considered to develop a cybersecurity Vendor Risk Management program.

This included covering policies and processes, roles and responsibilities, vendor management and cybersecurity vendor risk scoring, questionnaires, the tiering of vendors, what information you should be collecting as part of the review, risk management, and finally, the expected governance and reporting overlooking the program.

The next section took us into the details of integrating a process for your cybersecurity risk management program across the broader business. We started the section with a process map for how the review process should flow by assessing cybersecurity risk with your vendors. We then looked at ensuring an escalation process is created and what considerations should be in place. This led to a discussion about the cybersecurity incident process for your vendors when they become compromised, and ensuring that training is made available for all those involved in the cybersecurity Vendor Risk Management process. We then reviewed contract management because there will be a need for you to be involved with contract reviews from a cybersecurity and risk perspective. This included an overview of CLM and what is involved in the entire lifecycle of contracts. We then looked at some of the different types of contracts you can expect to see within your organization before finishing off the section with an overview of insurance requirements. We then finished off the chapter with some insights into the ongoing management and monitoring of your vendors, which included continuous monitoring, annual reviews, and BCP considerations.

In the next chapter, we will be reviewing proactive services. We will be covering all the activities that you can do proactively to help identify any known risk to allow you to mitigate any findings. To begin the chapter, we will provide an overview of why you should be executing proactive cybersecurity services within your organization. Next, we will review security testing, such as penetration testing, before moving on to incident response plans and why you need to have one in place for your organization. Next we will review tabletop exercises and provide more details about what is involved with these exercises. The chapter will close with details on other types of proactive services that you should be considering, such as threat briefs.

Join our community on Discord!

Read this book alongside other users, Cybersecurity experts, and the author himself. Ask questions, provide solutions to other readers, chat with the author via Ask Me Anything sessions, and much more.

Scan the QR code or visit the link to join the community.

`https://packt.link/SecNet`

11

Proactive Services

Proactive services may not be a formalized function within your cybersecurity program today, but there is a possibility you are doing some form of proactive services as part of your program. If not, you must look to implement proactive services within your program. There are several different types of them you should consider. One is the ability to identify any weaknesses and risks within your environment from testing types of activities and assessments – for example, penetration testing activities. A couple of extremely relevant proactive services include the need to ensure that you have a robust **Cybersecurity Incident Response Plan** (**CIRP**) in place along with the need to execute tabletop exercises. You've heard me state before that it's only a matter of "when" and not "if" a cybersecurity incident is going to occur. Statistics are not on our side, and the reality is that we need to be best prepared for when a major cybersecurity incident occurs. This will allow us to respond more efficiently and restore business operations much quicker. Having a well-defined proactive service program in place will allow for this.

The chapter will begin with explaining the importance of implementing a proactive services program, as this may not be a function you have defined yet, although you may be executing some proactive services within your organization already. We will briefly review the different types of proactive services that should be considered for implementation into your cybersecurity program. Then, we will shift our focus to cybersecurity testing activities and all of the different testing activities that will help identify any risk within your organization before threat actors are able to exploit them. We will also go into detail about the process of executing a penetration test.

In the section that follows, we will review the need for a CIRP to better prepare you for when a major cybersecurity incident or breach occurs. In addition, we will look at what is needed to implement a CIRP and all the details contained within one for your organization.

Following the CIRP creation will be the section that covers tabletop exercises. Once you have a plan in place, you need to test it and ensure that everyone is familiar with it and knows how to use it. Executing a tabletop exercise with a realistic scenario of a cybersecurity incident will allow this to occur. We will look at the different tabletop exercise types, with references to examples of how to run them. To finish off the chapter, we will take a further look at some of the other proactive services that weren't covered previously. This includes activities such as threat briefs, incident response type of training, threat hunting activities, and more. The following will be covered in this chapter:

- Why proactive services?
- Cybersecurity testing
- Incident response plans
- Tabletop exercises
- Other proactive services

Why Proactive Services?

First, let's take a high-level look at all sub-functions that should be addressed as part of proactive services. The following image captures much of what the Proactive Services function entails.

Figure 11.1: Sub-functions of the Proactive Services function

In short, the Proactive Services function executes activities that identify potential threats before threat actors find them – essentially, those services you execute before a cybersecurity incident occurs versus reactive services, which are those that occur after a cybersecurity incident, such as **Digital Forensics and Incident Response (DFIR)**. Proactive services also better prepare our teams to react more efficiently when an event occurs. At a high level, a lot of what we do as part of our cybersecurity program is to take a proactive approach as best as possible. Essentially, the essence of having a cybersecurity program sets the precedent that we are taking a more proactive approach with our cybersecurity posture. For example, within the program, multiple functions take a more proactive approach to cybersecurity, such as cybersecurity architecture, cybersecurity awareness, training, and testing, vulnerability management, vendor risk management, and so on. The reason why we need to specify a proactive services function is to cover all the other unique activities that need to be part of our program in order to be more proactive. It also sets a precedent that the program takes all the actions possible to reduce risk as much as possible for the organization. Another reason is ensuring that all proactive services are being executed as expected, along with following up on any remediation work. Once you start implementing proactive services, you will quickly realize that a lot of effort is needed to ensure that these activities are scheduled and executed, with remediation occurring.

In addition, you'll want to bring in external vendors to help execute many of these services to ensure there is no conflict of interest, and an independent party provides a second set of eyes with a formal report of findings. This will be extremely important, as you need to show evidence that you are taking all precautionary measures to review and check for any weaknesses within your environment. Like any audit, your internal teams shouldn't be the only resources looking for vulnerabilities and attesting to all controls; this requires external expertise to provide an additional level of due diligence. You'll also want to ensure that this function has the necessary dedicated project management resources assigned. There will be a lot of coordination needed with your internal teams, the business functions, and all the vendors you engage with to complete any of the proactive work on your behalf.

We have referenced many statistics from multiple different sources throughout the book so far. Without spending too much time on statistics for proactive services, I think it's important to highlight some of the more concerning statistics that support the need for the Proactive Services program we have referenced throughout the book. In 2023, IBM's *Cost of a Data Breach Report* revealed significant trends and costs associated with data breaches, highlighting increased security investments, the role of third parties in breach detection, and the financial impact of comprehensive incident response strategies.

IBM 2023 Cost of a Data Breach Report:

- As a result of a breach, 51% of organizations plan to increase investments in security.
- Only one in three organizations self-identified a breach. Third parties or attackers represent 67% of reported breaches.
- It was noted that organizations experienced an additional cost of $470,000 by not including law enforcement.
- 82% of breaches included cloud infrastructure.
- Greater levels of incident response planning and testing saved organizations $1.49 million when containing a data breach.

Source: `https://www.ibm.com/security/data-breach`

When looking at these statistics from a proactive mindset, why does it take a breach for investments to increase? This mindset needs to change. If only one in three organizations self-identify a breach, more needs to be done from a proactive perspective to implement better detection capabilities, add or execute more testing, and execute threat hunts, as examples. There may be several reasons why law enforcement wasn't engaged, but a good CIRP will ensure this is not overlooked, along with including up-to-date contact information.

82% of breaches included cloud infrastructure, which shows that a lot more proactive work needs to occur here, whether it is testing, configuration reviews, assessments, etc. For the last bullet, we can see that increased incident response planning and testing saved a substantial amount of money for those organizations that were better prepared. This only happens by being more proactive.

Another report from 2023, the *Qualys TruRisk Research Report*, underscores the ongoing challenges in vulnerability management and remediation.

2023 Qualys TruRisk Research Report:

- The average noted time to weaponize a vulnerability is 19.5 days, while the same vulnerabilities have a **Mean Time to Remediate (MTTR)** of 30.6 days and are only patched on average 57.7% of the time.
- In 2022, Qualys scanned 370,000 web applications, in which more than 25 million vulnerabilities were found. 33% of these vulnerabilities were classified as **Open Web Application Security Project (OWASP)** Category A05, which represents Misconfiguration.

Source: `https://www.qualys.com/forms/tru-research-report/`

As you can see from the statistics above, it takes less time to weaponize a vulnerability than it does to remediate a vulnerability, and by a lot. If we know this, why aren't we being more proactive with remediating vulnerabilities? The same applies to the second bullet; of the 25 million vulnerabilities identified, 33% were identified as one of the categories in the OWASP Top 10. This clearly indicates that we aren't doing due diligence as we deploy our web applications. We need to do better, and this is where having a proactive mindset is needed.

According to the Verizon 2023 *Data Breach Investigation Report* (https://www.verizon.com/business/resources/reports/dbir/), stolen credentials, phishing, and exploitation of vulnerabilities are the primary entry points for attackers. With the exploitation of vulnerabilities being a primary entry point, this should be a red flag for us to better understand these entry points and execute some additional testing services, as well as ensuring our vulnerability management program is modernized to proactively identify vulnerabilities within our environment. As for stolen credentials and phishing, we covered this in detail, with a lot of statistics to show the challenge we are dealing with, in *Chapter 9, Cybersecurity Awareness, Training, and Testing*, which covers the proactive types of services you should consider for your users.

The above is just a snippet of the data available to demonstrate the need for us to do better from a proactive perspective. Proactive services can reduce a lot of overall risk for your organization, and the cost to run these services will be minimal compared to the cost of a major breach. Some of the benefits of executing proactive services include:

- Identifying vulnerabilities before threat actors exploit them
- Being better prepared to respond to a major cybersecurity incident or breach
- Allowing you to better understand your risk profile and take action to reduce risk
- Providing more visibility and transparency into a cybersecurity program and the current state
- Helping with compliance requirements for any audits, certifications, frameworks, etc
- Providing significant savings by preventing a major cybersecurity incident or allowing for a speedier recovery

Showing a **Return on Investment (ROI)** for these activities can be difficult, however, hope is to prevent or reduce an actual major cybersecurity incident or breach from occurring. Because of this, you will never know the exact cost or true damage for your organization because these are prevention measures. Make sure however that you use documented statistics to show what others have lost due to a major cybersecurity incident or breach. For example, two very powerful numbers that can be referenced to help with your ROI from the *IBM 2023 Cost of a Data Breach Report* (`https://www.ibm.com/reports/data-breach`) include:

- $4.45 million was reported as the average cost of a data breach.
- Greater levels of incident response planning and testing saved organizations $1.49 million when containing a data breach.

The following are many of the proactive services that you should consider as part of your cyber-security program:

- Penetration testing
- Application testing
- Other testing, including physical, WiFi, **Active Directory (AD)**, etc.
- **Cybersecurity Incident Response Plans (CIRPs)**
- Incident response training
- Other proactive planning, including **Business Continuity Planning (BCP)** and **Disaster Recovery Planning (DRP)**
- Tabletop exercises
- Threat hunting activities
- Threat briefs
- A cybersecurity framework
- Assessments and configuration reviews
- End-user training, awareness, and testing
- And much more...

Some of the proactive services listed above are functions within the cybersecurity program, but it's important that you are aware that these are also proactive services to help reduce overall risk. As we go through the chapter, we will cover the above items in more detail and reference back to chapters where more details may have already been provided.

Cybersecurity Testing

Running cybersecurity testing activities within your organization shouldn't be an option; there should be requirements in place to ensure that cybersecurity testing occurs as part of your cybersecurity program. An example includes ensuring that a web application has undergone cybersecurity testing activities before being released into production. The same applies to mobile applications. To ensure that this happens, processes must be in place to ensure the correct scrutiny and testing of any application (or solution) occurs before go-live. This essentially ties back to the broader **Architecture Review Board** (**ARB**) process and the need to ensure that all cybersecurity requirements have been met before anything goes into production. This doesn't just apply to application testing but also to all other types of testing activity that allow vulnerabilities to be identified within your organization.

Types of Testing

Many types of cybersecurity testing can be executed within your environment. You may not be able to execute all, or there may be no need to execute some, but you'll need to determine what testing should occur within your environment and when. The following is a list of some of the cybersecurity testing exercises to be familiar with:

- Penetration testing
- Application testing
- Physical security testing
- Other (examples include user, network, cloud, etc.)

As mentioned earlier in the chapter, user testing is also part of cybersecurity testing, but we focus on this separately within *Chapter 9, Cybersecurity Awareness, Training, and Testing*. First, let's look at penetration testing, which everyone should consider as part of their cybersecurity program.

Penetration Testing

Penetration testing, or pen testing, is another method for identifying cybersecurity risks and an important component of the proactive services program. These tests should be executed at a minimum once a year within your environment. Penetration tests validate risk by performing specific testing activities against your organization's infrastructure, such as system hosts, applications, devices, and even your users to exploit known vulnerabilities. Tests are executed by skilled security professionals, referred to as ethical hackers, to try and replicate the activities of a malicious actor. This type of activity is better known as ethical hacking.

Penetration tests can be executed externally to simulate an outside threat trying to break in, or internally, to simulate an insider threat that has already breached your perimeter network. There are many different types of penetration tests, and they commonly cover the following areas:

- Systems and servers, including AD
- Web, APIs, databases, mobile applications, and standard applications
- Networks (internal/external/DMZ), including wireless
- Social engineering, such as phishing simulations
- Physical security tests against facility access and data center controls

Penetration testing can also cover software testing, which will most likely be executed in conjunction with your developers and programmers. We will cover this in more detail in the *Application Testing* section. There are a few different testing types that can be used as part of penetration testing, known as black box, gray box, and white box:

- **Black box testing**, also referred to as no or zero knowledge testing, is where the tester is provided with no information about your environment upfront. This best represents an actual hacker. For example, if you hired someone to execute a penetration test against your organization, you wouldn't provide any information; they would be reliant upon public information only.
- **Gray box testing**, also referred to as some or partial knowledge testing, is where the tester is provided with limited information about the environment. For example, when hiring someone to execute a penetration test, you may provide them with all your public IPs and URLs, which will allow the penetration tester to get off to a quicker start, allowing for a more efficient test to occur based on available time.
- **White box testing** is also referred to as complete or full knowledge testing. In this type of testing, the tester is provided with complete knowledge of the environment. For example, when you hire someone to execute a penetration test, you provide them with all the information needed to complete the test, such as IP addresses, URLs, credentials, architecture diagrams, source code, etc. This allows for a much more comprehensive test and is typically targeted more toward specific applications or systems.

Although there may be an appetite and an option to run penetration testing exercises internally as part of your cybersecurity program, it is highly recommended to outsource these activities to an external vendor. If you do have the resources to execute penetration testing activities internally, there is no harm in this, and it is certainly an added benefit to your program.

However, engaging a third party to execute an annual (at minimum) penetration test ensures a separation of duties to show added due diligence.

Most penetration testing vendors are equipped to handle small to large-sized environments and include experts who are very familiar with ethical hacking techniques, and following the rules of engagement to not inadvertently cause damage to critical systems. Many third-party companies offer penetration testing services, and a few examples include the following:

- **Secureworks**: https://www.secureworks.com/services/adversarial-security-testing/penetration-testing
- **Mandiant**: https://www.mandiant.com/services/technical-assurance/penetration-testing
- **Rapid7**: https://www.rapid7.com/services/security-consulting/penetration-testing-services/

These types of testing activities are also considered red team exercises in which they take an offensive approach toward an organization. The defensive side, on the other hand, is known as the blue team, which will primarily be your SOC team. When these two teams come together to collaborate on security testing, a purple team (red and blue) exercise is executed. For example, the red team will execute an attack on the organization and the blue team will attempt to validate whether they were able to detect the attack and respond to it. If the attack is detected, the blue team is successful in their role. If they don't identify the attack, they can address gaps for improvement. The idea of a purple team exercise is for both teams to collaborate and learn from each other to build a better defense for the organization.

 One way to truly test your SOC team's capabilities and response time is to not inform them of any penetration activities occurring within your organization. As the penetration tester identifies vulnerabilities within your environment, monitor your SOC team's performance and response. This provides some valuable data points and real-life types of incident response activities for your teams.

There are many tools available to help conduct your own tests within your environment if you have the expertise in-house to conduct internal testing to look for vulnerabilities. A few tools you may be familiar with include the following:

- **Metaspoilt**: https://www.metasploit.com/
- **Wireshark**: https://www.wireshark.org/
- **NMAP**: https://nmap.org/

As mentioned, if you do have the expertise and resources in-house, executing these types of tests will significantly help reduce risk for your organization. But don't forget to bring in a third party at least once a year to bring some diversity and provide a second set of eyes for your cybersecurity posture.

Executing a Penetration Test

Whether a penetration test is conducted internally or outsourced to a third-party company, you will need to ensure that a rigid process is followed as part of its execution. Simply deploying some tools and trying to uncover vulnerabilities without a plan and proper approval can be extremely risky and potentially disruptive to your operations. This will not go down well with a business. The penetration test needs to be carefully planned with the correct approvals in place before execution. There may be slight variations and different approaches, but the following diagram illustrates a standard process for executing a penetration test.

Figure 11.2: The penetration testing process

Let's review each phase:

1. The **Scoping and Planning** phase is where you will plan and define the scope of the test. This is where contracts will be signed and the **Rules of Engagement (RoE)** document is agreed upon.

2. **Reconnaissance** is the phase where information about the company or environment being targeted is gathered.

3. In the **Vulnerability Assessment and Scanning** phase, you will scan the environment and search for and identify weaknesses and vulnerabilities.

4. **Exploitation** is where you will attempt to exploit and breach an environment with any of the identified vulnerabilities.

5. **Reporting and Analysis** is the final stage, where you will receive a report with an overview and detailed analysis of the test. This can also include recommended actions for remediation.

As reviewed in the *Scoping and Planning* point, the RoE is a very important component before proceeding with any testing in any environment. Because of this, let's review the RoE in more detail.

Rules of Engagement

As part of your RoE document, you are going to need to ensure that you clearly document specific requirements so that everyone is fully aware of expectations. Within the RoE, you will need to ensure that you capture specific items within the document. This includes:

- The purpose and scope of the testing
- Who will be performing the testing, with contact details
- Contact details for internal personnel for any escalation
- On what dates and times the testing will be performed
- What infrastructure or systems will be included with IPs and URLs (test scope)
- What tools and technologies will be used to execute the test
- Incident handling for the SOC when vulnerabilities are identified
- Reporting details for both vulnerabilities that need to be shared immediately for remediation, and the expectations on the final report
- A signature required by leadership

The RoE document should be defined in clear language and approved and signed off by leadership before any work begins. It will also be important that any penetration activities follow your change control process, in order to ensure that the correct approvals are in place, unless you have executive leadership sign-off to execute the test without others knowing. The NIST SP 800-115 includes an RoE template that can be used as a starting point, as shown below. You can access the full template in *Appendix B—Rules of Engagement Template* of NIST SP 800-115, located here: `https://nvlpubs.nist.gov/nistpubs/Legacy/SP/nistspecialpublication800-115.pdf`.

TECHNICAL GUIDE TO INFORMATION SECURITY TESTING AND ASSESSMENT

Appendix B—Rules of Engagement Template

This template provides organizations with a starting point for developing their ROE.[42] Individual organizations may find it necessary to include information to supplement what is outlined here.

1.   Introduction

1.1.   Purpose

Identifies the purpose of the document as well as the organization being tested, the group conducting the testing (or, if an external entity, the organization engaged to conduct the testing), and the purpose of the security test.

1.2.   Scope

Identifies test boundaries in terms of actions and expected outcomes.

1.3.   Assumptions and Limitations

Identifies any assumptions made by the organization and the test team. These may relate to any aspect of the test to include the test team, installation of appropriate safeguards for test systems, etc.

1.4.   Risks

Inherent risks exist when conducting information security tests—particularly in the case of intrusive tests. This section should identify these risks, as well as mitigation techniques and actions to be employed by the test team to reduce them.

1.5.   Document Structure

Outlines the ROE's structure, and describes the content of each section.

2.   Logistics

2.1.   Personnel

Identifies by name all personnel assigned to the security testing task, as well as key personnel from the organization being tested. Should include a table with all points of contact for the test team, appropriate management personnel, and the incident response team. If applicable, security clearances or comparable background check details should also be provided.

2.2.   Test Schedule

Details the schedule of testing, and includes information such as critical tests and milestones. This section should also address hours during which the testing will take place—for example, it may be prudent to conduct technical testing of an operational site during evening hours rather than during peak business periods.

---

[42]   The structure of this template is intended to be illustrative. Organizations should organize their ROEs in whatever manner they choose.

B-1

Fiure 11.3: NIST SP 800-115 RoE template (page 1 of 3)

On a separate note, you also need to be aware of other environments that could potentially be impacted if those services have rules of engagement defined. For example, Microsoft has aan RoE document that you will need to review if you're testing an environment hosted in Azure. You can find the Microsoft RoE at `https://www.microsoft.com/en-us/msrc/pentest-rules-of-engagement`. The following screenshot shows the Microsoft RoE introduction and scope:

INTRODUCTION AND PURPOSE

This document describes the unified rules ("Rules of Engagement") for customers wishing to perform penetration tests against their Microsoft Cloud (defined below) components. In many cases, the Microsoft Cloud uses shared infrastructure to host your assets and assets belonging to other customers. Care must be taken to limit all penetration tests to your assets and avoid unintended consequences to other customers around you. These Rules of Engagement are designed to allow you to effectively evaluate the security of your assets while preventing harm to other customers or the infrastructure itself.

All penetration tests must follow the Microsoft Cloud Penetration Testing Rules of Engagement as detailed on this page. Your use of The Microsoft Cloud, will continue to be subject to the terms and conditions of the agreement(s) under which you purchased the relevant service. Any violation of these Rules of Engagement or the relevant service terms may result in suspension or termination of your account and legal action as set forth in the Microsoft Online Service Terms. You are responsible for any damage to the Microsoft Cloud and other customers data or use of the Microsoft Cloud that is caused by any failure to abide by these Rules of Engagement or the Microsoft Online Service Terms.

SCOPE

For the purposes of these Rules of Engagement, "Microsoft Cloud" is defined as including the following Microsoft products:

- Azure Active Directory
- Microsoft 365
- Microsoft Azure
- Microsoft Defender
- Microsoft Dynamics 365
- Microsoft Intune
- Microsoft Power Platform
- Microsoft Account

Figure 11.4: Microsoft's RoE

Make sure to review the section with the activities prohibited by Microsoft, as the list is quite lengthy. Activities on the list include not gaining access to any data that is not owned by your organization, not performing any kind of **Denial of Service (DoS)** type of testing, ensuring you are not using services that violate the **Acceptable Use Policy (AUP)**, etc.

After the test concludes, the next step will be to review the penetration test result and begin any remediation work needed.

Reviewing the Findings

The penetration test reports typically list any identified **Low**, **Medium**, **High**, and **Critical** vulnerabilities or findings. It may also include an **Informational** category that provides details on items of interest but that may not be of high concern or applicable to your current systems. Once the test concludes and the report is finalized, the work doesn't stop. It is just as important that you take the report provided and build a plan around the remediation of those items. Once you receive the report, you are going to need to centrally track all findings in one place. As part of a mature program, you will have multiple tests being executed throughout the year (most likely) with findings that will need to be remediated. First, you will need to document all findings into an issue tracker, and, depending on the severity of what you find, your centralized risk register. This will then allow you to efficiently work through the findings based on severity. Ideally, you want to work out of a digitized issue tracker where you can add all the details, assign resources, apply dates, and so on for completion. If you don't have a digitized issue tracker, you can always begin by using an Excel-based tracker to get you started. Something like the following can work.

Unique ID	Finding	Details	Test Type	Owner	Priority	Status	Completion Date	Resolution	Resolution Details
Rem_2024#1	Default admin username and password found on public site	When testing a public website (www.website name.com), it was observed that........	Penetration Testing	User responsible to remediate	Critical	Complete	July 1, 2024	Mitigated	Details of mitigation here
Rem_2024#2	Weak TLS ciphers	It is observed that the application is using TLS 1.1, which is........	Application Testing	User responsible to remediate	Medium	In-Progress	Expected completion November 30, 2024	Mitigation in progress	Details of mitigation here

Figure 11.5: Example tracker for all remediation items with any testing activities

> The risk register, which is part of the broader **Governance, Risk, and Compliance** **(GRC)** program, will be covered in more detail in *Chapter 14, Managing Risk*.

After the remediation plan is developed, the remediation owner and the cybersecurity team should work together to mitigate any critical or high-risk items immediately, and they should be held accountable by setting deadlines for completion. Once remediation and mitigation efforts have been completed, you will need to ensure you carry out retesting to validate that the controls put in place work as designed so you can close the risk. In some instances, you might accept a risk if the cost to remediate outweighs the value of the data or the type of information at risk. You will also want to ensure you are frequently reporting on the status of the remediation items to your leadership team so that they are aware of what the findings are and how they are progressing to completion.

With **High** and **Critical** (you could do this for all findings if preferred) findings, you may want to consider formally documenting each of them for review with your leadership team. These reports may also come in handy in the event of an audit, or there may be a need in the future to review how a specific finding was remediated. The following image provides an example of how you can document a vulnerability or finding:

Unique ID: Rem_2024#1	Title of Vulnerability Identified	Risk Accepted, Mitigated, Avoided, or Transferred
Severity: Low or Medium or High or Critical	Resolution	
Description: Description of vulnerability uncovered. **Date Complete:** June 1, 2024 **Owner:** User responsible to remediate **Test Type:** Penetration Test **Systems/Applications Effected:** List any systems or applications that were affected from this finding. **Recommendation:** Provide recommendations to remediate vulnerabilities.	Provided details on what steps were taken to mitigate the finding or provide further details if the risk was accepted, avoided, or transferred.	

Figure 11.6: Example of a report to document a single finding

The results from a well-executed penetration test highlight the importance of its inclusion in a mature cybersecurity program. The results provide great value in identifying risks to your users, systems, and company data. Ensure that you understand the technologies and concepts of penetration testing thoroughly to apply them efficiently within your organization.

 NIST has a very detailed document on security testing and assessment. You can view the *SP 800-115, Technical Guide to Information Security Testing and Assessment* publication at https://csrc.nist.gov/pubs/sp/800/115/final.

Hopefully, you are executing penetration testing today within your environments. If not, start by bringing in an external vendor to execute an external penetration test so that you can better understand the risk of your external attack surface.

Now that we've reviewed penetration testing, let's move on to application testing, which everyone should be executing against any new applications or new development work before they go into a production environment.

Application Testing

Another important consideration is application testing. This can include all types of apps, like web applications, desktop applications, or mobile applications. No matter what the application is, it must undergo security testing before being released into production. The same applies to any versioning updates. Ensuring that there are no known vulnerabilities within an application before going live is critical. To some extent, application testing can also be considered penetration testing, and the same principles above will technically apply to your application testing methodology.

Application testing will most likely be executed in conjunction with your developers and programmers. This can include code reviews, **Static Application Security Testing (SAST)** and **Dynamic Application Security Testing (DAST)**, misuse case testing, fuzz testing or fuzzing, interface testing against APIs and user interfaces, and more. One excellent resource used in this type of testing is **OWASP**. OWASP is a nonprofit organization that helps improve the security of software for individuals and enterprises. It provides a tremendous number of resources, such as tools, documentation, and a community of professionals, all looking to continually enhance software security. OWASP focuses on application and web application security, and you can find more information here on OWASP: `https://owasp.org/`.

One of the most useful resources from OWASP is the OWASP Top 10. The OWASP Top 10 provides a list of the 10 most critical cybersecurity risks identified as they relate to your web applications. The latest available version was published in 2021 and is presented here:

1. Broken Access Control
2. Cryptographic Failures
3. Injection
4. Insecure Design
5. Security Misconfiguration
6. Vulnerable and Outdated Components
7. Identification and Authentication Failures
8. Software and Data Integrity Failures
9. Security Logging and Monitoring Failures
10. **Server-Side Request Forgery (SSRF)**

Source: `https://owasp.org/www-project-top-ten/`

Another great resource is the *CWE Top 25 Most Dangerous Software Weaknesses.* It's another list that provides insight into the most common software weaknesses that can lead to exploitation. You can find more information on the list here: `https://cwe.mitre.org/top25/`. The latest update as of August 2024 is from 2023.

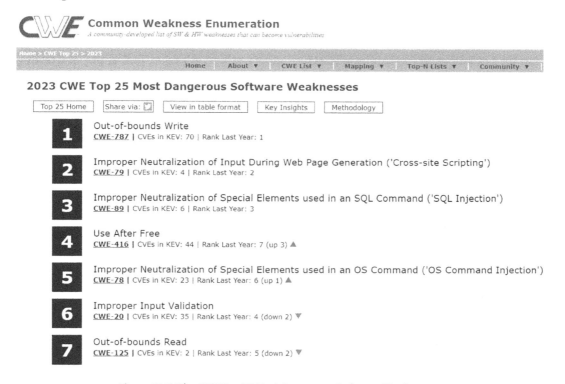

Figure 11.7: The CWE Top 25 Most Dangerous Software Weaknesses

Source: `https://cwe.mitre.org/top25/archive/2023/2023_top25_list.html`

Another consideration is testing your APIs to ensure there are no vulnerabilities as you release them into production. An API in some sense can provide the keys to the kingdom and must be efficiently protected. Therefore, it is critical that testing is complete to ensure best practices are adhered to, especially with passwords and private keys. OWASP also has a Top 10 list for APIs, *API Security Top 10 2023*: `https://owasp.org/www-project-api-security/`.

Make sure DevSecOps is incorporated into the development process and followed by all developers within the organization, which will help reduce risk with applications before they are released into production.

Including these requirements in your cybersecurity policy will help drive the need for stricter security with the development of applications, ensuring a process is in place to be followed. If you have a mature **ARB** process in place, ensure that security testing is required for all applications before going into production, and ensure at a minimum that **Medium, High,** and **Critical** findings are addressed before final approval.

Next, we will shift our focus to an area of testing that is sometimes overlooked but is just as important as your penetration and application types of testing. Let's take a look at physical testing and the importance of running these types of tests.

Physical Security Testing

It is important to highlight the need for physical security testing and to ensure that this is not overlooked as part of your overall testing program. When we think of physical security within an organization, the following should come to mind:

- Managing access to your buildings
- Managing access into your datacenter or any room with a higher level of privileges
- The need for surveillance cameras
- Perimeter security such as fencing and gates
- Security guards at your physical sites
- The installation of alarm systems with ongoing monitoring

The need for physical security is a critical component, designed to prevent anyone gaining unauthorized access to or extracting any assets or data from an organization. More importantly, it is used to protect the well-being of employees by preventing any harm to them from unauthorized personnel.

One of the more common physical security tests executed is to engage a third party who specializes in physical testing services and see how far they can penetrate your physical security controls. Once penetrated, the tester will then identify what information they are able to access or collect once on-site. The following are some examples of activities that may occur with a threat actor gaining access to your physical building:

- Gaining access to your building by bypassing any badge control systems.
- Accessing confidential or sensitive information within the building.
- Connecting to the internal network via the Ethernet to see what vulnerabilities exist.
- Once connected to an internal network, extracting any information that can be accessed.

- Gaining access to confidential areas of the business or gaining access to any rooms with infrastructure in them, for example, servers or networking equipment.

As part of the engagement, the tester will provide details on other findings, such as observing employees to see whether sensitive data is left on desks (including passwords), whether sensitive information is disposed of correctly, whether users allow tailgating to occur, whether there is any opportunity to social engineer employees, etc. Any physical access to a building can cause a substantial amount of damage to an organization, and it is important to understand what opportunities exist within your organization to improve physical security. Physical security tests will also allow you to build more awareness for your users on any identified weaknesses and risks with your physical security controls, as well as ensure that your policies and procedures are current and being followed.

There are many cybersecurity organizations that provide physical testing services as part of their portfolio. For example, one of the many services that Secureworks provides is a physical security testing option: https://www.secureworks.com/services/penetration-testing.

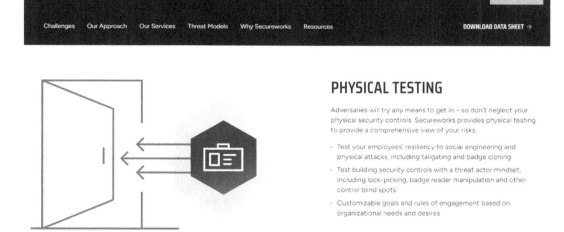

Figure 11.8: Secureworks' physical security testing service

 Nowadays, the ability to replicate someone's badge and use tools to act as a badge reader has never been so easy. These capabilities can be purchased and utilized by the consumer, for example, through the Flipper Zero: https://flipperzero.one/. Make sure you are aware of these tools and their capabilities.

Let's finish off this section by reviewing other types of testing activities that you should be aware of as part of your broader proactive services program.

Other Testing Activities

As a reminder, there is a magnitude of different types of testing activities that can be executed within your organization. Although some of these could also be classified within the penetration category, it is worth mentioning some other areas of testing activities you will want to consider for your organization:

- Testing your Wi-Fi network, especially if any major changes are being made or you are upgrading your Wi-Fi infrastructure. It will be important that the security on your wireless network meets the latest requirements to prevent a threat actor from accessing it. Other areas of the network, such as your firewalls, remote access equipment/systems, etc, should also be considered for testing.

- Another good use case for testing is against the devices issued to your users. For example, how easy is it to compromise one of your company laptops if they were to get lost or stolen? What about mobile devices being compromised? Ensuring that best practices for these devices are followed is critical.

- Other testing activities can include social engineering testing with your users, such as sending phishing campaigns, calling your service desk to try and extract information or gain access to an account, using smishing to try and extract information, and more. We covered social engineering in more detail in *Chapter 9, Cybersecurity Awareness, Training, and Testing.*

- With the ongoing adoption of cloud services, you'll want to focus on different testing methods against your cloud environments, especially as you continue to adopt more. One area that continues to be overlooked by organizations is that of cloud storage and ensuring that access has been set up correctly; you may want to consider some testing focused on your cloud storage, as an example.

- Although this falls within application testing, it is important to highlight the need to thoroughly test and analyze any source code that your organization produces before going into production. This should be a standard process built into the **Software Development Life Cycle (SDLC)**.

- Some other areas that can be considered for testing include any other specific infrastructure testing, such as your **AD** environment, **Domain Naming System (DNS)**, any database servers, etc.

- Although we covered this in more detail in *Chapter 8, Vulnerability Management*, it is important to remind you that there are many different types of vulnerability assessments and scans, such as web applications, external network scanning, cloud infrastructure, and so on, that can be used to help identify vulnerabilities within your organization. Vulnerability assessment and scanning is also part of the penetration testing process, as shown earlier in the section.

There are many different types of testing activities that can be executed against your environment. Make sure you are aware of all the available cybersecurity tests and that you execute them within your environment. Check with your cybersecurity vendors to review their catalog of available testing activities, as new types of testing will most likely be added over time. It should also be a best practice to run cybersecurity testing against any new service, infrastructure, or application before going live, ensuring that the risk of compromise is reduced as much as possible.

Now that we have covered cybersecurity testing in more detail, let's shift our focus to being best prepared for an incident through incident response planning.

Incident Response Planning

As we have continued to demonstrate throughout the book, there is an increase in advanced cyber-attacks, and they are becoming more sophisticated each day. The reality is that many companies aren't prepared for these levels of attacks, which can take days, weeks, or even months to recover from and get back to normal operations. Depending on the size of the business and the severity of the breach, the impact could be as damaging as a complete business shutdown, or worse. With that, it is critical that your organization fully understands the impact of a potential breach and how to best prepare to handle one. This is why we must spend time ensuring that a CIRP is in place for your organization. It is the tool that will help you navigate through a breach "when" it occurs. Your CIRP should include elements such as identifying the roles and responsibilities of the incident response team, and updated contact information of everyone involved, including vendors, incident response procedures, communication procedures, playbooks, and more. The CIRP is essentially your one-stop shop for anything related to a major cybersecurity incident or breach, and should be a requirement for all organizations nowadays, whether you are officially required to have one through regulation or not.

Although an incident response plan is intended to support being reactive by responding to an incident, not having one in place can be catastrophic. Hence, you need to be proactive and ensure that a robust and mature plan is in place for your organization. In *Chapter 7, Cybersecurity Operations*, we covered **Business Continuity Planning** (BCP) and **Disaster Recovery Planning** (DRP) in detail and briefly reviewed a **CIRP**.

In this section, we will go into more detail, specifically with the CIRP, and see what is needed to create a mature CIRP plan. As a reminder, the CIRP falls under the umbrella of broader BCP, as this program will need to be invoked in the event a major cybersecurity incident is declared. The same applies to DRP; there will possibly be a need to invoke this plan, depending on the type of cybersecurity incident. For example, a ransomware incident will most certainly invoke the need for DRP.

There are many resources available to help with the creation of your CIRP if you opt to create it in-house. A quick search on the Internet will return examples of cybersecurity incident response plans, along with templates available for reference. A great place to start is to review the *NIST SP 800-61 Rev. 3, Computer Security Incident Handling Guide*. This publication acts as a guide to help you better understand the incident lifecycle and provides you with the details needed to build a comprehensive plan. The guide can be found here: https://csrc.nist.gov/pubs/sp/800/61/r3/ipd.

 A couple of other relevant resources for incident management and handling include the SANS *Incident Handler's Handbook*, available at https://www.sans.org/white-papers/33901/, and the ISO/IEC 27035 series of documents, where you can find the foundational document, *Information technology — Information security incident management, Part 1: Principles and process* here: https://www.iso.org/standard/78973.html.

If you are creating an incident response plan for the first time, it can be a little intimidating if you have never built a plan like this before. Because of this, you'll want to engage someone to help with the creation of your initial incident response plan. There are cybersecurity vendors with services to help build your CIRP. Start by checking with your current vendors to see what options they have available. They may also have templates available for you to use to allow you to get started. Whichever way you go, I would recommend working with someone who is familiar with building a CIRP if you don't have one in place yet. If you do have one, it's always good to get a second set of eyes from a third party to ensure nothing is missing.

 One plan that provides some good guidance on what is included in a CIRP is the US Homeland Security **National Cyber Incident Response Plan (NCIRP),** which can be found here: https://www.cisa.gov/ncirp-background. Although this plan will most likely differ from what you create for your organization, it is a good plan to review as you gain a better understanding of how these plans are created.

Before we move into detail on building an incident response plan, it is important to ensure you have an incident response policy in place, or at a minimum, have language within your broader cybersecurity policy. As we have covered throughout the book, we must put a policy in place to ensure that users comply with expectations relating to cybersecurity. The same applies to the incident response plan; a policy must be put in place to ensure that everyone is aware of the plan and know they need to follow it in the event of a major cybersecurity incident. Referencing the CIS Controls, there is a template available for the incident response policy for you to look at: `https://www.cisecurity.org/insights/white-papers/incident-response-policy-template-for-cis-control-17`.

Figure 11.9: The CIS incident response policy template

As a reminder, we will cover policies in more detail in *Chapter 13, Governance Oversight*.

Building an Incident Response Plan

In this section, we will provide insight into what should be considered as part of your CIRP. With every organization being different, the need for different requirements depending on your country, industry, any regulation, and so on will determine what should be included in your CIRP.

The guidance here may not include everything you need to consider, but it will provide a foundation and a baseline of the minimum requirements that you should consider for your CIRP.

As with any business type of document or plan, you will need to ensure a formalized document is used for consistency. For your incident response plan, you will need to ensure that the title and company are listed with the document owner or sponsor and the document revision. You should also include the last time it was updated, with a sign-off from an executive leader or someone authorized to finalize the document for production. Make sure the document clearly states the classification, which should, at a minimum, be confidential, and ensure that the location of the document is clearly stated.

Introduction

Your introduction shouldn't need to be too detailed, but it would be good to provide an overview of what the CIRP is and what you can expect it to include. It should provide an overview of your organization and which group within the cybersecurity team is responsible for the CIRP. The introduction should also provide additional details on how the versioning will work, how often the document will be reviewed and updated, the criteria around why it has been classified as Confidential (or something else), and the importance of keeping the document secure. One other consideration is ensuring that the CIRP references back to the broader BCP plan for the organization and the location of that plan. It should also detail that the BCP plan will also be invoked when the CIRP is invoked, and that the correct personnel from the BCP must be included when the CIRP has been initiated for a major cybersecurity incident. Ensure that this section notes that human life is most important with any considerations with incident response.

Purpose and Scope

As with any purpose and scope for your documentation, you will clearly define the purpose of the CIRP, which is to provide the guidance, processes, and procedures for your users to formally navigate a major cybersecurity incident. The scope should be specific as to what the CIRP applies to. For example, if there are multiple organizations under a corporate umbrella, does the plan being created apply to all of them, or just a subset? Another example is the scope of your vendors versus your internal organization. Since you may need a dedicated incident response plan for your vendors, as we discussed in the previous chapter, the scope of your internal CIRP will not include external vendors.

Roles and Responsibilities

In this section, you will list everyone who will be involved with a major cybersecurity incident. This scope is not just limited to the cybersecurity team but will also include the broader organization as a whole, as well as all external parties who will possibly need to be engaged. Obviously, the cybersecurity and IT teams will be critical, as they will serve as primary resources within the incident. If you outsource some of your responsibilities to an **Managed Service Provider** (MSP) or **Managed Security Service Provider (MSSP)**, it will be critical that they are familiar with their responsibilities, as they will need to be engaged, depending on their role. From the broader organization, you are going to need to collaborate with the legal team, HR team, BCP team, and the marketing and/or communications team at a minimum.

The relationship with your legal team is very important as a cybersecurity leader. When any major cybersecurity incident is declared, you will need to immediately engage and work closely with your legal team to ensure that compliance is maintained and the correct processes are being followed. We cover more on the importance of the legal team in *Chapter 13, Regulatory and Compliance*.

From an external perspective, you will need to engage your cybersecurity insurance provider and vendors that need to support the investigation with impacted technologies, such as any software providers and **Internet Service Providers (ISPs)**. There's a possibility that local or government law enforcement will need to be engaged, like the FBI. As you work with your cybersecurity insurance provider, there will be a need to engage, at a minimum, a Breach Response Council and DFIR vendor. Extended support from your cybersecurity insurance provider may include the need for assistance with potential Bitcoin payments and/or negotiation-type skills if needed.

Be familiar with your cybersecurity insurance process to ensure the correct process and formalities are being followed. They will require you to engage a Breach Response Council that will provide guidance through an incident and digital forensics for a thorough investigation. We will cover cybersecurity insurance in more detail in *Chapter 14, Managing Risk*.

As you identify everyone who will be involved, you will need to ensure all contact information is collected and documented. At a minimum, you should have an email address and phone number for each contact. To ensure that everyone is aware of what is expected of them, you'll need to build a detailed roles and responsibilities matrix. It will be important there is no confusion about what is expected from everyone when a major cybersecurity incident occurs.

Make sure you are familiar with your local law enforcement offices, and look to reach out ahead of time to build relationships. For example, for organizations in the US, you can find your closest FBI office here with all the contact information: `https://www.fbi.gov/contact-us/field-offices`.

It will be important that both the incident manager along with a coordinator or project manager are highlighted, to ensure that processes are followed and everything is coordinated with all the roles that will be involved in the incident. This will get complex, as meetings will need to be scheduled and coordinated, some teams will be investigating and others will be making decisions, and continuous reporting and updates will be expected from your executive leadership team. It is important that the people in these roles know what is expected of them and that you have strong resources to fill them.

Communications

There are a couple of components you will need to track within the *Communications* section. First, you will need to ensure that you have all outbound communication channels documented. This can include everything from your internal users, contractors, executive leadership, board, customers, stakeholders, partners, vendors, and even the media, depending on the severity and any potential impact on the company's reputation. With each of these communication channels, you will need to ensure that you have a current contact list for each and that they are readily available when needed. All this ties back to having a good process in place for communications. This section will lean heavily on the crisis communication plan for your organization.

Another consideration for your outbound communication is the need to communicate in the event of a full system outage, like the possibility of ransomware. Because of this, you must consider an out-of-band type of communication that is available and accessible independently from your network. This way, you can still communicate in the event of a full system outage. In addition, make sure you have multiple communication options for those you need to communicate with, for example, text, voice, email, social media, etc. To accomplish this efficiently, you will need to consider an **Emergency Mass Notification System** (**EMNS**).

The second component that must be well documented is the logistics around how teams will collaborate and meet for a cybersecurity incident. What are the tools to be used, such as conferencing capabilities, dial-in bridges, collaboration platforms, etc? Like outbound communication needs, make sure capabilities are still available in the event of a full system outage.

For example, if you use Microsoft Teams within your organization and it becomes unavailable, what is the alternative? You may want to have either a separate Zoom or Webex account available for emergency purposes. This will also be a good idea in the event that a threat actor has compromised your network and is able to listen in to any phone calls or meetings. Also, consider creating a schedule ahead of time to provide updates to executive leadership and any other stakeholders that need to be updated. You will need to ensure that a war room type of capability is documented and well understood; this includes both in-person and virtually.

Incident Response and Recovery Process

The *Incident Response and Recovery* section provides the incident response team with the necessary steps and processes to follow when handling a major cybersecurity incident. To work through this process efficiently, it is recommended to follow the incident response lifecycle provided in the *NIST SP 800-61 Rev. 3, Computer Security Incident Handling Guide*: https://csrc.nist.gov/pubs/sp/800/61/r3/ipd.

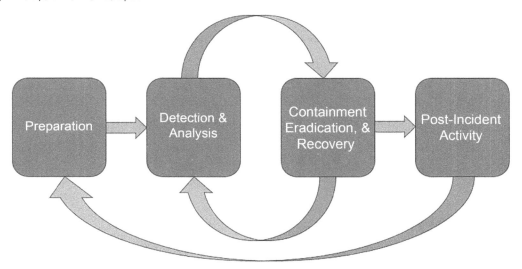

Figure 11.10: The NIST SP 800-61 Rev. 3 incident response lifecycle

As you can see, there is a four-step process that should be followed as you work through the incident investigation:

1. **Preparation**: This is where you will have all the information needed to ensure that you are best prepared to handle a cybersecurity incident. Essentially, the final copy of the CIRP will serve as your complete preparation activities.

2. **Detection & Analysis:** The detection phase will essentially be the process followed within your SOC to detect threats within the organization. Once escalation occurs, additional processes will need to be followed to ensure that the right resources are engaged. From an analysis perspective, this will include all processes to investigate an incident to understand the actions that will be needed for the next steps. It will be important that the plan includes parameters that define the different types of incidents that may need to be investigated. In addition, ensure that the classification of the breach has been defined, in which you will want to ensure that the risk matrix is included in the plan for review.

3. **Containment, Eradication, & Recovery:** In this phase, processes should be in place and followed to ensure that an incident is contained, preventing any further infection or damage throughout the network. It will be important to ensure that evidence, including audit trails and logs, are correctly collected and stored for review, along with providing any intel to the SOC so they can take the needed action to block any additional malicious attempts. Within the eradication phase, it is important that all remnants are removed from the network. This will require an external provider such as a DFIR to confirm any eradication. Recovery will cover everything needed to ensure that all systems are back up and running in the same state they were prior to the cybersecurity incident. This process may need to invoke DRP for successful recovery.

4. **Post-Incident Activity:** This section will primarily address any lessons learned from the incident and will be covered in more detail in the next section.

This section will become very detailed, and it is important that you spend the time needed to ensure that nothing is missing from this section, as you will need it to work through a major cybersecurity incident.

Lessons Learned

A component that is typically overlooked once an organization has recovered from an incident is the lessons learned. For the most part, there is always a sigh of relief when restoring operations after a major cybersecurity incident. In addition, the incident most likely created a lot of exhaustion, depending on how bad it was. Because of this, everyone rushes back to their normal operational duties to catch up and get back on with business. When this happens, the lessons learned always become a "we'll get to it later" activity. As hard as it may be, it is critical that time is spent reviewing the activities within the incident and understanding whether anything can be improved for the next time a major cybersecurity incident occurs. There is always room for improvement, and this comes from reviewing what can be done better from the lessons-learned review.

Another very important factor is ensuring that you fully understand the events and activities leading up to the cybersecurity incident so that they can be addressed, ensuring that the same cybersecurity incident doesn't occur again (although there will never be a guarantee).

A great white paper to provide additional guidance with your lessons learned has been provided by Secureworks. You can access their *Incident Response: Lessons Learned Template* here: `https://www.secureworks.com/resources/wp-incident-response-lessons-learned-template`. Within the template, you are provided with guidance on pre-engagement planning, lessons-learned formats, some sample interview questions, and a section that addresses reporting and debriefs.

Appendix

The final section of your plan should be the appendix, where you will host all additional supporting documentation for your CIRP. The items that you need to include and track within the appendix for your CIRP will depend on your organization. This section will easily get lengthy if you have a well-documented plan. At a minimum, you should include the following within your appendix:

- **Definitions**: There will be a lot of IT and cybersecurity-based language within your plan. With the plan needing to be used by non-technical IT and/or cybersecurity users, it is important that there are definitions to help ensure that everyone understands everything within the plan. The cybersecurity industry is well-known for the use of acronyms, so make sure all of them are spelled out.

- **Contact lists**: You may have this listed within the *Communication* section, but there is a chance that this will get extremely lengthy and probably shouldn't be within the main plan itself. Pulling all contacts into the appendix will keep the plan a lot cleaner.

- **Regulations**: With regulations and compliance constantly changing and being updated, it is important that you track any that may impact your organization. Think of any requirements to notify customers or the need to report an incident under **General Data Protection Regulation** (**GDPR**), the Securities and Exchange Commission, etc. You will need to collaborate with your legal team on these requirements.

- **Templates**: Include any relevant templates that will help as part of the incident process. This can include items such as communication templates, meeting note templates, evidence-gathering templates, schedule templates, reporting templates, timeline templates, etc.

- **Playbooks**: Here, you will track any specific documentation that helps and guides you through a specific type of cybersecurity incident. We will cover this in more detail in the next section.

- **Testing**: It is important that the plan requires testing to be conducted on an annual basis. An example of this includes the need to execute an annual tabletop exercise, as reviewed earlier in the chapter. It is important that this is well documented so that everyone understands the expectations for testing.

- **Training**: The CIRP is going to be quite a lengthy document, and it will not have been reviewed by everyone who needs to be aware of it. In addition, there may be uncertainty with parts of the plan. Because of this, providing training for everyone to attend and walking through the document should be a requirement. This will ensure that everyone is familiar with the plan and provides a channel for any questions to be addressed to ensure clarity.

- **Update process**: Make sure the process to update the CIRP is well documented and not overlooked. The document should be reviewed, at a minimum, annually and formally published as a new working version once reviewed. Ensure that the process includes where the document is to be stored securely and that all stored versions are current.

A couple of critical items to cover before we close the section are the need to ensure that the CIRP is secure and that only those that need access to it can access it. You also need to ensure that the document is available in the event the network is not available. When securing the document, bear in mind there have been instances where threat actors have accessed an organization's CIRP and used it against them – more specifically, with the cybersecurity insurance limits that you may have included in the plan. Although you do need to include a reference to your cybersecurity policy within the CIRP, you'll want to exclude any information that can be used by threat actors from the CIRP and leave it within the cybersecurity insurance policy only. This also means ensuring that the cybersecurity insurance policy is well-secured and that only a few have access to it and are aware of the policy limits.

 Make sure the access controls to your CIRP are thoroughly reviewed and audited frequently to ensure that there is no unauthorized access.

The second item is ensuring that the policy is accessible offline in the event all systems become unavailable; think of this as an air-gapped backup. Some options include having a printed copy or an e-copy stored in a secure location, such as a water proof and fireproof safe within your facilities. As always, there are a lot of considerations with these types of documents and plans to ensure they are secure.

Make sure you are following the best practices with the handling of your CIRP, ensuring that the plan is encrypted, whether online or on external/USB drives; that you have backups of your documents, including an air gap; ensuring access to the plan is only for those authorized, and that access is being audited.

Playbooks

As we noted in the appendix, you'll want to build playbooks for specific scenarios as part of your CIRP. Playbooks provide the steps needed to work through a specific cybersecurity incident. This allows for a more efficient investigation from whoever is working on the incident. The playbook will provide a consistent methodology for any analyst qualified to work through the steps for quicker response. For the most part, your SOC will already be leveraging **Standard Operating Procedures (SOPs)** for everyday incident investigation, which are somewhat similar to playbooks. There will be a defined step-by-step process that needs to be followed for both an SOP and a playbook. Because of this, the process of working through a playbook and an SOP will be a familiar process for those working through an investigation. You can build a playbook for any scenario that makes sense for your organization, although you will want to focus on playbooks for incidents that tend to be more common than others. Some examples of playbooks you will want to consider include:

- A ransomware playbook
- Email vulnerability playbooks for phishing, **Business Email Compromise (BEC)**, etc.
- A malware playbook
- A **Distributed Denial-of-Service (DDoS)** playbook
- A data loss playbook

As you build your playbooks, you will want to ensure consistency by using a standardized template. Like your CIRP, you will want to ensure that they are version-controlled, reviewed annually at a minimum, and formally approved to production for use.

You should also check with any of your vendors to see if they have any pre-prepared playbooks for any technology within your environment. For example, Microsoft has a few different playbooks available for reference here: `https://learn.microsoft.com/en-us/security/operations/incident-response-playbooks`. One of the available playbooks is for phishing incidents. In addition to the detailed instructions provided, they have both a playbook workflow and checklist available. You can access the playbook workflow here and download either the PDF or Visio format within the *Workflow* section: `https://learn.microsoft.com/en-us/security/operations/incident-response-playbook-phishing#workflow`.

The following image is page 1 of 4 from the PDF version of the playbook workflow.

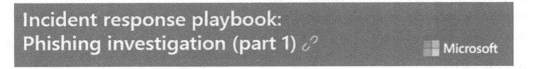

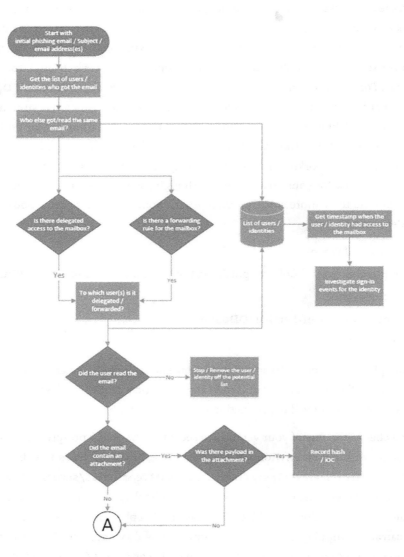

Figure 11.11: Microsoft incident response playbook (Phishing)

It is important that the playbooks created are easy to use and follow, allowing those investigating to run through the process more efficiently. Once they have been created, ensure that everyone is familiar with them and that they have been tested to ensure accuracy.

Now that an incident response plan has been put in place, the next step is to ensure that you are familiar with it. The best way to do this is by executing a tabletop exercise to test your response to a theoretical cybersecurity incident.

Tabletop Exercises

As already mentioned, statistics continue to show it is not a matter of "if" but "when" a major cybersecurity incident will occur. The better prepared we are to respond, the quicker we can get the business back up and running. One of the ways to better prepare for a breach is through executing tabletop exercises. Some of the benefits of running a tabletop exercise include the following:

- At a minimum, it is recommended to execute tabletop exercises as a best practice, but there are also audit and regulatory requirements that require these to be run.
- There may be a requirement from your cybersecurity insurance provider to help reduce costs with the policy.
- Providing the capability to test responsiveness to a realistic cybersecurity incident situation.
- Can help reduce impact and expenses by allowing a more efficient response.
- Providing better preparation for when an incident occurs.
- Ensuring that everyone is familiar with the plan.
- Brings all groups together into one room to allow everyone to become more familiar with roles and responsibilities.
- The exercise allows for improved communication, coordination, and collaboration between all parties involved.
- Identifying any gaps within the response plan in advance of an incident.
- Allowing improvements to be made to processes within the plan prior to an incident.
- They can be fun!

What Is a Tabletop Exercise?

In short, a tabletop exercise is the process of walking through a realistic scenario that resembles a potential real-life major cybersecurity incident or breach. Although this will not provide real-life experience, the process ensures that everyone is familiar with the CIRP and what can be expected when a major cybersecurity incident occurs. For the exercise, you will need to assemble the personnel (as identified in your CIRP) that will be involved in a major incident response. The tabletop exercise is designed to walk your key stakeholders through the process of a realistic incident scenario, asking questions that require the team to make timely decisions, troubleshoot, and communicate in a way that replicates an actual incident. A trained moderator will assess each response and provide feedback on strengths and areas for improvement. Tabletop exercises may include a wide range of personnel, which covers different real-world scenarios based on your industry, the size of the organization, and any potential regulation or compliance requirements.

Although we primarily focus on tabletop exercises within cybersecurity, it is important to be aware that there are other types of exercises. For example, the **Homeland Security Exercise and Evaluation Program (HSEEP)** provides a standard for exercises and references six different types of exercise within two different categories. Discussion-based exercises include seminars, workshops, and tabletop exercises. Operations-based exercises include drills, functional exercises, and full-scale exercises. If you want to learn more about all these exercises, you can access the *Homeland Security Exercise and Evaluation Program Doctrine* on this website: `https://www.fema.gov/emergency-managers/national-preparedness/exercises/hseep`.

Planning a Tabletop Exercise

When planning a tabletop exercise, you will first need to determine what type you will be running. There may be additional specific target groups within your organization, but at a minimum, you should execute the following types of tabletop exercises annually:

- **BCP Tabletop:** Although not necessarily cybersecurity-related, it is important that a BCP tabletop is executed annually for your organization. Everyone involved in the BCP must be familiar with the plan and what is expected of them. When a major cybersecurity incident is triggered, the BCP plan will need to be invoked also, since the CIRP falls under the broader umbrella of the BCP program. Other scenarios include active shooter incidents, hurricane preparedness, severe weather incidents, full system outages, and many more.

- **Technical Tabletop Exercise:** This tabletop will test your technical team's ability to respond to a cybersecurity incident. This exercise is mostly technical in nature and focuses on identifying whether an incident has occurred, investigating the incident, and the response provided from a technical perspective.

- **Executive Tabletop Exercise:** This exercise brings your executive leadership team together to test how they will respond to an incident. This will focus more on decision-making skills, how and whom to communicate with, the process and timing to engage cybersecurity insurance, whether to pay a ransom, etc.

- **Board Tabletop Exercise:** Although this will be similar to the executive tabletop exercise, you can combine the two if you can get both the executive leadership team and the board together at the same time. Regardless, it will be critical that the board runs through a major cybersecurity incident as there will be critical decision points that will need to be addressed. There should also be a focus on reputation and the need to address any public media concerns at this level. Making them aware of anything they can expect to make decisions on will be important so they can prepare ahead of time.

- **Supply Chain Tabletop Exercise:** With the increase of vendors becoming compromised, the supply chain continues to become more vulnerable. It's worth your time to work with those involved in managing your vendors, walking through a tabletop exercise that addresses the major impact within the supply chain, caused by a major cybersecurity event. As discussed in *Chapter 10, Vendor Risk Management*, there will be a need for an incident response plan dedicated to vendor cybersecurity incidents and breaches.

Once you have determined the type of tabletop exercise you would like to execute, the next step is to determine a theme. A theme for a tabletop exercise can be centered around anything relevant to cybersecurity. To make the tabletop exercise more realistic, you'll want to ensure that it is relevant to current trends within the organization, within your industry, or more generally, throughout the world. For example, if you are observing an ongoing increase in **BEC** threats within your organization, build the tabletop exercise around this as a theme. If there is an ongoing increase of ransomware throughout the world and organizations continue to become compromised, build your exercise around a ransomware event. Other scenarios can include insider threats, social engineering, data exfiltration, malware infection, zero-day vulnerability, **DDoS,** etc.

Now you have a theme, you will need to create an exercise based on that theme. If this is your first time, there are many resources available, including templates that you can leverage to build your tabletop exercise. Although not the most detailed, the CIS has some very basic tabletop exercises for reference: `https://www.cisecurity.org/insights/white-papers/six-tabletop-exercises-prepare-cybersecurity-team`.

Another great resource with more detailed tabletop exercises is available from CISA: `https://www.cisa.gov/resources-tools/resources/cybersecurity-scenarios`.

PUBLICATION, EXERCISE, TRAINING

Cybersecurity Scenarios

Cybersecurity-based threat vector scenarios including ransomware, insider threats, phishing, and Industrial Control System compromise.

Revision Date: February 27, 2023

RELATED TOPICS: CYBERSECURITY BEST PRACTICES, MALWARE, PHISHING, AND RANSOMWARE, INDUSTRIAL CONTROL SYSTEMS

◆————————————◆

Cybersecurity Scenario CISA's Tabletop Exercise Packages (CTEPs) cover various cyber threat vector topics such as ransomware, insider threats, and phishing.

For more information, please contact: CEP@cisa.dhs.gov

Resource Materials

⬇ Chemical Sector CTEP Situation Manual - November 2023
(DOCX, 7.49 MB)

⬇ Commercial Facilities CTEP Situation Manual - November 2023
(DOCX, 6.59 MB)

⬇ Communications CTEP Situation Manual - February 2024
(DOCX, 5.82 MB)

⬇ Critical Manufacturing CTEP Situation Manual - November 2023
(DOCX, 11.86 MB)

⬇ Cyber Insider Threat CTEP Situation Manual - September 2023
(DOCX, 4.64 MB)

⬇ Dams Sector CTEP Situation Manual - March 2024
(DOCX, 9.23 MB)

⬇ Early Voting CTEP Situation Manual - January 2024
(DOCX, 5.33 MB)

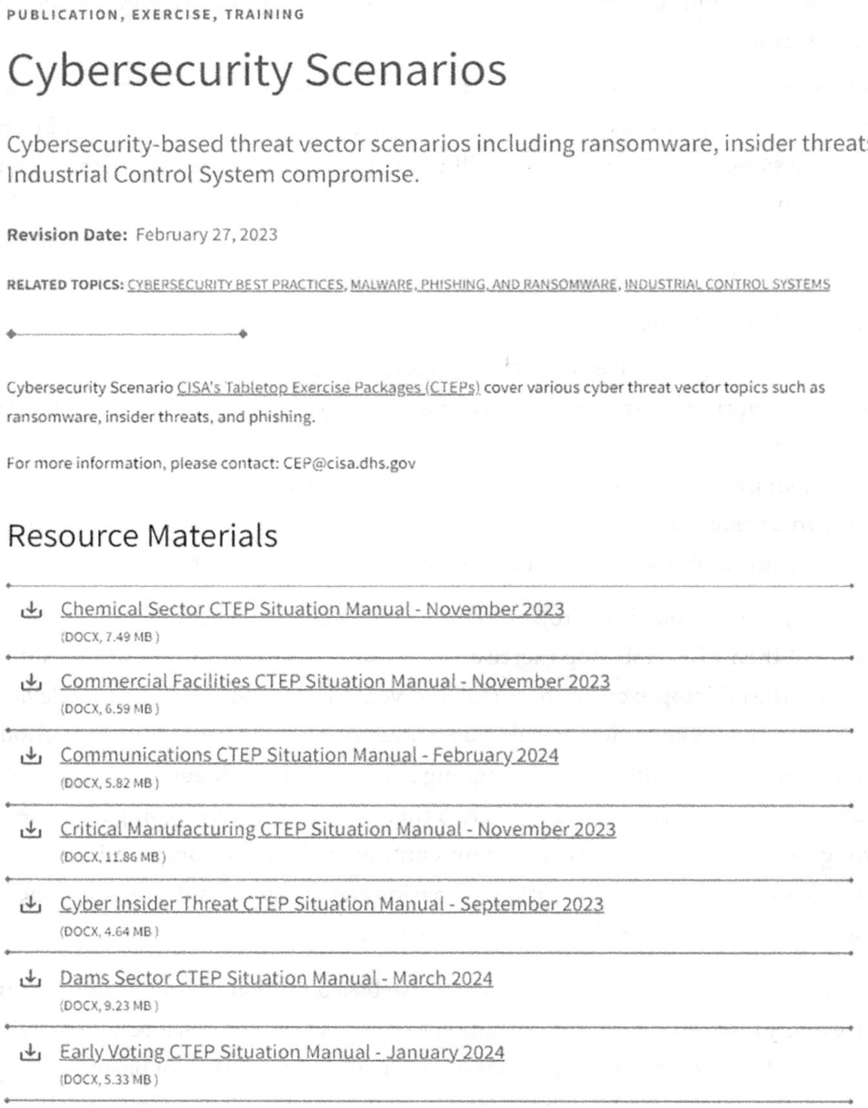

Figure 11.12: CISA tabletop exercise example scenarios

There are also additional types of scenarios, along with supporting documentation, available here: `https://www.cisa.gov/resources-tools/services/cisa-tabletop-exercise-packages`

One additional resource that will help you better understand the process of running a tabletop exercise is from the **National Cyber Security Centre (NCSC)**. They have an excellent guidance article that walks you through the process of creating and executing a tabletop exercise, which can be found here: https://www.ncsc.gov.uk/guidance/effective-steps-to-cyber-exercise-creation. The guidance walks you through nine distinct steps, as shown in the visual below.

Infographic summary

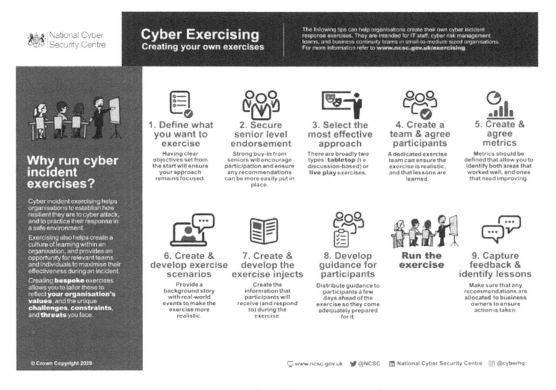

Figure 11.13: NCSC cyber tabletop exercise guidance

Source: https://www.ncsc.gov.uk/files/cyber_exercise_creation.pdf

One important decision you will need to make is whether to run the tabletop exercises internally or to bring in a third-party vendor to execute them. These types of exercises are not difficult to run, especially after you have completed a couple of them and you are familiar with the process. Running them internally should be feasible, but my recommendation would be to bring in a third party to execute the tabletop exercises, especially for the executive team and the board members, at a minimum.

The reason for bringing in an external vendor will be to allow a neutral party with no conflict of interest to provide their recommendations, based on their observations. If you and the team execute the tabletop exercise and make recommendations, there will be potential challenges made against those recommendations, and they may not be taken seriously. With a third party, you will have formally documented recommendations made by an external resource that will have been attested to, making it harder to challenge. However, there is no issue with running an internal tabletop exercise. Just make sure that you protect yourself, as you need to make progress with the recommended changes.

Executing a Tabletop Exercise

To execute the exercise, you are going to need to ensure that you plan ahead of time, especially when working with the executive leadership team or the board. Make sure you schedule the exercise in advance to provide ample time for everyone to review and be prepared. Ensure that everyone is aware of what the exercise is and what they can expect ahead of time. If they need to be aware of the CIRP, make sure it is made available before the meeting for review. Don't give too many details on the specifics of the exercise, as this will spoil the surprise. When executing the tabletop exercise, it is highly recommended that you schedule it in person if possible. There will be much better interaction when everyone is in the room, rather than users being remote on a conference bridge. There will be times when users may not be able to attend in person, and you will need a remote option. But make sure you push for in-person, as you will get a lot more out of the exercise.

For the exercise, there will be a dedicated moderator who will manage and execute it. They will be the ones to present, generate discussion, take notes throughout, provide the final report, and more. As you begin the tabletop exercise, you will want to start with introductions, especially since you will have multiple groups in the room. You'll want to provide an overview of the exercise and ensure that everyone knows what a tabletop exercise is, along with the purpose and scope of the exercise. Everyone should also be aware of the logistics and how the exercise will flow, along with any rules of engagement and expectations from everyone. Make sure that you also formally document attendance of the tabletop exercise, as this may need to be referenced in the future for any possible compliance requirements.

As you begin to work through the exercise, you will start by opening with the selected theme. For example, if the theme is based around a phishing event, you may open with the following scenario:

1. One of your users receives a malicious phishing email containing a link to a fake website, which looks to be a legitimate email from a trusted vendor.
2. The user clicks on the link in the email and submits their credentials to the fake website.
3. The user thinks nothing of the action and continues with their day.

This can also be thought of as the initial injection. An injection is the details provided to support a theme's storyline, in which the following injections enhance the storyline. You should keep the opening very brief and not go into any other details, as you will slowly build the exercise up to the malicious activity through your injections. Then, you will want to understand how the teams respond to the initial information provided. This is where you open up for questions and discussion. You will have a pre-defined set of questions that will help drive the conversation, allowing you to better assess the current state and any opportunities for improvement. For example, based on the initial injection around the theme of a BEC event, you may ask some simple questions such as:

*   Are there any actions that can be taken at this point?
*   Would you expect to be alerted to the above scenario?
*   How would an employee report the above activity?

During the discussions, the moderator will collect feedback and document opportunities. Once all questions have been reviewed, the moderator will insert a new injection that will enhance the storyline, which will lead to more questions and open discussion. Again, the moderator will continue to collect feedback and document any opportunities. The exercise will continue with more injections and supporting questions until the exercise is complete. Once complete, the exercise should finish with a debrief to review anything that went well and anything that didn't go well. This will conclude the exercise, after which the moderator will use their notes to build a final report for review.

To give you an example of a very simple tabletop exercise, the following exercise has been made available by CIS for reference.

This is one of the six exercises available from CIS that were referenced earlier and can be found here: `https://www.cisecurity.org/insights/white-papers/six-tabletop-exercises-prepare-cybersecurity-team`.

 Tabletop Exercises: Six Scenarios to Help Prepare Your Cybersecurity Team

Exercise 5

Financial Break-in

SCENARIO: A routine financial audit reveals that several people receiving paychecks are not, and have never been, on payroll. A system review indicates they were added to the payroll approximately one month prior, at the same time, via a computer in the financial department.

What is your response?

INJECT: You confirm the computer in the payroll department was used to make the additions. Approximately two weeks prior to the addition of the new personnel, there was a physical break-in to the finance department in which several laptops without sensitive data were taken.

OPTIONAL INJECT: Further review indicates that all employees are paying a new "fee" of $20 each paycheck and that money is being siphoned to an off-shore bank account.

Having this additional information, how do you proceed?

Discussion questions

- What actions could you take after the initial break in?
- Do you have the capability to audit your physical security system?
- Who would/should be notified?
- Would you able to assess the damages associated from the break in?
- Would you be able to find out what credentials may have been stored on the laptop?
- How would you notify your employees of the incident?
- How do you contain the incident?
    - *Optional Inject question:* How do you compensate the employees?

Processes tested: Incident Response

Threat actor: External Threat

Asset impacted: HR/Financial data

Applicable CIS Controls: CIS Control 4: Controlled Use of Administrative Privileges, CIS Control 16: Account Monitoring and Control, CIS Control 19: Incident Response and Management

Figure 11.14: CIS tabletop exercise example on a financial break-in

Now that we have reviewed how to execute a tabletop exercise, let's move on to the final reporting and remediation activities that should be executed.

Final Report and Remediation

Like any test, assessment, or exercise, you'll need to ensure that a formalized report is created, with all issues and opportunities identified. This report will serve as the findings that will need to be reviewed and addressed. In some sense, you can think of this as an audit against your tabletop exercise preparedness, in which the final report documents all areas for improvement that will need to be remediated. When receiving the report, you will typically find details around what was observed by the moderator in the exercise. Each observation will provide a high-level explanation of the findings, along with details to support the reason for documenting the findings. In addition, the findings should all have a recommendation provided to allow remediation to occur. Most importantly, each finding should be prioritized from Low to High. If the priority wasn't included, make sure this is added to help with your prioritization. The following is an example of what a report may look like, with some made-up observations.

Observation	Details	Recommendation	Priority
Cybersecurity Insurance Policy	It was identified that the organization currently doesn't have a cybersecurity policy in place.	It is recommended that options for a cybersecurity policy are reviewed, and one is implemented as soon as possible to ensure the appropriate support both technically and financially is made available in the event of a cybersecurity incident.	High
Cybersecurity Awareness, Training, and Testing	During a discussion on user testing through phishing simulations, it was identified that tests were only occurring once every quarter.	Although phishing simulations are being executed, it is recommended to increase the frequency to monthly to better prepare your users against the ongoing threat through phishing emails. In addition, make sure you are leveraging the latest types of phishing threats including the use of AI.	Medium

Figure 11.15: Example of observations on a tabletop exercise report

To close out the section, it is important to discuss the remediation actions. A lot of the time, the final report is reviewed, and no action is taken. You must ensure that a remediation plan is put in place and all findings are reviewed and actioned. You'll need to assess each finding in detail and determine next steps. Essentially, these items will fall within the issues tracker or risk register, depending on how you track risk. You will then need to determine if the observed findings will be mitigated, accepted, transferred, or avoided. Regardless of the decision, each finding will need to be formally documented within the GRC for tracking and compliance reasons. You should aim to ensure that all findings are closed out by the next equivalent exercise. Ensuring all remediation items are actioned and closed will allow for continuous improvement, which, in turn, will improve your incident response.

Other Proactive Services

In the final section, we will cover some other proactive services that should be considered as part of your proactive services function. As already mentioned, there are many proactive types of services that can be executed to help reduce risk for your organization.

Some of these will have been covered in more detail throughout other chapters, but we'll briefly cover them so that you are aware of everything you should be considering.

Threat Briefs

Threat briefs provide insights into past and current threats, including **Tactics, Techniques, And Procedures (TTPs)** for review and to allow you to be better prepared using the intelligence provided. These briefs can provide insight to your teams for review and a better understanding of how threat actors are being malicious. This will allow us to create better defenses by learning from those who have (unfortunately) been impacted. Threat briefs can be delivered from your current cybersecurity vendors if they have threat intelligence available to share. This can be in the form of notifications for you to review as they become available, or you may opt to have someone deliver a threat brief to the broader team. Ideally, you will want to receive this type of data as frequently as possible. However, this may not be realistic, so you will need to determine how often you can receive this information and how often you are able to have a threat brief hosted. Either way, the more information you can attain from threat briefs, the more proactive you can be by implementing the appropriate controls for known threats and TTPs.

Another method of receiving threat brief information is searching online for resources. Some vendors provide threat briefs publicly for your review. A couple of examples include Microsoft (`https://www.microsoft.com/en-us/security/security-insider/emerging-threats/`) and Unit 42 by Palo Alto Networks (`https://unit42.paloaltonetworks.com/category/threat-research/`).

Figure 11.16: Palo Alto Networks Unit 42 threat briefs

The information provided by threat briefs will be very valuable and allow preventive measures to be put in place against known adversaries, so make sure you don't overlook them; make them a requirement for your cybersecurity operations team.

Threat Hunts

We covered threat hunts in more detail in *Chapter 7, Cybersecurity Operations*. As a reminder, threat hunting is a technique used to determine if there is any active malicious activity occurring within your environment. More specifically, **Cyber Threat Intelligence (CTI)** data is used to detect any current activity being used to exploit your environment. Although considered more of a proactive activity to help identify if there is any current malicious activity occurring within your environment, these activities are typically run by your SOC, or blue team. Since this team has the most knowledge about your environment, access to the latest intel, and access to all your alerts, it only makes sense that they can be more efficient at running these types of activities. As threat actors continue to evolve, these activities only become more critical, and you should consider running constant threat hunts throughout your environment if possible.

Incident Response Training

Although we covered tabletop exercises earlier in the chapter, which, in some aspects, provide training or set expectations for when a cybersecurity incident occurs, incident response training can be looked at a little differently. This type of training is considered more technical for your primary incident responders, who will undergo the incident investigations. This will be geared primarily toward your IT service desk and SOC teams, serving as a basic level course to better understand how to identify cybersecurity incidents and how to handle an incident more efficiently. Advanced training will be targeted more toward your senior staff, who will be involved with any escalations for more detailed investigation and review. Check with your cybersecurity vendors to see if they have training programs available for your teams. There are also certifications available for incident response from some of the popular training and certification organizations. An example of some incident response training courses from SANS is shown in the image below. More details on these courses and what they cover can be found here: https://www.sans.org/cyber-security-courses/?focus-area=digital-forensics&q=Incident%20Response.

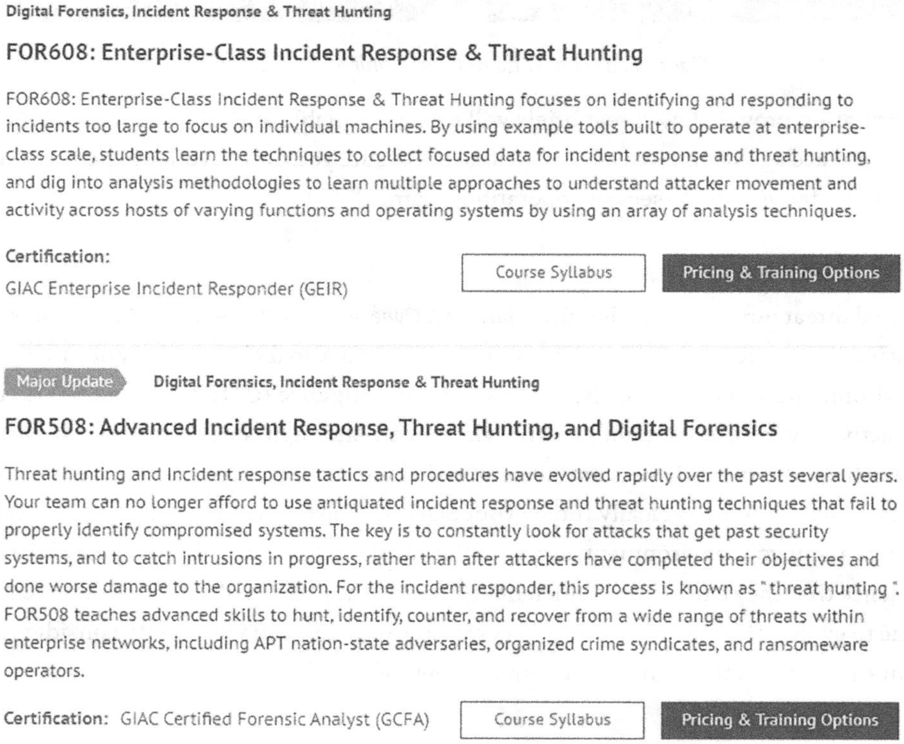

Digital Forensics, Incident Response & Threat Hunting

FOR608: Enterprise-Class Incident Response & Threat Hunting

FOR608: Enterprise-Class Incident Response & Threat Hunting focuses on identifying and responding to incidents too large to focus on individual machines. By using example tools built to operate at enterprise-class scale, students learn the techniques to collect focused data for incident response and threat hunting, and dig into analysis methodologies to learn multiple approaches to understand attacker movement and activity across hosts of varying functions and operating systems by using an array of analysis techniques.

Certification:
GIAC Enterprise Incident Responder (GEIR)

Course Syllabus | Pricing & Training Options

Major Update   Digital Forensics, Incident Response & Threat Hunting

FOR508: Advanced Incident Response, Threat Hunting, and Digital Forensics

Threat hunting and Incident response tactics and procedures have evolved rapidly over the past several years. Your team can no longer afford to use antiquated incident response and threat hunting techniques that fail to properly identify compromised systems. The key is to constantly look for attacks that get past security systems, and to catch intrusions in progress, rather than after attackers have completed their objectives and done worse damage to the organization. For the incident responder, this process is known as "threat hunting". FOR508 teaches advanced skills to hunt, identify, counter, and recover from a wide range of threats within enterprise networks, including APT nation-state adversaries, organized crime syndicates, and ransomware operators.

Certification: GIAC Certified Forensic Analyst (GCFA)   Course Syllabus | Pricing & Training Options

Figure 11.17: A couple of SANS incident response cybersecurity courses

Now that we have reviewed incident response training, let's review what a disclosure program is and what programs can be considered for your organization.

Disclosure Programs

Another proactive service to consider is that of disclosure types of programs. In short, these programs allow ethical hackers and/or cybersecurity researchers to disclose vulnerabilities to an organization in return for payouts, rewards, and/or recognition. A couple of well-known and adopted programs include a bug bounty program and a **vulnerability disclosure program (VDP)**.

A bug bounty program is one in which organizations incentivize ethical hackers to search for "bugs" or vulnerabilities within a defined scope of their services and/or applications. Bug bounty programs typically pay out rewards in the form of cash. Many organizations run bug bounty programs to allow the open community to disclose vulnerabilities to organizations before threat actors exploit them. One example is the Microsoft Bug Bounty Program, which can be found here: `https://www.microsoft.com/en-us/msrc/bounty`.

Figure 11.18: Microsoft Bug Bounty Program

On the other hand, a VDP is a program that enables anyone to disclose a vulnerability that they may have stumbled across in an organization's services and/or applications.

The idea around a VDP is to show that a company provides a process to allow vulnerabilities to be disclosed, along with providing the correct channels for cybersecurity researchers and/or ethical hackers to report any vulnerabilities without being concerned about any legal action.

One of the main differences between a bug bounty and a VDP is that a VDP will typically not pay out monetary rewards, although there may be some incentives or recognition involved, depending on the program in place. A simple search online for VDP will return many companies who offer this program.

In addition, there are platforms available to support your bug bounty and VDP needs. A couple of examples include Bugcrowd (`https://www.bugcrowd.com/`) and HackerOne (`https://www.hackerone.com/`).

Ransomware Best Practices

When it comes to ransomware, we need to be prepared as best as possible because this can be one of the most impactful threats to our organizations. With this being the proactive chapter, it seemed the most appropriate place to provide some best practices for both protecting and responding to ransomware events. There are many great resources available for ransomware preparedness and response. The following resources are excellent guides to support your ransomware preparedness and response needs:

- **CISA:** `https://www.cisa.gov/stopransomware`
- **NIST:** `https://csrc.nist.gov/projects/ransomware-protection-and-response`

Most, if not all of these items, have been covered throughout the book. Bringing them together in a checklist should provide better guidance on how to best protect and respond to a ransomware event. The first checklist shows the best practices for protecting against ransomware:

Best practices for protecting against ransomware

❑ Enforce MFA, use least privilege or just-enough privilege, and implement Privileged Access Management (PAM) and Privileged Identity Management (PIM).

❑ Patch and update all software and OSs (including network devices) to the latest supported versions.

❑ Ensure you are using the latest protection solutions including EDR or Extended Detection and Response (XDR).

❑ Implement next-generation network protection: firewalls, Intrusion and Detection Prevention (IDP), Intrusion Prevention Systems (IPSs), and so on.

❑ Implement network segmentation.

❑ Restrict the use of scripting to approved users.

❑ Secure your Domain Controllers (DCs).

❑ Block access to malicious sites.

❑ Only allow trusted devices on your network.

❑ Disable the use of macros.

❑ Only allow approved software to be used by your users.

❑ Remove local admin permissions.

❑ Enable advanced filtering for email.

❑ Use Sender Policy Framework (SPF), DomainKeys Identified Mail (DKIM), and Domain-based Message Authentication, Reporting, and Conformance (DMARC).

❑ Block the Server Message Block (SMB) outbound protocol and remove outdated versions.

❑ Follow best practices to harden your end-user and infrastructure devices.

❑ Protect your cloud environment with best practices, especially public file shares.

❑ Review your remote strategy and ensure outside connections into your environment are secure. If Remote Desktop Protocol (RDP) is needed, ensure best practices are deployed.

❑ For backups, maintain an offline backup or air gap, encrypt all backups, and validate recovery by testing regularly.

❑ Implement and focus attention on a well-defined Vulnerability Management Program (VMP).

❑ Implement a good cybersecurity and awareness program. Train users not to click on links or open attachments unless they are confident that they are legitimate.

❑ Build a mature Vendor Risk Management (VRM) program.

❑ Conduct annual pen tests and frequent threats hunts.

Figure 11.19: Best practices for protecting against ransomware

The next checklist shows the best practices for responding against ransomware:

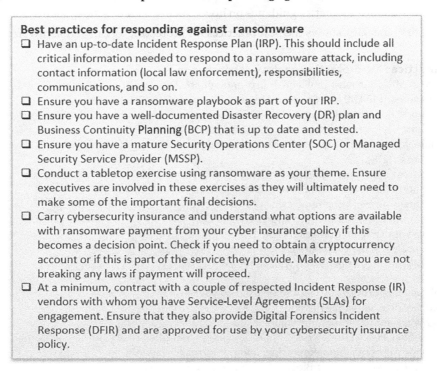

Best practices for responding against ransomware
❑ Have an up-to-date Incident Response Plan (IRP). This should include all critical information needed to respond to a ransomware attack, including contact information (local law enforcement), responsibilities, communications, and so on.
❑ Ensure you have a ransomware playbook as part of your IRP.
❑ Ensure you have a well-documented Disaster Recovery (DR) plan and Business Continuity Planning (BCP) that is up to date and tested.
❑ Ensure you have a mature Security Operations Center (SOC) or Managed Security Service Provider (MSSP).
❑ Conduct a tabletop exercise using ransomware as your theme. Ensure executives are involved in these exercises as they will ultimately need to make some of the important final decisions.
❑ Carry cybersecurity insurance and understand what options are available with ransomware payment from your cyber insurance policy if this becomes a decision point. Check if you need to obtain a cryptocurrency account or if this is part of the service they provide. Make sure you are not breaking any laws if payment will proceed.
❑ At a minimum, contract with a couple of respected Incident Response (IR) vendors with whom you have Service-Level Agreements (SLAs) for engagement. Ensure that they also provide Digital Forensics Incident Response (DFIR) and are approved for use by your cybersecurity insurance policy.

Figure 11.20: Best practices for responding against ransomware

Don't underestimate the impact of a ransomware event. You may be thinking that most of these items are obvious, but implementing these best practices will help both reduce the risk of a ransomware event happening and, if one does, allow you to recover much quicker than not being prepared.

Other

In addition to the proactive services we have just covered, there are many other proactive services that you can implement into your cybersecurity program to reduce risk. As already mentioned, most we cover in this book pertains to proactive services to some extent, including the vulnerability risk management program, the user awareness, training, and testing program, the vendor risk management program, and so on.

Ensure that you run ongoing vulnerability scanning to identify vulnerabilities as quickly as possible, including **External Attack Surface Management (EASM)**, and execute frequent configuration reviews along with assessment and baseline reviews of your infrastructure, including your cloud services. Don't forget to run audits within your environment and, more importantly, bring in a third party to audit you to ensure there is no conflict of interest. Other services include coaching-type services that provide you with the needed guidance as you navigate a major cybersecurity incident. Check with your cybersecurity insurance provider, as they may provide this. Look into breach and attack simulation services that provide continuous and automated testing to identify vulnerabilities within your environment.

Make sure you work closely with your cybersecurity vendors to understand any proactive services they have available as part of their portfolio. It will be important that you continue to review and plan to execute proactive services on an ongoing basis.

Summary

We covered quite a lot relating to proactive services throughout this chapter. Hopefully, you now realize the importance and need to ensure that this function is an active part of your cybersecurity program. These types of services can significantly help reduce risk in your organization, as you will proactively identify vulnerabilities before a threat actor can exploit them. They will also allow you to be better prepared as you complete theoretical exercises, helping to coach and guide you through the process of a major cybersecurity incident. Ensuring that you have a well-documented CIRP that provides all guidance will help speed up recovery and get the business back to a normal operational state, which can have the potential for significant savings. There is an overwhelming number of proactive services available, so it is important that you understand them and prioritize which ones will provide the most value to your organization, reducing risk. You won't be able to execute them all at once, so build a plan to work through each of the proactive services.

The first section covered the need for proactive services, along with statistics to support their execution within your organization. We also provided insight into many of the proactive services you should be considering. Then, we introduced cybersecurity testing and provided details about what is involved. We also reviewed different types of cybersecurity testing before going into detail on penetration testing. As part of penetration testing, we covered details on how to execute a penetration test, the rules of engagement, and reviewing the findings. We then covered application testing before finishing the section with insight into other types of testing activities.

In the section that followed, we covered building an incident response plan and everything that should be included in a CIRP at a minimum. This included the purpose and scope of the document, roles and responsibilities, communications, incident response and recovery, lessons learned, and everything that should be considered within the appendix. We then finished the section with a review of playbooks as part of your incident response planning. Then, we reviewed tabletop exercises and what exactly they are. This led to planning a tabletop exercise before going into details on how to execute a tabletop exercise. We then finished off the section by discussing the importance of the final report and ensuring that remediation of the findings within a report is complete. In the final section, we covered other proactive services, which included threat briefs, threat hunts, incident response training, disclosure programs, ransomware best practices, and some other proactive service considerations.

In the next chapter, we will cover **Operations Technology (OT)** and the **Internet of Things (IoT)**. We will begin the chapter with a deeper dive into each topic to gain a better understanding of what they are and where they are used. Then, we will review why securing these devices is so important, which will cover some statistics to show the current challenges with this technology. Because these technologies are considerably different from IT, we will cover the need for a dedicated program to ensure improved cybersecurity with these technologies. This will lead us to the need to protect these environments and what is involved, with them being unique. Finally, we will finish the chapter with an overview of how to respond to cybersecurity incidents with OT and IoT.

Join our community on Discord!

Read this book alongside other users, Cybersecurity experts, and the author himself. Ask questions, provide solutions to other readers, chat with the author via Ask Me Anything sessions, and much more.

Scan the QR code or visit the link to join the community.

`https://packt.link/SecNet`

12

Operational Technology and the Internet of Things

Operational Technology (OT) and the **Internet of Things (IoT)** are not necessarily part of everyone's portfolios. For the most part, specific industries will primarily leverage these types of technologies, such as manufacturing, power plants, water treatment facilities, and telecommunication companies. IoT will be an exception to an extent as many companies will most likely have some form of IoT within their organization today. However, I would expect it to be quite minimal compared to those companies that primarily depend on IoT technology. Whether you must manage any of these technologies or not, it is important you understand the principles and what is involved in securing these technologies because, you never know, you may find these technologies in your portfolio at some point. Although the same principles we have covered throughout the book will apply to securing these technologies, they do pose different and unique challenges that will need to be addressed as part of the program. It will be critical that both OT and IoT receive the same level of scrutiny and control as the rest of your cybersecurity program.

We will begin the chapter by providing an overview of OT and IoT, describing what each of them is, and ensuring you understand the distinctions between these technologies. Many of you may already be familiar with these technologies, but everyone must know what they are, what purpose they serve, and the difference between them, including how they differ from IT. We will also look at where each of these technologies are used. Next, we will review why protecting these technologies is so important, along with the differences in protecting these technologies versus traditional IT. Like other functions, we will also look at statistics as they relate to these types of technologies to demonstrate the challenges we face with threat actors exploiting these technologies.

Then, we will go into detail as to why a dedicated program is needed for these types of technologies. Because of the uniqueness and the purpose they serve, protecting these technologies requires different skill sets to your everyday IT protections. We will then investigate how you can best protect these technologies as part of your cybersecurity program. In short, the same principles that apply to your broader cybersecurity program will apply to your OT and IoT technologies. To finish off the chapter, we will review incident response as it relates to these technologies. The same processes and concepts will apply when responding to a compromise to these technologies. But it's important to ensure the response plans are built to account for the differences with these technologies as they will require different skill sets to work through a cybersecurity incident. The following will be covered in this chapter:

- What are OT and IoT?
- Why securing this technology is so important
- A dedicated program
- Protecting OT and IoT environments
- Responding to OT and IoT cybersecurity incidents

What Are OT and IoT?

First, let's take a high-level look at all the sub-functions that should be addressed as part of OT & IoT. The following image captures much of what the OT & IoT function entails..

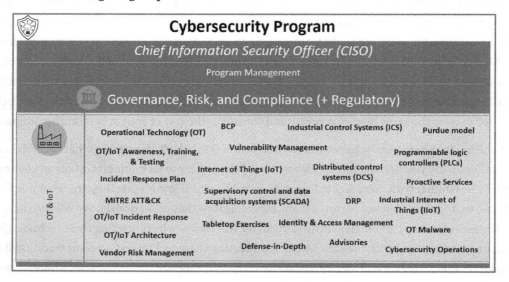

Figure 12.1: Sub-functions of the OT and IoT functions

As we have gone through the book, we have focused on everything related to **Information Technology (IT)** from a cybersecurity perspective. As a reminder, IT is everything used to process digital data within the organization. This is all the technology you are familiar with today as you complete your everyday tasks at work. Some examples of IT include:

- Laptops, tablets, and PCs
- Mobile phones and corporate phone systems
- Cloud systems and infrastructure such as servers
- Physical data centers and equipment
- Networks such as firewalls, switches, and Wi-Fi
- Web applications and business software
- Databases, data warehouses, and data lakes

OT and IoT differ significantly to IT and pose a complete set of new challenges as it relates to management and operations, including cybersecurity requirements. As we navigate through this chapter, we'll be providing a high-level comparison of what is all-inclusive in an OT and IoT program. Each of these technologies may already be part of the IT program or they may already be run within their own programs, depending on the size of your organization and the industry you are in. Either way, the content within this chapter will apply to both OT and IoT environments and, as a leader, it will be important you understand these environments and the requirements needed to secure them.

OT

OT comprises both the hardware and software used to manage, secure, and control technology within industrial systems and critical infrastructure. Industrial systems that may not necessarily be classified as critical infrastructure include manufacturing plants of consumer goods, electronics, vehicles, and food and beverages. On the other hand, some examples of where you will find OT within critical infrastructure include:

- Chemical sector
- Commercial facilities sector
- Critical manufacturing sector
- Dams sector
- Defense industrial base sector
- Emergency services sector

- Energy sector

- Food and agriculture sector

- Healthcare and public healthcare sector

- Nuclear reactors, materials, and waste sector

- Transportation systems sector

- Water and wastewater systems

Source: https://www.cisa.gov/topics/critical-infrastructure-security-and-resilience/
critical-infrastructure-sectors

For the most part, these are essential services that support our everyday lives, and even a minor disruption could be catastrophic. We need to ensure these critical systems used to support everyday life are highly protected.

As we navigate throughout the chapter, we will be referencing the NIST *Guide to Operational Technology (OT) Security: NIST Publishes SP 800-82, Revision 3*, which is an extremely comprehensive guide to OT and everything that should be considered to secure these environments: https://csrc.nist.gov/pubs/sp/800/82/r3/final.

OT comprises several components, including **Industrial Control Systems (ICSs)**, building automation systems, transportation systems, physical access control systems, and physical environment monitoring and measurement systems. One of the major components of OT from those listed above is ICS. Within the OT domain, ICS comprises hardware and software that manages and monitors control systems and processes within an industrial environment. Some of the more common control systems within ICS include:

- **Supervisory Control and Data Acquisition (SCADA) systems**: SCADA combines software and hardware that provides the capability to control, monitor, and analyze industrial processes through the collection of data from industrial devices like pumps, valves, sensors, motors, etc.

- **Programmable Logic Controllers (PLCs)**: A PLC is a control system used for automation in an industrial environment. Examples of where PLCs are used include traffic lights, elevators, production lines, and heating systems.

- **Distributed Control Systems (DCSs)**: A DCS is a computerized system that centralizes the controls and automation of plant processes for improved productivity and efficiency.

To provide a clearer picture, the following is how each is encompassed within the broader OT architecture:

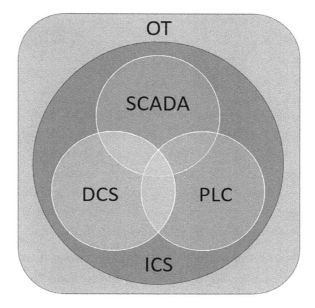

Figure 12.2: Components found within OT architecture

There are a lot more components involved with the OT architecture; those referenced are just some of the core components. If you have any involvement with OT or if you have any OT within your environment, make sure you are familiar with the architecture and all the components that are involved within an OT environment.

IoT

IoT comprises any network-connected device with sensors that exchange data over the Internet (or a public or private network) with other IoT devices and/or systems. IoT has grown exponentially in recent years and continues to grow, as devices are now being built for everything imaginable. Some of the many IoT devices available include:

- Wearable technology like smartwatches
- Connected appliances
- Smart cars
- Smart security systems
- Lightbulbs

- Health monitors

- Household appliances

- Smart entertainment systems like Alexa

- Landscaping equipment

- Automobiles

- Drones

There's a very high possibility you have interacted with, or are currently using, some type of IoT device within your personal lives today. They are everywhere and the reality is that anything electronic is IoT-enabled by default. As this growth continues, so do the cybersecurity concerns as we continue to become more dependent on it every day. With these devices, usability is typically the primary focus with a lack of cybersecurity standardization. Because of this, we must understand the lack of standardization with cybersecurity controls, and ensure IoT devices receive the same scrutiny as any other device within the cybersecurity program.

Another term to be familiar with is the **Industrial Internet of Things (IIoT)**. In some aspects, this is covered under the broader OT umbrella, but since we have IoT specified separately, it's important to note that IIoT does differ from IoT in that it focuses primarily on the industrial aspect of IoT. Some examples of where you will find IIoT devices being used include:

- Sensors (temperature, pressure, etc.)

- Industrial robots

- Smart meters

- Warehouse automation

- Vehicle tracking

- Power plant automation

- Machinery condition

- Autonomous equipment

As we have reviewed each of these technologies, it is important to acknowledge that they are all unique in their own ways, but they all need the same level of attention as they relate to cybersecurity. Although they are unique and each has separate management platforms, we are seeing some convergence between these technologies where they are becoming integrated with one another, such as the use of IT within the OT space.

To be more effective with cybersecurity, we need to see more convergence between all three of these technologies where we can follow one standard, the ability to centrally manage these technologies within one management platform, and for protocols to become standardized between the technologies. We are a long way off but the sooner a foundation is built to support this, the quicker we can become more efficient with the management of these technologies. In the current state, the following visual presents a high-level view of the current convergence between each of these technologies:

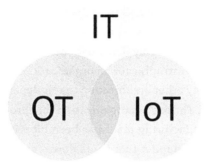

Figure 12.3: IT, OT, and IoT convergence

Unfortunately, full convergence is not something we can expect anytime soon with IT, OT, and IoT, which means we are currently challenged with the need to manage each of these technologies independently including cybersecurity. As mentioned, not every company will need to manage OT and IoT, which is why they are not included as one of the core functions. Another reason for the need to cover this section separately is the requirement for a different set of skills to efficiently manage these technologies. As a cybersecurity leader, it is important you are familiar with these technologies to ensure they are receiving the right attention, and the correct controls are put in place.

Now that we have reviewed what OT and IoT are, let us understand why it is important to secure them.

Why Securing This Technology Is So Important

Both OT and IoT technologies serve as and support many critical infrastructure operations in our modern world today. With OT, our everyday lives heavily depend on infrastructure that is critical to our current living ways. For example, the power plants that serve our electricity heavily rely upon OT.

Imagine a major cybersecurity attack against our power plants that prevents power from being produced for a prolonged amount of time. Think of the water treatment facilities that supply water to our homes; what if a threat actor infiltrated the environment and modified the chemicals that treat the water? What about an attack on a company that manages the transportation of an energy source highly depended on by millions of people, like fuel? The last incident will bring familiarity with the Colonial Pipeline, an incident that caused mass panic on the east coast of the US and made headlines throughout the world: `https://www.cisa.gov/news-events/news/attack-colonial-pipeline-what-weve-learned-what-weve-done-over-past-two-years`.

Let's shift our focus to the business side and the dependency on OT with the manufacturing of products. For the most part, everything we purchase or consume goes through some form of manufacturing process that will be highly dependent on OT to produce the final product. Any impact on this process can be catastrophic for an organization – not only the organization that owns the manufacturing but any other organization that is dependent on the manufacturing of materials or a specific product. Think of an organization that produces food but can't make the final product because the manufacturing plant has been hit with a major cybersecurity event. What about an attack on a chip supplier that shuts down its production lines causing a mass delay and a potential shortage of chips? One extremely unfortunate example is that of Clorox, whose production was majorly impacted by a cybersecurity incident causing a shortage in their products, which, in turn, impacted both supply and sales. It has been estimated that this single incident cost more than $49 million for the company: `https://www.securityweek.com/clorox-says-cyberattack-costs-exceed-49-million/`.

When looking at IoT, there are billions and billions of devices throughout the world that are now internet-connected. Once connected to the internet, they are susceptible to compromise, and many of these devices are being used for a magnitude of both personal and business purposes. Think of everything in your house that now has the capability to connect to the internet, specifically your security system and internet-enabled cameras. What if threat actors gain access to your system and/or cameras to gain access to your home or view live footage of your cameras, specifically, any internal facing cameras, which can be extremely concerning. What about the usage of IoT within the healthcare industry and the ability of threat actors to compromise and impact health-related devices that are crucial for someone's health? The outcome can be catastrophic. Another example is the auto industry; there are millions and millions of smart cars on the road today and they only continue to grow. Any compromise to these vehicles could potentially diminish their safe operations and result in significant damage, injury, or loss of life.

For example, in 2015, security researchers Charlie Miller and Chris Valasek were able to remotely exploit a Jeep Cherokee driving on the highway and disable the car's engine and brake systems: `https://www.wired.com/2015/07/hackers-remotely-kill-jeep-highway/`.

The reality is all the examples above are possible and there is nothing unrealistic with how advanced threat actors have become. We must take the security of OT and IoT environments very seriously as the compromise of these environments and devices can be life-threatening in some instances.

Now, let's take a look at some statistics to support why securing OT and IoT is so important.

OT Statistics

Looking into some statistics, the OT landscape paints a similar picture to the cybersecurity challenges within IT. In short, incidents are becoming more prevalent, and the impact can be substantial. Referencing a report released by Waterfall Security Solutions and ICS STRIVE, *2024 Threat Report OT Cyberattacks with Physical Consequences* showed the following concerns:

> The report is strict on its inclusion requirements, which require that the attack was deliberate, physical consequences were caused such as production outages, it involved specific industries (manufacturing, building automation, heavy industry, and critical industrial infrastructures, including transportation of people and goods), and the incident was made public.

- Half of all cyberattacks that caused physical consequences impacted manufacturing.
- It was noted that, in 2023, 68 attacks met the inclusion criteria, which impacted operations at 500+ sites, leading to a 19% increase from the previous year.
- Between 2019 and 2023, attacks have almost doubled on an annual basis.
- Incidents have cost some victims tens of millions to hundreds of millions of dollars.
- In approximately a quarter of those attacks reported publicly, it was stated that OT systems were directly impaired or manipulated by threat actors.
- Direct and deliberate impairment of OT systems and physical operations continues to occur from hacktivists and nation-states.

Source: `https://waterfall-security.com/ot-insights-center/ot-cybersecurity-insights-center/2024-threat-report-ot-cyberattacks-with-physical-consequences/`

From the same report, the visual below provides an extremely concerning trend where it is predicted to only get worse in the years to come. As shown, minimal incidents were reported in an almost ten-year period with a sudden spike on a year-to-year basis with OT-related cybersecurity incidents.

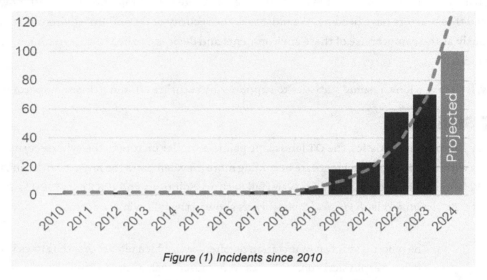

Figure (1) Incidents since 2010

Figure 12.4: OT incidents history from the Threat Report OT Cyberattacks with Physical Consequences

A great resource in addition to the report is the ICS STRIVE OT incident database (the data source of OT incidents referenced above), which can be found here: `https://icsstrive.com/`. For example, if you search for Maersk in the database, you will be provided with the details from the 2017 ransomware attack.

Maersk Ransomware Attack

September 19, 2020

A NotPetya attack disrupted operations for two weeks, blocking access to systems the company relied on to operate shipping terminals. The incident temporarily shut down the Port of Los Angeles' largest cargo terminal. The company lost $300 million in business disruption and equipment damage. Maersk had to undertake an almost complete infrastructure overhaul. They reinstalled 4,000 servers, 45,000 PCs and 2,500 applications over the course of ten days, a process that would normally have taken six months to implement.

Incident Date	Location	Estimated Cost
June 27, 2017	Global United States	$300M in business disruption

Victims
- Maersk

Type of Malware
- NotPetya

Threat Source
- Russia

References
- Ransomware: The key lesson Maersk learned from battling the NotPetya attack
- Cyberattack cost Maersk as much as $300 million and disrupted operations for 2 weeks

Industries
Transportation (Includes Logistics, Shipping, Maritime, Rail, Trucking)

Impacts
IT

Figure 12.5: Maersk incident as reported in the ICS STRIVE OT incident database

Another great report to reference on both OT and ICSs is the Dragos *2023 OT Cybersecurity Year in Review*, Source: https://www.dragos.com/ot-cybersecurity-year-in-review/. Some of the findings in the report for 2023 include:

- Dragos observed ten active threat groups within OT of which three of them were new.
- Network exploitable and perimeter facing was observed in 16% of advisories.
- Errors were found in 31% of the advisories.
- 53% of the advisories reviewed had the potential to cause both loss of view and loss of control.

- Over the last year, an increase of 50% was observed in ransomware attacks against industrial organizations.

- 28% more ransomware groups were tracked impacting ICSs/OT in 2023.

- There were 905 reported ransomware incidents tracked by Dragos, which was an increase of 49.5% from 2022.

- Manufacturing was the most impacted sector by ransomware incidents with 71%.

As you can see from the visual below taken from the Dragos report referenced above, by Dragos, 76% of ransomware incidents were within North America and Europe.

Figure 12.6: Dragos global map of OT ransomware incidents

IoT Statistics

Let's look at some IoT statistics to get a better understanding of the current threat landscape with this technology. Before we delve into some cybersecurity statistics, let's look at how many IoT-connected devices there are today with the predicted growth in the coming years.

IOT Analytics provides a great visual of the current and forecasted IoT device count from their *State of IoT Summer 2024* Market Report. You can view the image below in the article *State of IoT 2024: Number of connected IoT devices growing 13% to 18.8 billion globally* at `https://iot-analytics.com/number-connected-iot-devices/`.

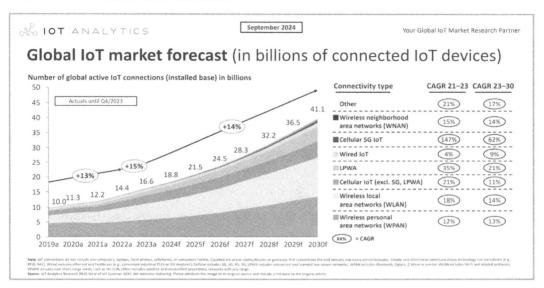

Figure 12.7: IOT Analytics visual of the current and forecasted IoT device count

As stated in the IOT Analytics article referenced above, 16.6 billion IoT endpoints were reported at then end of 2023. It has been predicted for 18.8 billion endpoints at then end of 2024 which equates to a 13% growth prediction.

That equates to over two devices per person on this planet. How do we even begin to secure the scale of this technology?

A report released by SonicWall, *2023 SonicWall Cyber Threat Report*, provided the following IoT statistics:

- SonicWall Capture Labs recorded 112.3 million instances of IoT malware in 2022, which was the first time the number of attacks was over 100 million. This is a staggering increase of 87% from 2021.

- Every industry studied within the report showed an increase in IoT malware. This includes healthcare, government, retail, education, and finance.
- Routers, cameras, firewalls, load balancers, and **Network Attached Storage (NAS)** were the highest targeted IoT devices.

Source: `https://www.sonicwall.com/resources/white-papers/2023-sonicwall-cyber-threat-report/`

Research released by Check Point noted that in the first couple of months of 2023, there was a 43% increase in the average number of weekly attacks targeting IoT devices per organization versus 2022. On average, 54% of organizations suffer from IoT-based cyberattacks every week. The report also noted an increase in attacks within all studied industries as well as an increase throughout all regions across the globe. You can read more on the research here: `https://blog.checkpoint.com/security/the-tipping-point-exploring-the-surge-in-iot-cyberattacks-plaguing-the-education-sector/`.

As observed, the need to secure both OT and IoT is a critical component of a cybersecurity program. The safety of OT equipment and IoT devices can cross the threshold of human safety, which cannot be ignored. There must be strict requirements in place moving forward to ensure the protection of these devices to ensure the safety of human life.

A Dedicated Program

Because of the uniqueness of OT and IoT technologies and the lack of full convergence, a dedicated program will be needed to ensure that these technologies are efficiently managed and secured as best as possible. With these technologies being significantly different from traditional IT, they need skill sets that are trained and specialized in managing them. This requires the need for a dedicated program to ensure they are gaining the attention they need, and that the correct controls are being implemented to reduce risk.

To keep consistency, it's recommended to follow the same structure for the OT and IoT program as we have recommended for the broader cybersecurity program for IT.

This way, you can ensure the program is comprehensive and doesn't contain any gaps.

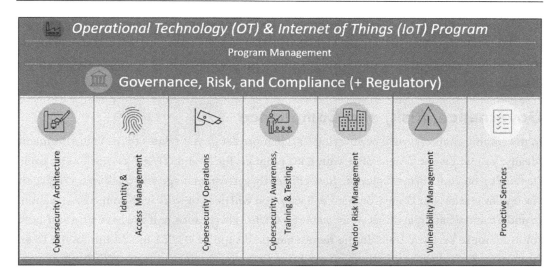

Figure 12.8: OT and IoT program recommendation

As we have journeyed through the book, we have covered all the major functions represented above (**GRC**, which is **Governance, Risk**, and **Compliance**, is coming in the following chapters) as they directly relate to IT. Because both OT and IoT bring a different set of challenges and oversight, you'll need to ensure you spend some time ensuring each of the functions represented above capture the requirements needed to efficiently manage your OT and/or IoT programs. At the same time, synergies will be needed between the cybersecurity program that supports the IT functions and you'll still need to centralize some of the capabilities and data in one place for more efficient governance.

Although there is a need for a dedicated program and different skill sets, it will be important that the OT and/or IOT program is not isolated from the overarching IT program. More and more convergence between IT and OT/IoT continues to happen, and it is only going to converge more in the future. To reduce risk as much as possible, there will need to be excellent collaboration between the groups to ensure IT protocols are secure as they manage and allow access to OT and/or IoT environments. There will need to be an understanding of the architecture between the two environments as they continue to cross into each other's areas. If the IT and OT/IoT teams' function is isolated, risk will only become more of a challenge to manage.

We did touch upon the management of OT and IoT as part of the core functions within the cybersecurity program if there was only a small footprint. This will likely be the case for IoT as it relates to the IT space.

But OT will most likely need the commitment of its own program regardless of how big or small it is. Within OT, you'll mostly find IIoT, which will fall under the same program and should follow the same concepts for management and oversight. Let's review each of the areas above in more detail as they relate to both OT and IoT.

Governance, Risk, and Compliance

In the coming chapters, we'll be covering GRC in more detail as it relates to the IT environment. Ideally, you will want to centralize your GRC program for IT and OT/IoT as much as possible. Depending on your current solution, how efficiently you can manage and track your OT/IoT environment versus your IT environment will vary. You will need to work with your current vendor to understand what capabilities are available for OT/IoT governance, or if you have a custom-built solution, make sure you build in the requirements to support OT/IoT in addition to the IT environment. Regardless of whether you can bring in the governance of your OT/IoT within your IT GRC program or not, you will need to ensure you have efficient GRC in place for your OT/IoT environments. Unfortunately, you aren't going to have a choice with this in the future, and the sooner you begin your journey with GRC for OT/IoT, the better position your program will be in.

When looking at GRC for your OT/IoT environments, like IT, your inventory management of all equipment, software vendors, and so on is critical. You must know everything applicable to your environment and have an accessible inventory of everything, including hardware type, firmware versions, software versions, manufacturer, and more. When advisories are released, you need to be able to quickly identify if you have any vulnerable hardware or software in your environment to remediate. The same applies to an active incident; the quicker you can identify anything impacted, the more efficiently you can respond to an incident.

Also, you mustn't overlook the need for well-defined policies, standards, processes, and procedures for your OT/IoT environments. For those of you with a large footprint, you will want to consider a standalone policy for these environments since they differ from IT and there will be unique language needed. In addition, ensure processes and procedures are well defined for all aspects of the program along with having well-defined standards that must be followed. Your risk profile for OT/IoT will be different from IT so you will need to ensure this is well documented and everyone understands the risk in these environments compared to IT. As we have alluded to several times, the risk with OT/IoT devices can have a significant impact on human safety. You'll also need to ensure a framework for your OT/IoT environments is being considered, including the need for a risk framework to be followed. You also need to ensure a risk register is being used to track your OT/IoT risk. It is highly recommended that all risk including IT is tracked in one place for more efficient reporting and visibility for your executive leadership team and board members.

Finally, the regulatory landscape for OT/IoT environments can be very different depending on where you are located in the world, any government-required regulations (whether national or local), and what type of industry you are in. For example, power plants will have their own set of regulations, water treatment facilities may have a different set of regulations, the medical field will have a different regulation, and so on. It will be important you work closely with your legal teams to ensure you remain in compliance. One more area that will need to be included is the need to execute audits, whether required or not; you will need to ensure you are bringing in outside expertise to review and validate that the controls you have in place are effective.

As a reminder, the GRC function is no easy feat, especially as you need to ensure your OT/IoT environments are efficiently being governed. Don't overlook your GRC requirements for OT/IoT as it will serve as the foundation to enable greater success with your program.

Cybersecurity Architecture

The architecture of an OT/IoT environment will not be the same as an IT environment, although the foundation and same principles should be applied across all environments. From a program perspective, the same the same **Architecture Review Board** (**ARB**) process covered in *Chapter 5, Cybersecurity Architecture*, should be followed for your OT/IoT architecture. In addition, it will be important you have an OT/IoT architect or representative on the ARB to help build the expected requirements and set the needed standards for these environments. As architecture gets deeper within OT, it will be important there is expertise that understands the requirements for ICS, SCADA, PLC, DCS, and so on as they are very specialized.

From an architecture strategy perspective (we covered ZTA in detail in *Chapter 5, Cybersecurity Architecture*), although zero trust should be considered for your OT/IoT environments, it may not be feasible in all areas because of older non-standard equipment. Because of this, ensure you are focusing on a Defense-in-Depth strategy in addition to **Zero Trust Architecture** (**ZTA**) where applicable. One of the reasons this architecture will be important is that there will most likely be older and potentially non-supported **Operating Systems** (**OSs**) that specific software is reliable on and cannot be updated because vendors lack support for a newer OS. This is not uncommon within OT environments. A couple of excellent resources to reference as you review your current architecture and look to reduce risk within these environments are:

- The CISA *Improving Industrial Control System Cybersecurity with Defense-in-Depth Strategies*, which can be found here https://www.cisa.gov/resources-tools/resources/ics-recommended-practices

- The *OT Cybersecurity Architecture* section in the NIST *Guide to Operational Technology (OT) Security: NIST Publishes SP 800-82, Revision 3,* which can be found here: https://nvlpubs.nist.gov/nistpubs/SpecialPublications/NIST.SP.800-82r3.pdf

As we have touched upon throughout the chapter, the convergence of OT and IT continues to become more prevalent. Traditionally, OT was much easier to secure because of the nature of segmentation and isolation, and these environments did not have any presence with any other networks. Today, they are becoming more interconnected with internal IT networks and expanding beyond the corporate firewall to be accessible via the Internet. This, as we know, significantly changes the dynamics and increases the risk ten-fold. With this, the network architecture of these environments becomes a top priority. Referencing the CISA Improving Industrial Control System Cybersecurity with Defense-in-Depth Strategies recommended practices document, the concept of implementing different zones with different levels to allow for segmentation and isolation between the different levels is critical. As shown below, you can see the level of detail needed to ensure a robust and secure architecture is in place, specifically as it relates to the IT and OT environments with the need for a **Demilitarized Zone (DMZ)**.

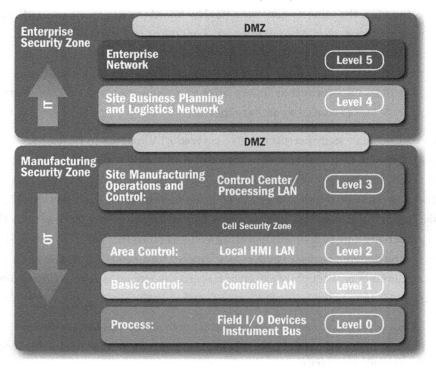

Figure 12.9: CISA Defence-in-Depth zone segmentation ICS architecture

Source: `https://www.cisa.gov/sites/default/files/2023-01/NCCIC_ICS-CERT_Defense_`
`in_Depth_2016_S508C.pdf`

 The NIST *Guide to Operational Technology (OT) Security: NIST Publishes SP 800-82, Revision 3* referenced above provides a few recognized models for reference for your OT network segmentation and isolation architecture: Purdue model, ISA-95 levels, and the three-tier IIoT system architecture.

In addition to the network architecture, make sure you have a clear understanding of all other architecture areas including identity, endpoint, data, application, infrastructure, and collaboration as they relate to OT/IoT. In addition, as mentioned in in *Chapter 5, Cybersecurity Architecture*, any engineering will need to be coordinated with the architecture function and you'll need to ensure you have the right skill sets with engineering requirements.

Identity and Access Management

The OT/IoT identity and access management landscape will differ greatly from your IT environment. Ideally, you will want to apply the same principles and recommendations to your OT/IoT environments as your IT environment, but you may be challenged with different requirements around the need for local accounts on OT/IoT equipment, older equipment and applications that are not capable of MFA, isolated systems, and others that aren't able to integrate with a centralized identity and access management system. You'll need to consider how users efficiently manage their access to systems without necessarily having direct access to password managers when accessing some systems, therefore, making it more challenging to create and remember longer more secure passwords.

Another consideration is the efficient governance and provisioning of access management for these environments. How do you ensure the auditing of access for each of the users accessing and managing the systems without using shared accounts? How do you efficiently provision all the access needed for resources and ensure all access is removed for any users exiting the organization? There may be a need for manual work with specific systems to manage access in which well-documented processes and procedures will need to be implemented with all required approvals occurring before any access is permitted. Identity and access management can quickly get very complex in this area as users may need many accounts for all the different systems, quickly making it very difficult to efficiently manage. Supporting these environments can quickly become challenging as the need to modify user passwords and support users having challenges logging in to systems will not be uncommon.

As you manage access to the OT and/or IoT environments, ensure you have implemented a privileged access workstation or jump server or workstation to prevent direct access to the OT and/or IoT networks. In addition, ensure access to any workstation used to access the OT and/or IoT environment is locked down and hardened.

To finish off, it will be important to ensure the identity and access management programs between IT and OT/IoT are collaborating closely if they are not integrated through a unified team. The more IT integrates into OT/IoT environments, access for users will be needed for the IT environment in addition to the OT/IoT environments. Because of this, it will be important that those managing the OT/IoT environments are aware of the different requirements for access to each of the environments.

Cybersecurity Operations

As in *Chapter 7, Cybersecurity Operations*, you will want to follow the same principles to your OT/IoT environment as your IT. Specifically, your threat detection is going to be different for your OT/IoT environments from your IT environment. In the following section, we go into more detail on the protection of OT/IoT with insight into some OT-specific malware along with MITRE ATT&ACK for ICS, and ICS advisories from CISA. All this is relevant specifically to OT. Incident response will also differ for OT/IoT and it will be important that processes are well-documented to handle any investigations and responses within these environments. You will need the correct resources within your SOC, or at a minimum, sufficient training provided to your cybersecurity analysts to be able to efficiently investigate alerts of concern within an OT/IoT environment.

One important aspect to define is the structure of your SOC as it relates to the OT/IoT environment. I wouldn't recommend implementing a dedicated SOC just for your OT/IoT environments unless there is a specific requirement. Assuming you already have a SOC setup for the IT environment and a SIEM system is in place, it would make sense to leverage the capabilities of the current SOC and build the capabilities to ingest the OT/IoT environment data sources into the current SOC for analysis and review. To do this, it will be important that you are able to specifically track the OT/IoT incidents separately and that the right resources are working through those specific incidents with priority.

You will also need to ensure you have well-written **Standard Operating Procedures (SOPs)** that can be followed for all potential OT/IoT-specific incidents. In addition, **Digital Forensics and Incident Response (DFIR)** will require different skill sets in the IT environment. You will need to ensure you have access to these skill sets whether in-house or externally through a vendor, which you will need to have on contract regardless.

One final area to mention, which was covered in *Chapter 7*, is how AI can benefit us from a protection perspective but impact us from a threat perspective. Make sure you are looking at opportunities to use AI to provide better protection and response within your OT/IoT environments. Discuss AI with your current vendors to see what they are doing with AI and how they are looking to embed AI within any of the current services being provided to you from them.

Cybersecurity Awareness, Training, and Testing

For your OT/IoT awareness, training, and testing program, you are going to need to ensure you are focusing on content related to OT/IoT technology. As covered in *Chapter 9, Cybersecurity Awareness, Training, and Testing*, we provided an in-depth outline of how to execute awareness, training, and testing for your users as they relate to IT. For OT/IoT, I would include this as a category within the broader program that provides data to those who manage and operate within these environments. These users will most likely have standard user accounts, which means they should be required to take standard training as it relates to IT in addition to the focused and specialized training as it relates to OT/IoT. You will need to check with your current awareness, training, and testing platform to see if they have anything specific on OT/IoT security as well as any options to test these users regarding OT/ IoT. You may need to be creative as most content will be built for an IT environment. Regardless, you will need to ensure this is not overlooked as part of your OT/IoT environments.

As you get more specific with your administrators, engineers, and/or operators for the OT/IoT, you'll want to ensure they have the latest training available and accessible. This will require investment on the organization's behalf to ensure these users are receiving the correct training so that risk is reduced as much as possible for OT/IoT environments. A great example of training for OT is available from SANS: `https://www.sans.org/job-roles-roadmap/industrial-control-systems/`.

The following poster provides the courses and tracks available from SANS.

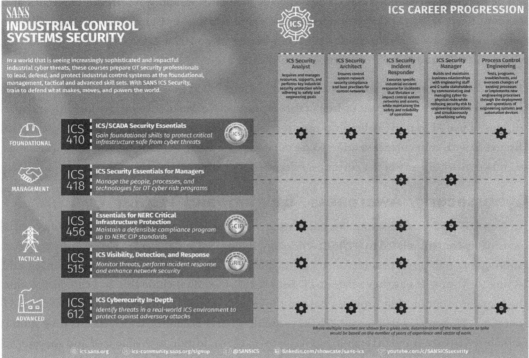

Figure 12.10: SANS ICS career training poster

Source: Click on **Download ICS Job Roles Poster** in the SANS link above.

Make sure you don't overlook this component for your users. The better educated they are in securing OT/IoT environments, the more secure your environment will become.

Vendor Risk Management

The overall vendor risk management program for OT/IoT vendors shouldn't be any different from the broader vendor risk management program covered in *Chapter 10, Vendor Risk Management*. Any OT/IoT vendors should receive the same level of scrutiny as any other vendor, and most will probably be considered important vendors as classified within *Chapter 10*. This will require the most thorough review ensuring any hardware and software being used within the environment have the correct controls in place. In the best practices section later in the chapter, an emphasis is placed on supply chain security.

Similar to what we covered in *Chapter 10*, it will be critical to ensure no hardware or software being used within your OT/IoT environments has been compromised by threat actors somewhere within the supply chain. In addition, ensure all contracts include cybersecurity language to hold the vendors accountable to maintain a high level of standards with cybersecurity and to ensure they are providing full transparency related to any cybersecurity event they are experiencing or with any vulnerabilities identified within their products and/or services.

Vulnerability Management

In addition to your vulnerability management program for your IT environment (covered in *Chapter 8, Vulnerability Management*) will be the need to ensure that OT/IoT-specific vulnerabilities are reviewed and addressed. However, it will be critical that all IT assets being used within an OT/IoT environment are fully assessed and in scope for the IT vulnerability management program. More so because some of these assets can possibly provide an entry point into the OT/IoT environments. For OT-specific vulnerabilities, you'll need to ensure you sign up for the CISA *Industrial Control Systems (ICS) Cybersecurity Advisories*, which will provide the needed details on all published vulnerabilities in the open. We will cover ICS advisories in more detail later in the chapter. You'll also want to ensure all vendors within your OT/IoT portfolio are required to notify you of any identified vulnerabilities immediately so you can address them right away. Make sure you have well-defined processes and procedures to remediate vulnerabilities in a timely manner once identified. And make sure your vendors have up-to-date contact information to notify you of any identified vulnerabilities.

Proactive Services

I believe by now you can see the bigger picture as to why OT/IoT requires its own dedicated function to ensure best practices and the required controls are being put in place. For proactive services, the same applies as covered in all other sections: ensuring the principles covered in *Chapter 11, Proactive Services*, are applied to OT/IoT. This includes executing the appropriate testing against your OT/IoT environments, especially penetration testing, to look for any vulnerabilities that can be exploited by threat actors. Tabletops and incident response plans will need to be customized specifically for OT/IoT environments, which we cover in more detail within the *Responding to OT and IoT Cybersecurity Incidents* section later in the chapter. Some other proactive services to consider include incident response training for OT/IoT environments, threat briefs relevant to the OT/IoT landscape, and executing configuration and/or baseline reviews of these environments.

Now that you have a better idea of some of the specific areas that you'll need to focus on within the OT and IoT programs, let's look at some specific areas that will need focus to reduce risk as much as possible within these environments.

Protecting OT and IoT Environments

It's important to understand that protecting your OT and IoT environments will differ from protecting your IT environments. Over the years, there has been a significant focus on protecting IT environments, and security products have specifically been built for IT. On the other hand, there hasn't been the same level of focus on OT and IoT over the years. However, this is changing quickly because of the potential impact that can be caused by these environments being compromised along with the reality that it is happening today. The good news is that vendors are realizing the need for increased protection and investments are being made to bring better security to the market for these environments. For example, Dragos has a platform for OT and ICS environments: `https://www.dragos.com/cybersecurity-platform/` and Microsoft has capabilities for IoT, OT, and ICS environments: `https://www.microsoft.com/en-us/security/business/endpoint-security/microsoft-defender-iot`. Make sure you are doing everything possible to reduce the risk of compromise with these environments including what we cover in this section.

OT Malware

In *Chapter 1, Current State*, we touched upon malware related to IT. Although malware within IT can have an impact on OT from some of the convergence of IT systems providing remote access and the management of systems, OT does have its own type of malware that have been created to directly impact these systems. Some examples of OT-specific malware include:

- Stuxnet
- Havex
- BlackEnergy2
- CrashOverride
- TRITON/TRISIS

Some of the malware from an IT perspective that can have an impact on OT include:

- EKANS
- LockerGoga
- BlackEnergy3

You can learn more about what OT malware is, including details on the malware referenced above, here: `https://www.ncsc.gov.uk/blog-post/what-is-ot-malware`.

MITRE ATT&CK

In *Chapter 7, Cybersecurity Operations*, you may recall the reference to MITRE ATT&CK, which represents real-world scenarios used by threat actors. As a reminder, the model allows cybersecurity experts to better understand how threat actors are infiltrating environments so they can assess and identify any gaps within their own environments. The specific model we referenced was the matrix for enterprise, which primarily references the IT environment. In 2020, MITRE released a matrix for ICS to represent the tactics and techniques used by threat actors targeting ICS because of the notable difference from IT. As of August 2024, the MITRE ATT&CK Matrix for ICS notates 12 unique tactics containing multiple techniques within each: `https://attack.mitre.org/matrices/ics/`.

ICS Matrix

Below are the tactics and techniques representing the MITRE ATT&CK® Matrix for ICS.

View on the ATT&CK® Navigator ↗

Version Permalink

Initial Access	Execution	Persistence	Privilege Escalation	Evasion	Discovery	Lateral Movement	Collection	Command and Control	Inhibit Response Function	Impair Process Control	Impact
12 techniques	9 techniques	6 techniques	2 techniques	6 techniques	5 techniques	7 techniques	11 techniques	3 techniques	14 techniques	5 techniques	12 techniques
Drive-by Compromise	Change Operating Mode	Hardcoded Credentials	Exploitation for Privilege Escalation	Change Operating Mode	Network Connection Enumeration	Default Credentials	Adversary-in-the-Middle	Commonly Used Port	Activate Firmware Update Mode	Brute Force I/O	Damage to Property
Exploit Public-Facing Application	Command-Line Interface	Modify Program	Hooking	Exploitation for Evasion	Network Sniffing	Exploitation of Remote Services	Automated Collection	Connection Proxy	Alarm Suppression	Modify Parameter	Denial of Control
Exploitation of Remote Services	Execution through API	Module Firmware		Indicator Removal on Host	Remote System Discovery	Hardcoded Credentials	Data from Information Repositories	Standard Application Layer Protocol	Block Command Message	Module Firmware	Denial of View
External Remote Services	Graphical User Interface	Project File Infection		Masquerading	Remote System Information Discovery	Lateral Tool Transfer	Data from Local System		Block Reporting Message	Spoof Reporting Message	Loss of Availability
Internet Accessible Device	Hooking	System Firmware		Rootkit	Wireless Sniffing	Program Download	Detect Operating Mode		Block Serial COM	Unauthorized Command Message	Loss of Control
Remote Services	Modify Controller Tasking	Valid Accounts		Spoof Reporting Message		Remote Services	I/O Image		Change Credential		Loss of Productivity and Revenue
Replication Through Removable Media	Native API					Valid Accounts	Monitor Process State		Data Destruction		Loss of Protection
Rogue Master	Scripting						Point & Tag Identification		Denial of Service		Loss of Safety
Spearphishing Attachment	User Execution						Program Upload		Device Restart/Shutdown		Loss of View
Supply Chain Compromise							Screen Capture		Manipulate I/O Image		Manipulation of Control
Transient Cyber Asset							Wireless Sniffing		Modify Alarm Settings		Manipulation of View
Wireless Compromise									Rootkit		Theft of Operational Information
									Service Stop		
									System Firmware		

Figure 12.11: MITRE ATT&CK Matrix for ICS

Dragos has taken the MITRE ATT&ACK Matrix for ICS model to the next level by adding mappings of techniques used by known threat actors into the matrix. They have also added additional techniques based on observations of these threat actors.

You can find the interactive map here: `https://www.dragos.com/mitre-attack-for-ics/`.

In addition to the tactics and techniques used by threat actors against ICS, MITRE has published a very thorough mitigation list to help prevent the techniques presented from being exploited. As of August 2024, a total of 52 mitigations have been provided for reference: `https://attack.mitre.org/mitigations/ics/`.

Home > Mitigations > ICS

ICS Mitigations

Mitigations represent security concepts and classes of technologies that can be used to prevent a technique or sub-technique from being successfully executed.

Mitigations: 52

ID	Name	Description
M0801	Access Management	Access Management technologies can be used to enforce authorization polices and decisions, especially when existing field devices do not provided sufficient capabilities to support user identification and authentication. These technologies typically utilize an in-line network device or gateway system to prevent access to unauthenticated users, while also integrating with an authentication service to first verify user credentials.
M0936	Account Use Policies	Configure features related to account use like login attempt lockouts, specific login times, etc.
M0915	Active Directory Configuration	Configure Active Directory to prevent use of certain techniques; use security identifier (SID) Filtering, etc.

Figure 12.12: MITRE mitigation list sample

It is highly recommended you make yourself familiar with the MITRE ATT&CK Matrix for ISC. This, in some respect, can serve as a foundation for your broader OT program to ensure there are no gaps with the recommended mitigations that MITRE provides. In addition, many participants and organizations who are actively involved and focused on OT and ICS technologies contributed to the publication of the ICS Matrix, making this an extremely relevant resource for reference.

ICS Advisories

In *Chapter 8, Vulnerability Management*, we reviewed the resources available from CISA. First and foremost, make sure you sign up for the **Known Exploited Vulnerabilities (KEV)** alerts to ensure you are receiving all known exploitable vulnerabilities for review with any ICS equipment or technology within your environment. In addition, you'll need to sign up for the **Industrial Control Systems (ICS)** Cybersecurity Advisories and ICS Medical Advisories alerts.

As a reminder, you can sign up for CISA KEV email alerts through the following link: `https://www.cisa.gov/about/contact-us/subscribe-updates-cisa`. Within the **Subscribe to Email Updates section**, click on **SUBSCRIBE**. Ensure you are subscribed to the ones highlighted at a minimum.

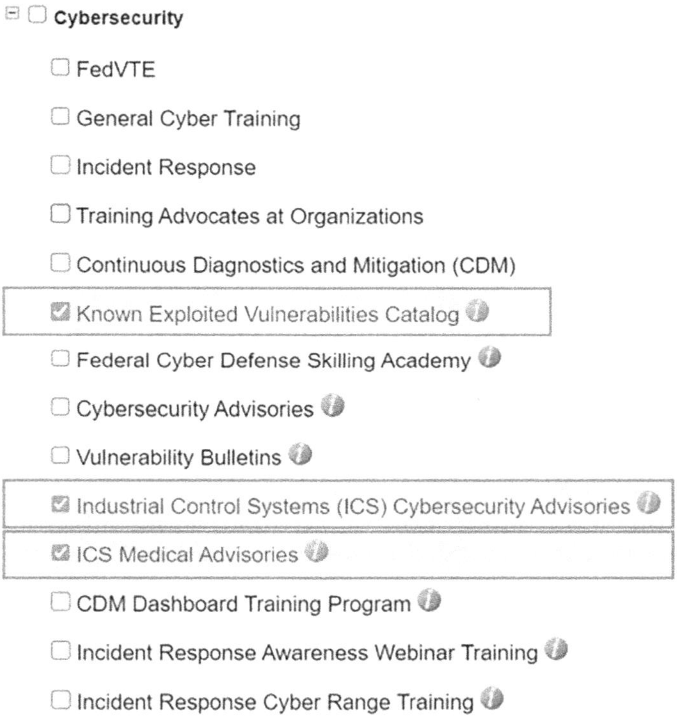

Figure 12.13: Recommended notifications from ICS

Though the ICS Cybersecurity Advisories will provide the primary details you need with any vulnerabilities, you'll want to ensure you are receiving the Known Exploited Vulnerabilities Catalog too. The KEV Catalog will primarily be related to IT but you'll want to be aware of it in the event anything from the IT environment is being used to access any of the OT and IoT environments. If you are not within the healthcare industry or working with medical devices, you won't need to receive alerts for the ICS Medical Advisories.

An example of an ICS cybersecurity advisory will look like the following:

You are subscribed to Industrial Control Systems (ICS) Cybersecurity Advisories for Cybersecurity and Infrastructure Security Agency. This information has recently been updated, and is now available.

CISA Releases One Industrial Control Systems Advisory

02/22/2024 07:00 AM EST

CISA released one Industrial Control Systems (ICS) advisory on February 22, 2024. These advisories provide timely information about current security issues, vulnerabilities, and exploits surrounding ICS.

- ICSA-24-053-01 Delta Electronics CNCSoft-B DOPSoft

CISA encourages users and administrators to review the newly released ICS advisory for technical details and mitigations.

This product is provided subject to this Notification and this Privacy & Use policy.

Having trouble viewing this message? View it as a webpage.

You are subscribed to updates from the Cybersecurity and Infrastructure Security Agency (CISA)
Manage Subscriptions | Privacy Policy | Help

Figure 12.14: CIS ICS advisory example

To access the advisory, you can click the link from within the email to navigate directly to the advisory received in the notification. Or you can navigate to https://www.cisa.gov/news-events/cybersecurity-advisories and, within **Advisory Type** on the left, select **ICS Advisory** to access all current ICS advisories.

Each advisory will contain the following:

- Advisory name, release date, and alert code
- Executive summary
- Risk evaluation
- Technical details: affected products, vulnerability overview, background, and researcher
- Mitigations
- Update history
- Vendor

It will be important to determine where these alerts will go. Similar to the broader cybersecurity program, these should be delivered to the SOC and the vulnerability management team for further review and next steps.

Best Practices

There are several best practice recommendations on the web for reference, but I want to focus on a couple of the more comprehensive ones available from NIST and CISA. Make sure you review both of them in detail to ensure all best practices are covered:

- NIST SP 800-82r3 Guide to **Operational Technology (OT)** Security, bullet points beginning on page 3 in the section named *In a typical OT system, a defense-in-depth strategy includes*: `https://csrc.nist.gov/pubs/sp/800/82/r3/final`
- CISA Cybersecurity Best Practices for Industrial Control Systems: `https://www.cisa.gov/resources-tools/resources/cybersecurity-best-practices-industrial-control-systems`

Focusing on the CISA best practices, ten specific areas are covered to provide a proactive approach to your OT security:

- **Risk management and cybersecurity governance**: Here, it is recommended to be aware of all threats that can impact your organization, ensure a full and complete inventory is maintained, ensure all applicable policies and procedures are in place, and ensure incident response procedures are in place that cover both IT and OT.

- **Physical security**: Ensure any field electronics are locked down and alerts are in place for any device manipulation, ensure your physical access only allows for authorized personnel, both logical and physical access to any OT equipment leverages MFA, and there are guards in place and barriers where needed.

- **ICS network architecture**: Ensure your networks are segmented, implement a multi-layer network model with the most critical communications within the most secure layer, implement one-way communication diodes where applicable, implement a DMZ for additional security, and ensure secure network protocols are being utilized.

- **ICS network perimeter security**: Ensure firewalls are in place to efficiently manage traffic between the ICS and IT networks, leverage geo-blocking where applicable, ensure all remote access is hardened and locked down, implement jump servers between your ICS network security zones, do not allow any remote persistent connections to control the network, and implement monitoring on all remote connections.

- **Host security**: Ensure patching is occurring as quickly as possible and a mature vulnerability management program is in place, ensure patches are tested in isolated test environments, leverage application whitelisting on human-machine interfaces, ensure all field devices are hardened, ensure no out-of-date software and hardware devices are being utilized, review and disable any unused ports and services on ICS devices, run backup and recovery exercises, and ensure both encryption and security is in place.

- **Security monitoring**: Make sure you have captured an ICS baseline of normal operations and network traffic, implement intrusion detection and prevention system capabilities for any out-of-norm network traffic, log and monitor all audit trails, and implement or leverage an existing SIEM solution for any OT-related activity.

- **Supply chain management**: Ensure cybersecurity is heavily weighed for your ICS vendor reviews, invest in secure ICS products, ensure you have the right cybersecurity language with contracts to cover incident handling, remote access requirements, general security requirements, and so on, ensure any cloud service providers are thoroughly reviewed with the correct certifications and audits in place, and ensure test labs are being used to check for malicious code with vendor software before being implemented.

- **Human element**: Ensure policies are in place and users are aware of them, ensure procedures are well documented for users to follow with managing ICS, implement training to help anyone involved with ICS to recognize any threats or compromises, and build a culture where IT and OT personnel are collaborating and supporting each other.

These best practices are provided on a great visual/poster, as follows:

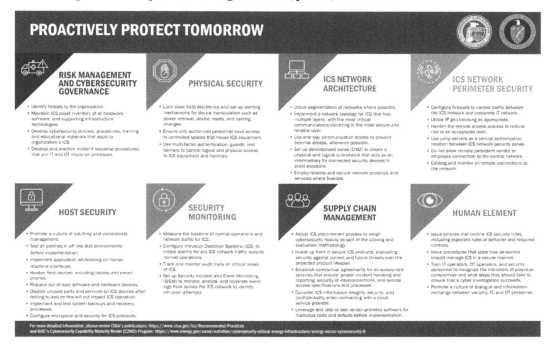

Figure 12.15: CISA ICS best practices

Source: `https://www.cisa.gov/sites/default/files/publications/Cybersecurity_Best_Practices_for_Industrial_Control_Systems.pdf`

In addition to the above, CISA has the following ICS recommended practices documents for reference available at `https://www.cisa.gov/resources-tools/resources/ics-recommended-practices`:

- Updating Antivirus in an Industrial Control System
- Improving Industrial Control System Cybersecurity with Defense-in-Depth Strategies
- Creating Cyber Forensics Plans for Control Systems
- Developing an Industrial Control Systems Cybersecurity Incident Response Capability
- Cross-Site Scripting
- Patch Management of Control Systems
- Securing Control System Modems
- Configuring and Managing Remote Access for Industrial Control Systems

- Department of Homeland Security: Cyber Security Procurement Language for Control Systems
- Mitigations for Security Vulnerabilities Found in Control System Networks

These recommendations and best practices, along with the mitigations available from MITRE, will provide for a very robust cybersecurity program for your OT environment to reduce risk as much as possible. These recommended mitigations and best practice documents are from reputable sources based on learnings and intel from real-world scenarios. Although we haven't specifically mentioned IoT, the same principles should be applied from this section, and for the most part, they will fall within the overall OT program unless your organization only handles IoT technology, in which case you will focus on the securing of IoT devices only.

Responding to OT and IoT Cybersecurity Incidents

In *Chapter 7, Cybersecurity Operations*, we covered incident response, BCP, and DRP in detail as related to IT. Although, at a high level, the processes will be the same, it's important to acknowledge that there will be some unique differences in the handling of incidents related to OT and IoT. Here, we will briefly cover the importance of BCP and DRP for your OT and IoT environments, the need to ensure your CIRP considers OT and IoT technologies, and you are running OT/IoT tabletop exercise themes.

BCP and DRP

The BCP plan for your organization will not change, whether you have OT/IoT or not. However, you need to ensure that you are familiar with the broader BCP for the organization if you are overlooking an OT/IoT program. If there is a major cybersecurity incident with any of your OT/IoT infrastructure, the BCP plan will also need to be invoked. As a reminder, we covered BCP in detail in *Chapter 7, Cybersecurity Operations*, by referencing *SP 800-34 Rev. 1, Contingency Planning Guide for Federal Information Systems*. This NIST publication is a great resource for your BCP and can be found at https://csrc.nist.gov/pubs/sp/800/34/r1/upd1/final.

If you are overlooking an OT/IoT program, make sure you know who runs the BCP for the organization so you can build a relationship ahead of time. It will be worth your time to build a relationship with this team and to ensure they are familiar with any specific response or recovery plans that will be invoked for a major cybersecurity incident within the OT/IoT scope.

In the same chapter that we covered BCP, we covered DRP in detail along with a reference to an example of a DRP plan. For the most part, the referenced plan will be applicable to the OT/IoT environments.

You will just need to ensure the specific technologies applicable to OT/IoT are referenced in detail as they will be unique compared to standard IT software and hardware. You must ensure that full recovery and/or redundancy is considered for SCADA, PLCs, DCS, and any other industrial equipment. As we covered in *Chapter 7*, it will be critical that a **Business Impact Analysis (BIA)** is complete with your OT/IoT environments. A key component of the BIA within a DRP plan is understanding the impact of OT/IoT and the **Maximum Tolerable Downtime (MTD)** that each can withstand before your business is negatively impacted. Two important factors that need to be understood as part of this analysis are the **Recovery Time Objective (RTO)** and **Recovery Point Objective (RPO)**.

One of the most important components of your DRP for your OT/IoT environments is the backup and recovery procedures, especially as ransomware continues to become more prominent. With OT/IoT being distinct compared to IT, you will need to ensure you have the right technologies in place to ensure successful and reliable backups are occurring for your OT/IoT environments. You will also need to ensure you have available hardware in the event of any damage to hardware through a cybersecurity event. This is where consideration of **High Availability (HA)** and failover types of capabilities also need to be taken into account, depending on the criticality of the software and/or hardware that ties back to your BIA.

Although we provided this list in *Chapter 7*, it is important that you fully understand what is needed to ensure reliable backups in the event of a worst-case scenario with your OT/IoT environment. You must implement your backup process to be resistant to a ransomware event. The following is a reminder of current modern-day best practices for your backups:

- Maintain multiple copies of your backups. An example is the 3-2-1 backup method that requires three copies of the backed-up data on two different types of media, with one being off-site.
- Implement an airgap with your backups. This is done by maintaining a copy of your backups offline and ensuring they cannot be accessed by any network or internet connectivity.
- Ensure all backups are encrypted and protect all private keys used to encrypt your backups.
- Make sure you utilize immutable backups to prevent them from being altered.
- Ensure application backups are being executed. You will need to coordinate with the OT/IoT teams to document backup procedures for these applications.
- Test your backups regularly.
- Ensure you have well-documented backup procedures, review frequently, and keep them up to date.

- Ensure you have a multi-layered access model to access your backups which includes phishing-resistant MFA, PIM, and/or PAM to ensure accounts are protected.
- Implement monitoring and alerts on all backup activity and access.
- Ensure auditing is in place for all backup activity.

As stated in the recommendations above, you must regularly test your backups to ensure you are familiar with the restoration process and, most importantly, that they work, and restoration is successful. At a minimum, you should be running a full recovery at least once a year to ensure the beginning-to-end process works as planned.

Incident Response Plan

It is important to note that an incident response for OT will differ from an IT incident. Because of this, you will need to create an incident response plan that is able to address the specifics within an OT environment. In *Chapter 11, Proactive Services*, we covered incident response in detail as related to an IT environment. For the most part, the foundations set within this section will certainly apply to the foundation of your OT incident response plan; there will just be some specifics within the plan that will differ from an IT approach. For example, the need for any DFIR will be distinctly different within OT and IT and you will need to ensure you have a retainer with an organization that specializes in OT forensics and incident response.

Some excellent sources to reference to gain a better understanding of the importance of an OT plan and the differences between IT include the following from Dragos:

- **Incident Response for Operational Technology (OT)**: https://hub.dragos.com/whitepaper/incident-response-for-operational-technology
- **An Executive's Guide to OT Cyber Incident Response**: https://hub.dragos.com/guide-an-executives-guide-to-ot-cyber-incident-response
- **Preparing for Incident Handling and Response in ICS**: https://www.dragos.com/resource/preparing-for-incident-handling-and-response-in-ics/

Some important points provided within these materials include the importance of taking into consideration the physical aspect of responding to an incident, and within an industrial environment, the concept of **Incident Management (IM)** must be understood. This is the management of an incident as related to the facilities themselves and equipment within the facilities, for example, the outbreak of a fire. If this fire was caused by a malicious threat actor because of intrusion into an OT system that was intentionally malfunctioned, both incident management and incident response (OT and possibly IT) will need to work collaboratively through the incident.

As already mentioned, the utmost priority in any incident response is human life, and this becomes more prevalent in an OT environment.

To give you a better understanding of some of the compromises that can be expected in an OT environment versus an IT environment, Table 2 beginning on page 8 of the *Incident Response for Operational Technology (OT)* from Dragos (`https://hub.dragos.com/whitepaper/incident-response-for-operational-technology`) provides some real-life consequences within an OT environment, which include:

- Plant damage
- Loss of production
- Impact on product quality
- Industrial safety event
- Environmental safety event
- Loss of system certification or assurance

Examples are provided with each of these consequences including the actual cyber events that caused them. For instance, examples from plant damage can include control system equipment damage, wear on elements like actuators, an increase in pressure on vessels and pipework, and the possibility of a fire or explosion occurring. With these examples, actual incidents were referenced with TRISIS and CrashOverride malware:

- **TRISIS malware:** `https://www.dragos.com/resources/whitepaper/trisis-analyzing-safety-system-targeting-malware/`
- **CrashOverride malware:** `https://www.dragos.com/resources/whitepaper/anatomy-of-an-attack-detecting-and-defeating-crashoverride/`

When writing the CIRP, it's highly recommended you consult with a vendor who specializes in this area when creating your initial incident response plan for your OT environment. If you do have one in place, have it reviewed to ensure nothing is missing based on current threats and the evolution of technology. As already stated, the foundation and outline referenced in *Chapter 11* will serve as a foundation for your OT incident response plan. In the Dragos *An Executive's Guide to OT Cyber Incident Response, a high level of nine requirements* is suggested for your OT plan:

- Roles and Responsibilities
- Risk Management, Triage, and Escalation Decision Making

- Selected IR Lifecycle (NIST, SNAS, PICERL, etc.)
- Categories of Incidents and Workflows
- Isolation Plan
- Communication Plan
- Regulatory and Legal Requirements
- Internal and External Resources and Contracts
- Supporting Forms and Documentation

Another thought is the prevalence of OT within your organization. If your organization has a large footprint with OT, then having a dedicated plan as referenced above will be needed. But, if your organization has OT but it's minimal, you may want to consider leveraging your current CIRP but expanding on the OT response requirements as attachments. If your organization specializes in IoT, there may be a need for specialized forensics and response for this equipment that will need to be considered.

Finally, it is important to note that with the convergence of IT and OT, and specifically the use of IT becoming more prevalent within the OT environments for remote access and management and so on, the need for incident response in both areas will need to be considered. Ensuring both areas are collaborating and familiar with each other's response plans (if they are separate) will be critical. Another reference that takes both of these items into consideration is a guide from Public Safety Canada, *Developing an Operational Technology and Information Technology Incident Response Plan*: `https://www.publicsafety.gc.ca/cnt/rsrcs/pblctns/dvlpng-ndnt-rspns-pln/index-en.aspx`.

Tabletop Exercises

We'll finish the chapter with a brief review of tabletop exercises and the need to ensure you are running them specifically for your OT/IoT programs. As we have covered throughout the *Incident Response Plan* section, the type of response needed for OT will differ for IT. Because of this, you will need to run a tabletop exercise themed toward an OT event to cover the differences from an IT-related cybersecurity incident. We covered tabletop exercises in detail in *Chapter 11, Proactive Services*, in which the principles covered will work for an OT-themed tabletop exercise. Within *Chapter 11*, we provided reference to some detailed tabletop exercises available from CISA: `https://www.cisa.gov/resources-tools/resources/cybersecurity-scenarios`. Within these exercises, there are many OT themes available for reference.

For example, the following covers critical manufacturing as a theme:

Figure 12.16: CISA tabletop exercise package for critical manufacturing

It is recommended to consult with a third party or external partner who has experience running tabletop exercises, especially if you haven't run one before. They will be able to bring experience to the tabletop, which will be important. In addition, having a third party, will provide a separation of duty and allow recommendations to come from a third party making it more viable, especially when the third party will be familiar with what is expected versus what your response may be.

Summary

As you have read through this chapter, the uniqueness and different challenges presented by both OT and IoT should be clear. These environments are within an industrial type of environment, which has physical consequences including potential impact on human safety, so the importance of protecting these environments is essential. Understanding that these environments create a complete set of different challenges in addition to the IT challenges is the first step to building a robust dedicated program for OT/IoT. Unless you have a small amount of OT/IoT, it is highly recommended you have a dedicated cybersecurity OT/IoT program built to replicate the same functions covered for IT. It's important to understand that the cybersecurity skill sets for OT will be different from IT so you'll need to ensure you are hiring the right personnel to build an efficient OT cybersecurity program. It's also important to note that there is already convergence between IT and OT occurring and it will only become more prominent. This will require close collaboration between the IT and OT teams. In the long term, we will most likely find more IT managing and operating OT and IoT environments.

In the first section, we provided a reminder of what IT is as a lead into what OT and IoT are and how they differ. This included providing examples of where you can expect to find both OT and IoT technologies. We also touched upon IIoT as related to industrial types of devices. In the next section, we took a deep dive into why securing OT and IoT is important. We began by providing examples of where you can expect to see cybersecurity incidents within each of these technologies and the impact they can have. This then led us to look into several OT-specific statistics and the challenges we face within the current state. We finished off the section with a review of IoT-specific statistics to show the current challenges faced by these devices.

In the next section, we reviewed the need for a dedicated program. In this section, we provided an overview of all the considerations for an OT/IoT program, which should include all functions within the IT cybersecurity program. This includes GRC, cybersecurity architecture, identity and access management, cybersecurity operations, vulnerability management, cybersecurity awareness, training, and testing, vendor risk management, and proactive services. We then looked at protecting OT/IoT environments, which included a look at OT malware, reviewing the ICS components of MITRE ATT&CK, and receiving alerts for ICS advisories, before finishing off with reviewing some best practices. In the last section, we covered BCP and DRP with regard to OT/IoT, which transitioned into incident response for OT environments. We then finished off the chapter with a review of the need for tabletop exercises with OT/IoT environments.

In the next chapter, we will be moving on to the third and final section of the book, covering bringing it together with the GRC program. The first chapter will be on governance oversight. Here, we will begin the chapter by looking at the importance of program governance as it relates to cybersecurity. Following this will be an insight into the program structure and governance of the cybersecurity program. Next will be policies, standards, process and procedures for your cybersecurity program, which is very important. This will lead to managing your executive leadership team through the governance program and communicating with both your executive leadership team and the board. This section will also cover reporting. The final section of the chapter will cover other governance considerations you should be aware of for your cybersecurity program.

Join our community on Discord!

Read this book alongside other users, Cybersecurity experts, and the author himself. Ask questions, provide solutions to other readers, chat with the author via Ask Me Anything sessions, and much more.

Scan the QR code or visit the link to join the community.

`https://packt.link/SecNet`

13

Governance Oversight

Governance, Risk, and Compliance (GRC) must be a requirement for all cybersecurity programs moving forward. This program is critical in that it is designed to bridge the gap between the cybersecurity program and the executive leadership team, including the board of directors if one exists. As a cybersecurity leader, you must provide full transparency to the executive leadership team and the board of directors moving forward. If not, you may find yourself liable for any cybersecurity incidents that occur if gross negligence is proven. It has never been more important to ensure that the executive leadership team and board of directors are aware of all cybersecurity risks within the organization. Even more important is ensuring that the executive leadership team and board of directors are taking accountability for cybersecurity risks within the organization. This cannot and should not fall on the shoulders of the CISO or cybersecurity leader alone. I have heard many times the term Cybersecurity Information Scapegoat Officer being used in place of CISO; this must change, and this is where GRC can help. The good news is that I've personally observed an uptick in roles being posted for GRC recently, which means organizations are identifying the importance of this role.

To begin the chapter, we will review the importance of program governance within the cybersecurity program. Here, we will go into more detail as to why governance should be a priority for your cybersecurity program, especially in today's world and with the ongoing threat of cyber groups and criminals. Following this section, we will take a closer look at the program structure needed as part of governance for your cybersecurity program. We will go into detail on some of the important components needed as part of your governance program, including the need for a GRC platform to efficiently track and document everything within the cybersecurity program. We will also review other items like an application inventory, asset inventory, vendor inventory, resource management, and more.

In the next section, we will be reviewing policies, standards, processes, and procedures in a lot more detail. We will provide an overview of each, so you understand the purpose, along with providing examples of each. We will also share examples of templates to help build your own policies. This will lead into the *Leadership Management and Communications* section, which will focus more on bridging the gap between your executive leadership team and the board of directors. This is where you will need to manage your stakeholder expectations through the GRC program. Here, we will review metrics and dashboards as part of managing those expectations. We will finish off the chapter with insight into other governance areas for consideration. This includes items like information protection, data loss protection, data discovery and classification, data disposal, and more. The following will be covered in this chapter:

- The importance of program governance
- Program structure and governance
- Policies, standards, and processes/procedures
- Leadership management and communications
- Other governance considerations

The Importance of Program Governance

First, let's take a high-level look at all sub-functions that should be addressed as part of the governance function. The following image captures much of what the governance function entails.

Figure 13.1: Sub-functions of the governance function

Although GRC are considered a unified program, we will be covering them separately because each serves its own unique purpose as it relates to the broader cybersecurity program, as described below:

- **Governance**: Governance is the overarching component of your cybersecurity program. This is the program that ensures alignment with the organization's objectives in addition to any compliance or regulation requirements. This is where you bridge the gap between the cybersecurity program and your executive leadership team, including the board of directors (if applicable), to ensure that full transparency and accountability sit at the correct level of the organization. This requires the need for excellent reporting and visibility into cybersecurity risk being tracked within the organization. You will also define your policies, standards, and processes/procedures for the organization within this program. Unfortunately, governance is not an option as we move forward with cybersecurity and must be included in your cybersecurity program whether it's a requirement (of a regulation) or not. We will review these items in more detail in this chapter.

- **Risk**: One of the most important components of the cybersecurity program is managing risk. Within this program, the focus is on identifying cybersecurity risk within the organization, analyzing that risk to determine the severity, and then determining how to address that risk, whether it be to mitigate, accept, transfer, or ignore. As part of this program, the need to translate cybersecurity risk into business terms and impact is critical. Here, you will also oversee the cybersecurity risk register for the organization. We will cover this in more detail in *Chapter 14, Managing Risk*.

- **Compliance (+ Regulatory)**: The need for compliance is only becoming more important as more breaches continue to occur. Within this program, the focus is on all applicable laws, any needed compliance, and regulatory requirements for your organization. For example, are you required to be compliant with GDPR, PCI-DSS, and so on? You will need to ensure all frameworks are in place and audited to allow for ongoing maturity. This includes the need for any other audits, such as SOC 1 or 2 types of audits. Data protection is also a priority as part of the compliance program. We will cover this in more detail in *Chapter 15, Regulatory and Compliance*.

With governance being the overarching program, it encompasses both risk and compliance, which, in turn, encompasses the broader cybersecurity program, as shown in the diagram below.

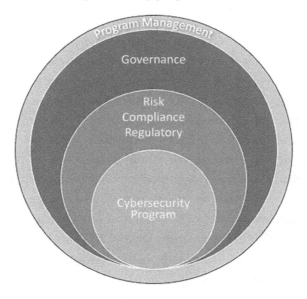

Figure 13.2: Governance in relation to risk, compliance, regulatory, and the broader program

With your governance program, it will be critical that you manage your executive leadership's expectations by providing full transparency into the cybersecurity program, along with all identified risks. By doing this, you are shifting the accountability of cybersecurity and the risks associated with it to the executive leadership team, or more specifically, the CEO and the board of directors if one exists for your organization. If you are not doing this today, the liability for a major cybersecurity event will fall on the shoulders of the CISO and the cybersecurity team as a whole. The reason this needs to be of the utmost priority for any CISO and/or higher-level cybersecurity leader is the reality that liability is already falling on cybersecurity leaders. A couple of prominent examples you may have heard of include:

- Uber's former Chief Security Officer was sentenced to three years' probation and fined $50,000 for failing to disclose a data breach: https://www.justice.gov/usao-ndca/pr/former-chief-security-officer-uber-sentenced-three-years-probation-covering-data

- The SEC filed charges against the SolarWinds CISO for fraud and internal controls failures: https://www.sec.gov/news/press-release/2023-227 (Case is still pending as of August 2024).

It is important to acknowledge that there have also been instances where the CEO has also been held liable. Although, it's possible that there were no cybersecurity executives in those organizations at the time, which is why the CEOs were held accountable. A couple of documented examples include:

- The CEO of Vastaamo was handed a suspended jail sentence for failing to protect customers' sensitive information: https://www.bitdefender.com/blog/hotforsecurity/ex-ceo-of-hacked-therapy-clinic-sentenced-for-failing-to-protect-patients-session-notes/

- The FTC finalized an order with the Drizly CEO for a data breach of approximately 2.5 million consumers' records: https://www.ftc.gov/news-events/news/press-releases/2023/01/ftc-finalizes-order-online-alcohol-marketplace-security-failures-exposed-personal-data-25-million

In no way am I defending any malpractice within the examples above, but they serve as an example of a new reality. The reality is that executives are being charged and held accountable for failing to efficiently protect their organization and any consumer data they are storing, and for a lack of transparency to the executive leadership team and regulators. These are serious consequences and cannot be taken lightly or taken for granted. This is why you must do everything you can to protect yourself as a cybersecurity leader, and this is where your governance program comes into play. It is important that no information is held back, moving forward, under any circumstances, and everything must be disclosed to your executive leadership team. Or, at a minimum, it should be accessible to your executive leadership team and the board of directors.

 This should also be a rude awakening for all executive leadership teams, specifically the CEO and the board of directors, as it delivers a clear message that no one is immune. The examples above should be more than enough to show the need to ensure the resources and capabilities are being provided to enable a mature cybersecurity program for any organization.

In addition to the examples of CISOs and CEOs being held accountable, some statistics available for reference include the Hyperproof *2024 IT Risk and Compliance Benchmark Report,* with over 1,000 respondents, which provided the following highlights:

- An astonishing 49% of the respondents have a difficult time identifying critical risks for remediation prioritization.

- Only 18% of the respondents have been able to align compliance and risk activities.

- 14% of the respondents are still using spreadsheets to manage IT compliance. This has increased by 40% year by year.

- The good news is that 83% have centralized their GRC program, which is up from 68% the previous year.

- IT risk is managed in silos, with departments, processes, or tools, by 19% of the respondents, which is down from 31% in 2023, although 70% of those managing risk in silos experienced a breach.

- In the previous 24 months, 59% experienced a breach, up from 42% in last year's report.

- 69% reported that they expect to spend more money and 60% expect to spend more time on IT risk in 2024.

- Respondents reported the NIST **Cybersecurity Framework (CSF)** as the most used framework.

Source: `https://hyperproof.io/it-compliance-benchmarks/`

Another report, by Navex, the *2023 State of Risk & Compliance Report*, which comprised over 1,300 global respondents, provided the following:

- Some good news, in 2023, more than 53% of the respondents reported their organization to be on the mature side with risk and compliance. This is up 15% from 2022, although there is still a long way to go for more organizations to mature.

- It was reported that 30% of the respondents experienced a data privacy/cybersecurity breach within the previous 3 years.

- Data privacy, protection, and security, along with regulatory compliance, were ranked as the highest priority against other compliance issues within the organization. Other compliance issues included harassment and discrimination, organizational culture, conflicts of interest, diversity, equity and inclusion, whistleblowing, reporting and retaliation, and bribery, corruption, and fraud.

- Of all the risk areas within an organization, both data privacy and IT/information security risks were reported as the highest priority by more than half of the respondents within both categories.

Source: `https://www.navex.com/en-us/resources/benchmarking-reports/state-risk-compliance/`

Although some of the data shows concerns and there is still work to be done, it is also clear that many are realizing the importance and need for a mature GRC program, and the trend is moving toward this. It is possible that many may be moving in this direction because they have no options based on regulation and/or other audit requirements. Either way, the shift we are seeing is positive and we need to ensure we continue to see a positive trend with the implementation and maturing of GRC programs.

Some other areas of importance (as they relate to governance) that we have already covered throughout the book, or will be covered in this chapter and the coming chapters, include:

- **Program management**: This is everything related to overlooking and running the cybersecurity program. This includes the need for good reporting and providing executive leadership and the board with full transparency and all known risks. We cover this in detail later in the chapter. Other areas of program management were covered within *Chapter 2, Setting the Foundations*.

- **Business enablement and alignment**: This is another important area that needs focus and attention under the broader governance program. Our roles are changing fast as we need to align with the business and enable success. It is not our place to tell the business what they should and shouldn't do, but to understand any risk and ensure the business is aware of that risk. This was also covered in more detail in *Chapter 2, Setting the Foundations*.

- **Change management**: Our modern-day worlds continue to change, and they are changing fast. A big component of this change is due to technology, which brings ongoing risk. As change occurs, we must work with our users on change management and keep them informed of what to expect. Transparency is key here. Again, this was covered in more detail in *Chapter 2, Setting the Foundations*.

- **Program frameworks and strategy**: It is imperative that your program has a framework in place today, and it's the governance component that needs to ensure this is a requirement, whether it's regulated or not. The same applies to your strategy. Without a well-documented and defined strategy, you will be managing chaos with no true end goal in place. We covered this in detail in *Chapter 4, Solidifying Your Strategy*, and we'll be touching upon frameworks in *Chapter 15, Regulatory and Compliance*.

- **Maturity models and assessments**: Here, you will need to ensure the current state is continuously reviewed to understand where maturity opportunities exist. This can only occur through the ongoing need to assess and audit to allow maturity to occur.

This, to some extent, falls within the need for a framework, as covered in *Chapter 4, Solidifying Your Strategy*, along with *Chapter 14, Managing Risk*, and *Chapter 15, Regulatory & Compliance*, coming up.

- **Financial management:** It is important you govern the financial component of the cybersecurity program. This includes everything as it relates to your ongoing operational costs, project costs, and personnel costs to run the program. You will continuously be challenged on budget spend so make sure you are tying back the program costs to the value-add of the cybersecurity program. This was covered in more detail in *Chapter 2, Setting the Foundations*.

- **Data governance:** The whole premise of cybersecurity programs is to protect your organization's (and users') data. Understanding the type of data being stored, where it lives, who has access to it, how important it is (classification), etc, are all critical components of data governance within the cybersecurity program. This will be covered in more detail in *Chapter 15, Regulatory & Compliance*.

- **Policies, standards, processes, and procedures:** One of the primary functions of the governance component of a GRC program is the overarching policies, standards, processes, and procedures that define what is expected of the users as they relate to cybersecurity. This will be covered in more detail later in the chapter, although we have touched upon this throughout the book.

- **Other considerations:** There are many other considerations that will be covered throughout this chapter, as well as both *Chapter 14, Managing Risk,* and *Chapter 15, Regulatory and Compliance*.

At this point, there shouldn't be a question in your mind whether the GRC program is needed or not. This program is a requirement with today's threat landscape, and the quicker you can implement and mature a GRC program, the better for both the organization and the cybersecurity team, especially yourself as a leader.

Program Structure and Governance

In this section, we will cover what is needed to ensure your GRC program is set up for success. The GRC program is going to need resources committed to ensure not only the success of implementation but also the ongoing operation to continue the maturity of the program as more requirements will continue to come from more audits, new regulations, increased risk, etc. The work within this program will not necessarily be technical and the resources will differ from those within a SOC, for example.

This program will work very closely with the leadership teams throughout the organization and will entail more documentation creation work – for example, the need to create and update policies, create retention schedules, review contract language, collect evidence, execute audits, build reports, etc. As you begin to mature this program, you will quickly realize that there will be a lot of effort needed and work that will take time to complete. Nothing should be outside the scope of GRC. The following is a reminder of where the GRC program falls within the broader cybersecurity program.

Figrue 13.3: GRC overarching the broader cybersecurity program

First, let's take a look at the strategy and roadmap for your GRC program. This will help set the foundation of your program to ensure greater success.

Strategy and Roadmap

The first detail you are going to need to address is to assess what stage your GRC program is at. For most, the real question may be whether a formal GRC program even exists. If one doesn't exist, you may already be executing some components of GRC within your cybersecurity program without even realizing it.

 As you build and formalize your GRC program, it will be important you receive support from your executive leadership team. If not, it will become challenging to formalize and mature your GRC program as it will require support from the broader business to inventory and classify all applications, ensure onboarding processes are being followed for new applications and vendors, that policies are being reviewed and published, and much more. Essentially, your GRC program requires close collaboration across the entire organization.

Regardless of the state, it will be important to understand where your GRC program is today and what is needed to either formalize a GRC program or enable the maturity of the program. If you recall, we covered the importance of a strategy in *Chapter 4, Solidifying Your Strategy*. Here, we covered ensuring the cybersecurity organization has a strategy defined at the cybersecurity function level with the need to define a strategy for each of the programs within the cybersecurity program. You will need to ensure you have your strategy defined and well documented, specifically for the GRC program. This only becomes more important if you need to justify building the program for the first time or if you need to mature it. For example, using the example from *Chapter 4*, the strategy may look something like the following:

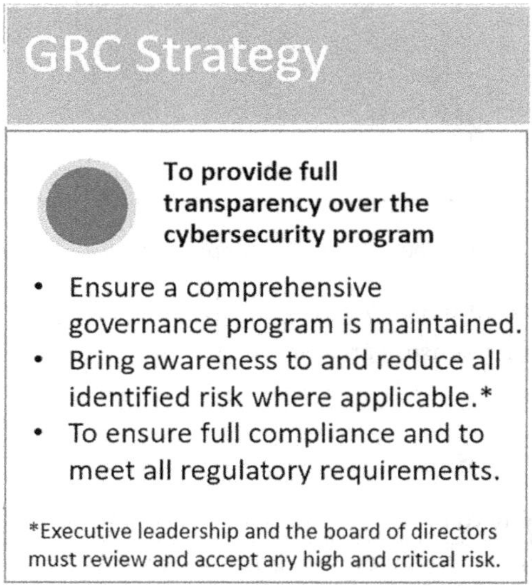

GRC Strategy

To provide full transparency over the cybersecurity program

- Ensure a comprehensive governance program is maintained.
- Bring awareness to and reduce all identified risk where applicable.*
- To ensure full compliance and to meet all regulatory requirements.

*Executive leadership and the board of directors must review and accept any high and critical risk.

Figure 13.4: GRC strategy example

In addition to the strategy for your GRC program, you are going to need to ensure you have a roadmap defined. This holds true whether you already have a GRC team in place or not. In *Chapter 3, Building Your Roadmap*, we covered the roadmap for the broader cybersecurity program, which included the GRC component. The reality is that the GRC function will not happen within the short-term timeframes unless there is already a GRC program in place today in which your roadmap becomes more mature. Many of the items covered within the short-term roadmaps comprise part of the GRC program, though. With this, much of your roadmap and effort may be evolved around formalizing the program and unifying all the components within the GRC program.

Bear in mind that some of the components of building a mature GRC program will be a very large undertaking, with many of them becoming ongoing work efforts, such as building a comprehensive application inventory with data classification, identifying business owners, and continuity requirements, ensuring you have a comprehensive set of cybersecurity policies that all users are aware of and attesting to, and implementing a framework and the ongoing need of audits to assess and allow for maturity, to name some examples.

 In addition to the GRC strategy, the governance program will need to ensure that the broader cybersecurity program strategy and roadmap are in place, being updated, and being followed.

Roles and Responsibilities

It will be important that the roles and responsibilities of the GRC program are clear and well-defined for all parties involved. GRC expands beyond the cybersecurity team to the broader organization and some specific key areas within the business. Some of these more prevalent areas include the executive leadership team, the board of directors, the legal team, the internal controls team, the auditing team, the privacy team, and risk management team, if any of these exist. These roles may be combined or fall within a few key areas within the organization. Either way, it's important to understand who is responsible for what.

This can be accomplished with the use of a RACI matrix if you are familiar with one. This is essentially a tool used to identify who is **Responsible, Accountable, Consulted, and Informed (RACI)** for a specific task or broader function. We will see some examples of GRC-related tasks later in this chapter and throughout the following chapters.

In *Chapter 2, Setting the Foundations*, we briefly covered roles and responsibilities for the broader cybersecurity program where a governance resource was mentioned as a more specialized role that will be needed within your cybersecurity program. This role is quickly becoming a norm within the cybersecurity program today. If you are a smaller organization, the governance role may be part of a broader role and responsibility, like a cybersecurity manager, for example. For a larger organization, the goal should be to build a team to support the ongoing demand of the GRC program in the long term.

Some of the roles you can expect to see within a GRC program, whether internal to the cybersecurity organization or external with a different team within the organization, may include:

- **GRC or governance, risk, and compliance director or manager**: As a GRC manager or director, you will be responsible for overseeing the entire GRC program and will most likely report directly to the CISO or the cybersecurity leader. You will be responsible for ensuring the GRC program has the resources it needs to be successful, allowing for efficient reporting to the executive leadership team.

- **Cybersecurity analyst**: The cybersecurity analyst can be thought of as a junior generalist resource who has knowledge of multiple areas and is able to support the GRC in many ways. This role will have a broader knowledge beyond GRC and will interact and collaborate with other teams throughout the cybersecurity program. They will typically report to a team lead or manager.

- **GRC or governance, risk, and compliance analyst**: A GRC analyst will be a junior role within the GRC program who has familiarity with all the GRC capabilities and tools. They will complete the day-to-day tasks required to run the GRC program. They will typically report to a team lead or manager.

- **GRC or governance, risk, and compliance administrator**: The role of GRC administrator will be a more senior role to the analyst who will carry more expertise in managing the tools that run the GRC program. They will typically administrate and configure the tools needed to support the GRC program. For the most part, they will report to a team lead or manager.

- **Auditor**: An auditor is someone who will assess controls that have been implemented as part of an assessment, regulation, certification, or any other type of compliance. The goal is to provide attestation to an audit once all controls have been confirmed. If any controls don't meet the requirements, the auditor will provide feedback and there's a possibility that attestation may not occur. Internal auditors will typically report to the risk or compliance organization if one exists. External auditors will be contracted through an audit firm.

If you recall, in *Chapter 2*, we also discussed outsourcing as a strategy for your broader organization. For the GRC function, I would highly recommend keeping this work in-house. This function needs to be close to home and you simply can't rely on a third party to efficiently overlook your GRC program for your organization. I'm not suggesting you don't outsource some of the busy day-to-day tasks, but it will be important that all management, oversight, and strategy-related work is kept in-house.

Obviously, the exception here will be the need to bring in external vendors for any audit type of work or any other activities that require a separation of duties.

It is important to note that the roles in the GRC program will differ significantly in some instances. The roles overlooking your GRC program will, for the most part, be a lot less technical than other areas of the cybersecurity function. In today's world, everyone needs to maintain some form of technical skill set to be successful, but the roles within GRC will be more business-oriented versus an engineer or development type of role. The role of those in the GRC program will interact with the broader business a lot more than other roles, specifically the legal and leadership teams. This role will need to understand the broader cybersecurity program at a higher level versus being a **Subject Matter Expert (SME)**. They will need to be knowledgeable about regulations, laws, audits, certifications, language within contracts, etc. However, since these resources manage audits and certifications, they will also need to understand the technical components as controls need attestation, etc.

One final topic to cover is that of certifications within the GRC space. With this being such a critical component for an organization, there are certifications available to ensure that those responsible for managing and operating GRC remain current. A couple of the more recognized certifications for GRC include:

- **ISC2 CGRC (Certified in Governance, Risk, and Compliance)**: `https://www.isc2.org/certifications/cgrc`
- Several certifications from ISACA including **CISA (Certified Information Security Auditor)**, **CRISC (Certified in Risk and Information Systems Control)**, and **CGEIT (Certified in the Governance of Enterprise IT)**: `https://www.isaca.org/credentialing/certifications`

Make sure you are aware of the available certifications and training for your staff and enable them to succeed. Provide the time and funds needed to allow them to learn, grow, and become certified. Make this a priority and ensure you are encouraging them to advance themselves. This will not only benefit them personally, but your organization will reap the benefits in the long term.

Organization Structure

As you look to build your organization structure for your GRC function, you'll want to ensure there is a direct report line to the CISO or the leader overlooking the cybersecurity program. This function cannot be embedded deep into the cybersecurity organization as it will not receive the support it needs to be successful.

This program requires full engagement and interaction from the CISO to ensure its success, especially since this will be considered the glue between the cybersecurity program and the executive leadership team.

In *Chapter 2, Setting the Foundations*, we covered defining the cybersecurity organization in more detail, where we provided an example of an organization chart. This chart included the GRC function reporting directly to the CISO as an example. This is exactly what the report structure should look like. Below is a partial view of the original organization chart from *Chapter 2* where the report line for the GRC function has been highlighted to the executive leadership team and the board of directors. As a reminder, this is just an example of what an organization chart can look like, and the other teams reporting to the GRC function don't necessarily need to be reporting to the GRC report structure, although they can be applicable to report to the GRC function.

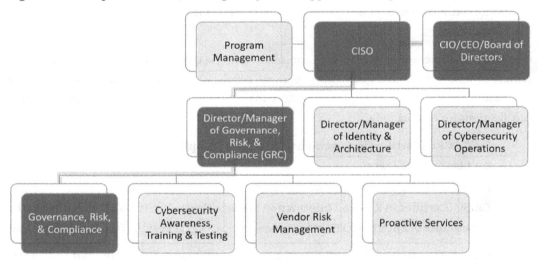

Figure 13.5: GRC within the cybersecurity sample organization structure

As you can see from the organization chart above, the report line of the GRC function does not stop at the CISO. It is important that this report line ties back to the executive leadership team, including both the CEO and the board of directors. It is critical this program reports up to the executive leadership team and board of directors with full transparency of the program, specifically, all risks that have been identified and are being tracked.

GRC Application

To ensure success for your GRC program, you are going to need a GRC platform to centrally document and track everything. You may be tracking components of your GRC program in other applications or spreadsheets, which becomes very challenging to manage efficiently long term. For example, are you tracking your audits in a spreadsheet? Are your assets in an independent standalone system? Is your risk register being tracked in a spreadsheet? If you are not managing your GRC components centrally, the management of your GRC program becomes inefficient very quickly when trying to link your risk back to your assets and applications. In the long term, this certainly won't provide the output needed for greater success and it is highly recommended you look to unify all the components of your GRC function in one platform. To help with this, there is the option of purchasing an off-the-shelf platform that will provide broader GRC functionality. Some examples of GRC tools include:

- ServiceNow: `https://www.servicenow.com/products/governance-risk-and-compliance.html`
- AuditBoard: `https://www.auditboard.com/`
- Archer: `https://www.archerirm.com/content/grc`

Another option is to build a custom in-house platform to meet your needs if the skill sets are available. Whatever the direction, make sure you are clear about the requirements to ensure there are no gaps in your GRC platform. Some of the benefits of implementing a unified GRC platform include:

- Centralized view of your entire cybersecurity program in one place.
- The ability for interdependencies between modules. For example, the vendor module links to the application module, which, in turn, can be linked to the risk register module for any identified risk.
- More efficient unified reporting with the ability for auto-reporting.
- Enhanced visibility with overall risk and compliance.
- The ability to efficiently track any required regulation.
- Providing improved transparency for the cybersecurity program.
- More efficient integration and automation capabilities.
- The ability to execute audits and track evidence more efficiently.
- Overall, the GRC platform will provide a more mature cybersecurity program for the organization.

At a minimum, the following core modules or components should be part of your GRC platform to ensure that a comprehensive view of your cybersecurity program is being represented in one place.

Vendors

We will start with the vendor module, where we will focus on centrally tracking all vendors being used within the organization. In *Chapter 10, Vendor Risk Management,* we covered the beginning-to-end lifecycle of your vendors and the importance of this program moving forward. Although there may be other systems to manage components of your vendors, like a cybersecurity scoring platform, contract system, risk management system, etc., it will be important to centralize the inventory of your vendors in one place. Integrating any of these data points from other systems will only benefit the GRC program by providing a single place for everything vendor-related.

Another important component of your vendor inventory is downstream dependencies. For example, as you build your application portfolio, they will need to be tied back to a vendor (unless built in-house). This will be important when an application has an identified vulnerability, as you will be able to quickly search the application portfolio, which, in turn, will provide the vendor information and any relevant information that may be needed, such as contact information. Some of the data points you will want to capture as part of the vendor module include the vendor's legal name, domain name, cybersecurity grade, the last time the vendor underwent a cybersecurity review, current contract status and renewal date, any relevant documentation from the cybersecurity review (SOC 2 Type 2, SIG questionnaire, etc.), supplier contact information, business contact for the vendor, and so on.

Applications

Next, you will need to ensure you are centrally tracking all your applications for the organization. Those of you who have been involved in this task will know it is not the easiest. As each business function purchases its own applications and processes aren't followed for onboarding, application sprawl becomes a big challenge. This, in turn, creates more risk as the need to track vulnerabilities, review vendors, etc., becomes very challenging. It will be important to embed your process into other existing processes to ensure there are no gaps. One of the more efficient processes will be working with the finance team to capture all vendors that are receiving payments from you.

Obviously, those purchasing directly with credit cards or in-house applications will not be captured here, so ensuring you have processes in place that business functions are aware of is critical.

For this module, it will be important that you are able to link the application (if applicable) back to the vendor module and the vendor from which the application is being subscribed. This is important as you need to track the beginning-to-end lifecycle of an application, including managing the risk for the vendor that has created/developed the application or service for use. As part of this module, you should be capturing the application name and description, vendor of the application, who in the business owns that application, status of the application, application user base, vendor/application contact information, data classification, application URL, whether publicly accessible, authentication type, any audit requirements, user access reviews, license information, application cost, renewal dates, hosting location, **Business Continuity Plans (BCP)** and **Disaster Recovery Plan (DRP)** information, support information, etc.

Risk Register

One of the most important modules is the risk register, which is designed to efficiently track all material cybersecurity risks within the organization. This is the module that will document all identified cybersecurity risks to be accepted, mitigated, transferred, or avoided. Once documented, it will be important they are reviewed by a cybersecurity council and/or executive leadership team and the board, depending on severity. Within the risk register, some of the data points that will need to be tracked include a risk ID number, date identified, risk title, what controls it links back to, application name or affected products, vendor name, risk description, risk identification, risk owner, risk type, probability and impact of risk, risk exposure, risk status, etc. We will be covering risk in more detail in *Chapter 14, Managing Risk.*

RACM

The **Risk and Control Matrix (RACM)** is the foundation to track all risks within an organization along with the controls that can be used to mitigate or reduce the known risk. It allows an organization to become more proactive by formally documenting all risks and predetermining the likelihood and potential impact of that risk. The RACM can become very complex and will take a lot of effort to implement efficiently. For the most part, you'll want to create your RACM around any framework, audits, or regulations you are required to follow (although there may be more risk that needs to be tracked in addition to the frameworks). For example, if you have a SOC 1 Type 2 requirement, you'll need to ensure all controls are captured and referenced back to a risk of not implementing the controls.

Think of the RACM as a control mapping exercise as you formally document all potential risks for the organization and tie them back to controls to reduce risk. The same applies to NIST, SOX, PCI-DSS, and so on. You'll also need to ensure that any duplicate controls from audits and regulations are tied back to a single risk and control to mitigate or reduce that risk. The RACM scope goes beyond cybersecurity to represent the broader risk for the organization. Although the scope here is for cybersecurity, it will be in the best interest of the organization to centralize the RACM for the broader organization and track all identified risks centrally.

Audits

In the audit module, you will be tracking all relevant audits for the organization. This can be anything from SOC 2 Type 2, NIST, ISO 27001, PCI-DSS, HIPAA, and more. In this module, you will need to provide the capability to manage each of the controls within each audit to be able to upload evidence and provide any data that shows how the control is being managed to reduce risk. There will need to be additional data points in the control such as a tracking number, control name and description, control owner, RACM reference, application, completed/due date, status, etc. Ideally, you will want to allow access to the controls to provide a central location for the control evidence, so you do not need to send documents back and forth. This will also create a more secure solution for your data. This can be thought of as the **Prepared-By-Client (PBC)** component of an audit. You'll also want to ensure any duplicate controls from multiple audits are tracked within a single control to reduce work effort and streamline the evidence collection. The audit section will also directly link back to the RACM and the controls with the predetermined risk profile. We will be covering audit types in more detail in *Chapter 15, Regulatory & Compliance*.

You'll also need to consider regulatory considerations depending on your industry and location throughout the world. This can be part of the audit module, or you may need a separate module to track any regulations that are required.

Inventory Management (Assets)

The inventory management or assets module may be part of a different platform that you manage. For example, if you have an **Information Technology Service Management (ITSM)** platform in place, does it also have **Configuration Management Database (CMDB)** capabilities? If so, you'll not want to duplicate any work for GRC as this is an enormous task within itself. But, it will be important that, at a minimum, you have the asset data available to the GRC platform so you can link back any risk to an asset within the organization. Your applications are also an asset so there'll be some synergies and cross-referencing between the two modules.

Your asset inventory data will vary depending on the asset type and how you would like to track it in the GRC versus any CMDB you may have in place already. We covered asset management in more detail in *Chapter 8, Vulnerability Management*.

Issue Tracker

The issue tracker is the place where you will formally track all issues identified from audits, assessments, tabletop exercises, tests, etc. Once an item has been added to the issue tracker, you will be able to assess and determine what action or next steps are needed for that issue. Depending on the issue identified, this may translate into something that needs to be entered into the risk register. In some way, you could argue as to why these items don't go directly into the risk register. But it's important to ensure analysis is complete for these items to determine what is needed to resolve them as it isn't necessary to add every finding into the risk register unless there is a material risk that needs to be reviewed and accepted. For example, a tabletop exercise may identify a process improvement. This doesn't need to go into the risk register; it needs to be tracked as an issue and then determine what actions are needed to close the identified issue. On the other hand, a penetration test that identifies a critical finding may need to be entered into the risk register.

Policy Management

Your policies are a critical component of your cybersecurity program, and you'll need to centrally track them for version control and, at a minimum, an annual review. Ideally, the fewer policies you need to manage, the better, meaning a broader policy that covers everything within the program will be easier to manage versus many smaller policies. This may not be feasible for every organization and, depending on the size of your organization, there may be several policies to manage. For example, you may have a general cybersecurity policy, a retention policy with a supporting retention schedule, a vendor risk management policy, etc. Either way, it is important to centrally house and track these policies with versioning and tracking capabilities as you keep them current for your users. Policies will be covered in more detail in the next section.

Business Continuity

Like your policies, you will need a centralized repository to track and house all your BCP, crises management, DRP, **Cyber Incident Response Plan (CIRP)**, and any other business continuity documents that will be needed for any event that triggers the BCP. These documents will also need to be reviewed and updated at a minimum on an annual basis, so ensuring you are able to track changes and version control them will be important as you need to publish them into production. *Chapter 7, Cybersecurity Operations*, covered business continuity in more detail.

Reporting

The last module we will cover is the reporting functionality. It is important that the platform is able to provide quality reporting based on all the data being collected and stored. This becomes critical for your output to the executive leadership team and board of directors as you need to translate the GRC program up to a higher level – especially the risk register module, which is where you will need to ensure the executive leadership team and the board of directors have visibility into what is in the risk register to assume accountability.

The suggested modules above are not all-inclusive of what you may find within a GRC platform. I would consider these items core to the GRC platform and things that I would ensure are included as part of your GRC platform, whether it is off the shelf or built in-house. As you look to mature your GRC program, it is important to look at ways to modernize and innovate in this area. There's a high possibility that much of your work is manual, even if you are running an efficient GRC program. This is most likely true as you execute audits or need to provide evidence as part of an audit or some other regulation. As you look to mature the GRC program, look at ways you can implement automation and enable continuous scanning capabilities. Is your current platform providing capabilities to automate the review and attestation of any of the controls required for your audits? This will prevent the need for manual work. And this will become more important as the need to execute more audits and meet more regulations continues to come our way. Make sure you are challenging your vendors to provide more advanced capabilities in this area to allow the shift from a traditionally manual process to more automated capabilities that allow your controls to always be up to date.

Policies, Standards, and Processes/Procedures

A major component of your governance function is the need for well-defined policies, standards, processes, and procedures. The policies, standards, processes, and procedures are the rules and guidelines your users must follow as part of working within your organization and it is critical they are being followed. In addition, policies should be reviewed and signed off by your executive leadership team to ensure enforcement for your users. Without this support, it becomes very difficult to enforce and cybersecurity will fail to get the attention it needs at an organizational level.

Hopefully, you have some form of general cybersecurity policy in place today, or, at a minimum, there is cybersecurity language within some of your primary company policies. If not, you must prioritize this today as they are needed to ensure a more secure and robust cybersecurity program. If you do have policies in place today, the reality is they will need to be updated on an ongoing basis. I find an annual review of cybersecurity polices typically evolves into multiple material changes and updates. This is the unfortunate reality in the ever-evolving threat landscape we live in today.

As you have observed throughout the book, we have touched upon the need for policies in most chapters. Your policies are what define the rules for your users to ensure they are not doing anything that increases risk for the organization. In a way, you can think of your policies as the set of rules and laws they must follow for working in your organization. When a user doesn't comply with your policies, they put themselves and the organization at risk. This is why it is important that your policies are well-defined and clear and that everyone is aware of them. Policy attestation can help with ensuring everyone is aware of them.

Defining Policies

A cybersecurity policy is the first level of formalized documentation for your organization's cybersecurity program, and it is mandatory. Policies are a critical component of your overall cybersecurity program, which requires sign-off and support from the executive leadership team to ensure success. Policies should be very broad and general, with no direct link to the technology or solutions within the organization. In general, they should not change often, but periodic review is critical. Some examples of general policies within an organization may include:

- Code of conduct and ethics policy
- Acceptable use policy
- Equal opportunity policy
- Workplace safety policy
- Compensation policy
- Change management policy
- Disaster recovery policy
- Time off, sick leave, or leave of absence policy

Although some of the policies above may cross into the cybersecurity policies as they may contain cybersecurity language, the following are some specific cybersecurity policies you can expect to see within an organization:

- Cybersecurity policy
- Privacy policy
- Information protection policy
- Vendor risk management policy
- Retention policy with supporting retention schedule

If you don't have any policies in place today, it is highly recommended you work with the appropriate teams to implement policies within your organization. From a broader organizational level, your general business policies will typically be owned by the HR department in collaboration with the legal team. Your scope will pertain to anything cybersecurity-related such as a cybersecurity policy. You will also want to work closely and partner with your legal team to ensure the cybersecurity policy is written accurately and receives sign-off from both the legal team and the executive leadership team. For your basic cybersecurity policy that your users will need to comply with, the following should be considered at a minimum:

- Acceptable use
- Security updates
- Encryption requirements
- Network requirements including firewall protection
- Password policy, **Multi-Factor Authentication (MFA)**, and biometrics
- Account management (including local administrative access strategy) and access controls
- Security protection tools including antimalware protection
- Data compliance and protection policies

- Data loss prevention and information protection (including data classification and retention)
- Incident response
- Physical security
- Cybersecurity training and testing
- **Artificial Intelligence (AI)** usage

Creating a policy for the first time is not easy, and you may find it beneficial to bring in a third party to help with the initial creation if nothing exists today or to have any current policies reviewed for recommendations. As you look at what you have in place today (if anything), it's worth reviewing how many separate policy documents you have with cybersecurity requirements to see if it makes sense to consolidate. As a user, it may be a little frustrating searching for a specific cybersecurity policy if there are multiple policies for cybersecurity. It may not be realistic to consolidate all policies, but it is highly recommended to create a master cybersecurity or information protection policy that encompasses all your primary cybersecurity policies.

To help get you started with your policies, there are many resources and examples available on the web for reference. A few that I have referenced in the past include the following:

- Within CIS Critical Security Controls Version 8, there are several policy templates available for download and review within the Policy Templates section at this location: `https://www.cisecurity.org/controls/v8`
- SANS also has many security policy templates available at this location: `https://www.sans.org/information-security-policy/`
- Another great repository of templates for reference has been provided by PurpleSec, which can be found here: `https://purplesec.us/resources/cyber-security-policy-templates/`

For the most part, all the templates provided above are individual policies for different components within the cybersecurity program. The following is the first page from the SANS Acceptable Use Policy template accessible in the above SANS URL.

Acceptable Use Policy
Last Update Status: *Updated October 2022*

Free Use Disclaimer: This policy was created by or for the SANS Institute for the Internet community. All or parts of this policy can be freely used for your organization. There is no prior approval required. If you would like to contribute a new policy or updated version of this policy, please send email to policy-resources@sans.org

1. Overview

Infosec Team's intentions for publishing an Acceptable Use Policy are not to impose restrictions that are contrary to <Company Name>'s established culture of openness, trust and integrity. <Company Name> is committed to protecting <Company Name>'s employees, partners and the company from illegal or damaging actions by individuals, either knowingly or unknowingly.

Internet/Intranet/Extranet-related systems, including but not limited to computer equipment, mobile devices, software, operating systems, storage media, network accounts providing electronic mail, WWW browsing, and FTP, are the property of <Company Name>. These systems are to be used for business purposes in serving the interests of the company, and of our clients and customers during normal operations. Please review Human Resources policies for further details.

Effective security is a team effort involving the participation and support of every <Company Name> employee and affiliate who deals with information and/or information systems. It is the responsibility of every computer user to know these guidelines, and to conduct their activities accordingly.

2. Purpose

The purpose of this policy is to outline the acceptable use of computer equipment and other electronic devices at <Company Name>. These rules are in place to protect the employee and <Company Name>. Inappropriate use exposes <Company Name> to cyber risks including virus attacks including ransomware, compromise of network systems and services, data breach, and legal issues.

3. Scope

This policy applies to the use of information, electronic and computing devices, and network resources to conduct <Company Name> business or interact with internal networks and business systems, whether owned or leased by <Company Name>, the employee, or a third party. All employees, contractors, consultants, temporary, and other workers at <Company Name> and its subsidiaries are responsible for

CONSENSUS POLICY RESOURCE COMMUNITY
© 2022 SANS™ Institute

Figure 13.6: SANS Acceptable Use Policy template

As we mentioned earlier, it's recommended to create a master cybersecurity policy that encompasses most of the cybersecurity-related items so there is one place for your users to go and reference.

Before moving on to the next section, it is important to point out the recent trend of AI availability and usage, as reviewed in *Chapter 7, Cybersecurity Operations*. As we discussed, you can't prevent users from accessing these technologies, so it's important you have a policy in place to ensure users are aware of the risks and how to use the technology appropriately. If you don't have an AI policy or any AI language within your cybersecurity policy today, it is highly recommended you implement this right away. For reference, SANS has a very good AI template available in their repository of templates.

Artificial Intelligence Use Policy
Last Update Status: *Updated January 2024*

Free Use Disclaimer: *This policy was created by or for the SANS Institute for the Internet community. All or parts of this policy can be freely used for your organization. There is no prior approval required. If you would like to contribute a new policy or updated version of this policy, please send email to policy-resources@sans.org.*

ARTIFICIAL INTELLIGENCE USE POLICY

I. Introduction.

YourCompanyName is committed to full compliance with applicable laws related to the use of artificial intelligence in the countries in which YourCompanyName provides products and services. Additionally, YourCompanyName is committed to the ethical use of artificial intelligence. This Artificial Intelligence Use Policy ("Policy") outlines YourCompanyName's requirements with respect to the adoption of all forms of artificial intelligence at YourCompanyName. Such artificial intelligence adoption includes use for business efficiencies, operations, and inclusion in YourCompanyName's products and services.

This Policy is applicable to all YourCompanyName directors, officers, board members, employees, contractors, representatives, affiliates, agents, and any person or entity performing services for or on behalf of YourCompanyName. The ResponsibleCorporateOfficer at YourCompanyName is responsible for the enforcement of this Policy.

II. Definitions.

"Artificial intelligence" or "AI" means the use of machine learning technology, software, automation, and algorithms to perform tasks and make rules or predictions based on existing datasets and instructions.

Figure 13.7: SANS AI policy template

Navigate to `https://www.sans.org/information-security-policy/` to download a copy of the *Artificial Intelligence Use Policy*.

Setting Standards

Standards come after the policies in that they define the specifics of the policy and are mandatory. They help enforce consistency throughout the organization and will provide details on how to follow a specific policy. Essentially, these provide the direction needed to support the policies. Let's look at some high-level examples of what a simple standard will look like for specific policies. The examples will be referencing a Windows endpoint environment:

- For a policy that states *All systems must be kept up to date with the most recent security updates*, the supporting standard may state something along the lines of *All Windows workstations will be configured using Windows Update for Business and Windows servers using WSUS or Azure Update Management. Update schedules will be defined and documented by the business use case.*

- A policy requiring encryption within your environment may have a supporting standard for Windows devices stating that *All Windows servers and end user workstations will be encrypted using BitLocker and/or Azure Disk Encryption.*

- A policy for device management or firewall requirements on end user devices may have a supporting standard with the following language: *A Windows firewall will be enabled and configured on all Windows end user devices and servers. Connection rules will be documented.*

- Your access management or password and biometrics policy may have a supporting standard along the lines of *A PIN and biometrics with Windows Hello must be used, and accounts will be required to use a password with a minimum of 16 characters. Passwords must contain a lowercase, uppercase, numerical, and special character and will only be changed in the event of a cybersecurity incident.*

- Your access management or MFA policy may have a supporting standard that states *MFA will be required for all users accessing all resources within the corporate enthronement.*

- Your access management, device management, or administrator policy may have a supporting standard along the lines of *There will be no standard user accounts assigned with local admin access on any Windows device.*

- Your device management or endpoint protection policy may state the following as a supporting standard: *All Windows end user devices and servers will be enabled with Microsoft Defender for Endpoint.*

- Your device management or compliance policy may have a supporting standard that states *Compliance policies for conditions such as device risk and minimum OS version will be assigned and enforced with Conditional Access on Windows devices.*

- Your information protection or data loss prevention policy may have a supporting standard with the following language: *Unified labeling with data loss prevention and information protection will be deployed to all Windows end user devices.*

Although there are a lot of examples above, the principle of the standards provided should be part of everyone's standards, whether Windows, macOS, or Linux. These are core endpoint requirements.

Bear in mind that these are brief and at a high level to demonstrate the difference between a policy and a standard. It's important to note that policies should be vague so they don't need to be updated frequently. Your standards documentation will be where the specifics are defined, and modifying these to meet the constant change and growth of technology becomes much easier than having to update your published policies.

Creating Processes and Building Procedures

Next in line are your processes and procedures once your standards have been defined. In *Chapter 7, Cybersecurity Operations*, we covered processes and procedures in more detail. Procedures were referenced as **Standard Operating Procedures (SOPs)**. As a reminder, a process is a step of defined tasks to achieve a specific outcome. This is essentially a document that formalizes the beginning-to-end lifecycle of accomplishing something, which is typically repeatable. This allows for consistency to ensure accuracy and compliance. Within technology and cybersecurity, they serve a very important purpose. On the other hand, a procedure is the step-by-step instructions used to accomplish a repeatable task or a specific step within a process. The instructions are intended to achieve a specific goal and assist with implementing the defined processes that tie back to your policies and standards. For example, let's take a scenario of deploying a Windows device within your organization, which becomes a repeatable process. A high-level example of a process to deploy a Windows device for a user may include the following steps:

1. Deploy a new device with Windows.
2. Ensure the device is connected to the internet.
3. Verify that the device configurations and applications have been installed.
4. Check whether the device is compliant.
5. Inventory and tag new devices.
6. Issue device to user.

Your procedures then become the specific step-by-step instructions to accomplish each of the defined steps above. For example, deploying a new device with Windows will have its own procedure and set of instructions to ensure Windows is installed correctly on that new device before moving to the next step in the process.

 There are many process mapping tools available to manage and track all your processes and procedures, which is important. You must ensure your processes and procedures are published and accessible for others to use. From a personal perspective, I've found Microsoft Visio a great tool to create higher-level processes.

Don't overlook implementing processes and procedures throughout the program. They serve a very important purpose to create consistency and reduce risk by standardizing how something should be accomplished.

Recommending Guidelines

Guidelines provide recommendations or best practices and are not mandatory requirements. They can be complementary controls in addition to standards, or even provide guidance where a standard may not apply. Guidelines will typically be created by the SMEs on the cybersecurity team who can provide guidance to the users for a better experience.

An example of a guideline might be: *Ensure you save and close all documents and programs before rebooting after receiving the latest Windows updates.*

Although not mandatory, guidelines provide a lot of value to users to help them to be more productive with technology. When building guidelines, it's important to think about how to efficiently provide the users with visibility and access to the guidelines. An effective communication plan is critical to ensure the users read and use the guidelines. The following are five ideas that will help with communicating your guidelines:

- Build a theme around your guideline communications, for example, smart cyber guidelines.
- Insert a section on guidelines in the company newsletters and/or communications.
- Link your guidelines back to a central repository for users to come back and access.
- Keep your guidelines short and to the point.
- Make your guidelines relevant to both professional and personal usage.

Now that we've completed the lifecycle of policies, standards, processes, procedures, and guidelines, let's take a look at the hierarchy from a high-level visual. As you can see, based on the pyramid, your policies will be at the top as there will tend to be fewer policies over your standards, processes, and procedures. Next is your standards, which, in turn, will generate your processes and procedures and any guidelines needed to support the organization.

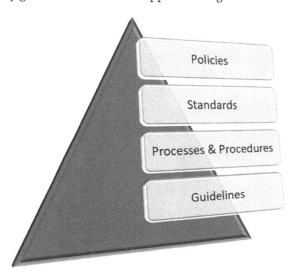

Figure 13.8: Policies, standards, processes & procedures, and guidelines hierarchy

Creating policies, standards, processes, and procedures may not be the most exciting task but it needs to be done. It's important that these are not overlooked and that time is spent ensuring you have well-written and clear policies in place, with well-defined standards, processes, and procedures for your organization to follow. This, in turn, will create a more resilient organization and aid in reducing risk.

Leadership Management and Communications

Now, we'll move on to what I would consider the most important component of your governance program. Here, you need to translate everything that is occurring within your cybersecurity program for the organization back to the executive leadership team and board of directors (if applicable). This won't be an easy task since the audience you will be communicating with will primarily be non-cybersecurity experts. Because of this, there will be a need for continuous education. Understanding your audience will be critical as you look to build your content and send communications and reports to these teams.

As you build your content, make sure you are providing relevant data to your executive leadership team and the board of directors. A baseline is provided below in the *Building Reports* section, but you may need to adjust to better support what is expected from the executive leadership team and board of directors. Don't just assume what you are presenting is what they want to see; ask for feedback and check if there are other data points they would like to be included. The executive leadership team and the board of directors won't care for the small non-essential details such as how many phishing emails were received or how many security events were triggered. All they want to know is how safe the organization is from cybersecurity threats. Because of this, keep the focus of your content around risk as it translates into business terms.

Communication Channels

When looking to report and communicate the cybersecurity program status, you need to take advantage of any channels that will allow you time to share the current state of the organization as it relates to cybersecurity. Although there is a realization that cybersecurity should be part of the executive leadership team and board of director's agenda, not every organization will be there yet. If you aren't receiving an invitation to these meetings, you need to reach out and ask the question as to why cybersecurity isn't on the agenda and whether there is an opportunity to get time on the agenda. If you work in an organization that isn't giving cybersecurity the attention it needs at the executive leadership and board level, I'd question whether this is a place I'd want to be working at. This will only put more liability on the CISO or cybersecurity leader's shoulders. If your organization is already accepting the need for cybersecurity to be on its executive leadership team's and board's agendas, make sure you are capitalizing on all available opportunities and communication channels.

Depending on your organization, the cadence of how your executive leadership meets, along with the board of directors (if one exists), will determine your communication channels. At a minimum, you will want to ensure you have a cybersecurity committee, executive leadership channel, and board of directors communication channel.

Figrue 13.9: Reporting hierarchy of the cybersecurity program

Based on the diagram above, the flow of each channel will feed into the next as you build your reports and present them. For example, the cybersecurity committee will cover content in detail, and some of the more relevant content will be presented to the executive leadership team. Finally, the most important and relevant content will be delivered to the board of directors. Let's look at each of the hierarchies in more detail.

Cybersecurity Committee

The first avenue into your upper management is the cybersecurity committee or council. You may not have a cybersecurity committee in place today but it's recommended you implement one if one doesn't exist. The idea behind the committee is to provide a forum where cybersecurity risk and other relevant items can be shared with the committee to review and make any necessary decisions. This committee will typically comprise personnel within the cybersecurity team like management, along with other key members within the organization such as other leadership and executive leadership members. Within this committee, a lot more content and reporting will be shared with the committee members where specific items can be discussed with, hopefully, decisions and direction provided by the committee. This will negate the need for all agenda items to have to go to your executive leadership team and only allow the focus of high and/or critical priority items to go to the executive leadership for review and any decisions needed.

This committee should meet, at a minimum, once a quarter, before the executive leadership and board of directors quarterly meetings, so any escalations from this meeting can be reviewed immediately without needing to wait.

However, it's not recommended to wait for any meetings to occur if there is any risk of substance that needs to be addressed; there should be a process to engage executive leadership and/or the board of directors immediately if needed.

If there are a lot of ongoing risks and items to discuss, you should consider meeting every other month or monthly. As part of this committee, some of the items that will need to be part of the agenda include:

- Quarterly executive dashboard
- Detailed metrics from each of the cybersecurity program functions – for example, reviewing user awareness, training, and testing metrics
- Overview of cybersecurity incidents
- Risk register
- Risks that need further review
- Projects charter
- A review of notable breaches
- Any cybersecurity policy-related items
- Other cybersecurity escalations

It will be important that the meeting is formally tracked with a charter, meeting notes are taken, and attendance is maintained. Certain decisions and risk acceptance will be made by this committee, and it will be important that it is formalized and documented for historical purposes if needed.

Executive Leadership

Your leadership team may meet as often as weekly, depending on your organization. This doesn't necessarily mean you are going to need to present weekly, but it will be important you are aware of the agenda and have a means of adding items to the agenda if needed, especially escalations. At a minimum, you will need to ensure you are reporting a detailed view of the cybersecurity program state to this group. Depending on when the cybersecurity council meets, you will want to ensure you are presenting to the executive leadership team after your cybersecurity council meeting. This will allow you to present the most relevant information to the group. For this meeting, the following agenda should be considered at a minimum for your executive leadership team:

- Quarterly dashboard
- Overview of cybersecurity incidents

- Risk register
- Risks that need further review

The idea is that the executive leadership team does not get overwhelmed with all the cybersecurity-related activity but is provided visibility into the more relevant items and those risk items that need to be reviewed and those that may need acceptance. Again, you must be transparent as the premise of this process is to ensure the executive leadership team and board of directors are accountable for cybersecurity risk as a whole.

Board of Directors

The final meeting you will need to ensure occurs is with the board of directors, if one exists. Getting time on this agenda may not be easy but it's important that time is made for cybersecurity. If you do get time, it will be very limited so you will need to ensure the content being delivered at this level is to the point and precise. Here, you will want to focus on a high-level view of the cybersecurity program, in addition to presenting any high or critical risk that requires the board of directors' review and any decisions to be made. For the board of directors, the following agenda should be considered at a minimum:

- Quarterly dashboard
- Major cybersecurity incidents
- Risks that need further review

Like your executive leadership meeting, ensure that full transparency is translated to the board of directors. They must be aware of the risk that has the potential to disrupt the organization. They need to take accountability for this type of risk by providing direction or, ultimately, accepting any high or critical risk on behalf of the organization. Without transparency, this cannot occur.

Moving from communication channels on to the next topic of building reports, it is important to ensure that your reports are tailored appropriately for each of the communication channels presented.

Building Reports

Here, you are going to create all your reports for the communication channels above. It will be important for you to have the most relevant and current data for each of your reports. These reports may not be the easiest to pull together, but once you identify how each of your reports will build out, make sure you document the process so it becomes easily repeatable each time you need to retrieve data.

Ideally, the reports should be dynamic and always up to date, but this may not be feasible for you in the current state. This is certainly something you need to strive for. At a minimum, make sure you can generate a quarterly report for the cybersecurity program that can be presented to each of the communication channels above.

Metrics

Before you can build any reports or dashboards, you are going to need to identify the metrics you would like to include in your reports and dashboards. There is no right or wrong answer as to what you include, but it's important to make sure the data is relevant and paints a clear picture of the current state and any risk that needs to be viewed. As mentioned in the introduction, make sure you are consulting with your executive leadership team and the board of directors on any expected metrics they would like to see as part of reporting. Since we have focused the program on the core functions throughout the book, it only makes sense to follow the same approach with your metrics as a baseline. This helps keep everything consistent with the cybersecurity program by reporting on each of the functions within the cybersecurity program. The following are considerations for metrics within each of the functions within the cybersecurity program:

- **Cybersecurity architecture**: For your cybersecurity architecture metrics, some data points should include the total number of projects or initiatives reviewed, how many have been approved, any pending further action, canceled items, any ARB items with security concerns or risk, and how many ARB reviews have been completed.

- **Identity and access management**: For the identity and access management function, some of the metrics to include are total active users, number of newly provisioned accounts, how many password and MFA resets occurred, number of privileged accounts, number of bot/service accounts, any compromised accounts, etc.

- **Cybersecurity operations**: For your cybersecurity operations metrics, some data points to include are the total number of incidents, total number of events, number of true positives versus false positives, number of incidents for different categories (phishing, system compromises, network, etc.), any **Service Level Agreement (SLA)** metrics, number of **Indicators of Compromise (IOCs)**, confirmed breaches, etc.

- **Vulnerability management**: Within the vulnerability management metrics, you should include your overall exposure score, how long it takes to patch your systems from when patches become available, your patching compliance rate, etc.

- **Cybersecurity awareness, training, and testing**: This is where you can show metrics relating to your user testing failure or pass rates, your training completion rate, departmental level completion rate, repeat offenders who fail, repeat offenders who don't complete training, etc.

- **Vendor risk management**: For your vendor risk management metrics, some of the metrics to track include an average score of your vendors, the total number of vendors being managed, how many vendors are below an acceptable grade, how many vendors are not within compliance, vendors that have had a recent breach, vendors currently under review, etc.

- **Proactive services**: Metrics for your proactive services should include penetration testing findings, other testing findings, tabletop exercise execution and findings, any proactive assessment findings, status on remediation of any findings, etc.

- **Operational Technology (OT) and the Internet of Things (IoT)**: For the OT and IoT metrics, you should look to provide metrics on all components covered above, but more specifically, you'll want to provide metrics around cybersecurity operations and vulnerability management with your OT and IoT environment.

- **Governance, Risk, and Compliance (GRC)**: This is where you will show metrics relating to any frameworks you have implemented, the outcome of any audits required for your organization like SOC 2 Type 2 and PCI-DSS, an overview of the risk register, etc.

There are many metrics that can be used for your reporting and dashboards. Make sure you aren't bringing too much noise but rather data that truly shows the impact of the cybersecurity program as a whole.

Dashboards

Now we have a better idea of some of the metrics that can be used, the next step is to generate your reports, or, as I prefer, dashboards that present your data in a more consumable format. For your reports and dashboards, you can be as creative as your imagination; there is no right or wrong way. As stated above, you are going to want to build detailed reports and/or dashboards for each of your functions as referenced above. These dashboards will provide more specific details for each of your functions. I'd highly recommend you aim to capture the metrics on a single page or slide if you are presenting in a PowerPoint type of presentation. Also, make sure you work with your vendors on providing relevant and quality reporting. For example, if you are using Security-Scorecard, they have *Board and Executive Reporting* capabilities as part of their service: https://securityscorecard.com/why-securityscorecard/board-executive-reporting/. The following is an example provided on their website of some of their reporting capabilities.

This specific example could be used as part of your own company's external attack surface monitoring or to show more details on important vendors that need further review.

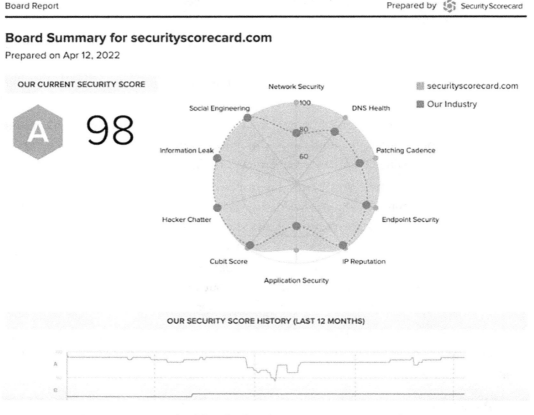

Figure 13.10: A sample of data for the cybersecurity program-specific functions

Once you have all your data for each of your functions, you'll need to compile a unified dashboard with the highlights from each function. This is not easy, and you may need multiple slides, but bringing them together on one dashboard will help with a holistic view of everything. Again, there's no right or wrong way to do this, and you should request feedback as you customize and evolve the dashboard to meet the executive leadership team and board of director's expectations. The following is a sample of a dashboard to give you some thoughts on how to build your own. Everything shown is sample data with some components referencing data from third-party platforms, as noted with asterisks.

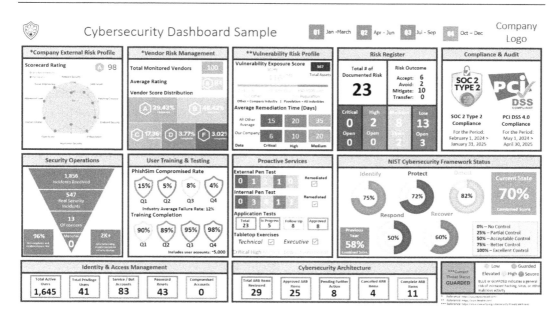

Figure 13.11: Cybersecurity sample dashboard

It may not be feasible to automate all the data you see in the dashboard above, but the ultimate end goal should be to integrate each of the data sources with a frontend application that can provide automation and up-to-date data for your dashboard. Even more impressive would be the capability to click on the data within the dashboard to take you to more detailed data for that specific function, like an interactive type of dashboard.

Now that we've reviewed the different communication channels and how to look at building metrics, next, we'll finish off the chapter with a look into some other governance considerations for your program, such as **Information Protection (IP)** and **Data Loss Prevention (DLP)**, insider threat, the data retention and information lifecycle, contract management, and resource management.

Other Governance Considerations

In the final section, we will briefly touch upon some other areas that need to be considered as part of the overall governance for your cybersecurity program. Although we didn't go into a lot of detail about the GRC application modules within the *GRC Application* section, it will be important for you to understand each of the modules in more detail and why they are important. As you look to centralize your GRC components, you will find the overall management of your cybersecurity program becomes much more efficient as you unify all data points.

This only becomes more advantageous as we need to share more with executive leadership and the board of directors.

We didn't cover risk, compliance, and regulations in this chapter as there is enough content for each to take up their own chapters, which will follow this chapter. Both risk and compliance tie very closely into governance as they comprise the lifecycle of your GRC program. It will be important that your governance, risk, and compliance all work hand in hand, especially if you are in a larger organization, as these may be run by separate teams. You must ensure that these functions aren't working in isolation and that each of the programs ties back to the broader governance oversight. In addition, think about how you can bring synergies with the broader risk management program within the organization if one exists. Managing risk in isolation throughout the business will only create more challenges. Work with your risk management and legal teams for opportunities to bring all risks under one platform; it may even make sense to leverage the GRC platform being used for your cybersecurity program.

As you continue to mature your GRC program, your vision should be a program that is fully unified by centralizing all components into one platform. Specifically, move away from individual spreadsheets used to manage your GRC components, and consider off-the-shelf applications, as discussed in the *GRC Application* section. There needs to be capabilities built in that allow for automatic and continuous scanning of your environment to be able to measure controls that tie back to any audits, regulations, or certifications that are required. We need to get out of the business of having to manually collect data as evidence for auditors to review; this needs to improve with automated capabilities for any audits, regulations, or certification requirements. Of course, there will always be a need for manual work, but being able to automate those easy repeatable tasks to confirm that controls are in place shouldn't be an unrealistic goal. As you strive for more automation and continuous scanning, this should begin to provide capabilities that allows for risk management to become more automated. This should translate into the GRC platform automatically identifying risk within the environment with the ability to send an alert to the SOC for review. Essentially, the GRC platform becomes your automated risk monitoring tool with capabilities to auto-validate controls or create an alert if any controls pose a risk. Some of this may not be realistic in the current state, but this should be the long-term vision of the GRC program.

Information Protection and Data Loss Prevention

IP and DLP are critical to your program and mustn't be overlooked. The information and data within our organization is what we need to protect and, to do this, we need to use capabilities such as IP and DLP. An important part of IP and DLP is the governance of the data and how we efficiently monitor and protect it. A program must be built around your data protection.

This will be covered in more detail in *Chapter 15, Regulatory and Compliance.*

Insider Threat

To some extent, this falls closely within the IP and DLP conversation we just had. Insider threats can cause a lot of damage to an organization, and insiders are able to go under the radar because they already have authorized access to data within your organization. It is important that capabilities are put in place that use a behavior type of detection along with IP and DLP capabilities to identify if data is being extracted that should not be. This is not easy, but capabilities are available to help better monitor insider threats within your organization. We reviewed insider risk in detail in *Chapter 6, Identity and Access Management.*

Data Retention and Information Lifecycle

Companies purposely collected more information than they needed in the past. This was primarily because data holds value, and the more they have, the more they can profit from it, especially our personal data. Unfortunately, this negatively impacts companies because when a breach occurs, more data is exfiltrated than what a company should have. Organizations need to do a much better job of disposing of data that is no longer needed or has been retained for longer than needed. This is where having a data retention schedule in place is needed, along with ensuring that the lifecycle of the data is managed correctly. Every organization should have a **Records and Information Management (RIM)** policy and retention schedule for all data. Another consideration is the disposal of physical documents as well as hardware. Make sure you have the correct processes in place to ensure physical documents and equipment are being disposed of correctly through an authorized vendor. This will be covered in more detail in *Chapter 15, Regulatory and Compliance.*

Contract Management

It will be important that good governance is in place for your contracts, especially in today's world, as it is recommended that all contracts contain cybersecurity language to allow for some protection and transparency with your vendors. As covered in detail in *Chapter 10, Vendor Risk Management*, you'll need to collaborate closely with your legal team, procurement team, and the business owner of the vendor. One of your biggest challenges from a governance perspective is ensuring all contracts are being reviewed. You will come across the business working directly with vendors without going through the correct protocols. This is where your policies and processes will need to be socialized and understood by the broader organization to ensure that risk is minimized with your vendors.

Resource Management

We touched upon resource management in both *Chapter 2, Setting the Foundations*, and *Chapter 4, Solidifying Your Strategy*, but it's important as part of being a leader that your resource allocation and the management of your resources is closely governed. As we have touched upon, in certain functions throughout the program, burnout is real and not something that should be taken lightly. This must be managed closely. In addition, we are all aware of the demand for both IT and cybersecurity as we support every part of the business. With the business continuously proceeding with projects and initiatives without our knowledge (unless good organization program management is occurring), we are expected to be able to provide support on an ongoing basis for the broader organization. With limited resources, this becomes challenging, and the work can quickly add up. As leaders, it is important to ensure you have the resources available without overworking your teams. As a reminder, we covered prioritizing well-being in detail in *Chapter 1, Current State*. Make sure you re-review this content.

Summary

Governance is not a small feat and to efficiently govern your cybersecurity program, you are going to need the right support and resources to make it a success. It is important that you are aware this is a program that will not be set up overnight. It will take time and dedication with support from your executive leadership team and the broader business functions. And the reality is, this program is required whether you like it or not. Without it, the liability for a major cybersecurity incident will fall directly on the CISO or cybersecurity leader in charge. One of the primary drivers for this program is to bring transparency to the executive leadership team and board of directors so that accountability occurs at the top, as it should.

To begin the chapter, we reviewed the importance of program governance and how governance encompasses risk, compliance, and regulation, which, in turn, overlooks the broader cybersecurity program. We also reviewed some real-life examples of where both cybersecurity leaders and CEOs are being held accountable for negligence with users' data. This led to reviewing some statistics related to GRC before finishing the section with an overview of some of the more prominent components of a governance program. In the section that followed, we covered in detail the program structure and governance. Here, we reviewed the importance of a strategy and roadmap for your governance program, a high-level overview of roles and responsibilities, and the governance organization structure, before finishing off the section with a detailed overview of the need for a centralized GRC application and the primary components that should be part of the platform.

Next, we took a deeper dive into policies, standards, processes, and procedures, which are considered some of the core principles of governance. We provided details on defining your policies, along with examples of where you can reference templates to get you started. Setting standards is what followed policies, along with reviewing processes and procedures in more detail, and the section finished with an overview of recommended guidelines. The following section, *Leadership Management and Communications* is what I consider one of the most important components of your GRC program and the primary driver around why the program needs to be in place. This is where you look to bridge the gap between the cybersecurity program and executive leadership team and the board of directors. Here, you will define your communication channels along with identifying the metrics needed to build your reports and dashboards to provide full transparency. We then finished off the chapter with a review of some other governance considerations as part of your GRC program.

In the next chapter, we will be covering the second component of the GRC program, managing risk. As we have alluded to many times, the foundation of the cybersecurity program is to manage and reduce risk for the organization. First, we will review further why risk is so important and the need to translate risk into business terms. We will then look at the different types of risk to provide a better understanding, which will lead to reviewing risk frameworks and assessments. Next, we will look at the importance of tracking risk, which will tie back to the GRC portal and the risk register module. We will then finish the chapter with a detailed review of the cybersecurity landscape.

Join our community on Discord!

Read this book alongside other users, Cybersecurity experts, and the author himself. Ask questions, provide solutions to other readers, chat with the author via Ask Me Anything sessions, and much more.

Scan the QR code or visit the link to join the community.

https://packt.link/SecNet

14

Managing Risk

Now that we have covered **Governance** in detail, we will review the **Risk** component of the **Governance, Risk, and Compliance (GRC)** program, along with the importance of risk. As we have stated multiple times, everything we manage as cybersecurity professionals is about risk. As you manage risk as a leader, it is important that you translate the technical component of risk into business terms so that a business can understand it from an impact perspective. It is not our job as cybersecurity leaders (or those on our team) to say no to the business. Our role is to assess the risk level of the identified risk and translate it into business terms for review. If a risk has been identified, it is then a business decision to determine whether it would like to accept the risk, look at ways to reduce risk, review whether the risk can be voided, or see if there is a way to transfer the risk before proceeding. Either way, the risk will need to be documented for reference in the future if needed.

We will begin the chapter with a deeper dive into risk, especially for cybersecurity. Here, we will cover the importance of focusing a cybersecurity program on risk and establishing the expectations that our role as cybersecurity professionals is in managing risk. We will also touch upon the fact that risk is much broader than cybersecurity and that, from an organizational level, there is a magnitude of risk that needs to be managed. Then, we will review the different types of risk that need to be managed, which expands beyond cybersecurity, so that you gain a better understanding of the broader landscape of risk. We will look at qualitative risk versus quantitative risk along with the different different risk mitigation options, which include risk acceptance, avoidance, transfer, and mitigation.

Then, we will review some of the different risk frameworks that your organization can consider. For example, the NIST Risk Management Framework is a great reference.

Then, we will look at tracking your risk, which is a very important component. We briefly covered this in Chapter 13, Governance Oversight, and the importance of keeping an up-to-date risk register with the latest identified risk.

The chapter will then finish with an overview of the cybersecurity insurance landscape and the importance of maintaining a cybersecurity policy for your organization.

The following will be covered in this chapter:

- Everything is about risk
- Understanding risk types
- Risk frameworks
- Tracking risk
- The insurance landscape

Everything Is about Risk

First, let's take a high-level look at all sub-functions that should be addressed as part of the Risk function. The following image captures much of what the Risk function entails.

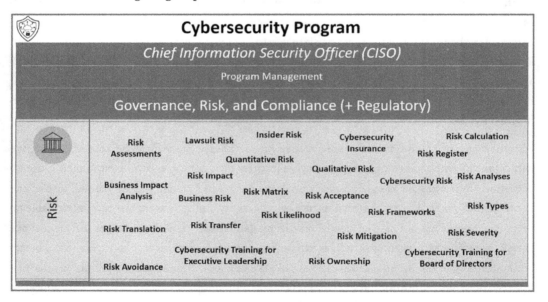

Figure 14.1: Sub-functions of the risk function

As we've continued to reference throughout the book, cybersecurity is all about risk management. At a time when the threat landscape has never been so advanced and active, risk management has only become more critical. It is important to acknowledge that risk will never be eliminated.

There will be some level of risk with everything in life; this is unfortunately the reality of the world we live in. Our role as cybersecurity leaders is to understand the level of risk present, also known as inherent risk. Once this risk is understood, we then need to advise on how to reduce the inherent risk as much as possible, typically through the use of controls. Once the controls have been put in place, it is important that we understand what risk is left over, otherwise known as residual risk, and at what level we are willing to accept this residual risk so that we can allow our organizations to operate as efficiently as possible? With residual risk, it is imperative that we continue to monitor it regularly to ensure that it is still within an accepted range for an organization.

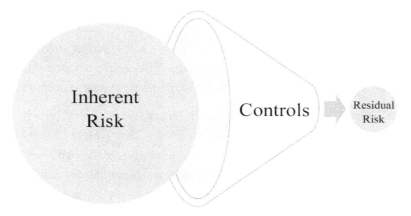

Figure 14.2: The process flow for inherent risk into residual risk

As we look to manage risk from a cybersecurity standpoint, it is important that all risks are formally documented and tracked, and all high and critical risks at a minimum are reviewed by the executive leadership team and board of directors. This is where the **Governance** component provides its value, as we reviewed in *Chapter 13, Governance Oversight*.

Risk Translation

Translating risk into business terms is a requirement for all CISOs and cybersecurity leaders. Unfortunately, this is not an option. This is not just important for those leading the cybersecurity program; this is a trait that everyone who works in cybersecurity needs to work on.

As you interact with a business on projects and support them on any cybersecurity issues, it will be valuable to ensure that any risk identified in a project or from a cybersecurity incident is described in a way that shows the impact it can have on the business. For example, let's say there is a new project where the finance team needs to send information to an external business partner. As part of the solution, they are looking to send data in a spreadsheet via email in an unencrypted format. Obviously, from a risk perspective, there are several faults with this method.

As a cybersecurity professional, you need to ensure that the business understands the risk of data easily being exfiltrated through this method, such as accidentally sending to the wrong email, a recipient saving a file on a personal device, or someone's mailbox becoming compromised and the data being accessed. Once this happens, threat actors can use the financial data to cause damage to the organization. Even worse, if this is consumer financial data, regulations and laws may come into play. Using examples of real-life situations can help with these conversations as well as demonstrate the potential financial, reputational, and business impact. In this situation, you'd want to remove the use of email and recommend a solution that provides the correct encryption, access controls, and auditing capabilities. It will be important that you can partner with the business as more secure solutions are needed to reduce risk when identified.

As stated, translating risk is very important. When reporting to your executive leadership team and the board of directors, ensure you translate risk in terms they are familiar with as you present to them. This may not be easy at first, but the more you become familiar with how your executive leadership team and board of directors operate, you'll be able to continue to fine-tune your data and presentations to better suit their expectations. However, as much as we do need to translate cybersecurity risk into business terms as much as possible, there is also a two-way street that needs to be addressed. The executive leadership team and board of directors need to understand the language of cybersecurity moving forward. This is no different from the executive leadership team and board of directors not understanding the language of financial risk. Today, cybersecurity risk poses as much risk as financial risk, if not more, and these conversations must happen at the executive leadership and board of directors' level, with an understanding of cybersecurity. As a cybersecurity leader, and as you (hopefully) engage and interact with the executive leadership team and the board of directors, take the opportunity to educate them as much as possible about cybersecurity principles and key items they should be more aware of. Make sure they are aware of the ongoing challenges and real-life examples so that they can take the initiative to research and educate themselves more on the matter. Furthermore, share the opportunity for them to become more versed with cybersecurity at their level.

For example, courses are becoming available specifically for executive leaders and boards of directors, which is a great sign. The MIT Sloan School of Management, for example, has a course for executive leaders and the board of directors, named *Cybersecurity Governance for the Board of Directors*. Making your executive leadership team and the board of directors aware of these opportunities will hopefully go a long way.

Figure 14.3: MIT Sloan School of Management Cybersecurity Governance for the Board of Directors

Source: `https://executive.mit.edu/course/cybersecurity-governance-for-the-board-of-directors/a054v00000qmgE1AAI.html`

In addition, and as we have already touched upon throughout the book, the broader organization and all users within your organization need to understand the language of cybersecurity and the risks associated with it. Cybersecurity is now a shared responsibility, and everyone within an organization must become more educated about the risk of cybersecurity and how to better protect themselves and the organization to reduce risk. This goes back to the need for a culture built around cybersecurity within your organization. Continue to educate your users on cybersecurity and the risk involved so it becomes second nature for them. We covered this in more detail in *Chapter 9, User Awareness, Training, and Testing*.

Risk Ownership

Risk, in general, is an organizational-level responsibility; more specifically, the executive leadership team and board of directors have the highest level of accountability. This is why they are in the positions they are in. And the same applies to cybersecurity risk; the accountability of cybersecurity risk must live at the highest level within an organization.

There shouldn't be any excuses moving forward that cybersecurity risk is something the executive leadership team and board of directors are not aware of. As much as it is the responsibility of the CISO and/or cybersecurity leader to bridge the gap between the cybersecurity team and upper leadership, it is also the responsibility of the executive leadership team and board of directors to ensure that they request insight into the cybersecurity program and are notified of any current risk. As we covered in *Chapter 13, Governance Oversight*, your GRC program is where this needs to be formalized for the organization.

The good news is that we are seeing data showing corporations identifying cybersecurity as a major risk at an organizational level. In *Chapter 2, Setting the Foundations*, we reviewed the *Allianz Risk Barometer 2023*, which identified the top risk for organizations. For 2024, the questionnaire included responses from 3,069 risk professionals across 92 different countries and territories. For the third year in a row, cyber incidents were listed as the top risk, and 17 countries listed it as their number one risk.

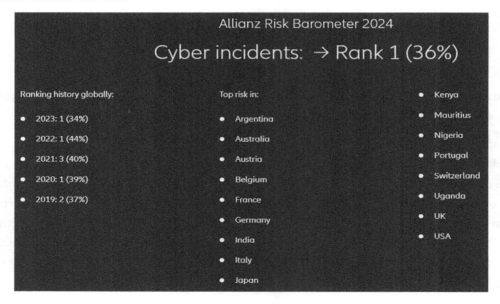

Figure 14.4: Allianz Risk Barometer 2024 for cybersecurity risk

Source: `https://commercial.allianz.com/news-and-insights/expert-risk-articles/` `allianz-risk-barometer-2024-cyber-incidents.html`

Something we continue to touch upon is the reality that the CISO, any cybersecurity leader, or anyone within a cybersecurity team for that matter must not accept any cybersecurity risk without following the approved processes in place. Any identified risk needs to be formally documented and the correct process followed for acceptance. This includes the need to ensure that risk is entered into your risk register. Again, this is part of your broader GRC program. For Low and Medium types of risk, processes should be easier to follow to allow risk to be accepted, preferably by the business function leads, and an inform to the executive leadership team. Any High and Critical risk must go to your executive leadership team and the board of directors for review. From here, they will need to determine how to address the identified risk, depending on options such as mitigation, avoidance, transfer, and acceptance. Ensuring that you fully document everything will be very important, as you may need an audit trail at some point in the future for reference.

To finish off this section, it is important to remind ourselves that it is not our job to say no to a business when they are looking to deploy new solutions or onboard new vendors and capabilities. As cybersecurity leaders, we need to review any new technologies and solutions as part of the broader **Architecture Review Board** (**ARB**) process, as discussed in *Chapter 5, Cybersecurity Architecture*. It is important to ensure that the business follows the defined strategy and that it complies with any policies, standards, or processes in place. The ARB process should capture and provide an avenue for these discussions to occur. If there is a need to proceed with anything that falls outside of compliance or the defined standard, processes should be in place to document and escalate to the appropriate resources within the organization, most likely the executive leadership team. If it is determined to proceed, the solution being deployed should be documented in the risk register with the appropriate approvals tracked, ensuring that accountability occurs at the highest level if an incident does occur. You will find many requests from the business that fall outside of the agreed-upon strategy, so these items must be addressed through defined policies, standards, and processes in place. Make sure you assess the risk profile at hand and ensure that the correct approvals are in place if needed; don't simply say no to the business.

Now that we have reviewed risk as it pertains to a business in more detail, including where the accountability for risk within an organization should live, let's shift our focus to the different types of risks.

Understanding Risk Types

In *Chapter 2, Setting the Foundations,* and *Chapter 10, Vendor Risk Management,* we touched upon some of the different types of organizational risk that must be accounted for within an organization.

Although we are focusing on cybersecurity risk within this book, broader risk management must be understood and managed appropriately. Here's a reminder of some of the organizational risks mentioned previously:

- Financial
- Market
- Compliance
- Legal strategic
- Supply chain
- Environmental
- Competition

These are just some examples of the many types of risk that your organization may have to understand and manage. This list will also differ based on the industry that your organization is located in. For example, if you're within an industry that manufactures, you may have risks associated with raw materials that are becoming depleted. If you are located within a specific country, your political risk may be different from organizations within other countries or regions. Although these risks will be out of your scope as a cybersecurity team, it is important that you understand the broader risk that impacts an organization and that you work closely with the individuals or teams that overlook some of the broader organization's risks. It is also important that the cybersecurity risk component is directly embedded into the broader risk program and tracked as part of all other risks within the organization. As a cybersecurity leader, you will naturally find yourself working with your risk management team if one exists. If one doesn't exist, the responsibility may sit with the Legal team, with which you'll already (hopefully) be collaborating closely. At the end of the day, risk management is a necessity for all organizations, and cybersecurity risk has quickly become one of the higher-priority risks to manage within an organization.

Let's take it down a level to cybersecurity risk, which includes everything and anything that can have an impact on an organization's technology, allowing for the compromise of any information or data from that organization.

This will typically occur through a major cybersecurity incident or a material breach within the organization, which can have a major impact on financial and reputational factors. As a reminder, in *Chapter 1, Current State*, we provided an overview of some of the techniques used in attacks against organizations. This, in turn, is the cybersecurity risk you need to be aware of within your organization. There is a lot of cybersecurity risk to be aware of, and the unfortunate reality is that it continues to grow. The following provides some of the more familiar cybersecurity risks you should be aware of:

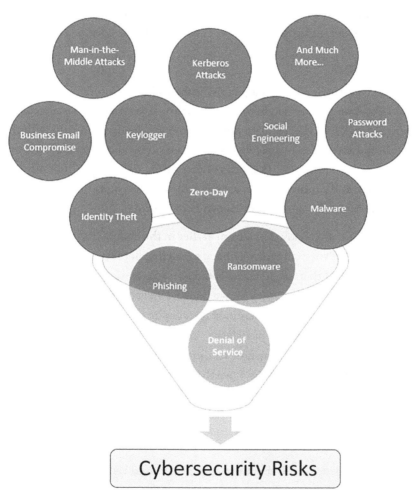

Figure 14.5: Cybersecurity risks

As you can see, the amount of cybersecurity risk shown in the diagram is quite substantial, and this is only a portion of the cybersecurity risk we must manage on a day-to-day basis, and it is constantly growing. It is important that you continue to review the current landscape to stay afloat with the latest risks that can impact your organization. As you look to focus on cybersecurity risk, it is important that you can measure and calculate the potential impact of that risk. In this chapter, we will focus on a couple of popular methods to best calculate risk, using qualitative and quantitative methods, which we will review next. Once that risk has been measured, you then need to determine how to address it by focusing on four of the more common management strategies: avoid, transfer, mitigate, or accept. These will also be covered later in this section.

Risk Calculation

One of the biggest challenges with managing risk is the ability to quantify it or be able to articulate the impact and likelihood of an event happening because of that risk. This is not easy and presents a very challenging task for cybersecurity professionals and leaders. To help with this, there are a couple of well-known methods typically used to assess risk and determine the importance of addressing specific risks. Primarily, the focus of measuring risk has traditionally been around a qualitative approach and continues to do so today. Moving to a quantitative approach would be preferred, but unfortunately, this method doesn't come easy.

Expanding on inherent and residual risk from earlier in the chapter. It is important to note that when working with risk, you will typically have a baseline deployment (of an application, software, device, service, etc.) that becomes accepted as a standard. Although there will always be some form of risk with anything deployed, we do need an acceptable baseline to measure from. If, for some reason, a vulnerability is identified on a deployed application, software, device, service, etc., we then have a new inherent risk that we need to calculate, which is risk beyond the accepted baseline. Once we determine how to address the risk (for example, an update needs to be deployed to mitigate the vulnerability), we then need to understand the level of risk that still remains, as we know as the residual risk. Once we understand the residual risk, we also need to ensure we are comfortable with that leftover residual risk.

Qualitative

First, let's dive a little deeper into the more commonly used risk measurement method when assessing risk, better known as qualitative risk. Qualitative risk evaluates potential threats based on likelihood and impact without using numerical data.

It involves describing and assessing risks through expert judgment and scenarios to understand their severity and implications. Essentially, it takes more of a subjective approach versus leveraging actual data, such as a mathematical-based approach. To determine this, the approach typically involves determining the likelihood vs. the impact:

Figure 14.6: Qualitative Likelihood x Impact Risk Analysis

There are many ways you can assess the likelihood and impact of something occurring. For example, in *Chapter 8, Vulnerability Management*, we briefly reviewed scoring qualitative risk as part of vulnerability tracking. Here, we referenced the OWASP risk rating methodology, found at this location: `https://owasp.org/www-community/OWASP_Risk_Rating_Methodology`. As part of their qualitative calculation for vulnerabilities, the following is used by OWASP:

- **Likelihood:**

    - **Threat agent factors:** Skill level, Motive, Opportunity, and Size

    - **Vulnerability factors:** Ease of discovery, Ease of exploit, Awareness, and Intrusion detection

- **Impact:**

    - **Technical impact:** Loss of confidentiality, Loss of integrity, Loss of availability, and Loss of accountability

    - **Business impact:** Financial damage, Reputation damage, Non-compliance, and Privacy violation

Each of these is translated into a score. For example, take *Privacy violation* from *Business impact* – within this, a score from low to high is assigned with the following scores for any **Personally Identifiable Information (PII)** disclosure:

- One individual = 3
- Hundreds of people = 5

- Thousands of people = 7

- Millions of people = 9

The same principle will apply to all other categories, with a similar scale of minimal impact/like-
lihood receiving lower numbers and maximum impact/likelihood receiving higher numbers. This
then ties back to the overall matrix that calculates the overall likelihood and impact, providing
an overview of the vulnerability of the risk being reviewed. As shown in *Chapter 8, Vulnerability
Management*, the following is the output of a risk calculation from the OWASP scoring template,
which can be accessed here: `https://wiki.owasp.org/index.php/File:OWASP_Risk_Rating_`
`Template_Example.xlsx`.

Risk: Full database theft from datacenter

Likelihood								
Threat agent factors					Vulnerability factors			
Skill level	Motive	Opportunity	Size		Ease of discovery	Ease of exploit	Awareness	Intrusion detection
4 - Advanced computer user	1 - Low or no reward	access or resources required	5 - Partners		3 - Difficult	3 - Difficult	4 - Hidden	3 - Logged and reviewed
			Overall likelihood:	3.375	MEDIUM			

Technical impact					Business impact			
Loss of confidentiality	Loss of integrity	Loss of availability	Loss of accountability		Financial damage	Reputation damage	Non-compliance	Privacy violation
2 - Minimal non-sensitive data disclosed	0 -	0 -	9 - Completely anonymous		1 - Less than the cost to fix the vulnerability	1 - Minimal damage	0 -	5 - Hundreds of people
Overall technical impact:	2.750	LOW			Overall business impact:	1.750	LOW	
			Overall impact:	2.250	LOW			

Overall Risk Severity = Likelihood x Impact					Likelihood and Impact Levels	
Impact	HIGH	Medium	High	Critical	0 to <3	LOW
	MEDIUM	Low	Medium	High	3 to <6	MEDIUM
	LOW	Note	Low	Medium	6 to 9	HIGH
		LOW	MEDIUM	HIGH		
		Likelihood				

Figure 14.7: The OWASP Risk Calculator Template

You'll come across many different versions of risk matrices; some will use different calculation
methods, some will have multiple tiers of impact vs. likelihood, some will have different input
methods, and so on. At the end of the day, they are subjective and may not provide the intended
accuracy, but they serve the purpose of providing a ballpark estimate of the expected level of risk
within reason.

You can also keep the model very high-level by using judgment to calculate risk, without the need
for any additional scoring to generate the likelihood or impact.

For example, the following five-tier scoring matrix can be used to quickly calculate risk. Applying a realistic scenario could include the announcement of a zero-day vulnerability with your public-facing firewall. The intel we've been provided with is that this vulnerability is already being exploited in the wild, and once exploited, access can be elevated to administration, allowing a full takeover of the firewall. With this knowledge, you can make the following determination:

- **Likelihood**: Most Likely
- **Impact**: Catastrophic

Based on the determination, the risk severity will be classified as **Critical** and require all hands on deck to work on mitigating the zero-day vulnerability. Again, every environment is different, but I would suspect this would be a **Critical** risk for most.

		Impact				
		Insignificant	Minor	Moderate	Major	Catastrophic
Likelihood	Unlikely	Low	Low	Medium	Medium	High
	Possible	Low	Medium	Medium	High	High
	Likely	Low	Medium	High	High	Critical
	Most Likely	Medium	Medium	High	Critical	Critical
	Guaranteed	Medium	High	High	Critical	Critical

Figure 14.8: Example of a five-tier risk matrix

As we have observed, qualitative risk evaluates the likelihood and impact of threats through descriptions and scenarios. Let's move on to see how quantitative risk is used to measure risk, with a more precision approach.

Quantitative

The other risk calculation method we are going to review is quantitative risk. This type of measurement will provide a lot more accuracy over qualitative but can be a lot harder to achieve. Instead of using a subjective approach like qualitative, quantitative uses a mathematical-based approach by leveraging data points, like financial information and statistics, to derive a risk. By using real-world data, you should be able to make better decisions on risk within an organization. For example, let's look at some examples based on statistics we've reviewed throughout the book.

From the 2023 IBM *Cost of a Data Breach Report,* it was reported that greater levels of incident response planning and testing saved organizations $1.49 million when containing a data breach.

Although this may not be 100% accurate for all organizations, this dataset is derived from 553 organizations affected by data breaches throughout 16 countries and regions within 17 industries. Based on this, the risk of not implementing a **Cybersecurity Incident Response Plan (CIRP)**, tabletop exercises, and incident response-based courses could cost us up to an additional $1.49 million with a major cybersecurity incident or breach.

Let's say we do our homework and receive quotes to create a CIRP, implement tabletop exercises, and provide incident response courses and certifications for $100K combined. In addition, we would need to onboard a dedicated resource to manage the incident response process, which will cost approximately $150K all-inclusive. The total cost to achieve greater levels of incident response and planning is $250K versus the actual $1.49 million in additional costs caused by a major cybersecurity incident. This, in turn, could provide the potential for an ROI of $1.25 million if there is a major cybersecurity incident. Based on the quantitative analysis provided, this should be an easy sales pitch to proceed with strengthening any controls around incident response and preparedness, thus reducing the risk. It is important to note that this will not eliminate the risk, and there will still be residual risk when a major cybersecurity incident or breach occurs.

Another example from the same IBM report stated that organizations experienced an additional cost of $470,000 by not including law enforcement. Like above, we need to understand what the financial impact is to ensure that law enforcement is involved when a major cybersecurity incident occurs. The good news is that there is no direct cost for law enforcement involvement, but you'll need to identify what is needed internally. Is there a lack of documentation? Does a CIRP need to be created? Are new processes needed? Do we need to dedicate a resource to collect the contact information of law enforcement? Do we need to build relationships with law enforcement local offices? Once you identify the gaps, you can calculate the monetary value to ensure that law enforcement is engaged. Once you do the math, you can then determine if implementing the gaps is worth the effort to reduce the risk. In this situation, it should be fairly simple to ensure that law enforcement is engaged for a major cybersecurity incident.

Another example pertains to the continuous awareness of statistics that show the human element continuing to be included in a high percentage of breaches. In the 2023 Verizon DBIR report, it was reported that the human element was included in 74% of breaches. In the 2024 Verizon DBIR report (https://www.verizon.com/business/resources/reports/dbir/), it was reported that the human element was part of 68% of breaches.

Other statistics show similar findings. With this data, you should be able to do a cost analysis against your cybersecurity program to better understand how much investment (including resources) is made within each of the functions. I bet, for most, your user awareness, training, and testing function receives the least amount of investment. Because of this, you should use statistics and analysis to justify why the need for more investment in your user awareness, training, and testing program.

This includes the need to justify users spending more time learning about cybersecurity, specifically, those areas that are at higher risk of causing a major cybersecurity incident to occur. If the human element is the biggest cause of major cybersecurity incidents, why aren't we investing more in this area?

The examples provided above are very basic examples of quantifying risk. To expand beyond basic risk quantification is a concept that is becoming more widely adopted known as **Cyber Risk Quantification (CRQ)**. CRQ essentially uses multiple data points beyond numerical and statistical inputs, such as reviewing current security controls, leveraging cyber intel, looking at vulnerability scores, etc. to provide more intelligence decisions around risk exposure to highlight the potential financial impact. This, in turn, allows cybersecurity risk to be presented more from a business point of view. More solutions are becoming available using a CRQ approach, providing the capability for leadership to manage the complexity of cybersecurity risk in a more meaningful approach. An example of a vendor providing CRQ services is Kovrr: `https://www.kovrr.com/cyber-risk-quantification`.

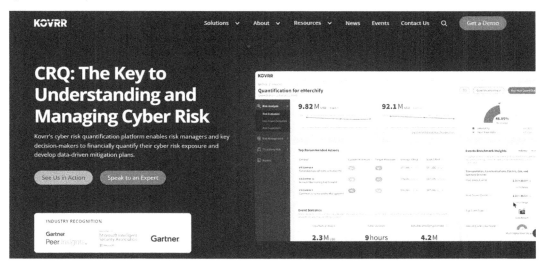

Figure 14.9: KOVRR's CRQ

As a leader, make sure you are familiar with and review CRQ to see how it can benefit your organization to more efficiently manage cybersecurity risk, and to provide the capability to translate cybersecurity risk into business terms.

Risk Mitigation

Next, we will look at the different mitigation options when managing your risk. As you are made aware of and need to manage your risk, you will need to determine how you want to handle it. As a risk needs to be reviewed, you will first need to determine the potential impact of it. This can be completed using the qualitative and/or quantitative methods we just reviewed, which will help, as you need to determine the next steps on how to reduce that risk if you decide to do so. As you look to handle risk, you will need to determine whether you avoid, transfer, mitigate, or accept.

Avoid

The first option is to avoid the identified risk. This typically means not proceeding with something when a risk has been identified. Take, for example, a new vendor you are looking to onboard within your organization. After a review of the vendor, it is identified that they come with high risk because of ongoing cybersecurity incidents and breaches. It has also been confirmed that they don't have a dedicated cybersecurity team and their cybersecurity practices are far from optimal. Ideally, you'd want to run a more realistic analysis of the cost of onboarding the vendor versus the impact and damage they may cause by compromising any of your data, or the potential impact on your business because they are not able to provide services during an outage from a major cybersecurity event or breach. Because of this, the decision should be made to not proceed with the vendor, thereby avoiding the risk. An alternative could be to look at other vendors or build the capabilities in-house.

Transfer

The next option we will review is transferring the risk. By transferring the risk, you move the risk and any financial impact from yourself to somewhere else. An example of this would be purchasing cybersecurity insurance. In the event of a major cybersecurity incident or breach, the cybersecurity insurance company assumes the risk to cover any costs and provides the needed resources to recover your business back to normal operation.

Mitigate

Next is risk mitigation, or risk reduction. This is essentially the ability to reduce any identified risk that poses a threat within your organization. The reality is that you aren't going to eliminate risk entirely, but if there is a higher level of risk identified, what can be done to reduce it?

An example may be a zero-day vulnerability that has been identified on one of your applications and it is considered high. Based on this, the cost to mitigate the vulnerability will be approximately $100K, but the cost of a major cybersecurity breach could add up to $1 million because the application stores sensitive information of your employees.

This breach also puts the reputation of the organization at risk. Based on this, you would proceed with implementing an update to the application immediately to mitigate the current vulnerability. Again, this doesn't mean that the risk is eliminated, as a new or more advanced vulnerability may be able to exploit the application in the future.

Accept

The final option available is risk acceptance; this is where an organization acknowledges the current risk at hand but accepts it as it is and takes no other action against it. This will typically be a decision taken when the cost to mitigate or reduce the risk outweighs the value of the overall asset with the identified risk. For example, you have an asset valued at $1K that has an identified vulnerability, but the cost to mitigate the vulnerability is $10K. With this information, the cost to mitigate the vulnerability outweighs the value of that asset, so the decision would be to accept the current risk, as it wouldn't make sense to invest $10K to protect a $1K asset. Obviously, this is a simple example, and there may be more data points to consider with a real-world example. However, the premise of the exercise is to demonstrate that it doesn't make sense to implement a control that costs more than the value of the asset with the identified risk.

Managing risk is not a simple task and can easily become very detailed and complex, depending on how accurate you would like to be with risk. We have barely touched the surface of risk management, merely providing a high-level overview of what is involved with risk calculation and risk mitigation.

Now that we have completed a review of the different types of risks, including what is involved in risk calculation and risk mitigation, we will review some different risk frameworks that are available for reference next.

Risk Frameworks

As we have continued to discuss throughout the book, the premise of the cybersecurity program is to manage and reduce risk related to cybersecurity for an organization. Essentially, our program can be thought of as a framework to an extent. If we take a step back from cybersecurity and look at risk from a broader perspective, it is critical that there is an overarching risk management program for the broader organization.

Your organization's size may determine whether you have a dedicated risk management function with a **Chief Risk Officer (CRO)** or whether this function falls within another group being a smaller organization. For example, does risk management for the organization fall within the CISO responsibilities, since our primary role is to already manage risk? Or does it reside with the Legal team? Regardless of where it lives, it will be important that there is some form of a centralized approach to manage and track all risks for the organization so that it's not distributed throughout.

And if risk doesn't sit within the cybersecurity function, there will need to be some strong synergies and collaboration between teams to ensure risk is being managed efficiently for the organization.

Earlier in the book, we discussed the importance and requirements for every organization to implement a cybersecurity framework. This essentially is a framework to manage risk more efficiently for cybersecurity. The following were mentioned as some of the more widely adopted cybersecurity-based frameworks:

- **International Organization for Standardization/International Electrotechnical Commission (ISO/IEC) 27001:2022**: `https://www.iso.org/standard/27001`
- **National Institute of Standards and Technology (NIST)** Cybersecurity Framework: `https://www.nist.gov/cyberframework`

In addition, an example of an IT-specific framework to manage and reduce risk includes the **Control Objectives for Information and Related Technology (COBIT)**: `https://www.isaca.org/resources/cobit`

> As a reminder, it is important to understand the difference between a framework and other similar items, such as controls, certifications, regulatory compliance requirements, acts, etc. When you think of a framework, the best way to conceptualize it is to ask, does it cover the entire scope of the cybersecurity program? For example, the NIST Cybersecurity Framework does, but the **System and Organization Controls (SOC)** doesn't. The SOC typically only covers a specific scope of your environment.

Before we look at some of the available risk management frameworks, it is important to ensure that you write and publish a risk management policy for your organization. This may not necessarily be the cybersecurity function's responsibility, but there will need to be collaboration and involvement with this policy, since cybersecurity currently holds one of the larger risks for organizations nowadays.

If you have a risk management team, this should be the responsibility of that team. If not, maybe your Legal team or internal controls teams will have some involvement in creating or overlooking the organization's risk management policy. Like all other policies, it will be important that users are aware of how risk is managed and handled within the organization, in addition to communication channels being made available to them to report any identified risk. The risk management frameworks mentioned below will provide the foundation for your risk management policy and what should be included.

Although we have reviewed cybersecurity frameworks throughout the book, we cover them in more detail in the next chapter. When focusing on risk-specific frameworks, there are some broader frameworks to consider for implementation. The idea behind a risk management framework is to help organizations better navigate the risk landscape across them, by putting a foundation in place that allows risks to be identified, tracked, and remediated. In *Chapter 4, Solidifying Your Strategy*, we briefly covered the ISO framework for cybersecurity. In addition to cybersecurity, ISO also has a framework for risk management. The *ISO 31000 for Risk Management* provides a framework that includes guidelines, principles, and processes needed to efficiently manage risk within your organization. It accomplishes this by using a process of identifying risk, and then analyzing and evaluating that risk once identified. This is followed by the process of treating the risk, which will then need to be monitored and communicated to others who need to be aware. You can learn more about ISO 31000 at `https://www.iso.org/iso-31000-risk-management.html` and directly purchase the guidelines at `https://www.iso.org/standard/65694.html`.

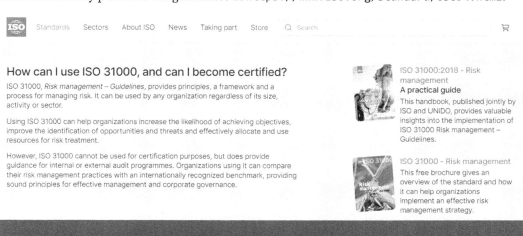

Figure 14.10: The ISO 31000 Risk Management Framework

The other cybersecurity framework we briefly covered in *Chapter 4* was from NIST. NIST also has a risk framework, better known as the NIST **Risk Management Framework (RMF)**. The NIST RMF process focuses on the integration of security, privacy, and supply chain risk management, and it can be applied to any organization that wishes to use it. You can access the NIST RMF here: `https://csrc.nist.gov/projects/risk-management/about-rmf`. The NIST RMF involves the use of seven steps as part of the framework:

1. **Prepare:** In this step, you will take the time to prepare an organization for the management of security and privacy risks while using the NIST RMF.

2. **Categorize:** Here, you will categorize all systems and information within your organization. This can be accomplished through the exercise of a **Business Impact Analysis (BIA)** for each of those systems and/or services.

3. **Select:** Following the categorization, you select the appropriate controls needed to protect the systems and information within your organization.

4. **Implement:** Once you have selected the controls, this step addresses the implementation of those controls for your organization's systems and information.

5. **Assess:** In this step, you will need to determine if the implemented controls meet the requirements to protect the systems and information within your organization.

6. **Authorize:** This is the step where accountability needs to come from the most senior leadership within the organization, determining if the current risk is accepted to proceed.

7. **Monitor:** The final step is ongoing monitoring and reporting to ensure the implemented controls maintain the accepted level of risk.

The current publication available from NIST is the *NIST SP 800-37 Rev.2*, which can be accessed by clicking on the link within the *Publication* section here: `https://csrc.nist.gov/pubs/sp/800/37/r2/final`.

NIST Special Publication 800-37
Revision 2

Risk Management Framework for Information Systems and Organizations

A System Life Cycle Approach for Security and Privacy

This publication contains comprehensive updates to the *Risk Management Framework*. The updates include an alignment with the constructs in the NIST Cybersecurity Framework; the integration of privacy risk management processes; an alignment with system life cycle security engineering processes; and the incorporation of supply chain risk management processes. Organizations can use the frameworks and processes in a complementary manner within the RMF to effectively manage security and privacy risks to organizational operations and assets, individuals, other organizations, and the Nation. Revision 2 includes a set of organization-wide RMF tasks that are designed to prepare information system owners to conduct system-level risk management activities. The intent is to increase the effectiveness, efficiency, and cost-effectiveness of the RMF by establishing a closer connection to the organization's missions and business functions and improving the communications among senior leaders, managers, and operational personnel.

JOINT TASK FORCE

This publication is available free of charge from:
https://doi.org/10.6028/NIST.SP.800-37r2

National Institute of Standards and Technology
U.S. Department of Commerce

Figure 14.11: The NIST RMF for Information Systems and Organizations

There are other risk management frameworks for reference, such as **Factor Analysis of Information Risk (FAIR)**, which assesses risk using a quantitative financial approach for cybersecurity and operational management versus the traditional qualitative method. You can learn more on FAIR here: https://www.fairinstitute.org/. Another notable risk management framework is the *COSO Enterprise Risk Management* framework, which you can also become certified in. You can learn more about the *COSO Enterprise Risk Management Framework* here: https://www.coso.org/guidance-erm.

Implementing a full-blown risk management framework may not be realistic for your organization at this time if one is not currently in place, especially since your focus will be building the cybersecurity program. If there is a risk management function or even an internal controls function within your organization, this may be where the broader risk management responsibilities live, although risk management falls within every function within an organization. If there is no risk management framework in place, make sure, at a minimum, that you follow the basics to efficiently manage risk within the cybersecurity program for the organization. The frameworks listed above should, at a minimum, provide guidance and the core concepts you need to identify risk, track it, and remediate it. While doing this, ensure that it is documented accordingly and shared through the correct channels so that executive leadership and the board of directors have full visibility, making them aware of the identified risk.

Moving on from risk frameworks, we will focus our attention on how to efficiently track risk for your organization and ensure that there is full transparency for the executive leadership team.

Tracking Risk

As we covered in *Chapter 13, Governance Oversight*, tracking risk is one of the most important components of a cybersecurity program, especially as we look to bridge the gap between the cybersecurity team, the executive leadership team, and the board of directors. It is important that when the risk is identified, it is formally added to the risk register. Once tracked, you will need to ensure that the identified risk is addressed, based on one of the mitigation methods covered earlier in the chapter. Even more important is ensuring that any high and critical risk is reviewed and acknowledged at the executive leadership and board level. If any decision needs to be made around a high or critical risk, it will be important that the correct processes are in place to help address these risks. Whatever direction is determined, it will be important that all activity and documentation are efficiently tracked.

As we discussed in the previous chapter, the risk register needs to be digitized and should be part of the GRC platform. With the risk register being part of the GRC platform, identified risk entered into the risk register can then be tied back to an application, vendor, and the **Risk and Control Matrix (RACM)** if one exists. This should also make it easier to report and provide a dashboard of all current risks being tracked within the risk register.

As a reminder, risk exists with everything within an organization, but the idea behind the risk register is to track identified risk beyond what is already acknowledged as risk with the standard controls in place. For example, you have an application that has been deployed following best practices and all cybersecurity controls in place.

You have completed an application penetration test, and nothing of any concern shows up on the report. Several months down the road, a critical vulnerability is identified in which PII could be accessed by an unauthorized user. This would trigger the need to add this to the risk register to bring visibility to the risk and immediately address it. As these types of risks are identified, they must be added to the risk register immediately for full transparency and to allow for tracking and updates to occur on status. It is also important to understand that judgment will need to be made at times when dealing with risk. Not everything is clear-cut, and your expertise, along with the expertise of those around you, will be needed when addressing risk. You will be challenged.

Implementing a Risk Register

As you look to implement a risk register for your organization, it will be important to understand if there are any capabilities available currently. For example, is there a broader risk management team that already tracks another risk, allowing you to inject the cybersecurity risk into a broader risk program? Is there a GRC tool being utilized for functionality that doesn't include the risk register? If so, is there an available module that can be purchased? If there is nothing available, you'll need to implement something immediately for the short term, whether it is a spreadsheet or something else to get you started. Ideally, you'll want to ensure this becomes part of the broader GRC platform in the long term. Regardless of the solution you implement, the following fields, at a minimum, should be part of your risk register:

- Unique risk ID
- Risk title
- Risk description
- Risk category
- Risk owner
- Date identified
- Date reviewed/modified
- Risk impact
- Risk likelihood
- Risk severity
- CVE ID (if applicable)
- CVSS score (if applicable)
- Impacted products
- Application/service owner

- Remediation steps

- Status

- Date remediated

- Additional information

 We covered the CVE ID and CVSS score in more detail in the *Vulnerability Overview* section of Chapter 8, *Vulnerability Management*.

Now that we've covered the details that should be included in a risk register, let's look at options to track your risk.

Digitizing the Risk Register

It is important that you digitize your risk register as a first option if applicable. As GRC becomes more prevalent for organizations, the risk register should be a default module as part of the platform. There are quite a few GRC platforms available that provide broader GRC capabilities, and it will be important you understand what the capabilities are of each if you are looking to deploy one. Ensure that you have defined your requirements, as each of the GRC platforms may have different capabilities. Regardless of what GRC platform you decide to proceed with, it must include a risk register to centrally track all your risks and tie them back to different modules and data within the platform. One example is Hyperproof's **Risk Register** within its **Risk Management** platform: `https://hyperproof.io/product/#risk-management`.

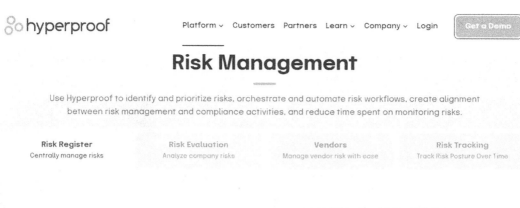

Figure 14.12: The Hyperproof Risk Register

Implementing a GRC platform or onboarding a vendor to provide capabilities may not be possible in the short term because of budgetary challenges and the time it takes to onboard a vendor. Because of this, there are other options available to quickly enable a risk register. One example is a risk register made available by SecureMetrics within the PowerBI platform: `https://appsource.microsoft.com/en-us/product/power-bi/securemetrics.sm_risk_register`.

This risk register does come at a small cost and requires you to have a Power BI license to deploy it. Although it does leverage a pre-defined Excel spreadsheet to manage risk behind the scenes, it does provides a **Risk Dashboard** visual on top of the data, along with the risk register in a more presentable format, as shown below.

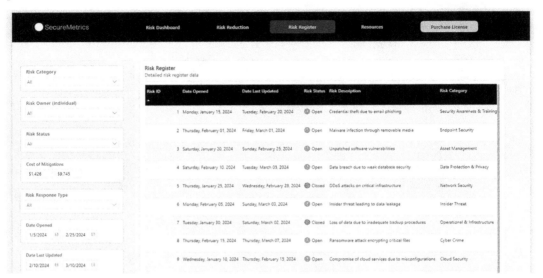

Figure 14.13: The SecureMetrics Power BI Risk Register

The final option is not ideal, but if you don't have anything else at your disposal, you'll need to deploy an Excel-based risk register, as you need to have something in place. This can be thought of as a temporary measure until you can get your GRC program, and the risk register, to a more mature state. Although not preferred, an Excel spreadsheet is better than having nothing in place at all, especially when we need to document everything to provide full transparency to the executive leadership team and the board. Simply use the fields referenced above within an Excel spreadsheet to begin tracking any identified risk for your organization to get started.

It is important that risk management gets the attention and focus it needs to ensure that everything is documented appropriately. This means that someone is responsible for the lifecycle of any identified risk. As a risk is identified, it must be entered into the risk register right away, determining the likelihood and impact of the identified risk. This will, in turn, provide the risk severity, which will help determine the next steps. As part of identifying the risk severity, it will need to be determined what the cost is to reduce the risk versus the value of the asset at hand. Once this has been determined, direction will be needed on whether to mitigate, accept, avoid, or transfer the risk.

Once all this has been determined, you'll need to ensure that any required actions are taken so that the risk can be resolved. Ensure that everything is documented and all documentation is retained when going through the lifecycle of managing risk.

Next, let's shift our focus from tracking risk to an in-depth review of cybersecurity insurance, looking at why it's important to consider purchasing a policy to support your organization as cybersecurity incidents occur.

The Insurance Landscape

Earlier in the chapter, we provided an example of transferring risk. This is primarily done by purchasing cybersecurity insurance for your organization, which should be a requirement for your organization. Before we get into more details on the policy, let's take a look at what the estimated cost of a breach may be to provide better context around why your organization should onboard a cybersecurity policy.

Estimating the Cost of a Breach

To provide more context into the need for cybersecurity insurance, Travelers' *The Hard Realities of Cyber Event* resource provides several examples of the estimated cost of a breach, with a breakdown of the costs involved. You can review the examples here: `https://www.travelers.com/business-insurance/cyber-insurance/claim-stories#`. For example, when a breach occurs, the following are some examples of costs you can expect to incur:

- Investigation costs such as forensics
- Crisis response costs
- Public relations costs
- Customer notification fees
- Call center costs
- Credit monitoring fees
- Legal fees
- Settlement costs
- Fines such as regulatory costs

It will not be possible to predict the accurate cost of a breach because there are so many variables and unknowns involved with a breach. The reality is that no breach is the same, and each poses its own unique set of challenges and, in turn, financial impact.

Fortunately, there are tools available to provide some guidance and understanding of what to expect, along with an estimated data point that can be used to set some expectations for your executive leadership team.

Breach Cost Calculator

For each of the examples provided by Travelers, the *NetDiligence® Data Breach Cost Calculator* was used to provide the cost of a breach. You can access the calculator here: `https://eriskhub.com/mini-dbcc`. Another example of a breach calculator is available from Arctic Wolf: `https://arcticwolf.com/resource/aw/calculating-the-cost-of-a-breach`. For example, let's select the following data inputs:

- **Industry**: Manufacturing
- **Location of organization**: United Kingdom
- **Number of employees**: 1,000
- **Annual revenue**: £3,000,000
- **Incident type**: Data breach
- **Root cause**: User action

The estimated cost of the breach for this type of organization totals £486,693, with the following breakdown.

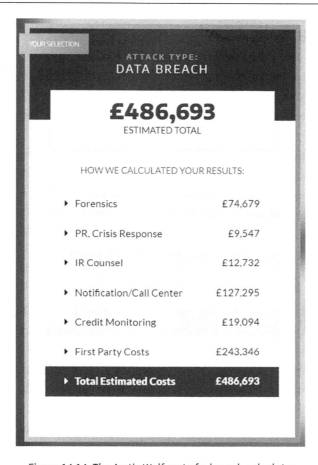

Figure 14.14: The Arctic Wolf cost of a breach calculator

Every breach is different, and there is no way you will ever predict the exact cost of any data breach. The breach calculator above provides a great estimate of what to expect. To put things into perspective, this example shows that you can expect the cost of a data breach to be approximately one-sixth of your annual revenue. That is substantial and will likely cause significant damage to your organization if you don't have any insurance to cover the risk.

Understanding Policy Exclusions

Over the years, the cybersecurity insurance landscape has continued to become more complex because of the continued increase in cybersecurity incidents that require cybersecurity insurance payouts. If you have cybersecurity in place today, it is highly recommended that you understand the details of your cybersecurity insurance. When reviewing and navigating your cybersecurity insurance, work closely with your legal team to ensure that you fully understand what is included in the policy, such as deductions, coverage specifics, and most importantly, any exclusions that may be in place. For example, Merck was originally denied its claim for the NotPetya cyberattack that occurred in 2017 because of an exclusion in the policy for anything related to an act of war. Fortunately, they were able to claim $700 million from the $1.4 billion in losses after years of dispute in the court system. This goes to show the complexity needed to navigate the insurance industry as we add cybersecurity insurance to our organizations.

Requirements for Cybersecurity Insurance

It is also important to be aware of the ongoing increase in cybersecurity requirements to purchase cybersecurity insurance today. Insurance writers are not handing out cybersecurity insurance to just any organization. They are getting a lot more stringent with their requirements, through questionnaires, surveys, assessments, and even to the extent of leveraging external attack surface management capabilities to score your organization. Based on the feedback and response you provide, in addition to the score they collect, will determine your coverage and, in a worst-case scenario, if a policy will even be issued. In addition, cybersecurity insurance providers will request minimum requirements to move forward with coverage. Every insurance provider will be different, and the requirements will continue to change; one example is Coalition, which shares five ways to meet your cybersecurity insurance requirements by ensuring that the following are implemented:

- Multi-factor authentication
- Cybersecurity training
- Maintaining good data backups
- Identity access management
- Enforcing data classification

Source: https://www.coalitioninc.com/topics/5-essential-cyber-insurance-requirements

Obviously, there is a lot more involved and required to implement a robust cybersecurity program, as we have covered throughout the book, but it's important to know that cybersecurity insurance providers are getting a lot more stringent with their requirements. Unfortunately, you can expect much higher premiums or even find yourself in a situation where you can't purchase coverage because of weak cybersecurity practices within your organization.

Coverage Types

As we briefly touched upon in *Chapter 10, Vendor Risk Management*, make sure you understand the specific coverages being purchased as part of the cyber liability insurance coverage. The following are some of the many coverages you can expect to see as part of a cyber liability insurance policy:

- Network security liability
- Privacy liability
- Data loss/breach
- Business interruption
- Media liability
- Extortion
- Reputation or brand protection

A great resource for your reference is the cybersecurity insurance guide from the **Federal Trade Commission (FTC)**. Although geared toward smaller businesses, the basic requirements should help you become more familiar with what should be included in your cybersecurity insurance, especially if you are going through the process for the first time. You can access the guide here: `https://www.ftc.gov/business-guidance/small-businesses/cybersecurity/cyber-insurance`. In addition, they have a simplified PDF version for reference and download, which can be accessed by clicking on the **Download/Print PDF** option within the link above.

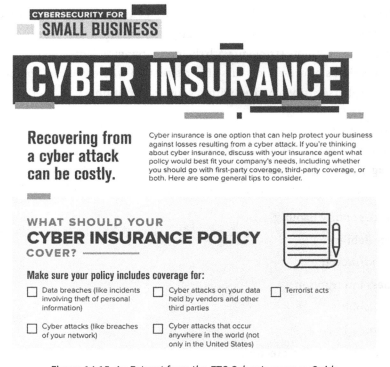

Figure 14.15: An Extract from the FTC Cyber Insurance Guide

The guide also expands on the need for both first-party and third-party coverage. As a reminder, first-party coverage will protect your organization directly from any losses related to your organization and the data being stored. The guide recommends including the following (with some additional items) as part of your first-party coverage:

- Legal counsel services
- Breach coaching
- Crisis management
- Public relations

- Lost or stolen data recovery
- **Digital Forensics and Incident Response (DFIR)**
- Cyber extortion
- Cyber fraud
- Customer notifications
- ID theft monitoring
- Call center services
- Cybersecurity-related fines, fees, or penalties
- Lost income because of a cybersecurity incident
- Negotiation services
- A Bitcoin wallet if executive leadership decides to pay

With third-party coverage, you will be protected against any losses from claims against you from a third party after a cybersecurity incident. The guide recommends including the following as part of your third-party coverage:

- Any litigation costs
- Costs relating to regulatory inquiries
- Any payments that may be needed to consumers
- Any costs related to lawsuits or disputes
- Defamation, copyright, or trademark infringement losses
- Any other potential damages or settlement costs
- Cost related to accounting

One final note about your cybersecurity insurance is to ensure that you are familiar with the process to engage it and that you follow the correct process to engage the resources needed. Your cybersecurity insurance may have a portfolio or panel of pre-approved vendors that you must use to file a claim. Make sure you are fully aware of any requirements so that when an incident does occur, you can engage the needed support as quickly as possible without compromising the receipt of any claims. Also, check with your cybersecurity insurance provider on the engagement process, and you may want to consider a direct contract with vendors that you prefer to use, specifically those that provide Breach Counsel and DFIR services. The quicker you can get your DFIR vendor engaged with an active incident, the better. Make sure you are aware of any **Service-Level Agreements (SLAs)** directly with the DFIR vendors also. As you are aware, time is of the essence, and every second counts when there is an active incident with threat actors within your environment.

In 2023, Munich Re estimated the cybersecurity insurance market was around $14 billion and expects it to increase to approximately $29 billion by 2027: `https://www.munichre.com/en/insights/cyber/cyber-insurance-risks-and-trends-2024.html`.

Summary

As one of the most important components of your cybersecurity program, risk management is a priority – especially in today's world, where we must show full transparency of all known risks to ensure that the right decisions are made at the right levels within an organization. Risk management is not a simple task to manage, and it is important that the right processes are put in place and everyone is aware of how to efficiently handle and manage identified risks. Risk management crosses the entire cybersecurity program, and everyone within the cybersecurity function must be fully aware of how to handle a risk as it is identified. The quicker a risk is identified, reviewed, documented, actioned, and resolved, the better for the organization. Remember that nothing is risk-free in this world, and the same applies to cybersecurity. There will always be a risk!

To start the chapter, we discussed in more detail why risks are so important and how everything we manage as cybersecurity professionals translates into a risk in some way. As we identify and manage this risk, it is important that it is efficiently translated into business terms so that the business, executive leadership team, and board can easily relate to and understand what that risk means to an organization. At the same time, it is important that the executive leadership team and board are familiar with cybersecurity basics, as the discussions around cybersecurity will be a more frequent topic of conversation. We also need to ensure that ownership of a risk sits at the top of the organization with the executive leadership team and the board of directors. Then, we reviewed the different types of cybersecurity risk you can expect to see in an organization. We then covered the difference between qualitative and quantitative risk before finishing the section by providing more detail about how to address risk by avoiding, transferring, mitigating, or accepting.

Next, we provided a list of some of the general frameworks, including ISO 31000, NIST RMF, FAIR, and COSO. We then covered the NIST RFM in a little more detail to provide an idea of how a risk management framework looks and what is involved. Then, we reviewed tracking risk and the importance of a risk register to track identified risks. Ideally, this should be through a centralized GRC platform, but at a minimum, make sure you track risks somewhere, even if they need to be in a spreadsheet temporarily until you can mature the process. This is critical as we look to oversee the cybersecurity program more efficiently for an organization.

We then finished off the chapter with insight into the cybersecurity insurance landscape and what should be included within a policy, related to both first-party and third-party coverage. It is also important that everyone should consider a cybersecurity policy for their organization based on the organization's current state.

In the next chapter, we will move on to the final component of both the GRC program and the final function of the cybersecurity program, regulatory and compliance. Here, we will take a deeper look into the evolving landscape of regulatory and compliance and the ongoing challenges in this area. We will then cover the need to ensure that you build positive relationships with the legal team because of the importance of their role within cybersecurity moving forward. Then, we will take a look at the importance of data protection for your organization, before going into greater detail on the need for frameworks and audits. We will finish off the chapter with a look into some other regulatory and compliance considerations that you should be aware of as a cybersecurity leader.

Join our community on Discord!

Read this book alongside other users, Cybersecurity experts, and the author himself. Ask questions, provide solutions to other readers, chat with the author via Ask Me Anything sessions, and much more.

Scan the QR code or visit the link to join the community.

`https://packt.link/SecNet`

15

Regulatory and Compliance

We have made it to the final component of both GRC and the cybersecurity program with Regulatory and Compliance. As a cybersecurity leader, you won't be able to avoid Regulatory and Compliance as part of your role. In addition, Regulatory and Compliance is a constantly evolving space with ongoing change, making this function very challenging to oversee. Unfortunately, the complexity of Regulatory and Compliance makes it almost impossible for any one person to know everything required, especially at the global level. Every country, region, state, and industry, for the most part, will differ in its regulations and compliance requirements. Your organization and where you provide services will determine how complex this becomes for you. For some organizations, there will be teams dedicated to ensuring your organization remains compliant. As a CISO or cybersecurity leader, you will need to work closely with the team that overlooks Regulatory and Compliance, most likely your legal department, if one exists.

To begin the chapter, we will look further into the evolving landscape of Regulatory and Compliance to gain a better idea of how vast and complex it has become for cybersecurity. As a cybersecurity leader, you will need to understand the different Regulatory and Compliance requirements for your organization, although you will not need to be an expert by any means. The expertise is where we move into the next section with building a strong relationship with your legal team. Your legal team will be much more knowledgeable in this area and will be the team to support and keep you educated on the latest Regulatory and Compliance requirements. In the following section, we will review the importance of data protection. This is essentially the primary responsibility of the cybersecurity function, to ensure the protection of information and data for the organization.

In the next section, we will be reviewing the need for frameworks and audits by covering them in more detail than previously covered throughout the book. There should be no cybersecurity program without a framework moving forward. This should be a requirement for every organization. In addition, you should be auditing yourself against the chosen framework for your organization. This is in addition to any other audits you are required to maintain. We will then finish the chapter with other Regulatory and Compliance considerations like privacy, data retention, e-discovery, etc.

The following will be covered in this chapter:

- The Evolving Landscape of Regulatory and Compliance
- Your Legal Team Is Your Best Friend
- The Importance of Data Protection
- The Need for Frameworks and Audits
- Other Regulatory and Compliance Considerations

The Evolving Landscape of Regulatory and Compliance

First, let's take a high-level look at all sub-functions that should be addressed as part of Regulatory and Compliance. The following image captures much of what the Regulatory and Compliance function entails.

Figure 15.1: Sub-functions of the Compliance & Regulatory function

It is important to work with both regulation and compliance hand in hand. At a high level, cybersecurity regulation is derived from the laws put in place to ensure you are implementing the requirements set forth and the best practices to protect the people, technology, data, and information within your organization. In addition, many regulations have an emphasis on the reporting and transparency of data breaches in a timely manner. On the other hand, the compliance component is to ensure any required regulation is being conformed to, in addition to other items such as audits, standards, policies, etc. As a leader, it's important that you validate you are meeting the requirements of any regulation that is required for your organization, especially since you could be breaking laws if you aren't meeting the requirements of any regulations, which in turn means someone will be held accountable. And as we've alluded to many times, you need to ensure that accountability doesn't fall on you as a cybersecurity leader. You must ensure accountability lives at the top of the organization with the executive leadership team and the board of directors.

Regulation within cybersecurity wasn't a need for many not too long ago. Today, regulation is becoming a requirement for many organizations and it's important we understand how this landscape is evolving. Regulation has become extremely complex, and it is not possible for any one of us to navigate these complexities alone. With regulation and cybersecurity, we are crossing into the boundaries of legal services more and more. Because of this, it's important you work closely with those who can provide the right guidance when dealing with regulations, especially as they change.

An excellent resource that provides insight into global privacy laws that we should have some familiarity with as part of being a cybersecurity leader has been made available from the **International Association of Privacy Professionals**, also known as **IAPP**: `https://iapp.org/`.

They provide an interactive directory in which you can click on a country to view what legislation, act, or law is in effect for that country: `https://iapp.org/resources/global-privacy-directory/`.

Global Privacy Law and DPA Directory

This tool identifies global data protection authorities and privacy legislation. Within each country listing, if available, there will be links to the DPA, privacy legislation and relevant IAPP resources. This tool is periodically updated and is for informational purposes and is not legal advice. Country information should always be verified via official sources. If you have comments about this tool, please share it with us at research@iapp.org.

Using the interactive map below, clicking on a country will display its information within the country directory. The map is an artistic rendering and there may be some inconsistencies with precise national borders.

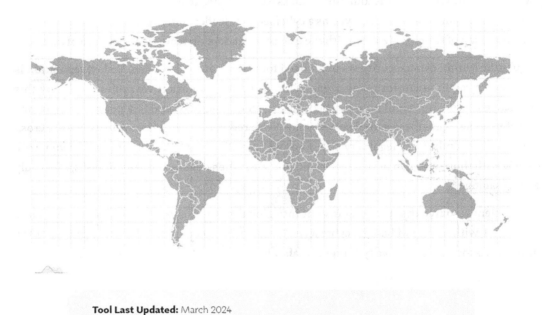

Tool Last Updated: March 2024

Figure 15.2: IAPP Global Privacy Law and DPA Directory

When you click on a country, you will be provided with additional details and links to each of the respective websites or documents with more details on the referenced legislation, act, or law in effect. The following is an example of clicking on Greece, a member of the **European Union (EU)**, which shows they are subject to the **General Data Protection Regulation (GDPR)**.

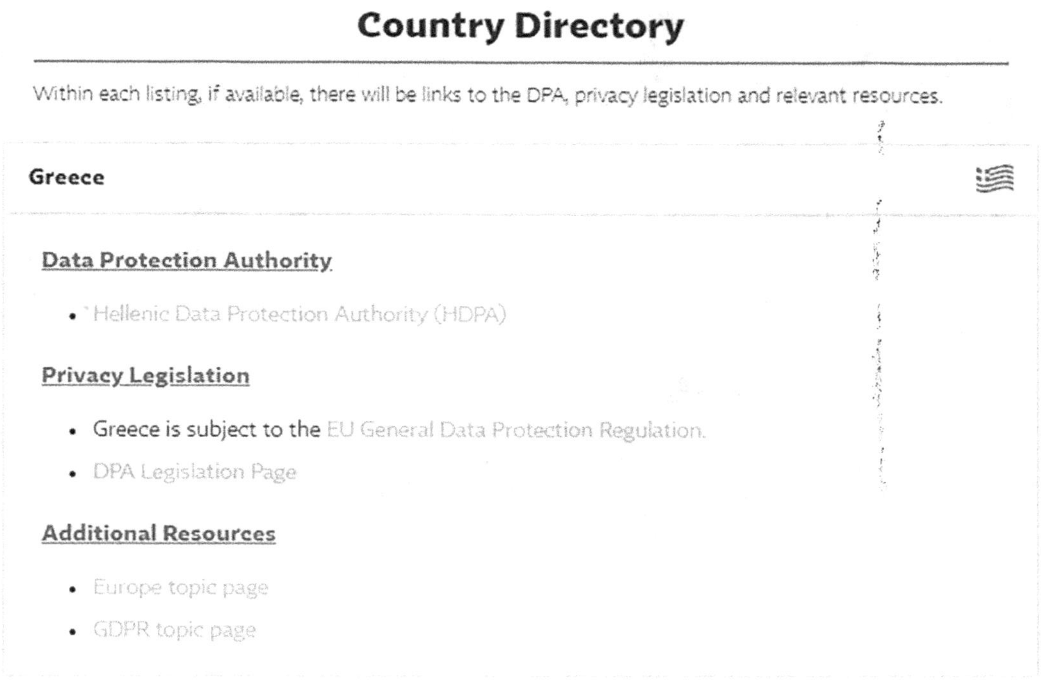

Country Directory

Within each listing, if available, there will be links to the DPA, privacy legislation and relevant resources.

Greece

Data Protection Authority

- Hellenic Data Protection Authority (HDPA)

Privacy Legislation

- Greece is subject to the EU General Data Protection Regulation.
- DPA Legislation Page

Additional Resources

- Europe topic page
- GDPR topic page

Figure 15.3: Greece's data privacy resources from the IAPP Global Privacy Law and DPA Directory

In addition to GDPR, which focuses on consumer privacy, the **European Union (EU)** has recently introduced legislation to ensure improved cybersecurity practices for organizations operating in Europe with the **Network and Information Security (NIS)** 2 Directive. You can learn more about NIS2 here: `https://www.europarl.europa.eu/thinktank/en/document/EPRS_BRI(2021)689333`.

For the US, there is also an interactive map to help navigate all the different privacy laws (if they exist) for each state, which can become very complex: `https://iapp.org/resources/article/us-state-privacy-legislation-tracker/`. You will be able to view details of all states that have a privacy law in effect by scrolling down the page. In the image below, you can see California, Colorado, and Connecticut listed with current privacy laws. For California, you can see that they have two acts in place:

- **California Consumer Privacy Act (CCPA)**
- **California Privacy Rights Act (CPRA)**

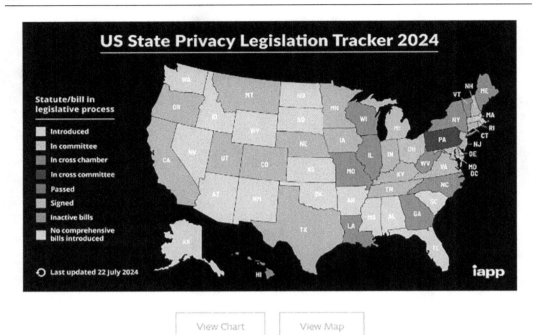

Figure 15.4: IAPP US State Privacy Legislation Tracker

In addition to the privacy laws shown above by IAPP, some other acts in the US to be aware of at the Federal level include:

- **Sarbanes-Oxley (SOX)**: In reaction to several major accounting mishandlings, the SOX act was put into effect in 2022. Although the act focuses primarily on the accuracy of financial statements and reporting to prevent fraud, there is a requirement around cybersecurity controls to protect financial information.

- **Health Insurance Portability and Accountability Act (HIPAA)**: The HIPAA act was passed in 1996 with several objectives focused on improving the healthcare system. One of these objectives focuses on ensuring the protection of healthcare information.

- **Federal Information Security Modernization Act (FISMA)**: FISMA is an act that was enforced in 2002, originally named **Federal Information Security Management Act**. In 2014, the Federal Information Security Modernization Act amended the 2002 version. This act requires federal agencies to implement and maintain a minimum set of security standards to protect government information..

- **Gramm Leach Bliley Act (GLBA)**: The GLBA was put into effect in 1999 to ensure financial institutions are protecting customers' nonpublic personal information.

We can expand even further with the latest **Securities and Exchange Commission (SEC)** rules for disclosure of a material cybersecurity incident along with the requirement of annual disclosure of cybersecurity risk management, strategy, and governance. You also have the very familiar **Payment Card Industry Data Security Standard (PCI DSS)**, which is a standard and not a law, but is certainly required for anyone handling credit card information.

The examples above are not all inclusive as there will be other regulations, acts, standards, or forms of legislation or laws in other countries that may have an impact on your organization. There may also be specific regulations that have an impact on your industry. And the more global your organization is, the more complex these requirements quickly become. If you are a global organization, you will need dedicated resources to navigate these complexities along with resources that are trained to understand the language of each of the regulations, acts, laws, or whatever it may be that impacts your organization. This is not something you can take lightly and certainly not something you want to assume risk for; you will want to ensure you are obtaining the correct guidance to allow full compliance when dealing with a major cybersecurity incident or breach

Although many of these regulations are recent, we can expect many more to continue to surface. In the US, there is still no Federal privacy regulation (as of September 2024) overseeing all states, although the **American Data Privacy and Protection Act (ADPPA)** has made the most progress to date: https://www.congress.gov/bill/117th-congress/house-bill/8152.

This means that we'll continue to see each state issue their own privacy regulations. As you can imagine, this will become extremely challenging to navigate, especially if you have employees and customers in multiple states that will have different requirements for reporting, notification, etc.

The Regulatory and Compliance area is going to require specialized skills to help navigate the complexity of these requirements. There's a lot to digest with Regulatory and Compliance and it's imperative that you have a comprehensive understanding of the requirements and the ongoing changes that continue to occur. You are going to need to have conversations with your legal team, your executive leadership team, and potentially the board of directors on these topics, so it will be important you are prepared with the necessary information.

Your Legal Team Is Your Best Friend

The reality is legal involvement within cybersecurity is becoming more of a requirement nowadays. And I only see this increasing as the regulatory landscape continues to evolve with more requirements because of the increase in ongoing breaches. Because of this, it is imperative you build immediate relationships with your legal team and General Counsel if they exist within your organization. If your organization doesn't have an in-house legal team or a General Counsel on staff, you will need to understand how these services are sourced and if there is an outside counsel being used. If an external service is being used for your organization, you will need to ensure engagement with the cybersecurity program. You will also want to ensure that the outsourced services have the right resources to support any cybersecurity needs. The same applies internally. You may not necessarily have the expertise in-house to deal with cybersecurity and privacy-related issues, so you may need to engage external legal services to fill any gaps you may have within the organization from a legal perspective.

The Legal Team's Responsibilities

As you can see, the involvement of legal expertise is becoming more prevalent within cybersecurity, and there are even instances where CISOs report to the General Counsel or head of legal services, which is making a statement. As you work with your legal services team, you may want to build a **Responsible, Accountable, Consulted, and Informed (RACI)** with all of the shared responsibilities for cybersecurity so there is a clear definition of who is responsible for what. It will be important that the legal team fully understands what is expected from them, as it relates to cybersecurity to ensure you are receiving the right level of support to be successful in your role.

Some of the areas you can expect involvement from your legal team include:

- Support with outside council if specialty services are needed
- Involvement with the cybersecurity insurance policy
- Support with cybersecurity-related regulation requirements throughout the world
- Support with any other cybersecurity law requirements for your organization
- Overseeing any breach notification requirements
- An active member of the cybersecurity council
- Support and sign-off on all policies such as the cybersecurity policy, records retention policy, and any other cybersecurity-related policy
- Involvement and support for the cybersecurity GRC program
- Oversight of all contracts and working with cybersecurity to ensure the appropriate cybersecurity language is within all contracts
- Collaborate with all contract reviews as red lines occur with contracts between your vendors
- Overseeing (along with the privacy team if one exists) all privacy-related requirements for the organization
- Overseeing (along with the risk management team if one exists) all enterprise risk management-related requirements for the organization
- Involvement with and a core team member of the CIRP
- Any other general legal services related to cybersecurity

There may be more items your legal team is involved in, or they may not necessarily be involved in everything shared above. My recommendation is to engage them as much as you can to ensure full compliance and to take advantage of their skillset as you cannot assume this responsibility for all these items yourself. Build full transparency with this team and collaborate with them as much as possible; it will only help as you look to reduce liability for yourself as a cybersecurity leader.

Obviously, the legal team's responsibilities are a lot broader and many of the items above aren't inclusive of cybersecurity. The legal's team responsibilities will spread across the entire organization and there will be specialty roles within the legal team outside of cybersecurity. This is why it's important that the correct legal resources are involved, whether it be in-house or outside counsel to ensure the right guidance is being provided for any legal-related issues. This is not an area you want to take any risks in. Clearly, the role of legal within cybersecurity is growing, and the list above is only going to grow bigger.

Data Breach Notification

To provide more detail about some of the expectations from your legal team, in the US, each state is unique and data breach notification requirements vary. This incredible resource from the *BakerHostetler* law firm provides an interactive map of notification requirements for each State if they exist: `https://www.bakerlaw.com/us-data-breach-interactive-map/`.

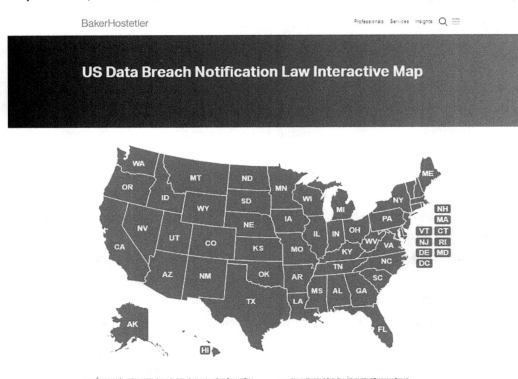

Figure 15.5: BakerHostetler interactive breach notification map

When you click on a State, you will be presented with some quick notes regarding any State requirements for breach notifications within that State. In the bottom left, you will have the option to **Download PDF** with more details, along with being provided the specific law requirements when you click on **Statute**. The following is an example of Colorado's requirements.

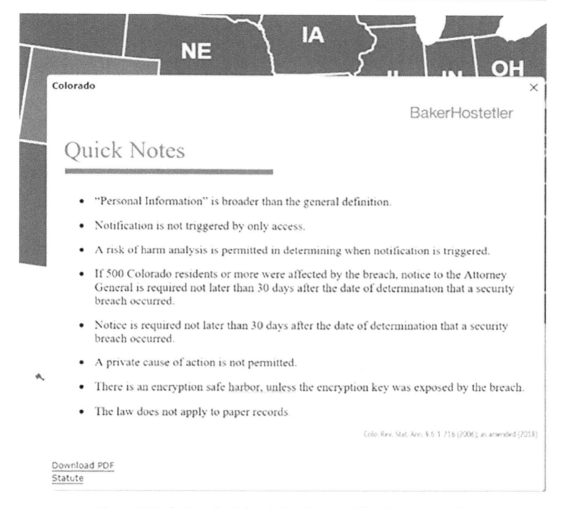

Figure 15.6: BakerHostetler Colorado State breach notification requirements

Obviously, this is only the breach notification requirement for Colorado. If you cover the entire US, you will need to check all State requirements. In addition, you may have different reporting requirements depending on your industry, such as HIPAA within healthcare (https://www.hhs.gov/hipaa/for-professionals/breach-notification/index.html). Or if you are publicly traded, you now have SEC requirements (https://www.sec.gov/files/rules/final/2023/33-11216.pdf) to comply with. If we take this a step further and you have a global footprint, for example, in Europe, you must comply with any breach notification requirements and other laws and/or regulations within Europe. GDPR is one example of this.

To provide additional details on what this looks like, we will mention another great resource from *BreachRx* (`https://www.breachrx.com/`). BreachRx has a *Global Cybersecurity & Privacy Regulations* webpage that provides details on breach notification laws and regulations required from around the world: `https://www.breachrx.com/global-regulations-data-privacy-laws/`. On this webpage, click on the **General Data Protection Regulation (GDPR)** link within the **Europe** section to access more details on the GDPR notification requirements in addition to other incident response guidelines for GDPR:

What's Required Under GDPR Incident Response Guidelines

If any of those trigger events occur, companies must issue a notification under GDPR incident response guidelines.

Notifying the GDPR Supervisory Authority

Organizations must **notify the appropriate GDPR supervisory authority within 72 hours after learning about the incident** (or include reasons for a delay past that timeframe).

The notification should detail:

- **The incident**, including the types and amount of data involved and the number of people associated with that data
- **Contact information** for someone at the company who can share more details (this should usually be the Data Protection Officer, a role required by GDPR)
- **Likely consequences** from the incident
- **Plans to address the incident** and reduce any negative effects

If companies do not have all of this information at once, they can provide the details in phases, but the initial notification should still come in the 72 hour window.

Notifying Affected Individuals

If the incident creates a situation that puts the affected **individuals at high risk, then organizations must also notify those people as soon as possible.** These notifications should clearly describe the incident and the type of information compromised.

GDPR sets deadlines of "without undue delay" for these notifications, which some interpret as 72 hours (the timeline used in other parts of the regulation) while others believe this part of the regulation is more tolerant of a wider timeframe. Either way, organizations should not delay, as courts have ruled in prior judgements that the deadline is "*not compatible with a time limit of several weeks or ... several months*".

Organizations do **not** need to send these individual notifications if:

1. The compromised data is protected in a way that renders it unusable to any attackers (for example, if it's encrypted and therefore unreadable).
2. The organization has already taken action that will reduce the fallout from the incident so that it no longer threatens individual safety or freedoms.
3. There are too many individuals affected that notifying every one of them would become a big burden, in which case GDPR allows companies to make a public announcement that notifies all of the people on which they have data.

Figure 15.7: BreachRx GDPR incident response guidelines
Source: `https://www.breachrx.com/global-regulations-data-privacy-laws/gdpr-guidelines/`

At this point, I think you see where I'm going. If your organization is publicly traded and impacts PII for users in all US States as well as Europe, you will be required to send dozens of different notifications to local authorities, customers, and/or employees of your organization. Strategically, you may use the most stringent requirements as your default and use the same communication with everyone else who falls in your scope.

Regardless, someone is going to need to keep up to date with all the requirements and monitor for any changes. The information here has been provided as guidance and your legal team will need to confirm what the actual requirements are from each of the States and/or countries. For example, your legal team should be accessing the requirements directly from the published source for confirmation of any communication and notifications like the GDPR at `https://eur-lex.europa.eu/legal-content/EN/TXT/?uri=CELEX%3A02016R0679-20160504`. A couple of articles within this regulation that reference notification and communication include *Article 33, Notification of a personal data breach to the supervisory authority*, and *Article 34, Communication of a personal data breach to the data subject.*

Directors and Officers (D&O) Liability Insurance

If you are a CISO, you will want to work with your legal team to understand if you are covered by insurance for any liability that may come your way. Based on some of the examples provided earlier in the book, CISO/CSO roles are being targeted in lawsuits for negligence. Traditionally, **Directors and Officers (D&O)** liability insurance provides coverage for directors and officers, but the CISO may not necessarily fall under this category. Because of this, you will want to ensure you are clear on any protections. Here is an insightful article that sheds more light on the matter: `https://www.iansresearch.com/resources/all-blogs/post/security-blog/2023/09/21/why-cisos-need-d-o-liability-insurance-coverage-now`.

As shown, it is important you build a positive relationship with your legal team. They, in turn, will become an extremely important partner of the cybersecurity program as regulations and laws continue to become approved and evolve. These regulations and laws are becoming very complex and the language within them needs to be clearly understood by someone trained in this area. Obviously, you need to be aware of the details, but it will be unrealistic to expect the CISO or cybersecurity leader to be able to keep up with the constant change of regulations along with the requirements to meet them. Now that we have covered the need for building good relationships and ensuring positive collaboration with your legal team, next, we will review the importance of data protection in more detail.

The Importance of Data Protection

One of the core principles for our existence as cybersecurity professionals is to protect the data and information within our organization. For the most part, threat actors are looking for financial gains. This is typically done through gaining access to data and information that can be used as extortion against an organization to pay the threat actors or to sell that information on the dark web.

Regulatory and Compliance

For example, Privacy Affairs has a *Dark Web Price Index 2023*, which provides an idea of what the average price is that data is selling for on the dark web during a specific timeframe. Obviously, these prices can fluctuate over time: `https://www.privacyaffairs.com/dark-web-price-index-2023/`.

Here is a sample of data from the link above with some of the credit card products available with an average purchase price.

Category	Product	Avg. dark web Price (USD)
Credit Card Data	Credit card details. account balance up to 5.000	$110
	Card.com hacked account	$75
	Credit card details. account balance up to 1.000	$70
	Stolen online banking logins. minimum 2.000 on account	$60
	United Arab Emirates credit card with CVV	$35
	Stolen online banking logins. minimum 100 on account	$40
	TDBank hacked account	$30

Figure 15.8: Privacy Affairs credit card data from the Dark Web Pricing Index 2023

Imagine your organization manages and stores credit card information and 10,000 records were compromised that contained credit cards with a balance of up to $5,000. The payday for this threat actor could land them up to $1.1 million from the sale of these records on the dark web. This is obviously concerning, and why we must take all measures to protect the information and data within our organizations from being accessed and exfiltrated by threat actors. Before you can do this, you first need to know where all your data exists, which ties back to the GRC program to ensure you have a comprehensive inventory of all applications and assets within your environment. Once you know where all your data is, it is critical you know the classification of that data to better understand the potential impact in the event any data is compromised.

Once classified, the correct controls will need to be implemented for this data to ensure it's protected efficiently. This is not an easy task, but as you need to meet specific regulations and ensure compliance, this becomes a crucial task.

As a useful reference, Microsoft has a high-level approach to how you should look to address information protection within your organization. Although the approach references Microsoft's Purview and the capabilities of that solution, the foundation can be applied outside of Microsoft's capabilities to best protect your data. You can view more details at `https://learn.microsoft.com/en-us/purview/information-protection`. The following principles are provided for your information protection foundation:

- Know your data by identifying the different types of data you need to protect
- Protect your data by creating and applying sensitivity labels across all data, services, cloud repositories, databases, and containers
- Prevent data loss by building and deploying DLP policies within your environment
- The final principle focuses on the governance of your data

Figure 15.9: Microsoft's Purview Information Protection capabilities

Source: `https://learn.microsoft.com/en-us/purview/information-protection`

Data Tracking

It is imperative that you know where all your data lives, not just within your organization but externally with any vendor or third-party entity that stores and processes data on your behalf.

If you don't know where your data exists, how are you meant to efficiently protect it? And even more concerning, if there is a major cybersecurity incident involving your data and you aren't aware of what that data is, how are you meant to efficiently respond? Knowing where all your information is falls under the broader governance requirements as it relates to the inventory/ asset management components of your cybersecurity program. Referring back to *Chapter 13, Governance Oversight*, we reviewed the importance of centrally tracking and managing all vendors and applications being used throughout the organization within a GRC application. Once in the GRC application, you will be able to view all dependencies of where your information exists throughout the organization and with your vendors. Once you know where all your data exists and understand the risk with that data, you can apply the appropriate controls for that data to best protect it.

We also covered asset management in *Chapter 7, Cybersecurity Operations*, and *Chapter 8, Vulnerability Management*. This continues to come up throughout the book because of the importance of knowing where all data is within your environment and who the owners of that data are. Although the scope throughout other chapters expands beyond data to hardware and other types of assets, you must ensure all data and information within your environment is being tracked as part of your inventory for your organization. In addition, as more regulations and laws become more prevalent, the need to know where all data exists will not be an option. You will be increasing your liability by not implementing the correct process and controls to ensure all data is being tracked and inventoried. When a major cybersecurity incident occurs, questions will need to be answered, and not knowing you had data that was compromised or exfiltrated will not go down well in a lawsuit.

Data Classification

Once you know where all your data exists, the next step is to classify that data based on the data type. Once classified, you will then need to ensure the appropriate protections are in place. For example, a general document that can be viewed by anyone will not need much protection depending on the scope of who can view it. On the other hand, an application with users' personal information must be protected with the correct controls such as encryption. Referring back to the GRC platform, this should be efficiently documented so you are aware of the risk with all data. This will then allow you to ensure the correct classification is applied to the data. This will not be an easy task and will require ongoing governance and the need to ensure compliance is in place. Ideally, you will want to have this process automated, but the reality is you may not be able to fully automate classification for all your data. For the most part, you should be able to easily classify most of the core environment, which includes mailboxes, document repositories, file shares, etc.

The challenge will be ensuring classification is being correctly applied to your application data and any SaaS applications that have been deployed for your organization.

When looking at the different classification levels, you may be familiar with the US government/military classification tiers that include Top Secret, Secret, and Confidential. The data type will depend on the classification. For example, the highest level of protection will need to be applied to Top Secret as access to this data could be catastrophic for the country. Looking at this from a business lens, the classification levels are typically different, although the same principles apply. There are many different variants available, and every organization will adopt what works best for their business and the data they handle. To provide an example of a standard approach, let's look at Microsoft's Purview default sensitivity labels as a reference:

- **Personal**: Data that is used for personal use only, no business-related data. Examples can include any information a user has downloaded that is personal like a class assignment from a college course.

- **Public**: Data that is OK to make available for the public to access and view. For example, anything that is typically published on an organization's public website, such as the executive leadership team.

- **General**: Data that is for internal business use only but may be shared with trusted parties such as vendors or contractors, typically with an NDA being signed. Examples may include contact information, policies, standards, company-specific information that is not intended for the public, etc.

- **Confidential**: This type of data can only be accessed by authorized personnel as it may cause damage to the organization. Examples include financial reports, contracts, customer information, forecast data, etc.

- **Highly Confidential**: This can be seen as the highest level of sensitive data that could cause significant damage to an organization if it gets into the wrong hands. Examples include PII and Sensitive PII, trade secrets, credentials, product information, etc.

Again, your organization will determine how you customize your labels. You may see some organizations use different labels like personal information, sensitive information, trade secrets, etc. in addition to or as a replacement for the examples provided above. You can view more details on the above default sensitivity labels at `https://learn.microsoft.com/en-us/purview/mip-easy-trials`.

Once you have determined your data classification labels, you need to determine what protection mechanisms are needed for each. For example, you will want to apply the strictest protections to your Highly Confidential data. This may include preventing anyone from outside the organization from accessing or viewing the data, ensuring the data is encrypted at rest and in transit, and access controls that require the strongest level of authentication such as phishing-resistant MFA or passwordless, as covered in *Chapter 6, Identity and Access Management*. For general classified data, access controls may include the need to ensure data is encrypted at rest and in transit with an approved list of recipients who can access the data. Essentially, when you configure your labels, they will then need to be applied to the data so you can then apply the correct controls that have been defined. This is a very high-level example, but it should give you an idea of how to ensure the protection of your data, especially the high-risk data within the organization.

Before we move into data loss prevention, it's important to ensure that your broad cybersecurity policy for the organization includes the requirements for data classification within your organization. The policy should clearly define each of the labels to be used within the organization and what the controls for each of those labels are. In addition, the policy should include examples of what type of data falls within each of the classifications and labels, as well as examples of data as it pertains to your organization.

Data Loss Prevention

Once your labeling has been defined and applied with the appropriate controls, the next step is to use capabilities that help prevent data loss from your organization. This is where **Data Loss Prevention** (DLP) can help. When looking at DLP, the capabilities include helping to identify, monitor, and prevent unwanted sharing and exfiltration of sensitive information from your organization. Referencing back to Microsoft's capabilities with Microsoft Purview, DLP (`https://learn.microsoft.com/en-us/purview/dlp-learn-about-dlp`) is a capability that works hand in hand with Microsoft's Purview sensitivity labels. By using Microsoft Purview, sensitive information can be prevented from leaking out of your environment by protecting the following services:

- Exchange Online, SharePoint Online, OneDrive for Business, and Microsoft Teams
- Word, Excel, and PowerPoint
- Windows 10, Windows 11, and macOS (latest three released versions)
- Non-Microsoft cloud apps
- On-premises SharePoint and file shares
- Power BI

Within the DLP engine, you can create policies that allow you to select which type of data you would like to protect. With the data type, you can select from many pre-defined templates that scan for **Personal Identifiable Information** (**PII**), finance, medical and health, and privacy types of information. You'll also need to understand how the same capabilities can be applied to applications or SaaS environments that may not be supported by Microsoft's Purview as an example. You may need to work directly with vendors to understand any native DLP capabilities that can be applied. Another consideration is that of insider threats and leveraging capabilities to detect the potential for any insider threats. Microsoft also has insider threat management capabilities as part of Purview (`https://learn.microsoft.com/en-us/purview/insider-risk-management`) to help identify the potential for data exfiltration through someone within your organization. We reviewed insider risk in more detail in *Chapter 6, Identity and Access Management*. Again, we are using Microsoft's Purview as an example of DLP capabilities but there are many solutions out there; just make sure they meet your requirements for your organization.

Before we finish the section, it is important to ensure mature governance processes are in place around your data protection requirements and data loss capabilities. Protecting against data loss will require resources dedicated to overseeing all the capabilities in place and the need to review any high-risk alerts or concerns with any potential data exfiltration. Ideally, you will want to automate as much as possible but there will be a need for your team to review and potentially need to engage with users or make executive leadership aware of certain situations. A lot of this essentially becomes ongoing operational work and can possibly become an extension of your SOC, in which you will have 24/7/365 monitoring of any data exfiltration activities that may be occurring within your organization.

The Need for Frameworks and Audits

We have briefly touched on frameworks throughout the book and the need to have one in place, especially in today's world with the increase in regulation and the need for greater compliance, as we have discussed previously. Once you have a framework in place, it is even more important to ensure the controls implemented from that framework are effective. This is where the need for audits comes into play, and it is critical that you validate that the controls that have been put in place are accomplishing what they were intended to do. It is important to ensure an external entity is auditing and attesting to your framework and controls in addition to any self-audited activities.

Validating Controls with Audits

Validating that controls are in place is a significant task in the cybersecurity program and one that should not be neglected. Ensuring the documented controls are enforced will help provide additional certainty and peace of mind. Having a second set of eyes to review anything you implement within cybersecurity is also needed these days. This doesn't necessarily mean an incident will never happen, but it does show that you are executing due diligence and doing what is right. This becomes very important with the liability issues that come with cybersecurity responsibilities.

 Remember, it's important to validate that your vendors also maintain the same level of protection in their environments that we covered in detail in *Chapter 10, Vendor Risk Management*. The more we move data to vendor-managed cloud and SaaS services, the more due diligence is needed to audit access and validate controls in the vendor's environment.

Although we have touched upon auditing in other chapters, it is important to know that auditing expands into many different areas throughout the organization beyond cybersecurity. Large organizations, and those publicly traded, will most likely have a dedicated auditing team to ensure compliance is being maintained throughout the organization for regulations and laws that apply to them. More specifically, the need for financial audits is nothing new in the corporate world. Depending on your organization, you may have a **Chief Audit Executive (CAE)** who can be thought of as the highest role overlooking the audit requirements for an organization. If there is an internal audit function within the organization, this doesn't negate the need for an external audit to occur. This will need to happen regardless and for the most part, the responsibility of the audit function will be to work and coordinate with any external auditors to get the information needed.

Audit Types

This is a reminder that auditing is the process that checks the intended controls are in place. At a high level, you will typically see two types of audits:

- An internal audit is one that is completed by an internal team employed by the organization to conduct audits within the business.
- An external audit is completed by a third-party company. The idea behind an external audit is to prevent any conflict of interest within the organization by using an independent party with no ties to the organization. Depending on your company's industry, this may be a legal requirement.

Commonly seen in an internal audit scenario, the auditing team reports directly to the CEO or the board of directors and not internally to the IT or cybersecurity function. This helps ensure accountability, prevents any conflicts of interest, and provides for a system of checks and balances.

There are many examples of completing an audit, and every country is different. Unfortunately, we won't be able to cover all of them in this chapter. As an example, let's review the **System and Organization Controls (SOC)** in more detail as a good example to reference. Even if you are outside the US, there's a chance you may have come across a SOC audit report. One of the more common practices and well-known reporting offerings is the SOC, not to be confused with the security operations center. The SOC services are part of the **American Institute of Certified Public Accountants (AICPA)** and **Chartered Institute of Management Accountants (CIMA)**. It's a good idea to become more accustomed to SOC and other auditing standards within the industry, especially as you look to adopt more cloud services. As you meet with vendors and subscribe to services, it's important that these audits and reports have been completed to ensure the controls are in place for your vendors.

SOC Reports

There are three different SOC reports to be familiar with:

- **SOC 1** reports focus on the financial aspect of an organization, ensuring the correct financial controls and reporting are in place. This report comes with two types of reports:

    - *Type 1* reports are based on the feedback and description of controls provided by management, completed at a specific time.

    - *Type 2* reports are more involved and look to test the effectiveness of the controls in place over a duration of time.

- **SOC 2** reports are focused more on the technology and security aspects of an organization to better protect users' data and reduce risk. The specific controls measured include security, availability, processing integrity, confidentiality, and privacy. Like SOC 1, there are type 1 and type 2 reports that follow the same format. SOC 2 reports are intended for internal use only and vendors will require a **Non-disclosure Agreement (NDA)** to be signed to view them.

- **SOC 3** reports take the output of SOC 2 reports and publish them in a more readable format that is less technical. This allows them to be made available for anyone to access and view.

You can view more information about the AICPA and CIMA SOC services at `https://www.aicpa-cima.com` and you can directly access the SOC 2 details at `https://www.aicpa-cima.com/topic/audit-assurance/audit-and-assurance-greater-than-soc-2`.

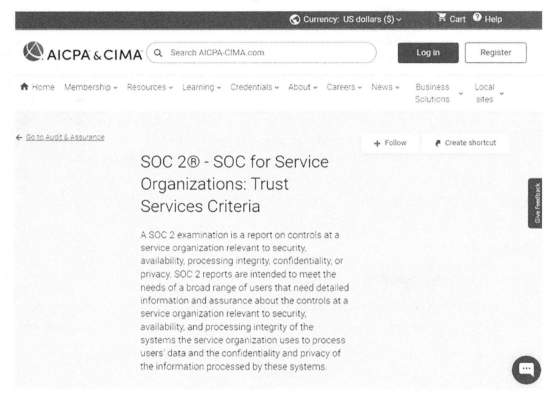

Figure 15.10: AICPA and CIMA SOC 2 information

 As already mentioned, the ISO/IEC 27000 is another common certification family to ensure best practices are implemented as part of your cybersecurity program. The ISO/IEC certification is used more widely internationally than SOC.

It is important as a cybersecurity leader that you are familiar with various audits and the process involved with auditing, especially as you need to engage an external auditor to complete an audit. There's a high possibility, you'll need to be engaged with other audits occurring throughout the business in addition to any cybersecurity related audits. For example, although SOC 1 is related to financial controls, it requires technical and cybersecurity-related evidence to show the protection of the financial data. For this reason, the cybersecurity team will need to be engaged to provide evidence for access controls related to the SOC 1.

Cybersecurity Frameworks

Currently, not all organizations are required to implement any form of framework within their environment. Although there is no requirement, all organizations should implement a framework today. This should not be an option for anyone in the current state. Your framework serves as the foundation of your cybersecurity program and allows both a measurement to be put in place and maturity to occur.

In *Chapter 4, Solidifying Your Strategy*, we covered the importance of a framework for your cyber-security program. More specifically, we reviewed frameworks that are created for cybersecurity programs. Two of the more prominent are:

- **International Organization for Standardization/International Electrotechnical Commission (ISO/IEC)** 27001:2022
- **National Institute of Standards and Technology (NIST)** Cybersecurity Framework 2.0

As you implement one of these frameworks, it is important that an audit occurs to ensure you are compliant. If you opt for the ISO/IEC 27001:2022, a process to be formally certified is available for this framework. This would be a process completed through a third party, thus providing that separation of duty and assurance that the required controls have been put in place. According to the *ISO Survey 2022* (https://www.iso.org/the-iso-survey.html), it was reported that over 70,000 ISO/IEC 27001 certificates throughout 150 different countries were issued (https://www.iso.org/standard/27001). For the NIST Cybersecurity Framework 2.0, there is no formal certification. Though this is recommended as a more feasible direction for anyone looking to implement a framework for the first time. Because of the recommendation, you can still proceed with implementing the framework and work with a cybersecurity vendor that will audit the NIST Cybersecurity Framework 2.0 for you. This will allow you to provide more assurance of the work you are doing to reduce risk for the organization. One of the benefits of the NIST Cybersecurity Framework 2.0 is the ability to measure your organization through a maturity model. This allows you to assess your current state and provides a baseline to work from. Of course, you'll want to bring in an external company to audit the initial assessment of the controls. Once you have this baseline, you can work through the controls to build a plan to ensure maturity is occurring with the controls.

In *Chapter 4, Solidifying Your Strategy*, we reviewed the NIST CSF 2.0 at a high level along with what a control looks like. Expanding beyond this, let's review how we can use the NIST CSF 2.0 as a maturity-based framework.

First, let's look at the controls for **Protect | Platform Security (PR.PS-01***)*, which can be found within the formal NIST CSF 2.0 publication at https://www.nist.gov/cyberframework. Expand **CSF 2.0 Resource Center** and click on **Download (PDF)**, open the PDF, scroll down to page 20, and look for the section **Platform Security (PR.PS)** as shown below.

- **Platform Security (PR.PS):** The hardware, software (e.g., firmware, operating systems, applications), and services of physical and virtual platforms are managed consistent with the organization's risk strategy to protect their confidentiality, integrity, and availability

    o **PR.PS-01:** Configuration management practices are established and applied

    o **PR.PS-02:** Software is maintained, replaced, and removed commensurate with risk

    o **PR.PS-03:** Hardware is maintained, replaced, and removed commensurate with risk

    o **PR.PS-04:** Log records are generated and made available for continuous monitoring

    o **PR.PS-05:** Installation and execution of unauthorized software are prevented

    o **PR.PS-06:** Secure software development practices are integrated, and their performance is monitored throughout the software development life cycle

---

Figure 15.11: NIST CSF 2.0 Platform Security controls

Next, let's look at an example provided in the *Quick-Start Guide for Creating and Using Organizational Profiles* for how to score and track each of your controls. The example in the guide references the PR.PS-01 control shown above and can be found by browsing https://www.nist.gov/quick-start-guides and clicking on **Download** in the **CSF 2.0 Organizational Profiles** section Open the PDF and scroll down to page 6.

CSF Outcomes		Current Profile				Target Profile
Identifier	Description	Practices	Status	Rating	Priority	Goals
PR.PS-01	Configuration management practices are established and applied	Policy: Configuration Management policy version 1.4, last updated 10/14/22. Defines the configuration change control policy [CM-1]. Procedures: System owners and technology managers informally implement configuration management practices. Change control processes are not consistently followed. The CIO specifies configuration baselines [CM-2] for the IT platforms and applications most widely used within the organization, but baseline use is not monitored or enforced consistently across the organization.	Configuration management is partially implemented within the organization. Some systems do not follow available baselines and other systems do not have baselines, so they may have weak configurations that make them more susceptible to misuse and compromise. Unauthorized changes may go undetected. Some changes are not tested or tracked.	3 *out of 5*	High	Policy: The Configuration Management policy requires configuration baselines to be specified, used, enforced, and maintained for all commodity technologies used by the organization. The policy requires change control processes to be followed for all technologies within the organization [CM-1]. Procedures: Each division of the organization has a configuration management plan [CM-9], as well as maintains, implements, and enforces configuration baselines [CM-2] and settings [CM-6] for their systems. Baselines are applied to all systems before production release. All systems are continuously monitored for unexpected configuration changes, and tickets are automatically generated when deviations from baselines occur. Designated parties review change requests and corresponding impact analyses [CM-4] and approve or deny each [CM-3].

Figure 15.12: NIST CSF 2.0 organizational profile example for PR.PS-01

As you can see in the example above, a rating has been added using a scale of 1-5 with 3 as the actual rating. This scale can be anything you choose but it's important that you have something that allows you to be measured. As a first step, you'll need to assess yourself before having a third party come in to complete an audit on the ratings.

Once the audit has been completed by a third party, you can then determine the score for each of the categories: Govern, Identify, Protect, Detect, Respond, and Recover. This in turn will allow you to calculate a total score for the assessment. Let's say the score for your first assessment is 2.6 with confirmation from an external audit. You now have a baseline to work from; you can assess each control and implement an improvement plan for each, so your score increases on the next assessment. However, it will be important to understand the risk of each control and ensure you are prioritizing appropriately and targeting the higher-priority controls first.

As you are aware from the constant reminders throughout the book, the need for a Cybersecurity Framework should not be an option. All organizations of all sizes need to implement a Cybersecurity Framework for their organization. This allows you to better understand your current state and build a baseline upon which you can mature from. If you are beginning your journey with a Cybersecurity Framework, it's recommended you focus on a maturity model-based assessment like the NIST Cybersecurity Framework that will allow for ongoing progression as you continue to improve your cybersecurity program. Your maturity journey will not happen overnight, and it will take time. Make sure you prioritize appropriately based on the methods outlined throughout the book.

Privacy Frameworks

Another type of framework you need to consider is a privacy framework. Again, your organization type, the industry it is within, location around the world, customer base, etc. will determine whether a privacy framework will help meet any privacy regulations or laws required for your organization. If you are required to follow any privacy regulations or laws, there's a high probability that you'll have a **Chief Privacy Officer** (**CPO**) who oversees the privacy program for your organization. If there is no CPO, there will most likely be ownership from the legal team for privacy-related items in addition to the CISO needing to be involved to support the program. Even if all three roles exist within an organization, there will need to be close collaboration on privacy-related issues for your organization. If your organization doesn't need to meet any privacy-specific regulations or laws, it is still a good idea for you to be familiar with a privacy framework to ensure you are abiding by best practices when protecting data within your organization. For the most part, every organization will have some form of sensitive PII based on employees working for them. This in turn requires the need for a high level of privacy to be maintained with this type of data.

Like all other frameworks, there are many available that can be adopted for privacy. Both ISO and NIST have privacy-specific frameworks that can be adopted within your organization.

The first is the ISO/IEC 27701:2019, also known as the **Privacy Information Management System (PIMS)**, which is an extension of the ISO/IEC 27001 and ISO/IEC 27002 to provide a broader coverage of privacy for your organization: https://www.iso.org/standard/71670.html.

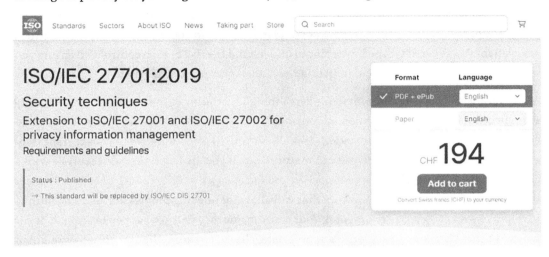

Figure 15.13: ISO/IEC 27701:2019 Privacy Information Management System (PIMS)

Like other frameworks, NIST also has its own version of a privacy framework, *NIST Privacy Framework: A tool for improving privacy through enterprise risk management, version 1.0*, which supplements NIST CSF 2.0 with some overlap. As you can see in the image below, there is a distinct overlap between cybersecurity and privacy. You can access the NIST Privacy Framework document at https://www.nist.gov/privacy-framework by expanding Privacy Framework in the top left and then clicking on **Version 1.0 (PDF)**.

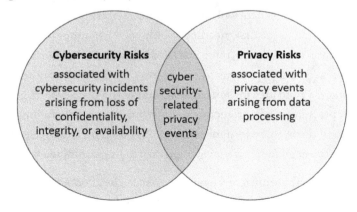

Figure 15.14: NIST cybersecurity and privacy frameworks relationship

Within the NIST Privacy Framework, there are five distinct functions: **Identify-P**, **Govern-P**, **Control-P**, **Communicate-P**, and **Protect-P**.

 The *-P* used in each of the Privacy Framework functions represents that these functions are from the Privacy Framework rather than the NIST Cybersecurity Framework functions.

When bringing both the NIST CSF 2.0 and Privacy Framework functions together, functions from each will support cybersecurity-related privacy risk. There are other privacy-specific frameworks available for review and we barely scratched the surface of what these frameworks involve. But it's important, as cybersecurity professionals, that we are aware of and understand privacy as we play a big part in protecting privacy for the organization.

Throughout the book, we have covered a lot of information about frameworks and their importance. From a cybersecurity perspective, there have been three distinct frameworks that help support the foundation of the cybersecurity program. The first is a cybersecurity-specific framework, which we covered in *Chapter 4, Solidifying Your Strategy*, and this chapter, which includes both the ISO/IEC 27001:2022 and NIST Cybersecurity Framework. The next framework we touched upon was risk management frameworks which we covered in *Chapter 14, Risk Management*, which includes the ISO 31000 for Risk Management and the NIST **Risk Management Framework (RMF)** among with some others. The final point we covered was on privacy frameworks, which included ISO/IEC 27701:2019 and the NIST Privacy Framework. Your priority should be the Cybersecurity Framework for your program as you focus on laying the foundation and maturing your cybersecurity program. The other two frameworks will most likely be a collaborative effort between other departments within the organization like legal, risk, and privacy if they exist.

Other Regulatory and Compliance Considerations

To finish off the chapter, we will look at some other items to consider as they relate to Regulatory and Compliance. As we've stated, this area of GRC can become very complex, and collaboration from other functions within the business will be needed as you look to maintain compliance with any regulations, frameworks, certifications, audits, laws, etc.

Privacy

Although we have covered privacy throughout the chapter in other sections, it is important that we directly address the need to ensure the privacy of the data we manage, store, and process is taken seriously with the utmost protection for that data.

Privacy laws and requirements to protect privacy data continue to become more stringent, and for good reason. Our data has been abused over the years and organizations have taken advantage of our data for better profit. This needs to change and the ownership of our data needs to lie in our hands and not with the organizations that collect it for profit. Our data should only be collected and used for legitimate reasons, and if it is no longer needed, it should be removed. Fortunately, regulations and laws are helping us get to a better place but it's not going to happen overnight. It will take time, but we will get there. As cybersecurity leaders, our role is to ensure we raise this awareness as much as possible and ensure our organizations are in compliance with any regulations and privacy laws that apply to us. Even if there are no requirements, we must ensure that our organizations are not retaining data that is not needed, especially privacy-related data.

One great example of where data privacy is taken seriously is within Europe and the **General Data Protection Regulation (GDPR)** that came into effect in 2018. GDPR is one of the stricter privacy regulations in effect today. The below-referenced checklist (partial) is taken directly from GDPR.EU (a website operated by Proton Technologies AG) and provides a great overview of the expectations and requirements for organizations that need to be GDPR compliant. The checklist is broken into four different categories: lawful basis and transparency, data security, accountability and governance, and privacy rights. If we look at the privacy rights section, the following is provided as part of the checklist.

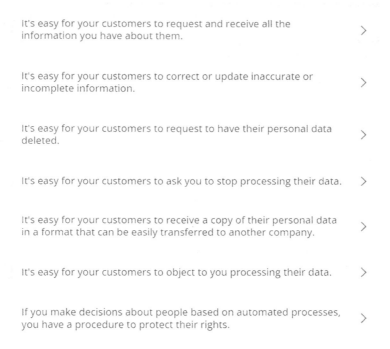

Figure 15.15: The GDPR.EU privacy rights checklist

Source: `https://gdpr.eu/checklist/`

Obviously, the work to meet the requirements from the checklist above will not be easy and it will take a large effort to get to the level required to become GDPR compliant. And the reality is, if you don't meet these requirements, big consequences can be expected. An example of a fine from GDPR includes a fine issued by the Irish **Data Protection Commission (DPC)** to TikTok for €345 million for a violation of children's privacy: `https://www.bbc.com/news/technology-66819174`.

Another expectation from GDPR is the need to know where all data is for your customers, more specifically to meet the requirement of removing customer data when requested. The same applies when being requested to correct or update customer data. If you aren't aware of where all this data is, you won't be able to remove it or modify it. If customer data is exfiltrated that should have been removed based on a customer request, you'll suffer the consequences. This is another example of the importance of the GRC platform and inventorying all your data by knowing where it all exists. An example of an organization providing the ability to request personal data to be removed looks like the following from Target: `https://www.target.com/guest-privacy/privacy-intake-form`.

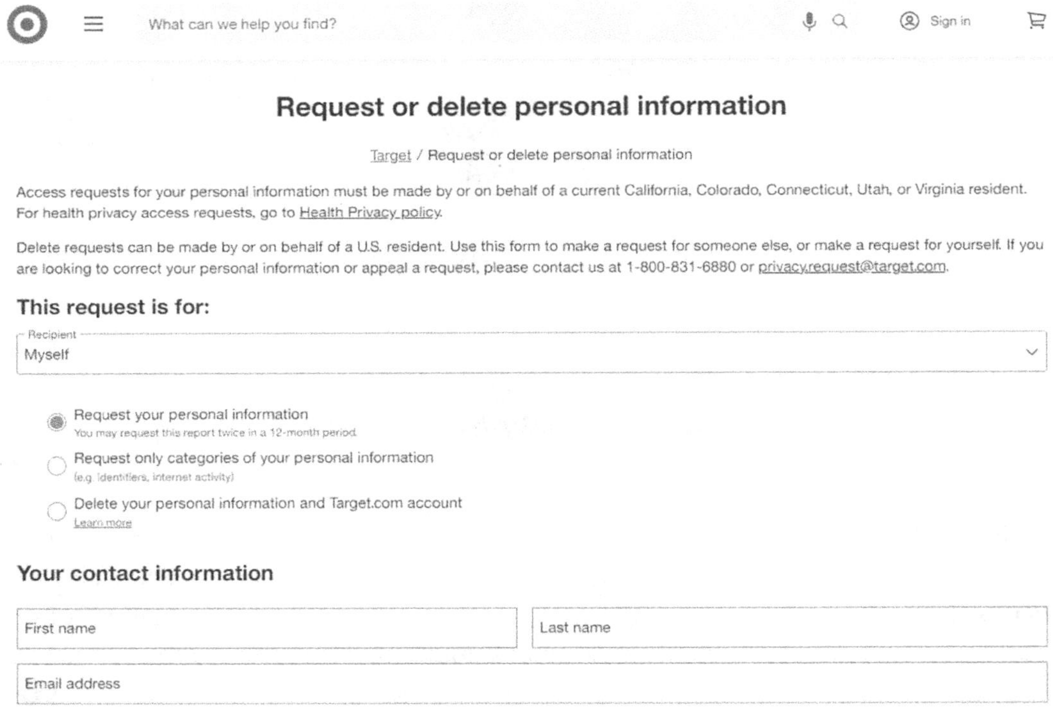

Figure 15.16: Target request form for data removal

The processes I have gone through to request data removal have not been the easiest to date and they are typically specific to where you are located. We need to get to a place where organizations allow anyone, no matter their location, to have data removed at their request. We have a long way to go but at least we are making some progress with the likes of GDPR and CCPA. In addition to requesting data removal, organizations should make the process of requesting for your account to be deleted much easier, although many larger organizations are making this an easier process. For example, if you want to close an account, the easiest way to begin is to do a *Google* search for *delete 'company name' account* and look for any articles from the company on how to delete an account. For example, searching for *delete Tesco account* provides an article with instructions on how to delete your Tesco account: `https://www.tesco.com/help/pages/tesco-account-faqs/closing-an-account/closing-my-tesco-account`.

If you can't find any instructions, reaching out to support will be the next best option. In reality, organizations should be removing accounts that have been inactive for a certain amount of time, like six months or a year. A lot of user compromises are coming from threat actors being able to easily access user accounts and I'd imagine a lot of these accounts are not even in use. Hopefully, organizations begin realizing that it's not about having many active accounts, but about doing what is right to better protect our data by reducing the footprint of the information they currently store on our behalf.

Another example of GDPR making an impact is a very visible one you have most likely come across when browsing the web. At some point, you will have seen a prompt or consent form on websites for using cookies and collecting our personal information when browsing.

An example from Marks and Spencer plc (`https://www.marksandspencer.com/`) is shown below by clicking on Manage cookies at the bottom of the screen.

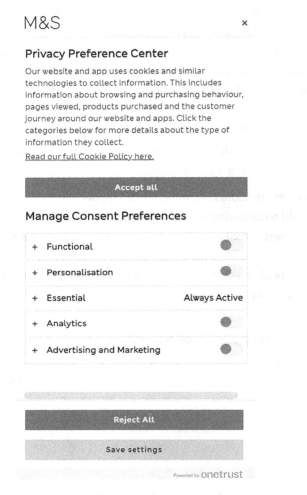

Figure 15.17: Marks and Spencer plc privacy preference center

Again, this is another great example of the progress we are making, but these consent forms clearly show how much our personal data has been abused over the years. Although I do appreciate the consent form being presented, organizations need to disable all data being collected by default and allow users to opt in if they wish, or at minimum, present the users with a **Reject all Cookies** option as an immediate option when the website appears. Some organizations make the consent form very challenging to work with and, in some cases, try to trick you into quickly accepting the consent form so they can track all data.

There are many websites that are doing this. An example below is from Starbucks (`www.starbucks.com`). As you can see in the image, you can't simply select **Reject all Cookies**; you are forced to click **Change cookie settings**, which presents you with another section where you can then select **Required Cookies**. Most will simply click **Agree** because of the annoyance of these pop-ups that appear on many sites we browse today.

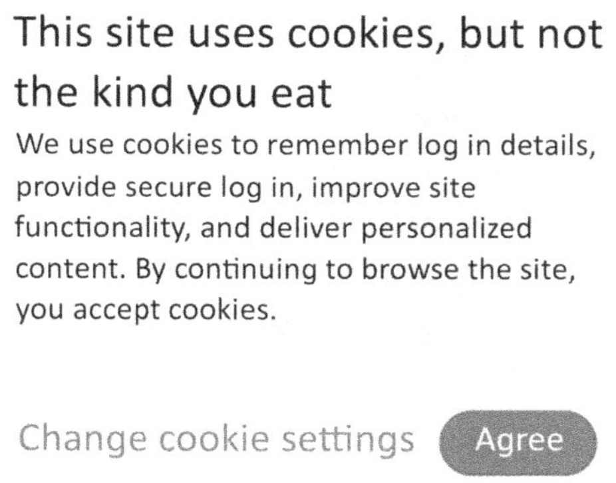

This site uses cookies, but not the kind you eat

We use cookies to remember log in details, provide secure log in, improve site functionality, and deliver personalized content. By continuing to browse the site, you accept cookies.

Change cookie settings   Agree

Figure 15.18: Starbucks cookie consent screen

This shows organizations are doing all they can to collect as much data as possible. But hopefully, we see this change as more regulation continues to focus on protecting our personal data and hold organizations more accountable for any neglect and mismanagement of our personal data.

Data Retention

As much as organizations like to collect user data, it is important that you only maintain it for as long as you need it. And this is only becoming more important as more organizations become compromised and more data continues to become exfiltrated. Organizations shouldn't be holding onto data if it's no longer needed unless there are regulations and laws that require that data to be retained. If you are holding onto data that is not needed, you could be holding yourself a lot more liable than needed if a major cybersecurity incident occurs.

As part of your data retention program, there will be a couple of primary components to support the governance side of data retention. The first is the need for a policy and the second is the retention schedule. First, as we've discussed throughout the book, the need for a policy to support your data retention requirements must be put in place. For data retention, this policy is known as the **Records & Information Management (RIM)** policy. The RIM policy will specify the purpose of the document, what is in scope for the policy such as specific records, who's responsible for what, disposal requirements, data storage requirements or reference to the policy that references these requirements, the retention schedule, and much more.

One of the most important components of RIM is the data retention schedule. This document will possibly become very lengthy and will require ongoing review and maintenance to ensure the schedule is current with the ongoing regulations and laws that continue to appear. This is an area that will need full engagement from the legal team along with each of the business functions that own data within their respective areas. It will be important that ownership of each of the documented classifications is captured and maintained. Tracking regulations for the retention schedule is not easy, especially at a global level. One great resource I have found very helpful in this area is filerskeepers (`https://www.filerskeepers.co/`). filerskeepers is a service offering that allows you to quickly build a retention schedule based on your organization's location and required regulations. They also keep the retention up to date to ensure you can keep your retention schedule current based on any changes in laws and regulations.

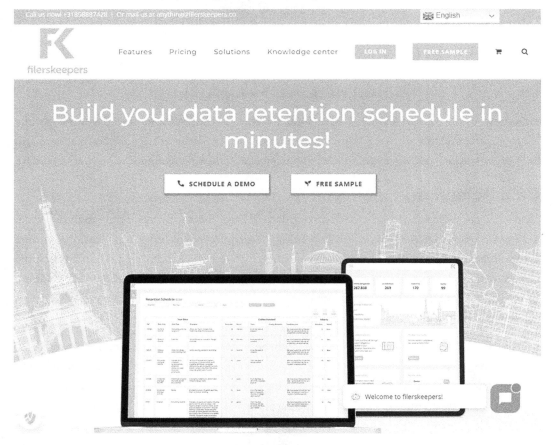

Figure 15.19: filerskeepers data retention

Implementing a RIM policy along with the supporting retention schedule is not a simple task. If you don't have one in place today, this should be something you are tracking to ensure one is implemented within your organization. Maintaining data for longer than needed only increases liability for your organization, especially when a major cybersecurity event occurs.

Data Disposal

Once any data being stored in your environment surpasses the agreed-upon retention, the next step is to ensure that data is disposed of. The first step is knowing where all of your data is. Once you know where all of your date is located, ensure you are able to efficiently audit the age of your data so it can be disposed of correctly when the retention is met. Once it is time for disposal, the disposal of that data must be done correctly. This gets very complex depending on where the data resides, what systems it is stored on, and the challenge of whether hardware where data has been stored will also need to be disposed of. Another example of disposal is end-of-life hardware. For example, physical devices such as laptops will need to have a certified disposal process in place to ensure any data on those devices is removed. An unfortunate example of where this wasn't done correctly was with Morgan Stanley, who contracted a vendor to remove data from the systems of a decommissioned data center. This didn't happen and customer data was found on hardware that was auctioned off: `https://www.cnbc.com/2023/11/16/morgan-stanley-fined-over-computers-with-personal-data.html`.

There will be many different scenarios in which you will need to manage the disposal of your data, whether electronic or on the hardware where it has been stored. In some instances, you will need to work with your vendors on their disposal policies and ensure that any data they process on your behalf complies with your retention schedule. This includes the termination of any services with vendors to ensure your organization's data is disposed of correctly. Some of these dependencies will require language within your contracts to support any data disposal requirements. For anything you need to manage internally with your organization, a great reference is the NIST SP 800-88 Rev.1 Guidelines for Media Sanitization: `https://csrc.nist.gov/pubs/sp/800/88/r1/final`.

Legal Hold

This is essentially the need to place all of a user's data on hold so it cannot be deleted in the event it is needed as part of a litigation. This functionality will require close collaboration with the legal team with the need for legal hold requests. This may involve more of the IT team than the cybersecurity team to support, but it will be important that you are aware of the process.

Ideally, the ability to place a user on legal hold should be with the legal team, but this may not be realistic depending on the technical capabilities available and to what extent legal hold is needed. When litigation comes into your organization, the legal team will need to notify the teams responsible for placing those users on legal hold unless they can do it themselves. Traditionally, legal hold has been centered around the user mailbox along with any local storage accounts where applicable. However, it is important to understand if a legal hold for a user requires other technologies to be placed on hold. This may include the need for SaaS, or other applications that the user has transacted in. Whether these capabilities are supported will need to be understood as you work with the legal team. In some instances, some applications or SaaS vendors may not have legal hold capabilities. The legal team will need to understand the risk with these environments with any potential litigation cases. On another note, it is important that users who are no longer required to be on legal hold have any holds removed immediately. You don't want to retain any data for users that isn't required. This will need to become an auditing task to ensure that the legal team is either informing the teams to remove the legal holds or removing them themselves. Either way, the legal team will be responsible for all legal hold oversight and will simply direct other teams with what their requirements are.

E-Discovery

E-discovery is a capability that works hand in hand with legal hold. For the most part, your legal team will utilize these capabilities. When there is an active legal case with your organization, there will be a need for the legal team to search for any content requested as part of litigation. This is another feature that requires the legal team to manage and have access to the tools to complete these capabilities. But like the legal hold requests, there may be complexities involved and a need to work with the IT and/or cybersecurity teams to collect data for a legal hold.

Another scenario for e-discovery may be as part of a cybersecurity incident where the SOC team requires capabilities to quickly search across users' data for specific data. Take, for example, an identified true-positive phishing campaign that could have the potential to cause a lot of damage to the organization if users click on the malicious link. A quick investigation shows the phishing email has been delivered to hundreds of users through a basic delivery report. To reduce the risk as quickly as possible, e-discovery capabilities should allow for a search of the phishing email and for it to be removed from the users' mailboxes immediately to prevent users from clicking on the malicious link. I'm not familiar with all e-discovery tools and capabilities, but Microsoft Purview does provide these capabilities within their e-discovery tool: `https://learn.microsoft.com/en-us/purview/ediscovery-search-for-and-delete-email-messages`.

Summary

Hopefully, you see the bigger picture as it relates to Regulatory and Compliance after reading through this chapter. Also, the reality is that this is an area where you are going to need a lot of support from others throughout the organization, more specifically the legal team. As regulations continue to be released and evolve, it is critical that someone in your organization stays current with any requirements from these regulations that have an impact on your organization. The bigger your footprint, the more regulations you are going to need to comply with, and it will be critical that you validate enforcement is occurring for any required regulation. To support this enforcement, you will need to ensure audits are conducted by a third party. You do not want to explain to anyone why you weren't in compliance with a regulation after a breach has occurred within your organization. Hence the importance of regulations and compliance within the cybersecurity program.

To begin the chapter, we reviewed the evolving landscape of Regulatory and Compliance and what is expected as we move forward in this area. As discussed, this is an area that continues to become more complex as we continue to see more regulation come our way and as current regulations and laws evolve and become more complex on a day-to-day basis. We also presented a list of many of the global regulations you may be familiar with or that you can expect to come across at some point. In the next section, we provided more insights into the importance of the legal team involvement as it relates to Regulatory and Compliance needs for your organization. We also took a closer look into some of the breach notification requirements throughout the US and the globe to provide a clearer picture of how complex this area has become. Thus, there is a need for legal experts who are familiar with regulations and laws to ensure your organization is in full compliance. This led to the importance of data protection for your organization. Knowing where your data is will allow you to appropriately classify your data, and finally, apply the correct controls to protect your data based on classification.

In the section that followed, we covered the need for frameworks and audits in more detail. Although we've covered Cybersecurity Framework throughout the book, we expanded the discussion on the maturity model of a framework that can serve as a foundation for improvement in your cybersecurity program. We also reviewed privacy frameworks in more detail, more specifically the NIST privacy framework and some of the cross-over into the NIST Cybersecurity Framework. This section also covered audits in more detail and the importance of executing audits, specifically bringing in an external vendor to complete audits and provide attestation for your frameworks. We then finished off the chapter with a look at some other Regulatory and Compliance considerations.

This included a deeper review of the importance of privacy for your organization, and a review of the need for data retention, which in turn requires a correct data disposal policy and processes. We finished off the section with a look at the legal hold and e-discovery needs.

In the final chapter, we will be sharing some final thoughts as we close out the book and everything that has been covered. We will begin the chapter by bringing it all together by defining the program, the core requirements, and finally the overarching GRC component for your program. This will lead into the next section, which will discuss the reality of your program as a journey. Your program will not be built overnight, and it will take time. This will take us into the top 10 considerations you should focus on with your program. We will also focus on what I consider my top three priorities as a CISO in the current state, although everyone will have different priorities. We will finish off the chapter with some thoughts and insight into what's next with cybersecurity as both technology and the threat landscape continue to evolve.

Join our community on Discord!

Read this book alongside other users, Cybersecurity experts, and the author himself. Ask questions, provide solutions to other readers, chat with the author via Ask Me Anything sessions, and much more.

Scan the QR code or visit the link to join the community.

`https://packt.link/SecNet`

16

Some Final Thoughts

We have made it to the final chapter where we will cover some final thoughts on everything we have covered throughout the book. The hope is that you have gained a better understanding of what is involved with building (or re-building) a mature cybersecurity program that will help meet the needs for today's ongoing and increasing cyber threats. As you have read through the book, hopefully, you have realized the depth and breadth involved with building and running a cybersecurity program. The goal is to share this knowledge with as many cybersecurity leaders (and experts) as possible to help with the building blocks for a more secure world from cybersecurity threats in the long term. My hope is that you gain some valuable takeaways from my experience over the years with building and overlooking a cybersecurity program.

In this chapter, we will begin with a brief review of all the sub-functions within your cybersecurity program. We will then shift into an overview of setting expectations that your program maturity is a journey, and the reality is that you are not going to have everything in place overnight. The importance of understanding where your program is and prioritizing what should be implemented first is important. Next, we will investigate 10 of the most important considerations with your cybersecurity program, looking at those items that should be considered higher risk and should be implemented with priority. We will also review what I consider my top three priorities as part of the top 10 considerations within the current state based on my observations. This is a question I get asked often. In the final section, we will cover some considerations for the future of cybersecurity.

The following will be covered in this chapter:

- Bringing It All Together
- Your Program Is A Journey

- Top 10 Considerations
- Some Observations And the Future

Bringing It All Together

Throughout the book, we have broken the content into 3 distinct sections to help shape the foundation of your cybersecurity program. The first of the 3 sections focuses on defining your program, essentially setting the foundation to allow for greater success with your cybersecurity program.

Defining the Program

For the most part, many large organizations will most likely have some defined program in place, especially those that have regulations. Even if you do have a well-defined program in place, it will be good to go through the exercise of reviewing where your program is today and ensuring it continues to be modernized to meet today's cybersecurity challenges. We began the section with some insights into the digital world we live in and the current challenges faced that relate to cybersecurity. As cybersecurity leaders, we must remain current and keep up to date with the most recent and evolving threats. We then reviewed the foundations needed to more efficiently support your cybersecurity program, such as the need to understand the services your organization provides and the business model in place. This includes the need to become a business partner, a very different role to the one that has been fulfilled by cybersecurity leaders in the past. Whether you have a cybersecurity program in place or not, you must continue to review the current organization structure and update it to meet modern-day threats, as you cannot afford to let your program become outdated. It is important that everyone is aware that the role of the cybersecurity organization is to manage cybersecurity risk for the organization, and that we aren't there to dictate what should or shouldn't be implemented for the business. Our role is to assess risk and make the business aware of that risk, and in turn, they determine whether to accept it or not.

This section also addresses in detail the need for building a robust roadmap for the cybersecurity program. To do this, you need to understand the current state of the cybersecurity program and, essentially, what maturity your organization is at with cybersecurity. This in turn will allow you to build a short-term roadmap with some quick wins, along with a long-term roadmap with more strategic items that will take longer to complete. If you are new to an organization, you will want to focus on an immediate short-term roadmap that allows you to gain a better understanding of the business. The final focus point within defining the program is the important requirement of having a strategy in place for your cybersecurity program. Not just a single strategy but a strategy for multiple items, including the need for a framework for your program.

Although frameworks are available for multiple areas, the need for a cybersecurity framework is undeniable. Ensure you have a strategy around your architecture, such as a zero-trust architecture model. You also need to consider a strategy for your application and vendor portfolio and the need to reduce your portfolio to eliminate complexity. The other strategy you need to define is your resource management strategy and how you are going to staff your team, along with in-house vs outsourcing considerations.

Once the program has been defined with the foundation in place and the strategy defined, we can now build the core components of the cybersecurity program by implementing each of the core building blocks to provide a well-rounded and robust program.

The Core

For the core of your cybersecurity program, we focus on seven unique functions that should be part of every cybersecurity program. If you oversee **Operational Technology (OT)** and **Internet of Things (IoT)** then you will have 8 unique functions as part of your core. In no specific order, the first is that of your **cybersecurity architecture**. This expands upon your core architecture strategy, for which we have reviewed a zero-trust architecture. Within the cybersecurity architecture function, this is where you will determine all the technologies needed to support your cybersecurity program, along with the engineering required to deploy and configure those solutions. It is also important that your team is part of the architecture review board to ensure compliance is being maintained with all solutions being deployed and vendors being onboarded. The next core function we reviewed for the program is that of **identity and access management.** In my mind, this is one of the more crucial functions with the shift from a traditional network-based security model to the identity becoming the new perimeter. It is important that you look to modernize your identity strategy for your organization, with the need to shift to a cloud-based **Identity Provider (IdP)** from a legacy-based directory store. Securing your identities should be one of your top priorities as threat actors continue to look for ways to compromise them and access your organization through compromising your user's identities. More concerning is the ease of access to passwords on the dark web allowing threat actors simple access to user accounts with simple passwords. You need to enable phish-resistant based MFA with the goal of becoming passwordless, eliminating the need for passwords altogether. In addition, you should be taking advantage of advanced identity protection capabilities with AI and machine learning.

The next function, which is probably one of the more traditional functions for most organizations, is that of **cybersecurity operations**. Everything from threat detection and prevention to incident response needs to be in place.

If you don't already have one, you need to have a 24/7 **Security Operations Center (SOC)** in place, whether it be in-house or outsourced to a **Managed Security Services Provider (MSSP)** or some other service to provide these capabilities. There are also more modern capabilities available, such as **Extended Detection and Response (XDR)**, that complement other traditional services within your cybersecurity operations. It is also important you are familiar with your organization's **Business Continuity Plan (BCP)**, especially your **Disaster Recovery Plan (DRP)** procedures and the **Cyber Incident Response Plan (CIRP)**. The next function is your **vulnerability management function**, which will work very closely with your cybersecurity operations function. Vulnerability management needs to be tracked, as its program requires resources to efficiently run. There are thousands upon thousands of vulnerabilities continuously haunting us, and it's this team's responsibility to track all known vulnerabilities and identify what needs updating within the environment. This includes the need to prioritize based on the risk of the known vulnerabilities. More importantly, it involves ensuring any known exploited vulnerabilities are addressed immediately. Here is where we stress the importance of asset management and ensuring a well-documented and comprehensive inventory of every digital asset within your organization is being tracked. If you don't know you have an asset, how are you meant to protect it? This unknown asset could be the entry point of ransomware into your environment that cripples your business operations.

Another extremely important and often overlooked function is your **cybersecurity awareness, training, and testing function**. Statistics continue to show that the human element is the most vulnerable, and for the most part, it is where major cybersecurity incidents stem from. With this being the case, why aren't we investing more in educating and training our users to be more aware and better prepared? Our programs need to focus on building a culture around cybersecurity, a place where cybersecurity isn't an afterthought, but at the forefront of everyone's minds and something that comes naturally as part of everyone's roles. Cybersecurity is everyone's responsibility nowadays and the cybersecurity awareness, training, and testing function is where this translates for the users. In the current state, another vital function that needs a lot of attention is that of **vendor risk management**. Vendors are becoming compromised more frequently and we are trusting them with more of our data. The traditional model of centralizing our data is long gone as it has become more distributed in cloud infrastructure and SaaS environments. Because of this, you need to focus on a robust vendor risk management program that includes a thorough review of each vendor. And more specifically, any high-risk vendors need full scrutiny of their cybersecurity posture. Your biggest challenge will be ensuring all vendors follow a process to be reviewed as you will find a lot of vendors being onboarded throughout the business without going through the correct process. From a cybersecurity perspective, you will need to understand contracts in more detail as cybersecurity language and insurance is a requirement today.

The last core function of your cybersecurity program is **proactive services**. We must take a more proactive approach with our program to reduce risk as much as possible. This includes the need to conduct different security tests within your environment, for example, penetration tests, application testing, configuration testing against **Active Directory** (**AD**), and more. More importantly, you need to take a more proactive approach to incident response. The reality is you are going to be dealing with cybersecurity incidents and you need to be better prepared to handle them. This is where tabletop exercises provide value, for both your technical and leadership teams. In addition, ensure you have a well-documented and updated CIRP that everyone is aware of. The quicker you can respond more efficiently, the quicker the business can continue to work. The final function we will cover may not be applicable to all but it is a requirement for anyone who needs to manage OT and IoT. Regardless of whether you manage OT and IoT or not, you should understand what this technology is and how it can be susceptible to threat actors. This technology is critical in supporting the core infrastructure of our countries, like power plants, water treatment facilities, traffic signals, healthcare equipment, and much more, hence the need for a dedicated program around this technology if it is applicable to your organization.

The next transition from the core is into the final section, which brings our entire program together to close the loop of a well-rounded and comprehensive program. Now that we have laid the foundation and the strategy, along with defining each of the required core components, we need to close the loop with the GRC component to ensure the program is meeting its objectives and remaining in compliance.

Bringing It Together

The next three chapters in the book cover an extremely important component of the cybersecurity program that can no longer be overlooked. The **Governance, Risk, and Compliance** (**GRC**) **program** is what provides oversight throughout the program and bridges the gap between the cybersecurity program and both the executive leadership team and the board of directors. This is the program that provides full transparency into all risks being managed within the cybersecurity program. The first component of GRC is the governance function, with the role of ensuring your cybersecurity program is appropriately structured and managed. Here, you will need to ensure that all relevant policies, procedures, and standards are in place for the organization as it relates to cybersecurity. This is the function that provides the avenue to your executive leaders and board of directors to ensure they are aware of all current risks being managed and to provide guidance or make any needed decisions around that risk. Ensuring you have quality reports and dashboards for the cybersecurity program will significantly help.

The next function focuses on everything risk-related. Your native language within cybersecurity needs to be risk moving forward, and you must figure out how to best translate that risk into business terms for your executive leadership team and the board of directors. You will need to understand the different types of risk applicable to your organization so you can efficiently manage them and ensure the appropriate decisions are being made with that risk. For example, does your organization accept, mitigate, transfer, or avoid? As you work with risk, you'll need a risk register to track everything. If you don't have a risk register today, implement one now. We then finish off the GRC program with regulatory and compliance. This has become a very important and complex area for cybersecurity, and you will not be able to oversee this yourself. You are going to need close collaboration with the risk management and privacy teams, if they exist, but most importantly the legal team. If you don't have an in-house legal team, you are going to need to work with an external law firm or outside counsel to provide guidance, and support your organization around the overly complex regulations we must work with today. In addition to needing to meet any regulation, it is just as important that we are maintaining compliance with any regulation. Even if there are no regulatory requirements, you need to ensure you are following a cybersecurity framework in which you should be ensuring you are compliant, specifically from an external independent source.

Now that we have come full circle with the entire cybersecurity program and reviewed everything involved, let's remind ourselves that building a cybersecurity program is no easy feat, and this journey will not happen overnight.

Your Program Is a Journey

One thing is clear, building your program is not going to happen fast. In fact, it will take months to years to become more mature if you are just beginning your journey. At the same time, you will never reach a destination with your cybersecurity program. There will always be new tactics and threats that evolve, requiring us to be dynamic and enhance our programs to meet the current state. In *Chapter 3, Building Your Roadmap*, we provided a lot of detail on how to best prepare for your journey. Everyone's journey is going to be different, depending on whether you are newly promoted to a cybersecurity leader within the organization or whether you are a new leader coming in from a different organization.

Assess the Current State

As you build your journey, there are three distinct areas you should focus on to better understand the current state that will allow a more successful program to be built and/or matured. The three areas are:

- **Learn Cybersecurity within Business**: Make sure you understand the organization's structure and learn the business as quickly as possible. This will help you better understand the culture and provide a better idea of how serious cybersecurity is (or is not) within the organization. For example, some questions you can ask other business leaders to help with this include: Are you aware of a formal cybersecurity program within the organization? Do you see cybersecurity as a priority? Has your team been involved in any cybersecurity incidents? Are you required to comply with any audits? Have you observed any cybersecurity gaps or risks throughout the organization?

- **Non-Technical Program Review**: Assess the current cybersecurity program from a non-technical perspective to identify what is currently in place and how mature the program is (or is not). For example, the following are some areas to review: Is cybersecurity training being executed, and are phishing simulations going out to the users? Is there a CIRP in place and are tabletop exercises being executed? Are cybersecurity policies, processes, and procedures in place for the organization? Does the organization currently hold a cybersecurity insurance policy?

- **Technical Program Review**: Spend time with the team to review everything from a technical perspective to help understand the current state of the cybersecurity program. For example, some areas to help get a better gauge of the environment include: Are there any vulnerability scans that have been run in the past for review? Are there any audit reports available for review, whether within the cybersecurity program or in the broader organization? Have there been any penetration tests against the environment? Is there a SOC in place?

Once you have a better understanding of the current state of the cybersecurity program, you can begin to build a roadmap to allow you to achieve your goal of building a more mature program.

Immediate Short-Term

Once you have completed your initial review, you'll need to define your short-term roadmap. Everyone's roadmap will be different and will need to cater to their own organization's needs and current maturity. This roadmap will follow your initial review of the environment and consists of a 90-day timeline. The idea here is that you look to make some immediate impact within the program. After your initial review, you may have identified some high-risk items that need to be addressed immediately. This roadmap will allow you the time to become better acclimated to the cybersecurity program. Ideally, you are going to need to formally document everything you would like to get accomplished within this timeline.

This will require good program management oversight and a project manager to support. We covered program and project management in more detail within *Chapter 3, Building Your Roadmap*. As a new leader, you will need to understand that accomplishments will not necessarily happen overnight, especially within larger organizations with more red tape and processes to navigate. You will need to be strategic with what can be accomplished within the specified timeframes.

Short-Term

The next roadmap will extend beyond the first 90 days with any hopeful quick wins or low-hanging fruit to the short-term roadmap, which will extend through the first 12 months of building or re-building your cybersecurity program. Here, you will begin working on some of those items that require additional time and resources to get movement. These items may not necessarily be as high a priority as the immediate roadmap items, but they are critical for the success of the overall program. It is important to remember that you will need to be dynamic as changes occur. New requirements or priorities will surface that weren't originally planned for and will need to be added to the roadmap as you progress.

Long-Term

The next roadmap that will need defining is the long-term roadmap, which shifts your focus to items that you plan to address in 1-3+ years for your cybersecurity program. You will observe the 1-3 year long-term roadmap starts naturally evolving into a stabilization phase with continuous improvement activities. Here, the program will hopefully be in a much more mature place, and you should feel more acclimatized with the program. There's also a high possibility that items you originally planned to work on in the long-term roadmap are either no longer applicable as they may have been addressed earlier, or they are no longer relevant. Again, you will need to be dynamic and adapt to the changing environment and requirements around you. In cybersecurity, this happens daily.

Cybersecurity Roadmap

The next step is to document your roadmap for the coming months and years. This allows you to follow a plan by tracking everything that is expected to be completed as part of your cybersecurity program journey. In *Chapter 3, Building Your Roadmap*, we provided a detailed list of examples of what you may include within each of the roadmaps. The following image from *Chapter 3* is a snapshot of how you may document your roadmap at a high level when sharing with a broader audience.

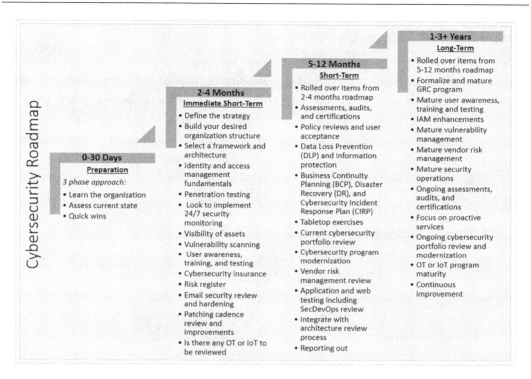

Figure 16.1: Cybersecurity roadmap sample

As you can see, your program is a journey with many components involved. Building a mature cybersecurity program will take time and you will need to ensure you are focusing on the higher-priority items first. The reality is that some of the components of your cybersecurity program will need to wait as you work on other priorities and ongoing operational items. In addition, technology is changing so fast and threat actors continue to evolve at such a great speed that you'll need to continuously review the program and adjust as needed. It will also be important that you review your roadmap on an annual basis and modify it as needed to support the coming year as you plan projects for the team.

A robust roadmap will only provide for greater success with your journey toward a more mature cybersecurity program. Now that you have read through all the chapters, you should have a much better idea of what is required to build a robust cybersecurity program and better understand the areas you need to focus on first. Your program may be in a good place with a lot of maturity, but the reality is, there is always room for improvement. As a reminder, your journey will have no final destination and will continue to evolve with many bumps and new challenges along the way. Your goal is to continue an upward trajectory of constant maturation as you continue to reduce risk for your organization.

The Top 10 Considerations

Let's move on to what I consider the top 10 considerations covered throughout the book for your cybersecurity program. These considerations are being recommended based on the current threat landscape (in the year 2024) and my personal experience. You may already be in a mature place with some of these, but it is still recommended you review and ensure you are applying the latest recommendations and technologies to each of these areas. In addition, we will begin the top 10 considerations with the *CISO Top 3 Priorities for the Year 2024*. Since I often get asked what my current top 3 priorities as a CISO are, I thought it would be a good idea to incorporate them with the top 10.

CISO Top 3 Priorities for the Year 2024

I often get asked what my top 3 priorities are as a CISO, probably a question most CISOs get asked. This is a question that you will most likely get different answers to from different CISOs depending on the industry they are in, how mature their program currently is, which country they operate in, what year is it, etc. As threat actors evolve and technology changes, our priorities will continue to change. As of the current state in 2024, the following are what I consider the top priorities to address, with the most important being addressed first.

User Awareness, Training, and Testing

Data continues to show that our users are most vulnerable as it relates to cybersecurity. The Verizon 2024 DBIR report (https://www.verizon.com/business/resources/reports/dbir/) stated that the human element was part of 68% of breaches, and although this is down from 74% the previous year, it still holds more than double the percentage of other factors involved in a breach. Other reports have shown this number to be even higher. Because of this, we need to ensure that we are providing the required investment to support the need to better educate our users. If our users are the primary threat vector for breaches, then we need to look at investing the most resources into the user awareness, training, and testing program. Traditionally, annual training and infrequent phishing simulations may have been sent. Unfortunately, this is not going to provide the results needed. We can't treat this program as a check box for compliance; it needs a lot more attention. The image below shows the level of detail you need to be thinking about with your user awareness, training, and testing program. At a minimum, you need to be running quarterly training with monthly phishing simulations. And your phishing simulations need to be AI-driven to provide the same reality of real-world phishing emails being received from threat actors.

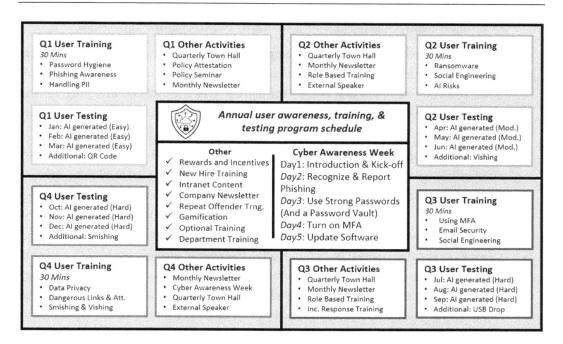

Q1 User Training
30 Mins
- Password Hygiene
- Phishing Awareness
- Handling PII

Q1 Other Activities
- Quarterly Town Hall
- Policy Attestation
- Policy Seminar
- Monthly Newsletter

Q2 Other Activities
- Quarterly Town Hall
- Monthly Newsletter
- Role Based Training
- External Speaker

Q2 User Training
30 Mins
- Ransomware
- Social Engineering
- AI Risks

Q1 User Testing
- Jan: AI generated (Easy)
- Feb: AI generated (Easy)
- Mar: AI generated (Easy)
- Additional: QR Code

Annual user awareness, training, & testing program schedule

Q2 User Testing
- Apr: AI generated (Mod.)
- May: AI generated (Mod.)
- Jun: AI generated (Mod.)
- Additional: Vishing

Other
- ✓ Rewards and Incentives
- ✓ New Hire Training
- ✓ Intranet Content
- ✓ Company Newsletter
- ✓ Repeat Offender Trng.
- ✓ Gamification
- ✓ Optional Training
- ✓ Department Training

Cyber Awareness Week
Day1: Introduction & Kick-off
Day2: Recognize & Report Phishing
Day3: Use Strong Passwords (And a Password Vault)
Day4: Turn on MFA
Day5: Update Software

Q4 User Testing
- Oct: AI generated (Hard)
- Nov: AI generated (Hard)
- Dec: AI generated (Hard)
- Additional: Smishing

Q3 User Training
30 Mins
- Using MFA
- Email Security
- Social Engineering

Q4 User Training
30 Mins
- Data Privacy
- Dangerous Links & Att.
- Smishing & Vishing

Q4 Other Activities
- Monthly Newsletter
- Cyber Awareness Week
- Quarterly Town Hall
- External Speaker

Q3 Other Activities
- Quarterly Town Hall
- Monthly Newsletter
- Role Based Training
- Inc. Response Training

Q3 User Testing
- Jul: AI generated (Hard)
- Aug: AI generated (Hard)
- Sep: AI generated (Hard)
- Additional: USB Drop

Figure 16.2: User awareness, training, and testing example schedule

It is important to remember that as you continue to build and mature your user awareness, training, and testing program, you need the capabilities to provide intelligence, dynamic learning, and automation. The ability of a solution to understand your users' behaviors and risks to allow content to be delivered that provides more value is key. Remember, humans are our greatest asset, but they are our biggest threat vector when it comes to cybersecurity. We covered this in more detail in *Chapter 9, User Awareness, Training, and Testing*.

Governance, Risk, and Compliance

Second on the list of the CISO Top 3 Priorities for the Year 2024 is the **Governance, Risk, and Compliance (GRC)** program. This is the program that brings the needed visibility to your executive leadership team and the board of directors. Nowadays, this is very important as we are seeing litigation being taken against organizations, and more specifically, in some instances, the CISO/CSO. Because of this, we need to ensure accountability sits at the top of an organization and that the executive leadership team and the board of directors are ultimately aware of any cybersecurity risk, and they are the ones accepting any risk if needed.

In the book, we covered each of the functions separately because of their importance:

- **Governance**: Governance is the overarching authority for the entire cybersecurity program to ensure everything is running as it should be to reduce risk while maintaining compliance. This is the program that bridges the gap between the cybersecurity program and any external stakeholders, such as the executive leadership team and board of directors (if applicable).

- **Risk**: Risk management pertains to how we efficiently assess and manage cybersecurity risk for the organization. Essentially, what are the threat vectors we need to be aware of and how do we address them to hopefully prevent a major cybersecurity incident from occurring? This also requires us to be able to translate any identified risk into business terms for the business to review. A risk register is critical as part of this function.

- **Compliance (+ Regulatory)**: Regulation is the laws put in place to ensure you are implementing the requirements set forth and best practices to protect the people, technology, data, and information within your organization. Compliance is to ensure any required regulation is being conformed to, in addition to other items such as audits, standards, policies, etc. It's important that you implement the required regulation but also validate you are meeting the requirements of that regulation.

Figure 16.3: Governance in relation to risk, compliance, regulatory, and the broader program

As you can see from the image above, the GRC function is what brings formality to the cybersecurity program. Governance along with risk and compliance oversees the entire cybersecurity program to ensure the program is meeting standards put in place and full transparency is being provided. This is critical as liability increases for organizations and, more specifically, cybersecurity leaders. We covered GRC in a lot more detail in *Chapter 13, Governance Oversight, Chapter 14, Managing Risk,* and *Chapter 15, Regulatory and Compliance.*

Vendor Risk Management

Next on the list is vendor risk management (aka third-party risk management). According to a report released by SecurityScorecard and the Cyentia Institute named *Close Encounters of the Third (and Fourth) Party Kind* (`https://securityscorecard.com/research/cyentia-close-encounters-of-the-third-and-fourth-party-kind/`), within the previous 2 years, over 98% of organizations have had at least one of their third-party relationships suffer from a breach. This data speaks for itself, and we need to be maturing our vendor risk management programs. We cannot allow vendors to continue to become compromised at the current rate. Vendors have so much access to our data that they need to be held to a higher standard. As you review and mature your program, it is critical items such as the following are addressed with your vendors to ensure they are taking cybersecurity seriously:

- Does the vendor have a cybersecurity program in place?
- Are there policies, procedures, and other documents in place?
- Does the vendor have certifications/audits like SOC2 Type 2 in place?
- Have there been any breaches with the vendor in the past?
- What are the vendors' external attack surfaces?
- Does the vendor execute testing like penetration testing?
- Do they have a BCP, DRP, Crisis Management Plan, and CIRP?

A example of the vendor risk management process may look like the following.

This is a high-level process flow you can reference and incorporate as part of your vendor's risk management process.

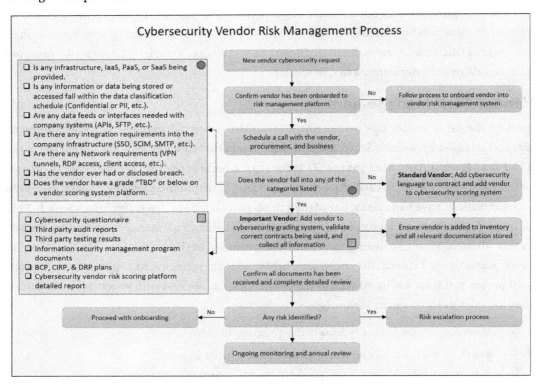

Figure 16.4: Cybersecurity vendor risk management process sample

Another important component of your vendor risk management processes is contract reviews. It is critical that the cybersecurity team is involved in contract reviews to ensure the correct language is being added, specifically the right to be notified when a major cybersecurity incident occurs, to ensure the vendor is implementing security controls per best practices, the right to access and review documents such as audits, certifications, and so on. You also want to ensure your vendors hold a cybersecurity insurance policy moving forward to provide protection in the event of a major cybersecurity incident. We covered vendor risk management in a lot more detail in *Chapter 10, Vendor Risk Management*.

The Remaining 7 Priorities

Next, we will move on to the remaining 7 considerations from throughout the book. These considerations are not in any particular order but do serve as some of the more important areas you should be reviewing within your cybersecurity programs.

Cybersecurity Framework

Throughout the book, we covered the importance of frameworks. Not just cybersecurity frameworks but other frameworks, including risk and privacy frameworks. For your cybersecurity program, the initial focus should be on implementing a cybersecurity framework as a priority. The two we covered in more detail were:

- **International Organization for Standardization/International Electrotechnical Commission (ISO/IEC) 27001:2022**
- **National Institute of Standards and Technology (NIST)** Cybersecurity Framework 2.0

Nowadays, having a cybersecurity framework should not be optional. All organizations of all sizes need to implement a cybersecurity framework for their organization. This allows you to better understand your current maturity and build a baseline that you can build upon. If you are beginning your journey with a cybersecurity framework, it's recommended that you focus on a maturity model-based assessment like the NIST cybersecurity framework 2.0 that will allow for ongoing progression as you continue to improve your cybersecurity program. Your maturity journey will not happen overnight, and it will take time. We covered cybersecurity frameworks in more detail in *Chapter 4, Solidifying Your Strategy*, and *Chapter 15, Regulatory and Compliance*.

Zero-Trust Architecture

It is important that you incorporate a foundational architecture with your cybersecurity strategy. The more familiar architecture for cybersecurity is that of **Zero Trust Architecture** (**ZTA**). The core of a zero-trust model will always fall back on the principle of never trust, always verify. Effectively implementing a zero-trust model requires a multilayered approach with your security strategy, along with the use of the most current and modern technologies available. As you look to strategize on a zero-trust security model, which will in turn become the foundation of your zero-trust architecture to best protect your environment, you are going to need to follow some form of deployment and maturity model. A couple of the better maturity models include that of Microsoft and the **Cybersecurity and Infrastructure Security Agency** (**CISA**):

- Microsoft Zero Trust: `https://learn.microsoft.com/en-us/security/zero-trust/`
- CISA Zero Trust: `https://www.cisa.gov/publication/zero-trust-maturity-model`

In *Chapter 5, Cybersecurity Architecture*, we covered ZTA in a lot more detail and we reviewed ZTA through the lens of both maturity models referenced above with an additional area, bringing the number of distinct areas to seven: identities, device and endpoint, data, applications and **Application Programming Interfaces (APIs)**, infrastructure, network, and collaboration. Don't overlook the importance of a ZTA for your cybersecurity program; you cannot rely on single layers of protection for your environment, you must implement multiple layers of controls to protect your assets.

Implementing Identity Protection and Privileged Access

In a world that has shifted outside the walls of the office to an anywhere-at-any-time access model, identities have become a high target of attention and are prone to weaknesses. They are a fundamental focus for attackers to gain access to your environment. Because of this, it is critical that your identities have multiple layers of protection and that preventative measures are in place.

Strong identity protection will require implementing account and access management capabilities and enforcing the principle of least privilege. A user must only be provided access to the specific data, applications, and systems that are necessary for their job role. Use **role-based access control (RBAC)** to streamline access and enforce strong passwords or adopt passwordless technologies. Encourage users not to use the same password more than once and provide an enterprise-grade password management tool to provide more efficient password management. Require **multi-factor authentication (MFA)** to access systems and implement conditional-based access controls that allow MFA to be bypassed from company-compliant devices for a better user experience. Enable biometric authentication when available and consider an end goal of working toward a passwordless authentication world.

Always enforce the principle of least privilege when assigning permissions to users. Furthermore, enhance access security for your privileged users by deploying the following solutions:

- **Privileged access management (PAM)**
- **Just-in-time (JIT)** access
- **Privileged identity management (PIM)**

These solutions provide a well-rounded privileged access administration program for both your traditional on-premises environment and your cloud environment. If you don't have any privileged management tools available, create a secondary account for these purposes and ensure you educate your users not to use the same passwords between accounts.

Here is a high-level list of some access management best practices to keep in mind when thinking about the scope of privileged access for your organization:

- Enforcing MFA should be at the top of the list and ensure you require MFA for all cloud-based accounts. Implement **Fast Identity Online 2 (FIDO2)** for the highest privileged accounts and at minimum Phish resistant MFA for all others. For on-premises servers, you will need to implement a third-party tool to enforce MFA.

- Deploy a PAM solution for your on-premises and IaaS infrastructure and a PIM solution for your cloud infrastructure.

- Use JIT access to assign permissions dynamically as needed to avoid permanently assigning accounts for privileged users.

- Have an effective account provisioning and deprovisioning process and ensure you automate disabling accounts when employees leave the company.

- Ensure auditing and monitoring of privileged accounts is occurring in your environment.

- Limit the number of administrators within your environment and follow the **Principle of Least Privilege (PoLP)** at all times.

- Make sure you have separate administrative accounts from regular users' accounts.

- Don't allow email and internet browsing from privileged systems.

- Enforce stricter password policies on administrative accounts.

- Limit the amount of emergency "backdoor" accounts and ensure they are being monitored 24/7.

Make sure approval processes are in place for any privileged activity that needs to occur.

With technology constantly evolving, so does the need to keep your access model and strategy up to date, especially as more workloads shift to the cloud. Make sure you are evolving your identity and access management programs to support the need to fight against current threats. We covered identity and access management in more detail in *Chapter 6, Identity and Access Management*.

Update Update Update

In *Chapter 8, Vulnerability Management*, we covered in detail the importance of keeping your environment up to date. To put things in perspective, the 2023 Qualys TruRisk Research Report (`https://www.qualys.com/forms/tru-research-report/`) stated the average noted time to weaponize a vulnerability is 19.5 days, while the same vulnerabilities have a **mean time to remediate (MTTR)** of 30.6 days and are only patched on average 57.7% of the time.

As you can see, there is a significant gap between the time it takes for vulnerabilities to be exploited and the time it takes to patch our environment. This needs to change, and it needs to change fast. Traditionally, your programs to update your devices would take a more passive approach of slowly deploying updates over weeks or even months in some instances to prevent potential impact on users or applications. Unfortunately, this mindset needs to change, and updates need to be deployed immediately, knowing that there will be a risk that an application or user may be impacted by a bad update. But you need to determine if the inconvenience of a bad update outweighs the cost of a major cybersecurity incident. Make sure you review your strategy around vulnerability management and update your assets to ensure you are applying updates as quickly as possible, the faster the better. It is also important to ensure you have robust asset management processes in place. If you aren't aware of an asset, then it most likely won't be getting any updates, providing an increased risk for compromise. We covered asset management in more detail in *Chapter 8, Vulnerability Management*. In addition, always be looking for continuous improvement, specifically around automation opportunities, to allow greater efficiency with vulnerability remediation.

SOC Modernization

If you have a SOC in place, it will likely be the most active function if set up correctly. This function should be operating 24/7/365 for your organization. If it isn't, you'll be incurring increased risk with the possibility of threat actors infiltrating your environment without anyone monitoring activity. Trends have shown that increased activity typically occurs during off-hours, weekends, and holidays. Your SOC team's primary responsibility is to detect and respond to cybersecurity incidents within the organization. This function serves as a critical component within your program and it's important that you invest the time needed to ensure your cybersecurity operations are running as efficiently as possible. For this to be an effective operation, a well-defined organizational structure for the SOC should be put in place along with firm processes. At a high level, the following are some of the areas and tasks for consideration that provide for more efficient operations:

- Ongoing protection against cyber-related activity
- Cybersecurity operations models: SOC, SOCaaS, XDR, and MDR
- Cybersecurity operations staffing models: in-house, outsourced (MSSP), or hybrid
- SOC organization structure and hours of operation
- Roles and responsibilities
- Log collection, analysis, and automation: SIEM, SOAR, on-premises, or cloud

- Incident management and response (ticketing system)
- Threat detection and prevention
- Processes and standard operating procedures
- **Service Level Agreements (SLAs)** and key metrics
- Knowledge management
- Governance including reporting, meetings, and leadership visibility

As you can see, there's a lot involved in running an efficient and modern SOC to meet today's threats. Make sure you are considering the latest capabilities for your SOC such as the option to leverage cloud-based SIEM and SOAR capabilities with the use of XDR throughout the environment. Just as importantly, make sure your processes and procedures are well-defined so incidents can be reviewed and addressed as quickly and efficiently as possible. We covered the SOC in more detail in *Chapter 7, Cybersecurity Operations*.

Incident Preparedness and Response

Having a well-defined BCP, crisis communication plan, DRP, and CIRP helps ensure that your organization is prepared for impactful events. If you don't have a CIRP in place today, it is highly recommended you prioritize this because it is not a matter of 'if', but 'when' an incident will occur. Your CIRP will allow you to be better prepared and guide you through a major cybersecurity incident when one occurs. Your CIRP should include elements such as identifying the roles and responsibilities of the incident response team, updated contact information of everyone involved, including vendors, incident response procedures, communication procedures, and playbooks. Once you have a CIRP in place, all parties involved in a cybersecurity incident should become familiar with the CIRP. You will also want to test the response of both your technical and executive teams with a tabletop exercise to address any gaps with your CIRP and better prepare the team for what decisions may need to be made in such a situation. You should be conducting both cybersecurity executive and technical tabletop exercises at the minimum on an annual basis. In addition, it is important you engage with a third party to execute your tabletops. Having an external vendor execute the tabletop and provide the report based on their observation will eliminate any possible internal conflict between teams along with providing a separation of duty. This will help hold executive leadership more accountable for remediating the findings from the tabletop exercise. We covered incident preparedness and response in more detail within *Chapter 7, Cybersecurity Operations*, and *Chapter 11, Proactive Services*.

Cybersecurity Testing

You must be running cybersecurity testing activities within your environment, if not, you need to look at bringing in these capabilities. There are many types of cybersecurity testing that can be executed within your environment. You may not be able to execute all of them right away, but you'll need to determine the priority of what testing should be occurring within your environment and when. The following is a list of some of the cybersecurity testing exercises to be familiar with:

- Penetration testing
- Application testing
- Network testing
- Infrastructure testing
- Physical testing
- User testing

Focusing further on penetration tests, these can be executed externally to simulate an outside threat trying to break in, or internally, to simulate an insider threat that has breached your perimeter network. There are many different types of penetration tests and, hopefully, you are executing them already. Some of the more commonly executed include:

- Systems and servers, including Active Directory
- Web, API, databases, and mobile applications
- Networks (internal/external/DMZ), including wireless
- Social engineering, such as phishing simulations
- Physical security tests against facility access and data center controls

Also, make sure you have a robust application testing program in place, especially with any applications that are going to be accessible over the internet. A couple of valuable resources for your application testing activities are the OWASP Top Ten (`https://owasp.org/www-project-top-ten/`), which provides awareness of the ten identified most critical cybersecurity risks as they relate to your web applications, and the CWE Top 25 Most Dangerous Software Weaknesses (`https://cwe.mitre.org/top25/`), which provides insight into the most common software weaknesses that can lead to exploitation. We covered cybersecurity testing in more detail in *Chapter 11, Proactive Services*.

Notable Mentions

With so many priorities required within cybersecurity at this time, it is hard to keep the list to 10 notable items. In addition to the top 10 considerations listed above, the following are some additional important items for the overall cybersecurity program that are worth highlighting.

Artificial Intelligence (AI)

I would be remiss not to bring up AI as a consideration within the cybersecurity space at this time. As we are all aware, AI and more specifically GenAI is gaining traction. We are still in the early days with GenAI, but it is already having a significant positive impact on cybersecurity. Unfortunately, the capabilities are also being abused by threat actors for their gain. We covered AI in more detail in *Chapter 7, Cybersecurity Operations*, and we will cover it more in the *Some Observations and the Future* section of this chapter.

OT and IoT

Although they are very important and relevant, since these technologies aren't part of everyone's organization, they weren't included in the top 10 considerations. Although, depending on your industry and how prevalent this is within your organization, it may be a top 10 consideration. Whether you manage any of these technologies or not, it is important that you understand the principles and what is involved in securing these technologies. Because you never know, you may find these technologies in your portfolio at some point. It is important to note that these technologies do pose different and unique challenges to IT technologies, which is why a dedicated program will be needed to address cybersecurity for OT and IoT. With OT, these are, for the most part, essential services that support our everyday lives, and even a minor disruption can be catastrophic. We need to ensure these critical systems used to support everyday life are highly protected. With IoT, many companies will most likely have some form of IoT within their organization today. Although, I would expect it to be quite minimal compared to those companies that primarily depend on IoT technology. When referencing the need for a dedicated program with IoT, companies like telecommunications, automobile manufacturers, power plants, etc. would be the use case. It is critical that both OT and IoT receive the same level of scrutiny and controls as the rest of your cybersecurity program. We covered OT and IoT in more detail in *Chapter 12, Operational Technology (OT) and the Internet of Things (IoT)*.

Program Strategy

You will not be able to execute a successful cybersecurity program without a well-defined strategy in place, especially nowadays as more scrutiny is coming our way from executive leadership, the board of directors, regulators, customers, consumers, etc. Without a strategy, you have no constructive way of obtaining the desired goals for your cybersecurity program. Your strategy is an integral part of the foundation for your cybersecurity program. Ensuring everyone is aware of the importance of the strategy and having a well-defined strategy will only provide for greater success. For your cybersecurity program, there should be at minimum four core strategies to support the program: an architecture strategy, framework strategy, resource management strategy, and product and vendor management strategy. In addition, there will need to be strategies defined for each of the functions within your program, for example, a strategy will need to be built around your cybersecurity operations function. We covered strategy in more detail in *Chapter 4, Solidifying Your Strategy*.

Stay Educated

Stay current on the ever-evolving threat landscape in today's world. It is important as a leader and a security professional that you are aware of and understand the complexities of current threats to ensure you are applying the appropriate remediations. There are many resources available to view the latest cybersecurity news. Make sure to follow the latest trends and understand current best practices. There is no way we could even begin to list all of them, but a quick Google search or interaction with ChatGPT will return many additional resources for review. The following is an example of a resource that provides over 50 blogs and websites for reference: `https://heimdalsecurity.com/blog/best-cyber-security-blogs/`. We covered staying educated briefly in *Chapter 1, Current State*.

Hopefully, this overview summarized some of the key takeaways from the book. The goal is to provide you with insights into some of the more important components to focus on. Obviously, we cannot cover everything from the book within the top 10 considerations, so it's important that you prioritize based on your current state and priorities.

Some Observations and The Future

In the final section, we will review some of the obvious areas that continue to progress, and we will briefly touch upon some areas of consideration that continue to evolve and some that have open-ended questions on what the future holds with these technologies. Curiosity and imagination should be triggered by several of the discussion points below.

Capabilities for anyone

It is important to note that nowadays, tools are available for anyone to use and carry out unethical and/or illegal activities. Not too long ago, it typically took a skilled threat actor to commit a cybersecurity crime. Today, anyone with access to a computer and the internet can easily commit crimes with tools being easily accessible. For example, ransomware is a lucrative business where the tools are made available for others to use and hold organizations at ransom. The dark web is a marketplace where you can purchase anything you can think of from a cybersecurity perspective, including credentials, access to organizations' networks and systems, 0-day exploits, users' data such as PII, and more. As GenAI capabilities continue to enhance at a rapid pace, threat actors are taking advantage of capabilities that prevent the need for them to be skilled at hacking. With GenAI, malicious code can be created, realistic phishing emails can be generated, and much more. Other examples include the ability to purchase tools like Flipper Zero that allows you to easily compromise access control systems, scan RFID chips, and much more. The idea of Flipper Zero, which is to allow ethical hackers and/or penetration testers to test hacking. The reality is, anyone can access and use this tool.

Figure 16.5: Flipper Zero Source: https://flipperzero.one/

The more accessible these tools and capabilities become, the more threat actors we will continue to see. On the flip side, these tools and capabilities provide benefits by allowing us to use them from a learning perspective to better educate ourselves.

Everyone Needs a Cybersecurity Program

This one is obvious. Moving forward, no matter the size of your organization or business, a security presence should be required. For a smaller business, security in the form of an outsourced model that leverages a MSSP might make more sense than hiring in-house. Having an MSSP available will give you the necessary resources to provide the expertise needed to handle security-related incidents. Larger organizations may opt for an in-house security team, but many MSSPs can cater to larger organizations as well.

Data Protection

It's commonly been said that "company data is the crown jewel of the organization." To protect data in an available-anywhere-from-any-device model, data protection needs to continuously be enhanced to prevent leakages. To do this, continue to grow and evolve your **data loss prevention (DLP)**, information protection, and insider threat programs using cloud-based technologies. Enhance your protection with **information rights management (IRM)** and data classification tools. Your organization's data should be automatically labeled and classified based on industry-standard privacy regulations, along with custom rules used to identify sensitive data unique to your business. Based on the data classification, there should be automatic protections applied that include the ability to enforce encryption, require authentication, block data from leaving your devices, and restrict copy and paste to non-protected apps.

Cloud First

Although there is still some resistance and not everyone may necessarily agree, shifting your technology and cybersecurity strategy to leverage cloud technologies will help you become more efficient with your cybersecurity capabilities. By adopting the use of next-generation technologies, you will be gaining the benefit of an environment that has little to no self-managed infrastructure. It allows for scalability and automation, makes use of AI, and incorporates behavioral analytics using big data.

A Passwordless World

Going passwordless is the future, and Microsoft, Apple, and Google are making a big push in this direction as an authentication strategy. The technology is already available, and you can go passwordless today. Unfortunately, it may not be easy to get to a passwordless world right away, but you do need to understand and begin this journey sooner rather than later. The methods that are used to provide a passwordless world are currently much more secure for your users.

With the elimination of passwords, authentication is improved by using something you already have, such as a phone or a security key, in addition to something you are or know, such as biometrics or a PIN. If you are not already familiar with **Fast Identity Online 2 (FIDO2)**, become familiar with it, as this specification is currently driving the passwordless initiative.

Digital Identity

As our personal digital identity continues to evolve, what is next? Are we heading toward a unified identity where a single item we carry represents ourselves? We already see this, as smartphones can supplement car keys, wallets, digital identification, and more. The concept of Web3 introduces a decentralized internet model where you control your digital identity, not the big tech companies that have profited tremendously by scraping your data and profiting from it. Web3 projects a fundamental change in how we transact as our digital selves in the future. An example of decentralized identity is Microsoft Entra Verified ID: `https://learn.microsoft.com/en-us/entra/verified-id/decentralized-identifier-overview`.

IoT

IoT has grown exponentially in recent years and continues to grow, as devices are now being built for everything imaginable. As this technology grows, cybersecurity needs to be brought to the forefront, not only within the enterprise but also within the consumer space. Devices, gadgets, lightbulbs, health monitors, household appliances, entertainment, landscaping equipment, automobiles, accessories, and drones are all examples of the types of internet-connected devices we can use today. Unfortunately, from a security perspective, usability is typically the primary design focus of these smart devices. Their internet connectedness, lack of security standards, and heavy usage in our daily lives expose a significant security risk. Hopefully, as we continue to evolve in this space, we will see the creation of a more universal standardization that can be followed with some form of certification showing whether a device meets the minimum security specifications for both enterprise and consumer usage.

OT

Services considered critical infrastructure, which includes OT, are another area that should be at the forefront of developing security-based technologies. Examples of critical infrastructure services include energy, emergency services, chemical and nuclear facilities, transportation, and the government sector.

These are essential services that support our daily lives, and even a minor disruption could be catastrophic. We need to ensure these critical components used in everyday life are highly protected, as the stakes are too high. Like IoT, hopefully, we see improvement in this space, with the creation of a more universal standardization that can be followed by some form of certification showing whether a device meets the minimum-security specifications for the enterprise.

AI and Deepfakes

As we've touched upon throughout the book, GenAI is taking the world by storm and every vendor is incorporating AI within their products and offerings. Although we are already seeing substantial benefits from using AI, there is the flip side where AI is already being used to benefit threat actors in many ways, with enhanced social engineering campaigns, the creation of malicious code with minimum effort, easier reconnaissance, more efficient automated attacks, the ability to attack organizations much quicker with capabilities like ChatGPT, and using deepfake capabilities with sound, images, and videos. As we expand on deepfakes, capabilities are now available for anyone to easily and accurately replicate one's voice. To take these capabilities even further, tools are available that allow us to easily generate images and videos of others with incredible accuracy. We are already hearing of instances where threat actors are spoofing high-profile executives like a CFO to trick those within the organization into transferring large amounts of money into fake accounts. The capabilities of these technologies are only limited by our imagination, and we are entering a new world of next-generation threats being generated by AI.

Autonomous Vehicles

Transportation services, from public to private to space travel, all involve internet-connected technology. These services are used by millions of people daily and any compromise could potentially diminish the safe operations of these vehicles and result in significant damage, injury, or loss of life. Like standard technology, this technology is susceptible to many different threats, including remote hacking such as the example in 2015, when security researchers were able to remotely control a Jeep Cherokee driving on the highway and disable the car's engine and brake systems: https://www.wired.com/2015/07/hackers-remotely-kill-jeep-highway/.

Microchips

Recently, although considered controversial, the topic of microchip implants has become more than just a conversation. Neuralink, a company of which Elon Musk is a founder, is conducting human trials on their brain chip implant already. Neuralink hopes to help people who suffer from neurological conditions by repairing certain cognitive and sensory-motor functions controlled by the brain.

Will this ever become a reality or requirement for humans? Perhaps we will live in a world where a microchip could become mandated in certain countries. Only time will tell.

Robotics

Another interesting area with significant technological advances is that of robotics and autonomy. What does the future hold, and where do we draw the line in terms of how intelligent robots can become? We all have most likely watched futuristic movies that entail robots becoming smarter than humans with the strength to overpower humanity. Could this ever become a reality? Could robots become programmed or compromised to do more harm than good? These are real conversations happening now and it's critical that we build a solid, core cybersecurity model that includes protection against these threats as robotic technology continues to evolve. There should be no failure of security in this space.

Other

Some other areas of consideration to be aware of include Web3 using blockchain architecture, cryptocurrency, the metaverse, and the expansion of smart devices being used for anything and everything. What about quantum computing and the impact it may have on current encryption algorithms? Even more concerning is cyber warfare, the ability to use cybersecurity as an attack method against other countries. As part of modern-day wars, we are seeing technology being used more and more, for example, the use of drones to gather intelligence, or even worse, as weapons. These devices are remotely controlled, meaning they can easily be susceptible to threat actors gaining control.

Summary

In this chapter, we provided an overview of what was covered within each of the three primary sections within the book: defining the program and the core, and bringing it together to bring us full circle with the cybersecurity program. We then reviewed more details as to why you need to focus on the program as a journey with no end state. This is when we discussed the method used through roadmaps to define your journey. We then reviewed what I consider the top 10 considerations from the book, including a CISO's top 3 priorities as of 2024. Each of these items included a reference to the original chapter, where you can review the material in more detail to gain a better understanding. We then finished off the chapter with a brief look into some considerations and final thoughts on more prevailing topics.

This chapter concludes the subjects in this book. We hope you enjoyed the content provided and were able to take away the necessary knowledge to help secure and strengthen your organization.

Join our community on Discord!

Read this book alongside other users, Cybersecurity experts, and the author himself. Ask questions, provide solutions to other readers, chat with the author via Ask Me Anything sessions, and much more.

Scan the QR code or visit the link to join the community.

`https://packt.link/SecNet`

packt.com

Subscribe to our online digital library for full access to over 7,000 books and videos, as well as industry leading tools to help you plan your personal development and advance your career. For more information, please visit our website.

Why subscribe?

- Spend less time learning and more time coding with practical eBooks and Videos from over 4,000 industry professionals
- Improve your learning with Skill Plans built especially for you
- Get a free eBook or video every month
- Fully searchable for easy access to vital information
- Copy and paste, print, and bookmark content

At www.packt.com, you can also read a collection of free technical articles, sign up for a range of free newsletters, and receive exclusive discounts and offers on Packt books and eBooks.

Other Books You May Enjoy

If you enjoyed this book, you may be interested in these other books by Packt:

Adversarial Tradecraft in Cybersecurity

Dan Borges

ISBN: 9781801076203

- Understand how to implement process injection and how to detect it
- Turn the tables on the offense with active defense
- Disappear on the defender's system, by tampering with defensive sensors
- Upskill in using deception with your backdoors and countermeasures including honeypots

- Kick someone else from a computer you are on and gain the upper hand
- Adopt a language agnostic approach to become familiar with techniques that can be applied to both the red and blue teams
- Prepare yourself for real-time cybersecurity conflict by using some of the best techniques currently in the industry

The Ultimate Kali Linux Book

Glen D. Singh

ISBN: 9781835085806

- Establish a firm foundation in ethical hacking
- Install and configure Kali Linux 2024.1
- Build a penetration testing lab environment and perform vulnerability assessments
- Understand the various approaches a penetration tester can undertake for an assessment
- Gathering information from Open Source Intelligence (OSINT) data sources
- Use Nmap to discover security weakness on a target system on a network
- Implement advanced wireless pentesting techniques
- Become well-versed with exploiting vulnerable web applications

Packt is searching for authors like you

If you're interested in becoming an author for Packt, please visit authors.packtpub.com and apply today. We have worked with thousands of developers and tech professionals, just like you, to help them share their insight with the global tech community. You can make a general application, apply for a specific hot topic that we are recruiting an author for, or submit your own idea.

Share your thoughts

Now you've finished *Resilient Cybersecurity, First Edition* we'd love to hear your thoughts! Scan the QR code below to go straight to the Amazon review page for this book and share your feedback or leave a review on the site that you purchased it from.

https://packt.link/r/1835462510

Your review is important to us and the tech community and will help us make sure we're delivering excellent quality content.

Index

V

Valimail
reference link 323

**Vendor Incident Response Plan (VIRP)
document 436**

Vendor Lifecycle Management 408

**Vendor Risk Management 403-406, 522, 660,
669, 670**
certification 413
current landscape 409, 410
cybersecurity risk 410, 411
hardware compatibility 413, 414
risk types 406, 408
supply chain risk 411, 412

Verizon 2024 DBIR report
reference link 666

**Verizon Data Breach Investigation
Report 14, 15**
reference link 15

Verizon DBIR report
reference link 596

virtualization-based security (VBS) 339

Virtual Local Area Networks (VLANs) 192

Virtual Private Network (VPN) 192

VirusTotal
reference link 288

vulnerability
overview 308-314

vulnerability alerting 317
external sources 319-321
SOC 318
threat intel 318
vendors 322

vulnerability disclosure program (VDP) 495

vulnerability management 95, 301
lifecycle 301, 302
sub-functions 300, 301
system 318

**vulnerability management,
considerations 337**
audits and assessments 343, 344
cybersecurity testing 342, 343
hardware vulnerabilities 337-339
ICS 344
IoT 344
network infrastructure 341, 342
OT 344
other activities 344, 345
virtualization infrastructure 340

vulnerability management, function 660

vulnerability management, program 523
building blocks 302, 303
meetings and status calls 305
metrics 305
modernizing 328, 329
policy and procedures 304
reporting 305
risk management 303
roles and responsibilities 304
SLAs 305
vulnerability tools 305
vulnerability tracking 304

vulnerability remediation 323-327
prioritization 327

vulnerability scanning 314, 316
External Attack Surface Monitoring
(EASM) 317

vulnerability tracking 323, 324
scoring 325

Download a free PDF copy of this book

Thanks for purchasing this book!

Do you like to read on the go but are unable to carry your print books everywhere?

Is your eBook purchase not compatible with the device of your choice?

Don't worry, now with every Packt book you get a DRM-free PDF version of that book at no cost.

Read anywhere, any place, on any device. Search, copy, and paste code from your favorite technical books directly into your application.

The perks don't stop there, you can get exclusive access to discounts, newsletters, and great free content in your inbox daily.

Follow these simple steps to get the benefits:

1. Scan the QR code or visit the link below:

https://packt.link/free-ebook/9781835462515

2. Submit your proof of purchase.
3. That's it! We'll send your free PDF and other benefits to your email directly.